Advances in Intelligent Systems and Computing

Volume 1114

The series "Advances in Intelligent Systems and Computing" contains publications on theory, applications, and design methods of Intelligent Systems and Intelligent Computing. Virtually all disciplines such as engineering, natural sciences, computer and information science, ICT, economics, business, e-commerce, environment, healthcare, life science are covered. The list of topics spans all the areas of modern intelligent systems and computing such as: computational intelligence, soft computing including neural networks, fuzzy systems, evolutionary computing and the fusion of these paradigms, social intelligence, ambient intelligence, computational neuroscience, artificial life, virtual worlds and society, cognitive science and systems, Perception and Vision, DNA and immune based systems, self-organizing and adaptive systems, e-Learning and teaching, human-centered and human-centric computing, recommender systems, intelligent control, robotics and mechatronics including human-machine teaming, knowledge-based paradigms, learning paradigms, machine ethics, intelligent data analysis, knowledge management, intelligent agents, intelligent decision making and support, intelligent network security, trust management, interactive entertainment, Web intelligence and multimedia.

The publications within "Advances in Intelligent Systems and Computing" are primarily proceedings of important conferences, symposia and congresses. They cover significant recent developments in the field, both of a foundational and applicable character. An important characteristic feature of the series is the short publication time and world-wide distribution. This permits a rapid and broad dissemination of research results.

** Indexing: The books of this series are submitted to ISI Proceedings, EI-Compendex, DBLP, SCOPUS, Google Scholar and Springerlink **

More information about this series at http://www.springer.com/series/11156

Tatiana Antipova · Álvaro Rocha
Editors

Digital Science 2019

 Springer

Editors
Tatiana Antipova
Institute of Certified Specialists
Perm, Russia

Álvaro Rocha
AISTI & University of Coimbra
Rio Tinto, Portugal

ISSN 2194-5357   ISSN 2194-5365 (electronic)
Advances in Intelligent Systems and Computing
ISBN 978-3-030-37736-6   ISBN 978-3-030-37737-3 (eBook)
https://doi.org/10.1007/978-3-030-37737-3

This Springer imprint is published by the registered company Springer Nature Switzerland AG
The registered company address is: Gewerbestrasse 11, 6330 Cham, Switzerland

Preface

This book contains a selection of papers accepted for presentation and discussion at the 2019 International Conference on Digital Science (DSIC 2019). This conference had the support of the Institute of Certified Specialists, Russia, AISTI (Iberian Association for Information Systems and Technologies) and Springer. It took place at Londa Hotel, Limassol, Cyprus, October 11–13, 2019.

DSIC 2019 is an international forum for researchers and practitioners to present and discuss the most recent innovations, trends, results, experiences and concerns in the several perspectives of digital science. The main idea of this conference is that the world of science is unified and united allowing all scientists/practitioners to be able to think, analyze and generalize their thoughts.

DSIC aims efficiently disseminate original research results in natural, social, art and humanity sciences. An important characteristic feature of the conference should be the short publication time and worldwide distribution. This conference enables fast dissemination so conference participants can publish their papers in print and electronic format, which is then made available worldwide and accessible by numerous researchers.

The Scientific Committee of DSIC 2019 was composed of a multidisciplinary group of 26 experts. Ninety-seven invited reviewers who are intimately concerned with digital science have had the responsibility for evaluating, in a 'double-blind review' process, the papers received for each of the main themes proposed for the conference: digital art and humanities; digital economics; digital education; digital finance, business and banking; digital health care, hospitals and rehabilitation; digital media; digital public administration; digital technology and applied sciences; and digital virtual reality.

The papers accepted for presentation and discussion at the conference are published by Springer (this book) and will be submitted for indexing by ISI, Scopus, among others. We acknowledge all of those that contributed to the staging of DSIC 2019 (authors, committees, reviewers, organizers and sponsors). We deeply appreciate their involvement and support that was crucial for the success of DSIC 2019.

October 2019   Tatiana Antipova
                                                                              Álvaro Rocha

Organization

General Chair

Tatiana Antipova   Institute of Certified Specialists, Russia

Honorary Chair

Álvaro Rocha   AISTI & University of Coimbra, Portugal

Scientific Committee

Alan Sangster   University of Aberdeen, UK
Altinay Fahriye   Near East University, Cyprus
Andre Carlos Busanelli   University of São Paulo (USP), Brasilia
  de Aquino
David Krantz   Arizona State University, USA
Elena Fleaca   Politehnica University of Bucharest, Romania
George Danko   University of Nevada, USA
Giuseppe Grossi   Nord University, Norway
Howard Frank   Florida International University, USA
Hu-Chen Liu   Shanghai University, China
Joao Vidal de Carvalho   Polytechnic Institute of Porto, Portugal
John Dumay   Macquarie University, Australia
Julia Belyasova   Public Social Welfare Centre, Belgium
Leonid Yasnitsky   Perm State University, Russia
Linda Kidwell   Nova Southeastern University, USA
Luca Bartocci   University of Perugia, Italy
Lucas Oliveira Gomes   Federal Court of Accounts, Brasilia
  Ferreira
Lucas Tomczyk   Pedagogical University of Cracow, Poland
Marina Gurskaya   Kuban State University, Russia

Contents

Digital Education

Digital Art and Humanities

Traffic Analysis in Data Networks: Design and Development of Monitor Software NMTraffic

Lidice Haz[1(✉)] [iD], Daniel Quirumbay[1], Jorge Isaac Avilés[2], and Ximena Acaro[2]

[1] Universidad Estatal Península de Santa Elena, La Libertad, Ecuador
victoria.haz@hotmail.com
[2] Universidad de Guayaquil, Guayaquil, Ecuador

Abstract. Data networks are part of the development of communication technologies. It is important to know the behavior of the data that is transmitted through the communication channels. Monitoring the traffic of a network becomes a task of administration and computer security. These tasks allow us to detect atypical behaviors in network traffic, possible threats and reduce security incidents. The purpose is to ensure the confidentiality, integrity and availability of the information assets. This work presents the design and development of a software to monitor and analyze the traffic of a network through the integration of WinPcap libraries. The project was developed applying the evolutionary prototype model focused on a rapid design that fits the needs of the user. The test scenarios allowed evaluating the functionality of the software and measuring the efficiency of the statistical reports generated by the application. The main objective is to optimize the consumption of bandwidth and network resources.

Keywords: Networking · Software monitoring · Network data analysis · Data transmission · Data traffic · Network scanning

1 Introduction

The technological development of organizations has modified their operations and operations, increasing their competitiveness. The information and communication systems are a fundamental part for the functioning of the processes of any organization. The core of the business is supported in computer systems that are integrated with telecommunications networks. The main objective is to optimize the development of productive activities that generate value in an organization [1].

The processing, storage and transfer of information are processes that are automated for each of the business areas. The uninterrupted use of information and communication systems increases the effectiveness and efficiency of organizations. The management and control of business processes are facilitated by a timely delivery of information, which allows senior management to make more effective decisions to achieve compliance with corporate responsibilities and objectives [2, 3].

© Springer Nature Switzerland AG 2020
T. Antipova and Á. Rocha (Eds.): DSIC 2019, AISC 1114, pp. 3–15, 2020.
https://doi.org/10.1007/978-3-030-37737-3_1

For an organization, the most important asset it has is information, therefore, physical and logical security mechanisms must be applied to safeguard said asset [4]. These mechanisms should make it possible to evaluate and improve the effectiveness and efficiency of organizational processes by analyzing the management and administration of the use of ICTs.

It is important to apply controls to monitor the activity of information and communication systems, given that currently these technological infrastructures are exposed as an easy target for espionage, crime and cyber-terrorism [5].

In this sense, it is necessary to ensure the correct functioning of the technological infrastructure, guaranteeing the security of communications and information assets. It is convenient that the IT management defines computer security procedures that include the automatic and periodic execution of tools that allow to analyze the traffic of the network. This, to optimize the use of the network and detect possible errors or failures in the communication systems; in addition, the final reports with the recommendations should be included in the reports of the internal audits addressed to senior management [3].

For the importance of controlling and measuring the performance of network resources through the analysis of their behavior is presented; therefore, different modes of operation of a sniffer and its implementation were studied in computer security tools available in the market [8, 9]. Finally, the design, development and functionality testing of the NMTraffic monitor software is presented as a tool to analyze network traffic, which integrates 3 important components: monitor mode, network traffic control mode, and network optimization mode band [10, 11].

2 Monitoring and Control of Network Traffic

The transfer of information between computer networks is an important activity that must be guaranteed to reduce delays in the transmission of data and ensure the quality of services to which users' access. The following describes some definitions on analysis, control and monitoring of data and network traffic [12, 13].

Analysis of data packages. Analyzing an object implies its decomposition into smaller parts to study its content. The data packages contain information that can be captured and analyzed. The capture and interpretation of data packets will be possible according to the level of security implemented in their transmission [14].

Network traffic analyzer. Are specialized computer programs in monitoring, analyzing and capturing frames or packets that pass through the data network. These programs allow the interception and recording of packet traffic passing over a data network, while the flow of data is moving within the network [15, 16].

There are various utilities such as:

- The conversion of the binary data to a readable format.
- Resolution of problems in the network.
- Analyze the performance of a network to discover bottlenecks.
- Intrusion detection in the network.
- Record of network traffic for argumentation and tests.

- The analysis of the operations of the applications.
- The discovery of defective network cards.
- Discover the origin of virus outbreaks or denial of service (DoS).
- Validate compliance with the company's security policies.
- As an educational resource in learning about protocols.

In the market there are different tools as mentioned in [17] that allow these processes to be carried out, such as:

- Tcpdump
- Nwatch
- Nmap
- Wireshark
- Ip traffic monitor.

3 Software Development NMTraffic

The project was developed applying the evolutionary prototype model focused on a rapid design that fits the needs of the user. The phases of the model and its results are described below.

3.1 Phase 1: Initial Specification

The requirements and limitations for the development of version 1.0 of NMTraffic were established. The main objective is to develop an application to monitor and analyze the network traffic and manage the different channels of wireless WIFI networks. For this, the following modules were defined:

- Administrator/Users: the software validates the login and installation from the administrator user as a security method.
- Equipment registration: the software administrator can register users who are connected to the local network by identifying the use of Ips addresses, their allocation and levels of packet traffic generated by each user in the network.
- User Control: the administrator can apply privileges to each user in the network. In this process, the login is used, which loads the available options according to the assigned privileges.
- Monitoring, WIFI channels or network traffic: the following types of monitoring are established.
  - Host scanning connected to the network, it is possible to enter a range of IPs or perform a scan automatically on the network to which it is connected.
  - Real-time monitoring with graphics, generates real-time graphics of network traffic.
  - Bandwidth, presents information about the bandwidth levels generated in real time.

- Channels and SSID of WIFI network, shows the frequency of the most used or congested channels by the users of the network.
- Sniffer mode, presents the traffic generated by the machines connected to the network, adding more detail in the information for subsequent audit.
- Sniffer mode per host, presents the traffic generated by a specific user of the network.
- Reports Module: there are several reports according to the information that you want to analyze, then the available reports are described:
  - General traffic per day, biweekly or monthly.
  - Traffic per host per day, biweekly or monthly.
  - Protocols most used by users.
  - Ports accessed by users.
  - Filter and/or general according to the scanned IPs and with information of the registered user.
  - URLs accessed by users

Limits of the Program

The project was developed in Microsoft Visual Studio 2015 programming language. It is compatible only with Windows operating systems. The software does not have the ability to block or manipulate traffic generated by other programs on the same computer. Therefore, it cannot be used in applications such as traffic limitation, and personal firewalls.

To perform the different monitoring an alpha WIFI antenna is used as a backup. This serves as a support to perform the scans because it provides better signal and better results are obtained. The scanning of the WIFI networks is done under the 802.11g standard in 2.4 GHz frequency, the 5 GHz frequency is not used, because some equipment or antennas do not have this frequency.

3.2   Phase 2: Development of the Application

For the construction of the application, the following technological tools were used:

- Windows 7 Operating System
- Microsoft Visual Studio 2015
- MySQL Workbench 6.0 CE Database
- WinPcap 4.1.3

Design of the Proposal

Next, the diagrams of the operation of the software components are described. Figure 1 shows the scheme of the application which corresponding the general process for scanning network traffic.

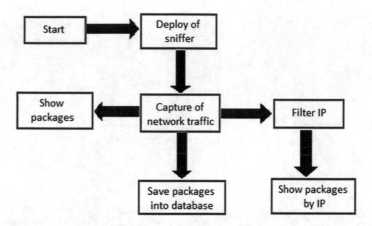

Fig. 1. Scheme of network traffic scanning.

In Fig. 2, the process of scanning equipment connected to the network through an ARP ping is shown, it is presented with a range filter of IPs.

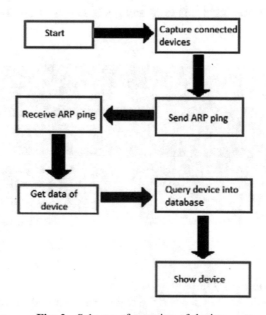

Fig. 2. Scheme of scanning of devices.

Prototype Design

In Fig. 3, the scheme of the main screen of the NMTraffic software is shown.

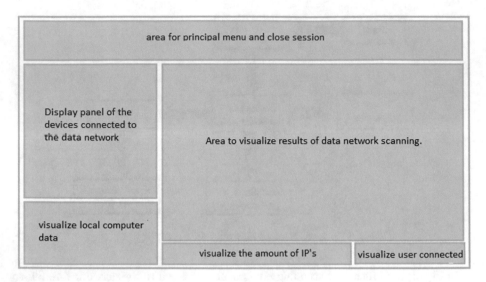

Fig. 3. Scheme of the main screen.

In Figs. 4 and 5, the login windows and main screen of the software are shown, respectively.

Fig. 4. Login window.

Development of the Prototype

The class diagram shown in Fig. 6 was the main scheme for developing the logic of the program and its coding. The use of external libraries such as WinPcap Pcap was necessary to perform network traffic scanning, and the SpPerfChart library for the design of the graphics in real time.

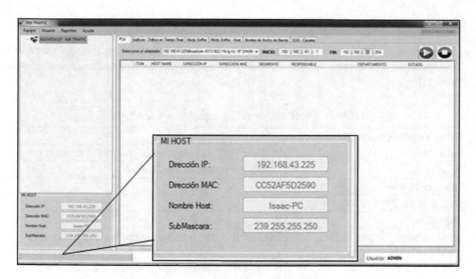

Fig. 5. Main screen with data equipment

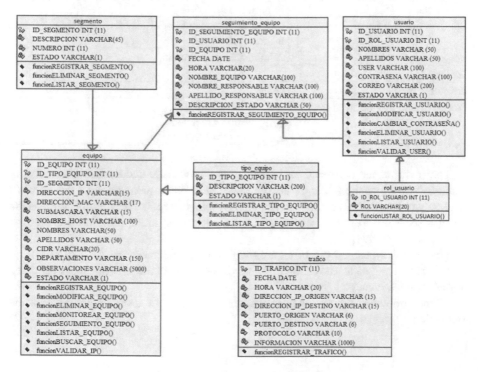

Fig. 6. Class Diagram: NM Traffic.

3.3   Phase 3. Implementation, Use and Evaluation

Different test scenarios were designed to verify the functionality of the software for each specified component. Then, in Table 1, the equipment operations scenario is described.

Table 1. Functionality tests - equipment operations.

Test: PC Scan Operations	
Objective	Perform start and stop operations of PC scanning
Description	Verify the operations performed
Complexity level	High
Case 1: Start PC Scan	
Input data	Expected output data
Select network adapter for the application	Display IP range obtained according to the network
Click on the start button	Show a list with all the equipment connected to the network
Case 2: Stop PC Scan	
Input data	Expected output data
Click on the stop button	Cancel scanning and not display devices connected to the network
Responsible involved	Administrators and users
Test results	
Errors obtained	None
Processes case 1 and 2 are corrects	_ X_ Correct execution _____ Wrong execution _____ Execution with errors

The performance results of each component were 90% successful. The software manages to capture the traffic of all the equipment that is connected to the network wirelessly or wired. The security mechanisms implemented in the program, such as the use of a password to start a session and data encryption, are appropriate for the first version of the prototype.

The tests were carried out in an area with multiple connections for 120 min for a week to obtain real data. The different views of the traffic of the network were validated in 90% of effectiveness.

In the following figures, the results are presented for each component of the application.

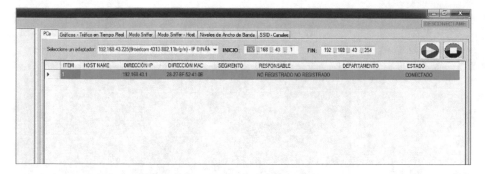

Fig. 7. Scanning of devices.

Figure 7 shows the scanning of equipment by selecting a network adapter in a range of IPs by default. In Figs. 8 and 9, the automatic scanning of the detected devices in the network is shown, in graphic and list form, respectively.

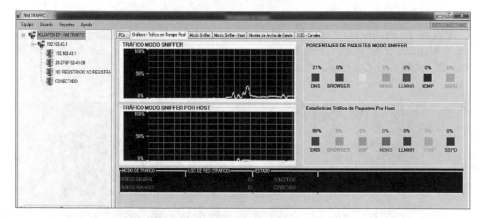

Fig. 8. Network traffic plot.

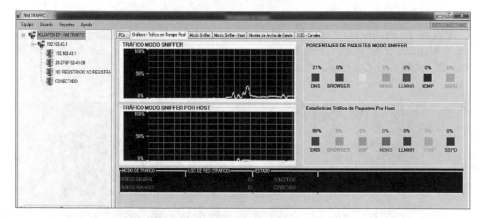

Fig. 9. Network traffic list.

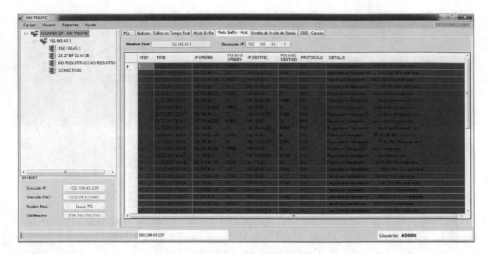

Fig. 10. Host traffic.

Figure 10 shows the scan of traffic for a specific host, selected from the previous list, the information that is described corresponds to the date, origin and destination IP, port of origin and destination, and a detail of the traffic analyzed.

Fig. 11. Bandwidth level.

Figure 11 shows the information of the load and download levels of the bandwidth in Mbps. This information was validated with an internet speed test obtained from a URL.

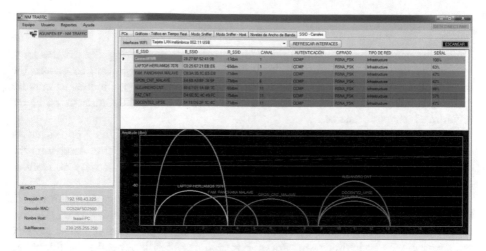

Fig. 12. WIFI scanning.

Figure 12 shows the scanning of all access points or Wi-Fi wireless networks. The information is presented in graphic form and in a list.

TECH SOLUTIONS C.A.
LISTADO DE PÁGINAS ACCEDIDAS
JPEREZ
08-15-2018 10:11

FECHA	HORA	DIRECCION IP ORIGEN	INFORMACION	
14/02/2017	20:20:04	192.168.100.1	www.facebook.com	
16/02/2017	13:6:37	192.168.43.225	Petición en Navegación:	teredo.ipv6.microsoft.com.
16/02/2017	13:6:37	192.168.43.1	Respuesta en Navegación:	teredo.ipv6.microsoft.com.
16/02/2017	13:6:37	192.168.43.225	Petición en Navegación:	teredo.ipv6.microsoft.com.
16/02/2017	13:6:37	192.168.43.1	Respuesta en Navegación:	teredo.ipv6.microsoft.com.
16/02/2017	13:6:37	192.168.43.225	Petición en Navegación:	teredo.ipv6.microsoft.com.
16/02/2017	13:6:37	192.168.43.1	Respuesta en Navegación:	teredo.ipv6.microsoft.com.
16/02/2017	18:19:21	192.168.43.225	Petición en Navegación:	teredo.ipv6.microsoft.com.
16/02/2017	18:19:21	192.168.43.1	Respuesta en Navegación:	teredo.ipv6.microsoft.com.
16/02/2017	18:19:21	192.168.43.225	Petición en Navegación:	teredo.ipv6.microsoft.com.
16/02/2017	18:19:21	192.168.43.1	Respuesta en Navegación:	teredo.ipv6.microsoft.com.
16/02/2017	18:19:21	192.168.43.225	Petición en Navegación:	teredo.ipv6.microsoft.com.
16/02/2017	18:19:21	192.168.43.1	Respuesta en Navegación:	teredo.ipv6.microsoft.com.
16/02/2017	18:19:27	192.168.43.225	Petición en Navegación:	teredo.ipv6.microsoft.com.
16/02/2017	18:19:27	192.168.43.1	Respuesta en Navegación:	teredo.ipv6.microsoft.com.
16/02/2017	18:19:27	192.168.43.225	Petición en Navegación:	teredo.ipv6.microsoft.com.
16/02/2017	18:19:27	192.168.43.1	Respuesta en Navegación:	teredo.ipv6.microsoft.com.
16/02/2017	18:19:27	192.168.43.225	Petición en Navegación:	teredo.ipv6.microsoft.com.

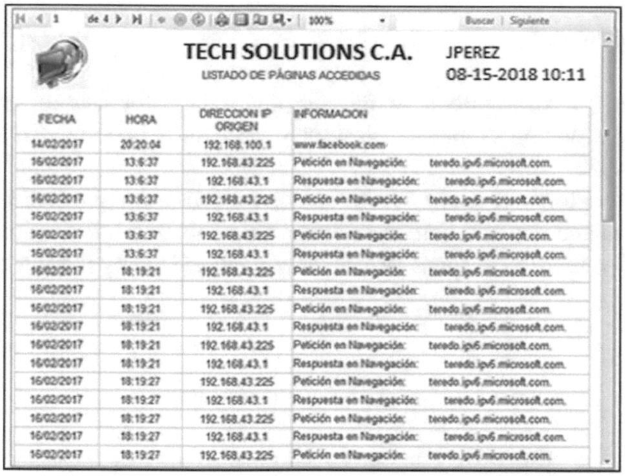

Fig. 13. Reports generation.

Figure 13 shows the generation of a report, specifically the navigation list or pages visited. This information can be downloaded in pdf or xls format.

4 Conclusion

Technological network infrastructures require constant monitoring due to the large amount of data that is transmitted by these channels. It is important to ensure the quality of network services. Using specific tools with adequate techniques to monitor and analyze the traffic of the network facilitates the search and solution of problems regarding their performance.

The administrators of technological networks are aware of the importance of ensuring and maintaining high performance in the network architecture. Analyzing the data traffic periodically with specific tools for this purpose is an activity adhered to computer security.

The NMTraffic software is a tool that works in windows desktop environments. It allows to monitor the traffic of the network and observe the bandwidth levels all in real time. It is possible to generate different reports for auditing purposes. It allows to carry out scans in wire and wireless networks. For scanning of wifi networks the frequency of 2.4 GHz is used.

For the development of the software it was necessary to apply the concepts on the TCP/IP model in order to know the behavior of the transmission of messages and identify the packets that circulate through the network and the communication protocols used, as well as the operation of programs sniffer.

5 Recommendations

For future work, it is important to migrate the project for its function on multiple platforms. The scanning of the wireless networks must be extended to support the frequency 5 GHz.

It is important to make network administrators aware that the tool is used only for ethical purposes and not as a sniffer to obtain unauthorized information.

The use of hacking tools must be known and authorized by the top management of the company.

References

1. Soomro, Z.A., Shah, M.H., Ahmed, J.: Information security management needs more holistic approach: a literature review. Int. J. Inf. Manag. **36**(2), 215–225 (2016)
2. Alsmadi, I., Xu, D.: Security of software defined networks: a survey. Comput. Secur. **53**, 79–108 (2015)
3. García, A., Ysabel, M., Espinoza, I., David, R., León Tenorio, G.M.: Modelo de gestión de riesgos de TI basados en estándares adaptados a las TI que soportan los procesos para contribuir a la generación de valor en las universidades privadas de la región Lambayeque (2016)
4. Pérez, M.T., Palomo, M.A.: Soluciones administrativas y técnicas para proteger los recursos computacionales de personal interno-insiders. Innovaciones de Negocios **4**(8), 357–376 (2017)

5. Agüero Jiménez, D., Calimano Meneses, Y.: Estudio de viabilidad de una herramienta software para monitorización de tráfico IP en Windows Phone. Rev. Cubana Cienc. Inform. **7**(1), 24–31 (2013)
6. Barrionuevo, M., Lopresti, M., Miranda, N.C., Piccoli, M.F.: Un enfoque para la detección de anomalías en el tráfico de red usando imágenes y técnicas de computación de alto desempeño. In: XXII Congreso Argentino de Ciencias de la Computación (CACIC 2016) (2016)
7. Hsu, S.T., Lee, C.P., Yao, P.C.: U.S. Patent 9,660,959. U.S. Patent and Trademark Office, Washington, DC (2017)
8. Shu, Z., Wan, J., Li, D., Lin, J., Vasilakos, A.V., Imran, M.: Security in software-defined networking: threats and countermeasures. Mob. Netw. Appl. **21**(5), 764–776 (2016)
9. Yang, J., Wang, L., Lesh, A., Lockerbie, B.: Manipulating network traffic to evade stepping-stone intrusion detection. Internet Things **3**, 34–45 (2018)
10. Kekely, L., Kučera, J., Puš, V., Kořenek, J., Vasilakos, A.V.: Software defined monitoring of application protocols. IEEE Trans. Comput. **65**(2), 615–626 (2016)
11. Yassine, A., Rahimi, H., Shirmohammadi, S.: Software defined network traffic measurement: current trends and challenges. IEEE Instrum. Meas. Mag. **18**(2), 42–50 (2015)
12. Solomon, T., Zungeru, A.M., Selvaraj, R.: Network traffic monitoring in an industrial environment. In: 2016 Third International Conference on Electrical, Electronics, Computer Engineering and their Applications (EECEA), pp. 133–139. IEEE, April 2016
13. Qi, H., Du, Z., Ge, L.: Computer Network Management Software (2017)
14. Hazarika, B., Medhi, S.: Survey on real time security mechanisms in network forensics. Int. J. Comput. Appl. **151**(2) (2016)
15. Kraitsman, R., Milstein, A., Raff, A., Matot, D.: U.S. Patent No. 9,270,690. U.S. Patent and Trademark Office, Washington, DC (2016)
16. Sanders, C.: Practical Packet Analysis: Using Wireshark to Solve Real-World Network Problems. No Starch Press, San Francisco (2017)
17. Bhosale, D.A., Mane, V.M.: Comparative study and analysis of network intrusion detection tools. In: 2015 International Conference on Applied and Theoretical Computing and Communication Technology (iCATccT), pp. 312–315. IEEE, October 2015

Tools and Metrics for Reputation Assessment in a Digital Environment

Alexander Ovrutsky[1] [iD], Alexandra Ponomareva[1,2(✉)] [iD],
Alexander Ponomarev[3] [iD], and Maxim Ponomarev[3] [iD]

[1] South Federal University, Universitetsky, 93, 344006 Rostov on Don, Russia
alexandra22003@rambler.ru
[2] Rostov State University of Economics,
B. Sadovaya. 69, 344002 Rostov on Don, Russia
[3] Russian Academy of National Economy and Public Administration,
Pushkinskaya Street, 70, 344002 Rostov on Don, Russia

Abstract. The article represents the study of the specifics of the formation, management and evaluation of the reputation ratings effectiveness in the digital environment. As a methodological basis for the presented research, the concept of social representation was used, says that the elements of mass and individual consciousness are emotional-evaluative and behavioral components harmonized with the information field. The concept of "reputation", its structure is discussed. The importance of reputation rating has been proved, and factors that positively and negatively affect reputation in the communication field are described. The essence of reputation management is disclosed. Digital services with built-in reputation score and metrics are considered as tools for reputation management. The description of digital-image reputation assessment tools is presented – the special services that allow assessing the current reputation in real time, their functionality was described. The existing metrics for the effectiveness evaluation of advancement in a digital environment are examined and the indicators to judge the reputation directly and indirectly are identified. The results of the expert survey on the reputation assessment problem in practical marketing and communication activities are presented; the most popular reputation assessment services, metrics used by practitioners for reputation assessment are described.

Keywords: Reputation · Digital tools for reputation rating · Reputation assessment metrics

1 Introduction

Reputation is an important component of the subjects in socio-economic space, formed by the attitude of the target audiences towards them based on the interaction of the consumers' system of values and socio-valued concepts that these subjects represent by their marketing and communication activities. Digital environment does not affects significantly on the content of the reputation concept, but radically transforms the method of its formation, development, preservation, protection. The speed of the marketing and communication processes occurring in the digital environment

© Springer Nature Switzerland AG 2020
T. Antipova and Á. Rocha (Eds.): DSIC 2019, AISC 1114, pp. 16–29, 2020.
https://doi.org/10.1007/978-3-030-37737-3_2

predetermines the importance of continuous monitoring of the status of reputation, of its assessment in dynamics. There are a large number of online services that allow you to track marketing and communication activities in the digital environment in real time, accurately assess its effectiveness. At the same time, the tools and metrics for reputation rating are not sufficiently systematized, and methods for using them to the needs of regional markets are poorly developed.

In our research, we refer to the scientific work of such authors as Shoven, Ryaguzova, Belevskaya, Brik, Weber, Erickson Sheri, Stone, giving an idea of the "reputation" phenomenon from the standpoint of sociology, psychology, jurisprudence, communication theory and reputation management [1–5]; Kravchuk, Khalilov discussed on the effectiveness of communication marketing, which highlight the types of efficiency, describes methods and techniques for evaluating the effectiveness of the communication process [6, 7]. We also refer to practical publications by Ivanichev, Koshik, Gudkov et al., reflecting the experience of using services and metrics for evaluating the effectiveness of marketing and communication activities in the digital environment [8–12, 14].

To solve this problem, we discuss the theoretical description of the "reputation" concept based on the social ideas' concept, considering the reputation assessment in the system of research of the effectiveness of communication marketing, conduct a structural and functional analysis of existing tools and metrics for evaluating marketing and communication activities in the digital environment and identify of them, which can be used to assess the reputation, develop tools for expert survey, conduct research and identify the specifics of existing tools and metrics of reputation assessment in the regional market.

In this way, first, the "reputation" concept (Sect. 2.1) will be reviewed and the reputation assessment will be represented in the communication marketing effectiveness system (Sect. 2.2), then the objectives, methods and tools of the expert survey on tools and metrics of reputation rating will be described and analyzed (Sect. 2.3), at the end of the article it will be possible to find the results, conclusions, the rationale for the novelty of the study, a description of further ways of its development.

2 Brand Reputation: From Concept to the Analysis of Practical Tools and Metrics for Market Performance Rating

2.1 The Concept of "Reputation"

In recent decades, "reputation" has been the object of active sociology, psychology, law, cultural studies, journalism, marketing, and communicology studies based on the formation of a scientific interdisciplinary field, in the center of which are such concepts as reputation, reputation management, reputational efficiency, reputational scandal, reputational crisis reputational damage, reputational losses, reputational discourse. At present, one can observe the process of the formation of such new scientific categories as "reputation economy" and "reputation society". In this way, not only the "reputation" concept is being studied, but also the other related concepts.

The relevance of the topic is predetermined by its public demand: a series of doping, offshore, political scandals that have caused significant reputational damage to their subjects indicates the demand on knowledge about the structure of the reputational scandal, ways to minimize reputational damage and restore reputation. In modern works, reputation is considered not only as a result of communication practices, but also as a key goal of the strategic management of enterprises, individuals and society as a whole.

The term "reputation" is of Latin origin, its original meaning – "thinking", "reflexion", "calculus", "calculation". In this way, etymologically, its rational basis is its semantic basis. Complementary, close concept for the concept of "reputation" is the image. Differentiation of these concepts is possible according to the following parameters: stability/variability (image - the construct is more stable, reputation is more susceptible to changes, external influences), the result of the impact on consumers (the result of designing the image is the image in the minds of consumers, the result of consumers interacting with the reputation - their consumer actions, communication reactions), means of formation (the image is formed primarily by means of branding and advertising, the reputation is mainly event marketing product). In general, practice shows that a synthetic, rather than a differentiating, approach is more productive in shaping and developing the image and reputation of goods, services, individuals, enterprises and organizations. If you summarize the existing approaches to the definition of the reputation essence, you can present them as follows: it is a system that synthesizes social representations, cognitive constructs and public opinion based on assessments, forming confidence in the object of reputation. The disadvantage of this approach is that it does not present a mechanism for reputation forming.

We are interested in an approach in which reputation is studied as a binary phenomenon. With this approach, reputation appears as "a psychological and economic resource that allows reputation objects (organizations, person, country, specific social groups) to convert emotional stakeholder patterns that affect reputation (acceptance, inclusiveness, goodwill, confidence, etc.) into economic benefits (competitiveness, brand premium, willingness to exchange, loyalty, etc.)" [15, p. 48]. However, stakeholders appear as reputational representations carriers. In this way, the resource reputation is represented in the objects of reputation management, the representation of reputation is reflected in the quality of social representations in the individual consciousness of stakeholders, as well as in the mass consciousness of consumers. Since reputation is a phenomenon being formed, one can make a conclusion about its communicative-managerial essence and means of creation, namely: events and media texts that form the reputational discourse. In the reputational discourse, elements of reputation materialize; therefore, the structural-functional study of reputational discourse gives an idea of the essence of reputation. The tools that form the reputational discourse include public relations, journalism, advertising, event marketing, social marketing, branding, etc.

The method of reputational discourse analysis developed by us includes highlighting the structural elements of reputation (self-appraisal/self-esteem, admiration/love, trust, empathy [16]), factors that form the elements of reputation, and discursive methods of influencing target audiences that form elements of reputation (see Table 1).

Table 1. Structure and methods of reputation design [17]

№	Reputational elements	Reputation factors	Discursive methods of influence on the target audience, forming the reputational elements
1	Self-appraisal/ self-esteem	Performance	Publications of ratings, participation in competitions and exhibitions, information on the implementation of a quality management system
		Quality of goods and services	
2	Admiration/love	Innovativeness	Formation of the image of the creative organization
		Working conditions	Internal corporate press, information about the quality of personnel management, the formation of internal loyalty
3	Trust	Corporative management	Information transparency, top management publicity, conceptual and systematic external and internal communications, the formation of external loyalty
4	Empathy	Social responsibility	Sponsorship, philanthropy, social marketing activities
		Leadership	

The approach to the analysis of the reputation discourse allows you to technologize its creation, differentiate design methods and describe the stages of reputation formation.

2.2 Reputation Ratings in the System of the Communication Marketing Effectiveness

The category of efficiency is universal and is usually understood as the efficiency of resource usage in sight of company performance. The concept of efficiency is inextricably linked with the concept of effect, that is, the result of activity. Efficiency is related to the process, and the effect - to the result. It is the idea of the relationship of the process and the result that underlies the category of "effectiveness". From an economic point of view, reputational efficiency is the ratio between the effects of reputation and the effort and cost of its creation. Reputational effectiveness is associated with the effectiveness of communication marketing. High reputational efficiency means maximum customer satisfaction with the brand. Reputational efficiency is achieved through the organization of continuous communication interaction with consumers. Reputational efficiency for an enterprise is an external, difficult to measure indicator. Formation of reputational efficiency may be based on product improvement, but over time it becomes clear that this is not enough. The conceptual evolution of marketing shows how the views of practitioners and theorists have changed over time on how to achieve reputational effectiveness. The production marketing concept prescribed to do this by improving production, the product concept - by improving the product, marketing - by intensifying sales, traditional marketing - by harmonizing the product with the needs of the target markets. The concept of socio-ethical marketing implies the impossibility of forming reputational effectiveness without simultaneously solving social, environmental, ethical, cultural and other problems existing in society, and the concept of marketing interaction is based on the idea that achieving reputational effectiveness is based on the interaction of an enterprise not only target markets, but also with partners, government, public organizations, etc.

The concept of reputational effectiveness links together the economical and communication effectiveness of marketing. Reputational effectiveness means harmonization in the implementation of multidirectional goals of an entrepreneur (customer satisfaction and profit generation) and is the key concept of the modern marketing philosophy.

If we talk about the practical aspect, then there are no universal indicators of reputational efficiency that could be applied in any sphere of economic activity in any situation. The category of reputational efficiency in the digital environment appears as a complex indicator formed by factors and criteria, the set of which in a situation of measuring efficiency is predetermined by the specifics of the actual communication and economic tasks of an enterprise. When developing a system of indicators for assessing reputation in an enterprise, the following can be used:

- The principle of KPI usage (Key Performance Indicators) dictates the creation of a set of metrics that demonstrate the level of goals and objectives achievement in the area of reputation formation (managerial approach);
- The principle of conversion: involves tracking the level of the relationship of quantitative indicators of the next stage to the quantitative indicators of the previous stage of marketing activity (communication and behavioral approach);
- The principles of ROI (Return on Investment) and profitability: dictate the description of the ratio of invested financial resources and obtained reputational results, transformed into economic indicators, for example, improving reputation, can lead to an increase in the value of shares of an enterprise (economic approach).

Analysis of the metrics used in the digital environment, particulary in SMM, made it possible to identify those that directly or indirectly allow us to assess reputation. It turned out that they can be divided into three groups:

1. General metrics for the public SM account: the number of subscribers, the increase in subscribers for a certain period, the number of people who unsubscribed from a public account for a certain period, the number of public/fast views, the public/fast coverage (the number of people who at least once contacted the public account/fast) organic, public/post coverage paid, public/post coverage is viral.
2. Metrics of interaction with the audience of the public SM account: the activity of the audience (the number of users who committed a certain number of actions (1,2,3, etc. actions), the level of attractiveness (Likes/Followers * 100%), the level of sociability (Comments/Followers * 100%), distribution ratio: (Shares/Posts (number of posts) * 100%), engagement ratio (number of likes + reposts + comments/total number of subscribers, check-ins, hashtags, negative user reactions, user-generated content (posts, photos, videos, uploaded by subscribers))).
3. Public interaction with the external environment: the conversion to the site, the number of leads, the digital environment mentioning.

Analysis of metrics that can be used to assess reputational effectiveness shows that they are based on the principle of KPI and on the principle of conversion. Using the ROI and metric-based profitability principle is almost impossible.

Let us consider tools/services for communication monitoring in a digital environment, the functionality of which allows you to assess reputation (Table 2).

Table 2. The functionality of assessing the reputation of tools/monitoring services in a digital environment (developed on the basis of [8, 9]).

Tool/monitoring service	Functionality of brand reputation assessing
Medialogia http://www.mlg.ru/	Media citation; positive/negative; article size, band number, audience (PR value); photo availability; main/minor role; mentioning an object in the title; determining the tone of the messages, the most interesting information guides and influential authors
Agorapulse https://www.agorapulse.com	Tracking references to other pages (by keywords, hashtags, links), Statistics and analytics, reporting on key indicators: subscribers, engagement, clicks, etc.
Babkee http://www.babkee.ru	Semantic analysis of texts and their tonality. Detailed reporting on the authors of references (age, region, contacts, etc.)
Mention https://mention.com/en/	Tracking brand references in a billion of sources (social networks, forums, blogs, etc.). Responses to comments on Twitter, Facebook and Instagram directly through the Mention interface
SemanticForce http://www.semanticforce.net/ru/	Search for brand references and reviews on Twitter, FB, VK, OK, YouTube; monitoring for reviews, forums, HR-portals, photos in Instagram and presentations in SlideShare. Filtering by language, geography, tonality and other parameters
StarComment https://starcomment.ru	Search for mentions in Facebook, Instagram, VK, OK, Twitter, YouTube, Google Play, TripAdvisor, Flamp. Alerts for new messages on e-mail, in Telegram and Slack. Filtering by keywords, geography, links and other parameters
Talkwalker https://www.talkwalker.com	Monitoring social networks, blogs, news sites and forums (187 languages supported). Definition of tone of a brand mention with an accuracy of 90%. Real-time monitoring and access to two-year historical data. Analysis of not only textual, but also visual content (due to image recognition technology)
Scan http://scan-interfax.ru/	Press clipping, annotations, media digest, newsletter, press report, thematic analytical note, media image analysis, psychological media portrait, PRV, MO, Tonality, SPI, etc.

(*continued*)

Table 2. (*continued*)

Tool/monitoring service	Functionality of brand reputation assessing
Integrum http://www.integrum.ru/	Information about business reputation, industry analysis, tracking the mentioning and nature of publications (positive/negative), identifying opinion leaders in social networks, with the possibility of setting personal monitoring, to track further account activity in the network
IQBuzz http://iqbuzz.ru/	The frequency of references, Tonality of references. Distribution of positive and negative messages; Main themes. Grouping messages by specified topics, their subsequent monitoring. Search for reviews, trends, opinion leaders from 10,000+ sources, including LiveJournal, VK, YouTube, Instagram, Twitter. Automatic determination of tonality and its personalization. Statistics collection (gender, audience age, etc.), KPI tracking (coverage, involvement, etc.)
Katyusha https://katyusha.info/mobile	Information about companies and individual entrepreneurs, including foreign owners of Russian firms, financial statements, insolvency information, financial analysis and interrelationships between companies and individuals, transcripts of television and radio programs, arbitration court rulings, bank reviews and agency analysis materials
Jagajam https://beta.jagajam.com/ru/products/basic-analytics	Identifying the most popular posts, calculating the best time for publications. Analysis of pages by the growth of subscribers, the frequency of publications, engagement and a dozen of other SMM-metrics
BrandAnalytics https://br-analytics.ru	Definition of tonality with an accuracy of 80–90%, filtering spam and irrelevant messages. Notification of serious events in the information field (negative explosions, posts from opinion leaders, etc.)
Socialbakers http://www.socialbakers.com	Evaluation of the effectiveness of content, advertising and the whole strategy of SMM promotion in comparison with competitors. Search for user content (UGC), identifying leaders of opinions, determining the tone of comments
Youscan http://www.youscan.ru	Automatic identification of trends, opinion leaders and major sites
Socialstats http://socialstats.ru	Sorts posts by popularity, authors and users - by activity. Calculation of engagement, core index and other SMM metrics
Popsteps https://popsters.ru	Evaluation of the effectiveness of posts by involvement, taking into account the format, volume of text, date of publication. Sorting and filtering publications for easy detailed analysis

The analysis of the functions shows that the existing tools allow to assess the current effects on reputation in real time, evaluating their types, intensification or weakening, but do not allow to describe it meaningfully. A meaningful description of reputation is currently not possible with the help of monitoring services of a digital environment and requires special primary research based on a methodology combining qualitative and quantitative methods.

2.3   Expert Survey: Brand Reputation Assessment Tools and Metrics

An expert survey was conducted in January 2018. Previously, we formed a pool of experts, which included the heads of PR-departments of government and large commercial structures (banking and engineering, trade, restaurant business, the media). The initial list included 20 surnames, both men and women. With a slight excess of the latter, which reflects the gender imbalance in this field of employment (the excess of women). All experts worked in Rostov-on-Don. The city is the center of the Southern Federal District, one of the eight federal districts that constitute the Russian Federation. In that way, our selected experts met the requirements of an expert survey, namely, they had special knowledges in PR, extensive experience and management skills in this area, and their opinions were reference for analyzing the PR market in the South of Russia.

We carried out the pilot phase, during which the questionnaire passed its approbation, the changes and clarifications were made in the wording of the questions. In the final version, the expert questionnaire included three questions, two of which were tabular and the answers to which suggested scaling (frequency scale). The average time to complete the questionnaire was seven minutes.

The expert form offered anonymous filling, no contact information was required. The questionnaire was sent by mail. In some cases, an appointment with the expert was previously arranged to participate in the study; in other cases, the expert received a form with a cover letter. The questionnaire was aimed at studying the opinions of experts on the relevance of reputation management issues for modern PR practices, and also offered to comment on the use and frequency of using analytical systems and SMM-profile assessment metrics in PR practices of reputational management.

Only half of the initial expert list agreed to participate in the study and filled out the questionnaire. The main reasons for the refusal were the lack of time and experience with the analytical system and public assessment metrics in the field of reputation management. In other words, not all PR practices of organizations in Southern Russia include modern Internet technologies for working with reputations, and partly operate in their classic pre-Internet form (press releases, press conferences, briefings, corporate media, etc.) which, in our opinion, reduces their effectiveness.

The first question of the expert questionnaire concerned the relevance of reputation management for modern PR practices. All experts referred these questions to the most relevant (three experts) or important, along with other issues (seven experts). In other words, issues of reputation management are already included in the main issues on the agenda of a modern PR-specialist in Southern Russia, however, while most experts assess them as "just important", leaving maximum priority for other tasks.

The second question of the expert questionnaire concerned the assessment of the frequency of using 17 analytical systems within the framework of the reputation management. Experts noted only nine systems as used, with the Medialogia analytical system leading (nine expert evaluations). At the same time, attention is drawn to the fact that its use is not assessed as permanent, but is characterized by a "case by case" assessment. The remaining systems, most likely, are not widely used in PR practices for managing the reputation of specialists in the South of Russia (Table 3).

Table 3. Expert assessment of the frequency of analytical systems usage

№	Tools	Permanent use	Rare use
1	Medialogia	3	6
10	IQBuzz	1	2
8	Scan	2	–
13	BrandAnalytics	2	–
3	Babkee	1	–
18	The other TNS-media_____	1	–
5	SemanticForce	–	1
6	StarComment	–	1
15	Youscan	–	1

The third question of the expert questionnaire was proposed to assess the SMM-profile metrics on a 4-point scale. Where: 1 - often used in professional activities; 2 - sometimes used; 3 - not used; 4 - difficult to answer (Table 4).

Table 4. Expert estimates of metrics

№	Metrics/experts	1	2	3	4	5	6	7	8	9	10
I	*General SMM-profile metrics*										
1	Number of subscribers/followers	1	1	1	1	1	1	1	3	1	1
2	Increase of subscribers/followers for a certain period	2	3	1	2	2	1	1	3	1	1
3	Number of unsubscribers for a certain period	3	3	2	4	2	1	2	3	1	1
4	Number of views of profile/post	1	1	1	1	1	1	1	3	1	1
5	Organic coverage (the number of people who at least once contacted with profile/post)	4	1	1	3	1	1	2	3	2	1
6	Paid profile/post coverage	4	3	1	3	1	1	3	3	2	1
7	Viral profile/post coverage	4	1	1	3	1	2	1	3	2	1

(continued)

Table 4. (*continued*)

№	Metrics/experts	1	2	3	4	5	6	7	8	9	10
II	*Interaction metrics with profile audience*										
1	Audience activity (number of users who performed a certain number of actions (1.2, 3 actions)	1	2	1	1	1	1	1	3	2	2
2	Level of attractiveness: Likes/Followers * 100%	1	2	1	3	1	1	1	3	1	2
3	Level of sociability: Comments/Followers * 100%	1	2	1	3	2	1	1	3	1	2
4	Spread rate: Shares/Posts (number of posts) * 100%	1	1	1	3	1	1	2	3	1	3
5	Engagement ratio: number of likes + reposts + comments/total number of subscribers	1	2	1	3	1	1	1	3	2	3
6	Check-ins	3	2	2	3	3	2	3	3	4	4
7	Hashtags	2	2	1	3	1	2	1	3	1	3
8	Negative user reactions	3	1	1	2	1	1	1	3	4	1
9	User content (posts, photos, videos uploaded by subscribers)	3	2	1	1	1	1	1	2	3	3	2
III	*The interaction of account with the external environment*										
1	Conversion to the site	2	1	1	1	1	1	1	3	4	1
2	Number of leads	2	2	1	2	1	1	3	3	4	3
3	Account mentions	2	1	1	3	2	1	2	3	4	2

Statistical processing included building a matrix of ranks. To accomplish this, all the metrics were presented with a single list with a continuous numbering, and expert assessment 4 - "I find it difficult to answer" for the correctness of the calculation was replaced by rating 3 - "not used" (Table 5).

Table 5. Rank matrix

Metrics/experts	1	2	3	4	5	6	7	8	9	10	Sum of ranks	d	d^2
x1	4	4.5	9	3	7.5	8.5	6	10	4.5	5.5	62.5	−37.5	1406.25
x2	10	18	9	7	16.5	8.5	6	10	4.5	5.5	95	−5	25
x3	16	18	18.5	14	16.5	8.5	14	10	4.5	5.5	125.5	25.5	650.25
x4	4	4.5	9	3	7.5	8.5	6	10	4.5	5.5	62.5	−37.5	1406.25
x5	16	4.5	9	14	7.5	8.5	14	10	11	5.5	100	0	0
x6	16	18	9	14	7.5	8.5	18	10	11	5.5	117.5	17.5	306.25
x7	16	4.5	9	14	7.5	18	6	10	11	5.5	101.5	1.5	2.25
x8	4	12.5	9	3	7.5	8.5	6	10	11	12.5	84	−16	256
x9	4	12.5	9	14	7.5	8.5	6	10	4.5	12.5	88.5	−11.5	132.25
x10	4	12.5	9	14	16.5	8.5	6	10	4.5	12.5	97.5	−2.5	6.25
x11	4	4.5	9	14	7.5	8.5	14	10	4.5	17	93	−7	49
x12	4	12.5	9	14	7.5	8.5	6	10	11	17	99.5	−0.5	0.25
x13	16	12.5	18.5	14	19	18	18	10	16.5	17	159.5	59.5	3540.25
x14	10	12.5	9	14	7.5	18	6	10	4.5	17	108.5	8.5	72.25
x15	16	4.5	9	7	7.5	8.5	6	10	16.5	5.5	90.5	−9.5	90.25
x16	16	12.5	9	3	7.5	8.5	14	10	16.5	12.5	109.5	9.5	90.25
x17	10	4.5	9	3	7.5	8.5	6	10	16.5	5.5	80.5	−19.5	380.25
x18	10	12.5	9	7	7.5	8.5	18	10	16.5	17	116	16	256
x19	10	4.5	9	14	16.5	8.5	14	10	16.5	5.5	108.5	8.5	72.25
∑	190	190	190	190	190	190	190	190	190	190	1900	–	8741.5

The resulting statistics allowed us to calculate the Kendall coefficient of concordance for the related ranks case. It was W = 0,24. This indicator shows a weak degree of experts consistency, which, in our opinion, is a consequence of various professional tasks of the experts of the surveyed group, as well as the lack of a group expert assessment regarding the metrics used and the reliability of their results.

However, for individual indicators it is possible to draw certain conclusions. Sic, the total ranking indicator allows us to identify the two most significant metrics in expert assessments (x1 and x4). These are: 1. The number of account subscribers and 2. The number of views of the account/post. Most likely, these metrics are the most common to use, are considered by experts to be the most reliable and indicative parameters that are understandable to non-specialists and are guided by customers and specialists in related fields of communication. The remaining numerous metrics are specific, we do not yet have the character of universal indicators and have significant limitations (Table 6).

Table 6. Analysis of the studied factors significance

Factors	Sum of ranks
x1	62.5
x4	62.5
x17	80.5
x8	84
x9	88.5
x15	90.5
x11	93
x2	95
x10	97.5
x12	99.5
x5	100
x7	101.5
x14	108.5
x19	108.5
x16	109.5
x18	116
x6	117.5
x3	125.5
x13	159.5

3   Results

The study leads to the following conclusions.

1. Expert evaluation shows that reputational management issues are considered as important and priority ones for regional PR practices of the South of Russia.

2. In PR practices, automatic monitoring of texts on the Internet is widely used to diagnose reputational threats and other parameters of reputation management. There are several analytical systems on the market, however, the leader is the Russian system Medialogy. In other words, the following analytical parameters are considered the most popular in this area: media citation; positive/negative texts; article size, band number, audience (PR value); photo availability; the main/secondary role of the person; mentioning an object in the title; the definition of the tone of the messages, the most interesting infopovody and influential authors.

   At the same time, even in large public and private organizations, PR practices are retained without the analytical systems usage and online reputation management metrics, which significantly reduces the effectiveness of professional activities. The main reasons for this situation, in our opinion, include the delay in educational programs for advanced training and retraining in the field of PR, due to both the internal processes of reforming Russian education and the high rate of innovations occurring in the digital environment.

3. A large number of existing Internet services, allowing the user to track marketing and communication activities in a digital environment in real time and evaluate its effectiveness, has not yet affected the scope of their use in real PR practices. Experts indicate that they use no more than 1–2 analytical systems and most often such use is not systemic. This confirms our assumption that reputation assessment tools and metrics are not sufficiently systematized, and methods of using them in relation to the needs of regional markets are poorly developed.

4. The specifics of tools and metrics usage for assessing the online reputation of the brand and the client are related to the use of KPI (Key Performance Indicators) principles, which measure cognitive, emotional, evaluative, and behavioral aspects of stakeholder responses. All metrics can be grouped into three blocks: 1. General SMM-account metrics; 2. Interaction with the audience of the account metrics; 3. Account interaction with the external environment. Metrics «The number of public subscribers» and «the number of views of the account/post» were the most reliable and versatile.

5. In our opinion, differences in expert assessments show the transitive state of the regional PR market, the change of methodological approaches to professional activity, generational (age) and professional breakdown, when marketers who can design and manage various communicative by the phenomena.

4 Conclusion

The purpose of the study was to identify the specifics of managing online reputation, in particular, assessing the effectiveness of reputation in a digital environment. Theoretically, the concept of "reputation" is considered, its structure, the importance of reputation assessment is proved, factors positively and negatively affecting reputation in the communication field are described, digital tools of reputation assessment are recorded and described. In empirical terms, expert assessments were made regarding the specifics of managing online reputation, and conclusions were drawn about the features of the regional PR market. In particular, the relevance of issues of reputation management for regional PR practices has been proved. Regional PR practices are characterized by the extensive use of automatic monitoring of texts on the Internet and the use of KPI principles; the market is at the stage of changing work methodologies, generational and professional scrapping; the tools and metrics of reputation assessment used by specialists are not systematized, there are problems of delayed vocational education and professional retraining.

The limitations of the study are due to the lack of quantitative data that could reliably indicate specific trends in the regional PR market, as well as a situation where experts, when the main expert criteria coincide (experience, leaders of large divisions, etc.), represent different segments of the economy, perform various professional tasks, which could be the reason for the lack of a group expert assessment regarding the metrics used.

Further research should be supplemented with quantitative methods (surveys), expanding the geographical boundaries of research (comparative interregional or international studies), include in research tasks an analysis of educational programs of basic vocational education and a system of advanced training and retraining in the direction of PR for compliance with the professional tasks of building and managing online - reputation. In general, the topic of forming and managing online reputation is recognized by us as relevant and heuristic.

References

1. Shoven, P.: Sociology of reputation. Domest. Notes **1**, 85–99 (2014)
2. Ryaguzova, E.: The reputation of the individual as the credibility of another, vol. 14, 1st edn, pp. 71–76. Saratov University Tidings (2014)
3. Belevskaya, E.: Reputation as a sociocultural phenomenon: mechanisms and mutual influence of culture and reputation. In: Directions of Modernization of the Modern Innovative Society: Economics, Sociology, Philosophy, Politics, Law. Materials of the International Scientific and Practical Conference, vol. 1, pp. 37–40 (2015)
4. Brick, E.: The ratio of the right to protection of business reputation and the right to freedom of speech and information. In: Modern Problems of Law-Making and Law Enforcement. Materials of the All-Russian Student Scientific and Practical Conference, pp. 100–102 (2016)
5. Weber, M., Erickson Sheri, L., Stone, M.: Corporate reputation management: Citibank's use of image restoration strategies during the U.S. Banking Crisis. J. Organ. Cult. Commun. Confl. **15**(2) (2011). https://www.thefreelibrary.com/Corporate+reputation+management% 3A+Citibank%27s+use+of+image+restoration...-a0263157563. Accessed 10 Feb 2019
6. Kravchuk, M.: Methods for evaluating the effectiveness of advertising tools (2019). http:// www.cossa.ru/articles/155/35288/. Accessed 07 Feb 2019
7. Halilov, D.: Social Media Marketing. Feber, Moscow (2013)
8. Ivanichev, I.: Monitoring of social networks for reputation management: 10 services that can do it. https://texterra.ru/blog/monitoring-sotsialnykh-setey-dlya-upravleniya-reputatsiey-servisov-kotorym-eto-pod-silu.html. Accessed 07 Feb 2019

9. Ivanichev, I.: How to evaluate SMM promotion in numbers: 27 services of statistics and analytics of social networks. https://texterra.ru/blog/kak-otsenit-smm-prodvizhenie-v-tsifrakh-servisov-statistiki-i-analitiki-sotssetey.html. Accessed 07 Feb 2019

10. Ivanichev, I.: KPI in SMM: 30+ social marketing effectiveness metrics. https://texterra.ru/blog/kpi-v-smm-metriki-effektivnosti-marketinga-v-sotsialnykh-setyakh.html. Accessed 07 Feb 2019

11. KPI indicators in social networks. http://adindex.ru/publication/tools/2013/09/24/102503.phtml?utm_source=facebook&utm_medium=pub&utm_campaign=kpi. Accessed 07 Feb 2019

12. Gudkov, S.: Key performance indicators for an online store. https://gaap.ru/articles/Klyuchevye_pokazateli_effektivnosti_KPI_dlya_internet_magazina/ (дата обращения 30.01.2019)

13. Industrial standard. http://www.akarussia.ru/knowledge/industrial_standarts Accessed 07 Feb 2019

14. Koshik, A.: Metrics in social media. http://www.icontext.ru/blog/metriki-v-sotsmedia-perevod-stati-avinasha-koshika/. Accessed 10 Feb 2019

15. Ovrutsky, A.V.: Reputation as a psychological and economic resource. In: Economic Psychology and Behavioral Economics in the Context of Global Social and Economic Changes, Materials of the All-Russian Scientific Conference, pp. 47–51 (2014)

16. Global RepTrak 100: The world's most reputable companies (2013). https://www.rankingthebrands.com/PDF/Global%20RepTrak%20100%20Report%202013,%20Reputation%20Institute.p

17. Ovrutsky, A.V.: Reputation. Reputation discourse. Reputational damage. Psychologist (4), 10–18 (2016). http://e-notabene.ru/psp/article_19631.html. Accessed 10 Feb 2019

Analysis and Detection of a Radical Extremist Discourse Using Stylometric Tools

Tatiana Litvinova[1,2(✉)] ⓘ and Olga Litvinova[1] ⓘ

[1] Voronezh State Pedagogical University,
86 ul. Lenina, Voronezh 394043, Russia
centr_rus_yaz@mail.ru
[2] Plehanov University, Stremyanny lane 36, Moscow 117997, Russia

Abstract. The amount of radical extremist texts published online is constantly growing with researchers as well as agencies working hard to design tools to detect and remove this type of content. However, a great deal of this research effort is being focused on English-language materials. Meanwhile, in the last few years extremist groups including radical Islamic ones have started to produce content in languages other than English and, more specifically, in Russian which has been increasingly employed. Qualitative analysis of this content is important but not sufficient. We show the possibility of using stylometric tools allowing different types of texts analysis to be performed in order to analyze and detect radical extremist content. Using a meticulously designed dataset consisting of real-world extremist forum texts on two topics and texts by common Internet users on the same topics, we were able to show that the authors from extremist forum remain consistent in their style irrespective of the topic and their texts could thus be detected using a variety of methods. We conclude by the underscoring the importance of combining qualitative and quantitative analysis of a radical extremist discourse for better understanding of radical minds and development of counter-extremist tools.

Keywords: Extremist texts · Stylometry · Corpus linguistics

1 Introduction

Despite the fact that social media has literally taken over our day-to-day lives and has emerged as a powerful tool, it is not always employed with the best of intentions. Therefore more often than not it becomes a platform for groups and individuals who use them in order to promote violent ideologies, cause confusion and instill fear in general public. One of such groups are extremist Muslims advocating jihad (for discussion about term "Islamist extremist" and related terms see [1, 9]). The necessity of analyzing their discourse and constructing tools for automatic detection of such texts is quite obvious. However, the work in this area to date has focused almost exclusively on English language data. Meanwhile, it seems urgent to examine similar texts written in other language including Russian as the terrorist groups banned in the Russian Federation are conducting their propaganda in Russian. E.g., an article compellingly titled "ISIS speaks Russian. The Islamic State speaks Russian – a new market for ISIS

© Springer Nature Switzerland AG 2020
T. Antipova and Á. Rocha (Eds.): DSIC 2019, AISC 1114, pp. 30–43, 2020.
https://doi.org/10.1007/978-3-030-37737-3_3

propaganda" [8] says that 2014 and 2015 have become turning years in ISIS' attention towards Russia - both government, officials, as well as potential fighters. An analytical review of the activities of the ISIS (banned in the Russian Federation) in the year of 2018 conducted by The Meir Amit Intelligence and Terrorism Information Center, showed that most of the informal media products of ISIS supporters appear in foreign languages including Russian[1]. Manual inspection of social media content is time- and labor-consuming, that is why it is crucial to develop automatic tools for detection of radical extremist content. In this paper we seek to show the potential of simple stylometric tools in analyzing and detecting radical extremist content by the example of texts from Islamist extremist forum. Our understanding of stylometry is broad: we define stylometry as the computer-assisted quantitative study of a writing style.

2 State-of-the-Art

Computer analysis of Islamic extremist texts currently involves the use of different lexicon-based tools like Linguistic Inquiry and Word Count (LIWC), content analysis software[2] which calculates the percentage of words in the entire text that matches the words in the predefined grammatical and semantic categories [13] as well as complex corpus-linguistics tools, such as Wmatrix which automatically assigns semantic codes to words in the corpus [14]. Most studies which employ these tools, however, are based on English-language data partly due to the fact these tools were designed and broadly validated on English texts. As Baker and Vessey notes, the work in the field of linguistics analysis of extremist texts to date has focused almost exclusively on English language data which 'means that the extent to which such materials pose a threat to national security "cannot be properly understood" [1]. Another problem involves using lexicon-based tools per se: tagging of words especially in non-English texts can lead to numerous misclassifications.

It is only very recently that researchers started to analyze such texts in a multilingual perspective. As a rule, these works do not rely on predefined lexicon but rather researchers compile and analyze frequency lists, find collocates, etc. For example, Baker and Vessey [1] preformed contrastive analysis of English and French ISIS media texts. They used AntConc software to derive keyword lists and then examined the top 500 keywords for each list (collocate, cluster and concordance analysis was conducted). A good examples of the work on detection of jihadi hate speech in multilingual setting are [4, 5] where multilingual (mostly English, about 40%, and Arabic, 30%) Twitter data were analyzed with different Natural Language Processing (NLP) tools to perform keyword analysis, classification (with character trigrams as features), including cross-domain experiments. Russian-language Islamist extremist texts were analyzed with WordSmith Tools software package to identify most frequent words and word clusters, build concordances and identify collocates [12]. To date, we are unaware of the works dealing with classification experiments on Russian-language

---

[1] https://www.terrorism-info.org.il/en/isiss-media-network-developments-2018-future-courses-action/.

[2] https://liwc.wpengine.com/ (last accessed 09/01/2019).

texts by radical extremists. In this paper we seek to classify texts as extremist/non-extremist using simple stylometric tools. as well as to apply other stylometric techniques to this type of content. What sets our work apart is not only with material but the methodology as we show how different stylometric tools can be instrumental in analysis and detection of radical Islamist extremist content.

3 Materials and Methods

3.1 Corpus Description

The material of the study are texts of the KavkazChat dataset which contains texts from a Russian-language forum with a focus on jihad in the North Caucasus. It is part of Dark Web forum portal which is part of the Dark Web collection [2]. This forum was defined as extremist by Russian authorities (currently, the site is blocked) and researchers who developed the Dark Web forum portal. KavkazChat dataset contains 699,981 posts written by 7,125 members in the period 3/21/2003–5/21/2012. These posts are organized into 16,854 threads (topics). For the ongoing analysis we have selected two threads dedicated to the discussion of two severe terrorist attacks for which Caucasus Emirate leader Doku Umarov claimed responsibility: the 2010 Moscow Metro bombings and the 2011 Domodedovo attack. For a detailed review of the first thread see [12]. The second thread dedicated to the discussion the deadly terrorist attack at the international arrivals terminal of the Domodedovo Airport on January 24, 2011. All the texts by extremist forum authors were compiled into one txt file per thread. Citations and usernames were removed. As the second thread is smaller (21453 tokens), we only extracted the first 21453 tokens from the first thread to avoid data imbalance.

These threads contain a lot of extremist sentences. A typical example goes like this: *Разрушайте прав. здания и забирайте от туда всё материально ценное - это всё награбленное у вас же самих* ("Blow up the government buildings as anything that is of any value in there is whatever has been stolen from you"). The danger is that even those people who came to the forum to find out more about the Islamic ideology, became tolerant to the actions of terrorists: *Теракт этот, вывел меня на ваш форум, где я не стал горячиться, а решил просто поговорить с людьми, понять вас, и честно, мой взгляд перевернулся на 180 градусов, я конечно не считаю что уничтожение мирных граждан, можно признать как необходимой мерой, но и вас я понимаю* ("This terrorist attack is actually why I joined the forum where I kept my cool and just chatted with some people and honestly, it has totally changed my perspective. Well, I don't really think that slaughtering innocent people is necessary but I think I kind of relate to you as well"). We will further refer these datasets as **Extremist_Moscow** and **Extremist_DMD**.

What we did next was to compile two datasets which consist of the texts by common Internet users who commented on the news about these same attacks. These corpora were as large as the datasets Extremist_Moscow and Extremist_DMD derived from the Kavkazchat. The first dataset contains comments on the material on the

Moscow bombing posted by an online newspaper "The Village"[3] and by the Radio Svoboda[4]. The second dataset compiles comments from users for the news on Domodedovo attack posted by the Radio Svoboda[5] and Komsomolskaya Pravda[6]. We will further refer these datasets as **User_Moscow** and **User_DMD**.

Overall, the volume of datasets is 81796 tokens. We have prepared two versions of our datasets in accordance with the tasks at hand: first, datasets were divided into 10000-word chunks **(Corpus10000)** and secondly, into chunks of around 1000-words (we preserved the sentences boundaries) (we further refer to it as **Corpus1000**).

Using datasets containing texts where authors discuss similar, but different events can help us to answer the questions: are there any general patterns of radical extremist discourse and, if so, can we detect segments of this discourse by means of stylometric tools, including cross-topic scenario?

3.2   Methods

We have performed several experiments using different techniques: (1) cluster analysis; (2) classification; (3) oppose; (4) rolling stylometry. All the experiments were performed with R package Stylo [6]. This package combines sophisticated state-of-the-art algorithms of text analysis with a user-friendly interface and are widely used for stylometric studies. In stylometry, the commonly accepted type of features (style-markers) is frequencies of the most frequent words (MFWs), even if there is no consensus on how many MFWs should be used [7]. As our study explores the potential of stylometric tools for analysis and detection of Islamist extremist texts, we limited ourselves to the use of these simple, yet widely used and effective features.

4   Results and Discussion

4.1   Cluster Analysis

One of the techniques implemented in Stylo package and the one widely used in stylometry is cluster analysis, namely hierarchical cluster analysis (HCA). This is an exploratory statistical technique that identifies group samples by similarity with dendrograms used to visualize the results. In HCA, individual data points are joined to the closest ones in a step-by-step (hierarchical) procedure until one large cluster containing all the data points are created. We used this technique to find similarity in our corpus. To reveal natural structure in our data, we did not specify the number of clusters prior to the analysis. Namely, we would like to reveal the reason for clustering text (topic or author group: extremist or non-extremist) on different ranges of MFWs (from 100 to 1000 with increment 100). We have performed our first experiment on Corpus1000

---

[3] http://www.the-village.ru/village/people/people/88720-v-moskovskom-metro-proizoshel-vzryv#comments.

[4] https://www.svoboda.org/a/1996131.html.

[5] https://www.svoboda.org/a/2285715.html.

[6] https://www.vrn.kp.ru/daily/25625.5/792227/.

(72 text samples). At the first stage, we used very basic settings: 100 MFWs as features, Classic Delta as a distance measure (Fig. 1). As Fig. 1 shows, two large clusters are clearly seen: one consists predominantly of texts from extremist forum, and the other of texts by users. Although the division is far from perfect, let us remind the reader that this result was obtained with only 100 word as features, which are mostly function words, and short segments of texts (around 1000 words).

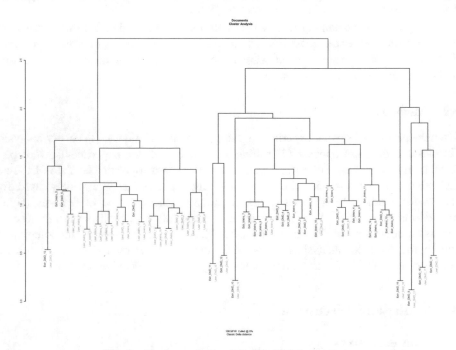

Fig. 1. Cluster analysis with 100 MFWs and Delta

The results with other distance measures and different numbers of MFWs (up to 1000) are quite similar, which shows that there are differences between an extremist and non-extremist discourse on different levels of MFW range (both function and content words). Other exploratory techniques (e.g., multidimensional scaling, Fig. 2) show the same pattern: texts by the authors from the same group are closer to one another (or have shorter distances) in the graph (i.e. more similar linguistically) than those by the authors from different groups irrespective of the topic although there is an overlap for some samples. We have a plausible explanation for this: our close reading of the user texts revealed that some texts by the users contain calls for aggression of the Muslims and are extremist by their nature.

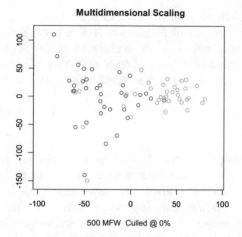

Fig. 2. Multidimensional scaling with 500 MFWs (red dots are extremist texts, green dots are user texts)

Next, we have performed a cluster analysis on Corpus10000. The results with large segments of texts are better in terms of division of texts by the group of the authors. We went further to validate the results of the cluster analysis using the Bootstrap Consensus Tree (BCT) technique. BCT enhances the explanatory power of visualization with a procedure of validation inspired by advanced statistical methods: it runs lots of cluster analyses with different numbers of MFWs and shows a radial graph with clusters emerging from the centre of the screen. The consensus strength of 0.75 was chosen, which means that visualized linkages appear in at least 75% of the clusters.

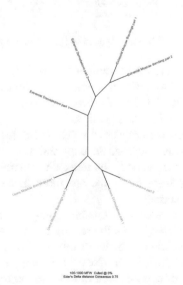

Fig. 3. Bootstrap Consensus Tree with 100-1000 MFWs

We performed the analysis with different distance measures, but the results were similar: texts by the extremist authors are more similar to each other than those by non-extremist authors. There are, however, differences in the further division of texts: whereas two topic-dependent clusters of texts by the users are clearly seen, texts by extremist related to the discussion of the Domodedovo terror attack are more divergent stylistically (Fig. 3).

4.2   Classification

Since we have established that the texts by the extremist and non-extremist authors differ in the use of the most frequent words, we next move on to the classification phase using different classification methods as implemented in Stylo package: Delta, k-Nearest Neighbors classification, Support Vectors Machines, Naive Bayes, and Nearest Shrunken Centroids [6, 11]. Our aim is to check the possibility of the classifiers to predict the group membership of the authors. Once the number of samples per class is equal (36 samples per group), random baseline is 50%. We ran classification experiments on **Corpus1000**.

To assess the quality of the trained model, we performed stratified cross-validation, or a number of swaps between the training and the testing sets preserving the representation of classes from the training set. Namely, we performed stratified 6-fold cross-validation with different number of MFWs (from 100 to 1000 with increment 100) and the different classifiers mentioned above. The results obtained with the various classifiers are similar. For the sake of exemplification, below (Fig. 4) we present the results obtained with Naive Bayes classifier with a varying number of MFWs.

100⊕0	200⊕0	300⊕0	400⊕0	500⊕0	600⊕0	700⊕0	800⊕0	900⊕0	1000⊕0
Min. :66.67	Min. : 75.00	Min. : 75.00	Min. :58.33	Min. :66.67	Min. :58.33	Min. :83.33	Min. :75.00	Min. :75.00	Min. :58.33
1st Qu.:75.00	1st Qu.: 77.08	1st Qu.: 75.00	1st Qu.:68.75	1st Qu.:68.75	1st Qu.:68.75	1st Qu.:83.33	1st Qu.:77.08	1st Qu.:77.08	1st Qu.:70.83
Median :79.17	Median : 83.33	Median : 79.17	Median :75.00	Median :79.17	Median :75.00	Median :83.33	Median :83.33	Median :83.33	Median :87.50
Mean :77.78	Mean : 84.72	Mean : 83.33	Mean :75.00	Mean :76.39	Mean :75.00	Mean :84.72	Mean :81.94	Mean :81.94	Mean :80.56
3rd Qu.:83.33	3rd Qu.: 89.58	3rd Qu.: 89.58	3rd Qu.:81.25	3rd Qu.:83.33	3rd Qu.:81.25	3rd Qu.:83.33	3rd Qu.:83.33	3rd Qu.:83.33	3rd Qu.:91.67
Max. :83.33	Max. :100.00	Max. :100.00	Max. :91.67	Max. :83.33	Max. :91.67	Max. :91.67	Max. :91.67	Max. :91.67	Max. :91.67

Fig. 4. Results of Naive Bayes classifier with 100-1000 MFWs

We can see that depending on the number of MFWs, the mean accuracy of the classifier ranges from 75 to 84.72% with stable results (not less than 75%) obtained in the range 200–300 and 700–900 MFWs. This fact shows that one should carefully select the number of features for classification.

It is notable that a general pattern was revealed during manual analysis of the output of different classifiers: the classifiers mistakenly assigned user texts on Domodedovo topic to the extremist, and vice-versa. This reflects a general trend previously founded with BST: extremist texts from this topic are more heterogeneous stylistically and therefore more difficult to classify.

In the next step we performed leaveoneout cross-validation where each sample is used once as a test set (singleton) while the remaining samples form the training set. In case of leaveoneout cross-validation, better results were obtained with a higher number of MFWs (75% with 1000 MFWs and Nearest Shrunken Centroids). These experiments show the possibility of classification of the texts on the same topic as written by extremist or non-extremist authors based on such simple features as MFWs.

4.3 Oppose

In an analysis of extremist discourses, a researcher is typically interested in their linguistic characteristics. To reveal differences in the word usage between the extremist and non-extremist authors, we used the Stylo function **oppose**. It performs a contrastive analysis between two given sets of texts using Burrows's Zeta in its different flavours, including Craig's extensions [3, 10]. This function splits input texts into equal-sized slices and checks the emergence of particular words over the slices. The number of slices in which a given word appeared in the subcorpus A and B is then compared using the criteria chosen by the user.

Unlike keyword analysis widely used for corpora comparison [5, 12], Zeta ignores frequencies of the words and concentrates on their consistency [3, 10] which is especially useful for our case as it is our goal to get rid of the effect of individual voices and to reveal the characteristics of the texts by the group of authors. The script generates a list of words significantly preferred by a tested author, and another list containing the words which are noticeably avoided.

First, we have performed this comparison for texts on the two topics (the Domodedovo and Moscow terrorist attacks) separately and then compile all the texts written by extremist forum authors in one file and all the texts by a non-extremist author in another file. We applied the same settings for all the three stages of the analysis: text slice length was set to 1000 words, text slice overlap was set to 0. We chose to avoid low-frequency words by setting a rare occurrences threshold to 5. Zeta filter threshold was set to 0.1

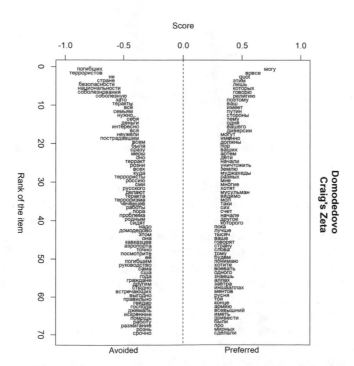

Fig. 5. Words preferred by the extremist forum authors in Domodedovo attack discussion

(default value) to eliminate words of weak discrimination strength. We use Craig's extensions of Zeta as the oppose method. Zeta analysis excludes the extremely common words and concentrates on the middle of the word frequency spectrum which is considered to be a powerful but simple method of measuring differences among authors [10] (in our case, we use it to measure the differences among groups of authors).

Figures 5, 6 and 7 show the results of the comparison (words are ranked based on the strength of differences).

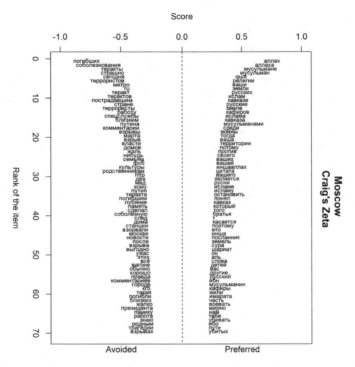

Fig. 6. Words preferred by the extremist forum authors in the Moscow attack discussion

We can see that the lists of words preferred by the extremist and non-extremist authors from two topics have a lot in common. While discussing the terrorist attacks, extremist forum authors refer to religion (category "Allah and worship" as introduced in [1]: аллах "Allah", мусульманин "Muslim", религия "religion"), Negative Othering (different forms of pronoun *Ваши* "your", *кафир* "unbeliever", *против* "against", *русня* "Russians"), war and death (*войны* "war", *воевать* "to fight", *убивать* "to kill", *убитых* "the killed"), *land* (*Кавказ* "Caucasus", *земли* "land"). On a general level, the rhetoric in extremist speech is religiously polarizing, with vitriolic references to unbelievers, apostates, crusaders, and so on which is similar to findings made on multilingual twitter data [4, 5]. The adverb *so "поэтому"* is typical for both lists of words preferred by extremist forum authors, which indicates explanations of terrorist actions. The users irrespective of the topic name killed people as *погибшие*

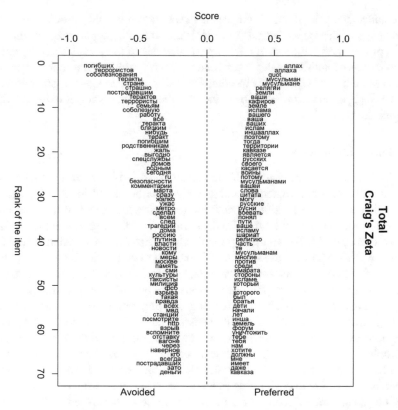

Fig. 7. Words preferred by the extremist forum authors in the discussion of both terrorist attacks

("dead"), and those people who have done those terror attacks are named as *тер-рористы* "terrorists" (word avoided by extremist forum authors). They express their feelings (*страшно* "scared", *соболезную* "condolences", *ужас* "terror", *жалко* "sorry"). There are also some differences, however. The name *Путин* ("Putin") avoided by the extremists in their discussion of the Moscow terrorist attack becomes one of the preferred ones in the discussion of the Domodedovo terrorist attack. In the Moscow metro discussions the users choose a lot of words related to law-enforcement and authorities (*ФСБ* "Federal Security Service", *КГБ* "KGB", *Путина* "Putin") (for details see [12]), while in the discussions of the Domodedovo explosion there are fewer differentiating words of this semantic group and they are different in nature (*руко-водство* "ruling authorities"), but more words related to nationality (*кавказцев* "Caucasians", *чеченцев* "Chechens", *русские* "Russians" – the last word is preferred by the extremists in the metro attack discussion). The results of the frequency contrastive analysis show the extremists picking up what users have done when they were blaming the Moscow metro attacks on the authorities. However, according to the oppose analysis (as well as close reading) national issues are more frequently brought up in the Domodedovo attack discussion.

A detailed analysis of the differences between extremist and non-extremist texts is beyond scope of this paper but this brief look into the results generated by the oppose function proves the usefulness of comparative analysis of the texts written by extremist and non-extremist authors under topic control. Such analysis, although rare in current literature (a common approach is to compare extremist texts against general language corpora, see [1] as an example) helps to identify the characteristics of extremist rhetoric more precisely.

4.4   Rolling Stylometry

Earlier in this paper, we performed the classification experiments of the texts as written by an extremist or a non-extremist author which can be useful in assisting the expert (forensic linguist etc.) in making the decision regarding the nature of the texts etc., but in the real-world scenario we typically deal with a lot of non-extremist texts and few extremist ones (see [5] for similar opinion) which leads to an imbalance in data and therefore to the difficulties associated with building classifiers. To solve the task of detection of extremist texts among "normal" discourse (i.e. an extremist user on a general forum or a chat), we propose to imply rolling stylometry technique which is employed in analysis of collaborative writing. Rolling stylometry [7] uses the concept of moving window in combination with standard supervised classification techniques. Unlike standard procedures, aimed at assessing style differentiation between discrete text samples, this method, supported with compact visualization, attempts to look inside a text represented as a set of linearly sliced chunks, in order to test their stylistic consistency. The ultimate goal of this technique is to detect stylistic takeovers, which is exactly our goal. Three flavors of the method have been introduced in Stylo package: (1) Rolling SVM, relying on the support vector machines classifier, (2) Rolling NSC, based on the nearest shrunken centroids method, (3) Rolling Delta, using the classic Burrowsian measure of similarity. We used all the three classifiers for our experiment.

To perform this analysis, we need a reference corpus (i.e. training corpus) which represent the style of two group of authors and a test file. We put texts by users and extremist forum authors dedicated to the discussion of Moscow metro bombings into a reference corpus (two 10000-word length files from User_Moscow dataset, two 10000-word length files from Extremist_Moscow dataset). The test file to be chunked automatically into equal-sized segments consisted of the texts from User_Domodedovo dataset (split into one file). Then two fragments from the Extremist_Domodedovo dataset were inserted into the test file. The first fragment (5 randomly chosen chunks from Corpus1000) was inserted after 6324 first words of the test file, the second fragment (5 randomly chosen chunks from Corpus1000) was inserted after 18041 words of the test file. In Fig. 8 a ground-truth division of the test file into 5 parts (1st part written by users, 2nd part – by extremist forum authors, etc.) is marked with a vertical dashed lines with corresponding section names. Sections attributed by the classifier to the extremist forum authors marked red, those attributed to users are green. Graphical output of classifiers is slightly different: while NSC output gives only one ultimate answer, SVM output keep fewer probable candidates slightly in the shadow (Fig. 9). Rolling Delta ranks the possible author: the most probable author goes first, i.e. the bottom stripe indicates the first ranked candidate (i.e. the most probable), then comes the second (Fig. 10). When classification results are consistent across a number of chunks, the stripes tend to be unicolored rather than patchwork-like.

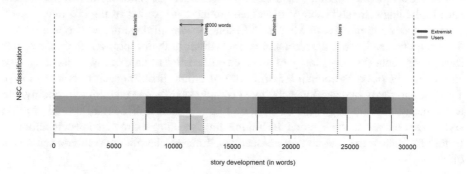

Fig. 8. Test file analysis with Rolling NSC and MFW=500, window size: 2,000 words, sample overlap: 500 words

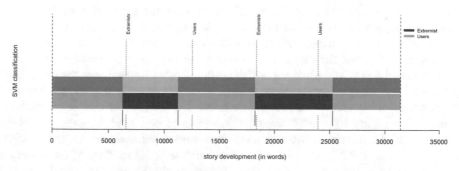

Fig. 9. Test file analysis with Rolling SVM and MFW=200, window size: 5,000 words, sample overlap: 4500 words

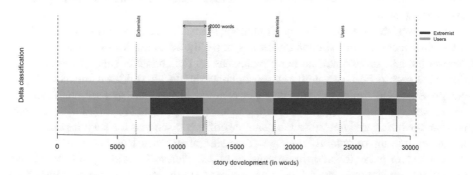

Fig. 10. Test file analysis with Rolling Delta and MFW=200, window size: 5,000 words, sample overlap: 500 words

Regardless of the number of MFWs (in the range 100-1000 with increment 100) and length of slices (we tested slices from 1000 to 5000 with 500-word increment) and different overlap values (from 500 to 4500 with 500-word increment), all the classifiers detect a few style breaks in the dataset. Although the result is not without flaws, a clear shift in the linguistic style can be discerned around the position of the takeovers. Even with a rather small number of MFWs (200) some meaningful results were obtained. Let us remind the reader that the train and test sets differ in their topic, which confirms our hypothesis about the consistency of some characteristics of an extremist discourse (see similar results of cross-domain classification of online jihadi speech obtained in [5]). The last part (texts by users) was the most complicated for the classifiers probably due to similarity of some fragments of the texts to the extremist discourse. Overall, rolling stylometry proves to be a good technique for detecting extremist speech although further research with different types of features, different classifier and slicing settings is obviously needed.

5   Conclusions and Future Work

Our study has shown the possibilities and prospects of the application of stylometric tools for analysis of texts from an extremist forum. Different stylometry techniques are shown to be useful for analysis of extremist rhetoric. Control for topic and author group allowed us to perform the experiments which were indicative of the linguistics characteristics of the extremist discourse which are detectable irrespective of the topic. Both supervised and unsupervised methods should be employed for better understanding of the extremist mind and development of efficient methods for detection of extremist content. The rolling stylometry technique is shown to be useful for detection of the fragments of the extremist discourse even in a cross-topic scenario. Overall, visualization techniques widely used in stylometry facilitate understanding of the extremist discourse and a better analysis of the experimental results. Combination of distant and close reading of extremist texts with the visualization of the obtained results will enrich both theoretical understanding of the extremism and practice of forensic linguists and state agencies who are due to detect and remove harmful content including radical extremist content.

Our research has some obvious limitations related to the small volume of analyzed texts. Our focus on the similar topics has not only advantages described in Dataset section but also disadvantages: our conclusions on the characteristics of the extremist discourse should be taken with caution as they should be validated on other topics. Nevertheless, the aim of the present paper is not to identify the general characteristics of the radical Islamic discourse, but rather to show the potential of different stylometric tools in analyzing and detecting the examples of this discourse for such a complicated language in terms of natural language processing analysis as Russian.

Our future plans include the expansion of our dataset by adding new topics and exploring new features beyond the most frequent words as well as a closer inspection of the texts by the most prolific extremist authors.

Acknowledgment. Funding of the project "Speech portrait of the extremist: corpus-statistical research (on the material of the extremist forum "KavkazChat")" from RF President's grants for young scientists (no. MK-5718.2018.6) is gratefully acknowledged.

References

1. Baker, P., Vessey, R.: A corpus-driven comparison of English and French Islamist extremist texts. Int. J. Corpus Linguist. **23**(3), 255–278 (2018)
2. Chen, H., Reid, E., Sinai, J., Silke, A., Ganor, B. (eds.): Terrorism Informatics: Knowledge Management and Data Mining for Homeland Security. Springer, New York (2008)
3. Craig, H., Kinney, A.F. (eds.): Shakespeare, Computers, and the Mystery of Authorship. Cambridge University Press, Cambridge (2009)
4. De Smedt, T., Jaki, S., Kotzé, E., Saoud, L., Gwóźdź, M., De Pauw, G., Daelemans, W.: Multilingual Cross-domain Perspectives on Online Hate Speech. CLiPS Technical Report 8, CTRS-008 (2018). https://www.uantwerpen.be/images/uantwerpen/container2712/files/perspectives-hate.pdf
5. De Smedt, T., De Pauw, G., Van Ostaeyen, P.: Automatic Detection of Online Jihadist Hate Speech. CLiPS Technical Report 7, CTRS-007 (2018). https://arxiv.org/ftp/arxiv/papers/1803/1803.04596.pdf
6. Eder, M., Rybicki, J., Kestemont, M.: Stylometry with R: a package for computational text analysis. R J. **8**(1), 107–121 (2016)
7. Eder, M.: Rolling stylometry. Digit. Sch. Hum. **31**(3), 457–469 (2016)
8. Fainberg, A.: ISIS speaks Russian. The Islamic State speaks Russian – a new market for ISIS propaganda (2016). https://www.ict.org.il/Article/1598/The-Islamic-State-speaks-Russian#gsc.tab=0
9. Frans, W.: There are radical Muslims and normal Muslims: an analysis of the discourse on Islamic extremism. Religion **43**(1), 70–88 (2013)
10. Hoover, D.L.: The Craig Zeta Spreadsheet. Book of abstracts of Digital Humanities 2010 (2010). http://dh2010.cch.kcl.ac.uk/academic-programme/abstracts/papers/html/ab-659.html
11. Jockers, M.L., Witten, D.M., Criddle, C.S.: Reassessing authorship of the 'Book of Mormon' using delta and nearest shrunken centroid classification. Lit. Linguist. Comput. **23**(4), 465–491 (2008)
12. Litvinova, T., Litvinova, O., Panicheva, P., Biryukova, E.: Using corpus linguistics tools to analyze a Russian-language Islamic extremist forum. In: Bodrunova, S. (eds.) Internet Science. INSCI 2018. Lecture Notes in Computer Science, vol. 11193. Springer, Cham (2018)
13. Pennebaker, J., Chung, C.: Computerized text analysis of Al-Qaeda transcripts. In: Krippendorff, K., Bock, M.A. (eds.) A Content Analysis Reader. Sage, USA, pp. 453–465 (2009)
14. Prentice, S., Taylor, P., Rayson, P., Hoskins, A., O'Loughlin, B.: Analyzing the semantic content and persuasive composition of extremist media: a case study of texts produced during the Gaza conflict. Inf. Syst. Front. **13**, 61–73 (2010)

Man - Machine Knowledge Mediation: Overview of Deep Learning Methods for Natural Language Processing

Ekaterina Isaeva[1][(✉)] and Vadim Bakhtin[2]

[1] Perm State University, Perm 614990, Russia
ekaterinaisae@gmail.com
[2] Perm National Research Polytechnic University, Perm 614990, Russia

Abstract. In this paper we suggest that efficient processing of natural language should be considered as a complex procedure, which comprises such stages as knowledge encoding by means of natural language units, its digitalization, and machine processing, which make up man – machine knowledge mediation. For this reason, methods chosen for natural language processing should be efficient for both knowledge encoding and knowledge retrieval, as well as for working with textual data of different length, since knowledge can be compressed either to words or to larger syntactic units.

The paper provides an overview of modern trends in Natural Language Processing, namely methods for deep learning. We have studied them as potential add-ons to the TSBuilder program based on the Rosenblatt's Perceptron initiated for automated term system building. Such methods as word embedding, convolutional, recurrent, and recursive neural networks are viewed through the prism of their application for the refinement of term extraction and categorization, as well as subsequent neural network training. Statistics on these methods' representation in scientific databases is also provided.

Keywords: Term extraction · Categorization · Word embedding · Word2vec · WordRank · Convolutional neural network · Recurrent neural network · Recursive neural network · Mediation · Natural language processing

1 Introduction

In traditional linguistics most of tasks were performed manually. It took years to examine texts, identify and determine language units, do sampling and categorization. Now with the development of corpus and computational linguistics, this work is measured in minutes and even in seconds. Still not all the problems under the remit of modern linguistics can be solved with computer tools available nowadays or not all the results can be as sound as those manually obtained. Particularly, our research team aims to elaborate a reliable methodology for special knowledge mining. The problem is not a new one but there still has not been found a universal method that would substitute man in the task of term system building of new developing areas with massive terminology borrowing, interference, and reconsideration of the term content. We suggest approaching this problem from the cognitive sciences' perspective to facilitate human-like reaction and reasoning in special knowledge mining.

© Springer Nature Switzerland AG 2020
T. Antipova and Á. Rocha (Eds.): DSIC 2019, AISC 1114, pp. 44–52, 2020.
https://doi.org/10.1007/978-3-030-37737-3_4

As part of ontology, knowledge is understood as a system of concepts, which preserves such features as networking, logics, and hierarchy. Underpinning human thinking, reasoning, communication, and intellectual development knowledge is subject for transferring, precepting, reconstruction and so on. To this extent originally an ideal notion knowledge becomes material, in particular expressed in signs, such as natural language (NL), which allows for its manipulating in some media [1], such as discourse, in the form of a text (written or audio). In order to be handled, namely, precepted and understood the text needs to be suitable for the recipient to perceive it (shared language, background knowledge consistence). So, the process of knowledge acquisition includes several states: NL designation (coding), textualization, text reading, text understanding, conceptualization. The process usually requires two participants: the one who prepares knowledge for understanding and the one who acquires knowledge from the text.

If we consider this process in terms of NL processing (NLP), we should note that machine learning methods operate numerical data, so, in order to solve text analysis problems, like classification or translation, using machine learning, textual data needs to be converted to a numerical form. Thus, we acknowledge additional stages of text processing (encoding in a machine-readable way), text mining (knowledge extraction), and reverse NL encoding. We entitle these additional steps, which are to settle or reconcile differences in NL and machine code and ultimately in NL understanding (NLU) and NLP, Man - machine knowledge mediation.

Various neural approaches to Man – machine knowledge mediation are focused around the problem of neural network modelling. In this paper we focus on those most applicable for updating and modification of our TSBuilder program [2], developed for terminology identification and categorization as automated steps for term system building. Roughly, the program is based on search trees for identification and a multilayer Rosenblatt's Perceptron for categorization. We are going to use the perceptron as the foundation for the further neural network add-on, for it is known that "a standard neural network consists of many connected <...> neurons" [3], activated by previous neurons depending on their weights. Nowadays a lot of literature is available on the problem of NLP enhancement by means of deep learning and application of different types of neural networks based on a multilayer perceptron. We provide an overview of the cutting-edge methods for NL mediation to be used in machine learning.

2 Background Knowledge on Deep Learning in NLP

2.1 Word Embedding in NLP

One of the most time-tested and approved neural NLP methods is Word embedding, which assumes that words' meanings are revealed in contexts consequently, words, which occur in similar contexts are likely to have identical or related meanings. The first mention of word embedding or distributional vectors in NLP and text mining found among Scopus publication date back to 2007. In their article "Evaluation of an unsupervised ontology enrichment framework" the authors from Technical University of Cluj-Napoca (Romania) used distributional vectors for encoding "contextual content information" aiming to "enrich the hierarchical backbone of an existing ontology, i.e. its taxonomy, with new domain-specific concepts <...> based on hierarchical self-organizing maps <...> by populating the existing taxonomy with the extracted terms and attaching them as hyponyms for the intermediate and leaf nodes of the taxonomy" [4]. Next year the authors continued reporting on their achievements in the area of unsupervised machine learning for categorization and taxonomy development (4 papers in conference proceedings).

From 2010 new players in the field of word embedding for NLP started appearing, introducing new initialization schemes and training technics for multi-layer neural networks, analyzing such issues as "induced word embeddings for extreme cases <...> providing insight into the convergence issues" [5], "fast spectral method to estimate low dimensional context-specific word representations from unlabeled data" [6], or "an application of matrix factorization to produce corpus-derived, distributional models of semantics that demonstrate cognitive plausibility <...> and [are] highly interpretable" [7].

The peak of publication activity on this topic is registered in 2018. Among the results of the Scopus search by "word embedding" and "distributional vector" as the keywords included into the name, the abstract and the keywords of the paper, restricted by the spheres of NLP or text mining, which comprise 903 papers[1], there are contributions of researches and developers from China (255 papers), the USA (136 papers), India (68 papers), the UK (44 papers), and others. The fields where word embedding and distributional vectors are applied for NLP vary from Computer Sciences (49.8%) to Decision Sciences (5.5%), Arts and Humanities (2.9%), Biochemistry (1.3%), and others.

The range of problems solved with word embedding in linguistics (which is of special interest in this paper) is quite representative. Particularly, in [8] the problem of multiword expressions recognition with the application of word embeddings is considered in the framework of the development of "Russian thesaurus RuThes, which contains various types of phrases, including <...> multiword terms", while in [9] the authors use diachronic word embeddings for the analysis of literal and metaphorical word senses. They determine "source domain words and target domain words <...> from the nearest neighbors" and derive conceptual metaphors.

Remarkable results have been achieved with the application of "distributed representations with terminology extraction". The algorithm includes the assessment of "the specificity of a word <...> to the target corpus by leveraging its distributed representation in the target domain as well as in the general domain, the <...> [adaptation of] this distributed specificity as a filter, and <...> [its application] to the corpus" [10]. The authors claim that the filters are compliant with other terminology identification methods and simplify the training procedure.

In terminology extraction another challenge is knowledge retrieval for polysemic terms. To solve this problem the study of word embedding as the key for finding term – concept correspondences should be mentioned. The researchers have developed a method of "short text conceptualization based on new short text similarity", and suggest "representation for short text based on concepts instead of terms using BabelNet as an external knowledge [to avoid] <...> multiple matches [by] <...> assigning a term to a concept" [11].

---

[1] *Estimated in 30 March 2019.*

Particularly representative are Scopus statistics of Word2vec and WordRank popularity for NLP purposes. The first one constructs a numerical vector, based on word frequency in source text, for every unique word in this text in such a way that this vector is the same for the words with identical spelling, and is different for words which spell differently.

Among 796 results obtained for "word embedding" and "distributional vector" keywords 470 papers mention Word2vec, which gains popularity from 2014 (6 papers) to 2018 (224 papers). The tasks are connected with spam and malicious attacks regulation, management of disaster risk advice, sentiment analysis, databases management, speech recognition and production, and others.

The WordRank method, is to a certain extent similar to the Word2Vec method. It also calculates numerical vector for each unique word it the text. However, their main difference is that this method takes into account the context of words and evaluates collocations (i.e. co-occurrence of the words in the text), therefore homonyms possess different vectors. According to Shihao et al. Word Rank is resistant to noise in data and "efficiently estimates word representations via robust ranking" [12] and obviously, this method processes text more accurately than Word2Vec, because of its ability to rely on the word's context.

According to Scopus statistics, this is not a widely spread method, for it is represented only by 6 instances in the period from 2005 to 2018 but is worthwhile considering if the task is connected with the NL belonging to different knowledge domains where homonyms and polysemy are of utmost importance.

Since the Word2Vec method is less complicated, it can process text more than 10 times faster than WordRank, on the other hand, the WordRank method more accurately describes the source text, thus, the machine-learning models trained on the WordRank-generated numerical vectors, show greater precision in solving NLP problems.

2.2   Convolutional Networks in NLP

Another method which is of interest in terms of NLP is the usage of convolutional networks, which have already proved their efficiency for visual data analysis (more than 30000 papers in Scopus), and since 2015 have been gaining momentum in text mining (49 papers). The method is employed for such tasks as text, sentence, and word classification, semantic analysis of noisy text streams, sentiment analysis and others.

In [13] it is suggested that convolutional network can contribute to achieving balance between two opposing aspects, namely results' accuracy and computational costs. The authors introduce "hierarchical gated convolutional networks with multi-head attention <...> that has word-level and sentence-level [and reuses] parameters of the model in each sentence". It should be noted that attention mechanism is a technique inspired from the need to allow the decoder part of the [neural network]-based framework to use the last hidden state along with information (i.e., context vector) calculated based on the input hidden state sequence" [14]. Then, the authors "apply gated convolutional network on both levels". Additionally, "a multi-head attention mechanism is employed [to evaluate the relevance of the word] for better construction of document representation" [13]. Another example is the application of a deep convolutional neural network for knowledge mining performed on the Persian corpus.

The development comprises character-based convolutional layers, which increase accuracy in extracting text features [15].

2.3   Recurrent Networks in NLP

Equally interesting is to examine how effective are recurrent networks for processing sequential information. The mechanism consists in computing "every instance of an input sequence conditioned on the previous computed results [with subsequent representation of] these sequences <...> by a fixed-size vector of tokens which are fed sequentially (one by one) to a recurrent unit" [14]. The method is frequently used for short-text classification, e.g. in [16] a group of developers "propose a bidirectional long short-term memory <...> recurrent network" either with a knowledge dictionary adopting the attention mechanism "to guide the training of network using the domain knowledge in the dictionary" or without a domain knowledge dictionary, in which case the authors "present a multi-task model to jointly learn the domain knowledge dictionary and do the text classification task simultaneously" [16].

Directly related to our project of term system development is the development of one-stage algorithm "that identifies terms via directly performing sequence-labeling with a BILOU scheme on word sequences" [17] unlike other systems, which carry out this task in two stages of identifying candidate terms and refining the results. With the focus on consecutive word and phrases representation there has been developed a framework of Word-Phrase-Entity language modelling, in which Recurrent Neural Networks "do not operate in terms of lexical items (words), but consume sequences of tokens that could be words, word phrases or classes such as named entities, with the optimal representation for a particular input sentence determined in an iterative manner" [18].

2.4   Recursive Networks in NLP

Another type of neural networks applied for NLP is recursive networks. This method is widely used for visual data recognition, decision taking, forecasting, and other tasks, what is proved by more than 7 thousand papers found on this topic in Web of Science. Publications on the application of recursive networks in the field of NLP are not so numerous, meanwhile, the method proved its feasibility for "parsing, leveraging phrase-level representations for sentiment analysis, semantic relationships classification, <...> sentence relatedness" [14], etc.

Recursive networks rely on recurrent ones collected together and able to "address trees" [3] for modelling sequential language data. This method provides data representation in a form of nodes comprising the representations of all their children nodes. Due to these features recursive networks are appealing perspective for multiword units' identification and classification. To prove this, we refer to the project on the development of a recursive network model for semantic role labelling introduced in [19], where the authors compare traditional shallow models with theirs, which "does not dependent on lots of rich hand-designed features <...> [and] is able to add many shallow features" [19].

3 Database Mediation for NLP Through Deep Learning

While working on the project "Special Knowledge Mediation by means of Automated Ontological and Metaphorical Modelling" we have collected a database of terms of various lengths, for each term there is its class (the category to which it was assigned in the classification), the context from which the term was derived and the translation into Russian (for terms in English). We have constructed a multilayer perceptron for processing scientific texts in natural language. Our algorithm identifies sequences of words of various lengths as terms. Those of sequences that are recognized as terms are used for analysis with the perceptron. The perceptron obtains a term from several words, receives a classification for the words of which the term consists, analyzes the term and classifies the entire phrase. Further, the algorithm was modified to classify terms of unlimited length. For this purpose, a term preprocessing level was built, which allowed to minimize any term to a specific mask of length equal to two and classify the latter with an existing perceptron. Now the training of the cascade of neural networks is underway.

In the nearest future, we plan to use various architectures of neural networks and their combinations in cascades to increase the accuracy of the terms' classification. We also plan to use some of the technologies to improve the accuracy of terms' identification.

1. We are determined to use word embedding to get vectors of words that will be included in the term. From the available options, mentioned above, namely Word2vec, FastText, and WordRank, we are inclined to choose the latter, since there homonyms can receive different vectors). This will provide the following opportunities: First, the model will be able to build terms not only from words that are cognate to those that are already in the database, but also from those whose vectors are close to the ones in the database. Secondly, the word vectors will become important additional information when classifying. For the terms in the database, it will be necessary to calculate their vectors and add them to the base, and then calculate the common vector for the compound term. This can be achieved with various algorithms, mentioned above.
2. As before [20], it is supposed to use a convolutional network to bring the potential term of unlimited length to a standard form for subsequent classification by the next level of the neural network cascade. In [11], the use of a convolutional neural network with word level parameters and sentence level parameters is described. The approach is worthwhile applying for our purposes to optimize the search for terms.
3. We also assume that instead of the classifying perceptron, a recurrent neural network, that is, a neural network with feedback, can be used to increase the classification accuracy employing the knowledge of the classification results of previous terms in the processed text.

In the framework of the current project the following tasks should be fulfilled:

1. to teach our handmade convolution on the existing base, which has phrases and their classes. The data is sufficient for machine learning, since it represents a marked dataset;

2. to measure the accuracy of the dataset for testing;
3. to integrate embedding vectors into the identification and classification procedures, test different algorithms for obtaining vectors, compare and choose the best algorithm;
4. to improve the convolutional neural network to bring to a unified mask, taking into account the use of word vectors and ideas obtained in [11];
5. to get results on identification accuracy;
6. to replace the perceptron with an experimental recurrent network;
7. to measure the accuracy of the cascade obtained;
8. to compare with the first option.

4   Results

The paper presents the overview of methods for deep learning based on the Rosenblatt's perceptron, which proved to be efficient for splitting NL dataset in two groups but not sufficient for a manifold categorization. For this reason, a study into such neural network methods, as word embedding, convolutional, recurrent, and recursive networks in terms of their robustness for NLP, has been undertaken.

As a result, we have determined that the usage of convolutional networks is reasonable for terminology identification task to achieve more precision and extend the method on multiword terms identification. This will allow to achieve word standardization and reduce computational costs.

To solve the problem of poor representativeness of the training dataset and overcome the cognit-word barrier, on the one hand and ambiguation caused by homonyms, on the other, the method of word embedding suggested by WordRank developers is found reliable, for it allows knowledge retrieval based on the term's context, thus term's vector is built with reference to its contextual meaning.

Recurrent networks are considered in this paper as an effective tool for terms classification, which are more effective than the perceptron at this stage. The method provides perspective of words to be represented as consecutive units and to be assigned a category as a whole entity. The feedback provided by the network of this type will also contribute to the program upgrade as concerning classification accuracy and computational costs.

5   Conclusion

The paper designed as a comparative study of the methods efficient for Man – machine knowledge mediation contains an overview of deep learning methods, which are gaining popularity in the last decade. Based on the scientific databases, namely Scopus and Web of Science statistics, we revealed the most approved methods for tasks similar to ours, as well as the ones that are breakthrough and demand developer's creativity to be implemented for NLP. The former are methods connected with word representation in the form of a numerical vector or the method of Web embedding, and the latter are convolutional, recurrent, and recursive neural networks, which proved remarkably effective for visual data analysis but are only coming into use, meanwhile seem effective in NLP.

Proceeding from the findings obtained in this study the present state of the TSBuilder program has been evaluated and further improvements have been outlined. The updates will help to streamline Man – Machine mediation in NLP.

Acknowledgements. The reported study was funded by RFBR according to the research project №. 18-012-00825 A.

References

1. Isaeva, E.: Metaphor in terminology: finding tools for efficient professional communication. Fachsprache **41**(Sp. Issue), 65–86 (2019)
2. Isaeva, E., Suvorova, V., Bakhtin, V.: Supervized machine learning: Computer-aided development of a specialized dictionary. Autom. Doc. Math. Linguist. **50**(3), 104–111 (2016)
3. Alshahrani, S., Kapetanios, E.: Are Deep learning approaches suitable for natural language processing? In: Natural Language Processing and Information Systems, Nldb 2016, Salford, England (2016)
4. Chifu, E.S., Chifu, V.R.: Evaluation of an unsupervised ontology enrichment framework. In: ICCP 2007 Proceedings IEEE 3rd International Conference on Intelligent Computer Communication and Processing, Cluj-Napoca, Romania (2007)
5. Son, L.H., Allauzen, A., Wisniewski, G., Yvon, F.: Training continuous space language models: some practical issues. In: Proceedings of the Conference EMNLP 2010 - Conference on Empirical Methods in Natural Language Processing, Cambridge, MA, United States (2010)
6. Dhillon, P., Foster, D.P., Ungar, L.H.: Multi-view learning of word embeddings via CCA. In: 25th Annual Conference on Neural Information Processing Systems 2011 Advances in Neural Information Processing Systems 24, NIPS 2011, Granada, Spain (2011)
7. Murphy, B., Talukdar, P.P., Mitchell, T.: Learning effective and interpretable semantic models using non-negative sparse embedding. In 24th International Conference on Computational Linguistics - Proceedings of COLING 2012. Technical Papers, Mumbai, India (2012)
8. Loukachevitch, N.V., Parkhomenko, E.: Recognition of multiword expressions using word embeddings. In: Communications in Computer and Information Science, Moscow, Russian Federation (2018)
9. Jia, Y., Zheng, Y., Zan, H., Wang, Z.: Analysis of literal and metaphorical senses based on diachronic word embeddings. In: Proceedings of the 2017 International Conference on Asian Language Processing, IALP 2017, Singapore (2018)
10. Amjadian, E., Inkpen, D., Paribakht, T.S., Faez, F.: Distributed specificity for automatic terminology extraction. Terminology **24**(1), 23–40 (2018)
11. Bekkali, M., Lachkar, A.: An effective short text conceptualization based on new short text similarity. Soc. Netw. Anal. Min. **9**(1), 23–40 (2019)
12. Shihao, J., Hyokun, Y., Pinar Y., Matsushima, S., Vishwanathan, S.V.N.: WordRank: learning word embeddings via robust ranking. In: EMNLP, pp. 27–41 (2016)

13. Du, H., Qian, J.: Hierarchical Gated Convolutional Networks with Multi-Head Attention for Text Classification. In 2018 5th International Conference on Systems and Informatics, ICSAI 2018, Nanjing; China (2019)
14. Deep Learning for NLP: An Overview of Recent Trends. https://medium.com/dair-ai/deep-learning-for-nlp-an-overview-of-recent-trends-d0d8f40a776d. Accessed 30 Mar 2019
15. Ghasemi, S., Jadidinejad, A.H.: Persian text classification via character-level convolutional neural networks. In: 2018 8th Conference on Artificial Intelligence and Robotics, IRANOPEN 2018, Qazvin, Iran (2018)
16. Cao, S., Qian, B., Yin, C., Li, X., Wei, J., Zheng, Q., Davidson, I.: Knowledge guided short-text classification for healthcare applications. In: Proceedings - IEEE International Conference on Data Mining, ICDM, New Orleans, United States (2017)
17. Kucza, M., Niehues, J., Zenkel, T., Waibel, A., Stüker, S.: Term extraction via neural sequence labeling a comparative evaluation of strategies using recurrent neural networks. In: Proceedings of the Annual Conference of the International Speech Communication Association, INTERSPEECH, Hyderabad, India (2018)
18. Levit, M., Parthasarathy, S., Chang, S.: Word-phrase-entity recurrent neural networks for language modeling. In: Proceedings of the Annual Conference of the International Speech Communication Association, INTERSPEECH, San Francisco, United States (2016)
19. Li, T., Chang, B.: Semantic role labeling using recursive neural network. In: Chinese Computational Linguistics and Natural Language Processing Based on Naturally Annotated Big Data (CCL 2015), Guangzhou, China (2015)
20. Bakhtin, V., Isaeva, E.: New TSBuilder: shifting towards cognition. In: 2019 IEEE Conference of Russian Young Researchers in Electrical and Electronic Engineering (EIConRus), St Petersburg, Russia (2019)

Digital Economics

Universal Basic Income as an Innovation in Social Policy and Public Finance: SWOT Analysis

Svetlana Tsvirko(✉) 📵

Financial University under the Government of the Russian Federation,
Leningradskiy prospekt 49, Moscow 125993, Russian Federation
s_ts@mail.ru

Abstract. The article focuses on the features of universal basic income (UBI). The method of research is SWOT analysis. The objective of the study was to analyze strengths, opportunities as well as weaknesses and threats that are connected with UBI. The main possible advantages of UBI revealed in the paper are as follows: poverty reduction; possibility to lead a decent life, no matter what happens; insurance against financial disasters in light of the threat of a decline in employment due to the growing automation of processes; possibility for people to spend more time with their families or volunteer activities; reduction of corruption opportunities. The most serious drawbacks of UBI identified in the paper are as follows: burden on public finances; possible tax increases; reduction of other social programs, fraught with public discontent; dismantling of labor and socially useful activities; polarization of society; loss of skills and the inability to return to the labor market. The analysis provides a basis for decision makers to exploit opportunities and minimize threats of this new approach in social policy and public finance.

Keywords: Universal basic income · Public finance · Social policy · Welfare · Poverty · Inequality · SWOT analysis

1 Introduction

The study of issues related to the universal basic income (UBI) is of current interest, that can be explained by massive socio-economic changes and problems that are observed in different types of countries – both developed and developing. So far poverty is a great issue in the global economy and solving this problem is an indispensable requirement for sustainable development. According to World Bank, about 46% of the world lives on less than $5.50 a day. Over a quarter of the global population lives on less than $3.20 per day and around 10% lives on less than $1.90 a day [1]. While many countries have made great strides in reducing monetary poverty, they still lag in crucial areas such as infrastructure, education and security. If we expand our analysis, "multidimensional" view reveals a world in which poverty is a much broader, more entrenched problem, underlining the importance of investing more in human potential. The World Bank's multidimensional measure looks at deprivations in areas that include income, education, electricity, water and sanitation. When viewed through

T. Antipova and Á. Rocha (Eds.): DSIC 2019, AISC 1114, pp. 55–64, 2020.
https://doi.org/10.1007/978-3-030-37737-3_5

this multidimensional lens, the share of poor globally is approximately 50% higher than it is when defined by income or consumption alone. Progress in reducing poverty has been very uneven. Over the last two decades much of the decline in extreme poverty has been due to impressive growth in Asia, particularly China and India. However, the number of people living in extreme poverty has been increasing in Sub-Saharan Africa and extreme poverty is increasingly becoming concentrated in that region. Shared prosperity—or annual growth in the income of the bottom 40% of people—has increased in many countries, however, in some countries in Europe & Central Asia and Sub-Saharan Africa slow growth or even declines in shared prosperity were seen [1]. As it is stated in the Global Goals for Sustainable Development, there is the need for creation of sound policy frameworks at the national, regional and international levels, based on pro-poor and gender-sensitive development strategies, to support accelerated investment in poverty eradication actions [2].

One of the possible approaches to fight with poverty and its negative effects and consequences is the universal basic income. UBI is a periodic payment unconditionally delivered to all in the country (community). According to one of the most famous proponents of the idea of UBI Philippe Van Parij, UBI is defined as «income paid by a political community to all its members on an individual basis, without means test or work requirement» [3, p. 4]. The characteristics of payments that can be considered the UBI are unconditionality, inclusiveness and regularity of payments.

An important reason for the popularity of the idea of UBI in developed countries was the fear that the labor market would be threatened by automation, that can force people out of their jobs. There is some evidence that the labor market will face a threat from artificial intelligence (AI) in particular and information technology in general. The level of intelligence that robots demonstrate already now disturbs the society. According to the research performed by BCG, when asked about the implications of AI for the economy and society, citizens expressed significant concerns about the availability of work in the future (61% agree) [4]. One of the publications, that caused a wide public response was the publication of Oxford Martin School, where it was stated, that around 47% of total US employment is in the high risk category due to computerization [5]. So, the threat of loss of income makes people turn to the idea of small but guaranteed income.

There is significant amount of publications devoted to the UBI – both in western economic literature and in Russia. Among them we can name publications by such authors, as Ackerman, Alstott and Van Parij [3], Andersson [6], Coady, Prady and Francese [7, 8], Widerquist [9], researchers from Kela (The Social Insurance Institution of Finland) [10], Gontmakher [11], Istratov, Belyanov [12], Kuznetsov [13], Kuznetsov [14] Tsvirko [15], Watson, Guettabi, Reimer [16] *and others*. A number of international organizations have issued their working papers devoted to UBI. For example, OECD is an opponent of UBI in advanced economies as it is revealed in [17]. As for IMF, it has a number of publications, devoted to UBI, for example in Fiscal Monitor in October 2017 it discussed UBI as one of the policies that can support inclusive growth under the conditions of rising inequality and slow economic growth in many countries [18]. It should be stressed that the IMF does not give direct answer about UBI: in a special announcement the representative of IMF said that IMF is not advocating for or against UBI [19]. There is a specialized website devoted to the concept of UBI [20];

initially it was created in 1986 as the international expert network BIEN (Basic Income European Network), but subsequently extended its activities to the whole world. This network serves as a link between individuals and groups committed to or interested in basic income, and its aim is to foster informed discussion on the topic.

However, it seems that there is no holistic approach to the study of UBI, which is really necessary to increase efficiency in implementation of this new approach. As far as we know there are no papers with SWOT analysis of the UBI, that can be a contribution to more integrated assessment of opportunities and risks of UBI as an innovation in social policy and public finance.

In this context the aim of this study is to analyze possibilities and drawbacks of UBI. For achieving the aim of the research, it is necessary to show some examples of the realization of this concept in practice, briefly explain the methodology of research in the form of SWOT analysis, clarify the essence of UBI in the framework of SWOT analysis and finally reveal limitations of research and potential for future studies.

2   Methodology of Research and Input Data on Experiments in the Sphere of Universal Basic Income

Analysis of different countries' experiments with UBI can produce relevant data and provide some understanding of the necessity and possibility to implement UBI as an innovation in social policy and public finance. We will use SWOT analysis for this research.

SWOT analysis is a method of strategic planning, consisting in identifying factors of the internal and external environment for the system. Factors are divided into four categories:

- Strengths,
- Weaknesses,
- Opportunities,
- Threats.

Strengths (S) and weaknesses (W) are factors of the internal environment of the object of analysis (that is, what the object itself is able to influence); opportunities (O) and threats (T) are environmental factors (that is, those that can affect the object from the outside and are not controlled by the object).

The objective of the SWOT analysis is to provide a structured description of the situation regarding which a decision needs to be made. A SWOT analysis is effective in conducting an initial assessment of the current situation, but it cannot replace the development of a strategy or a qualitative analysis of dynamics. This is a universal method that is applicable in a wide variety of areas of economics and management. It can be adapted to the object of study at any level (product, enterprise, region, country, etc.). This method is considered to be flexible with a free choice of the analyzed elements depending on the goals set and it can be used both for operational assessment and for strategic planning for a long period.

As for input data, we know about some UBI-like experiments, including cases of the USA, Canada, Finland, Italy, India, Kenya, Uganda. It is interesting that the first attempt to introduce a basic income at the national level was made in the United States.

In the spring of 1968 a petition was submitted to Congress on the need to introduce a system of "guaranteed income and security", that was supported by some well known economists, such as James Tobin, Paul Samuelson and others. In 1969 President Nixon suggested law on the introduction of basic income to the Congress, but it was rejected twice in the Senate. The idea of introducing basic income was contained in the electoral program of presidential candidate from Democratic Party George McGovern in 1972, but he was not elected [11]. Despite the fact that the idea was not implemented at the federal level in the USA, there are examples of the application of UBI at the regional level. For example, in 1976 the Alaska Permanent Fund was created, from which each resident of the state is paid annually (usually $ 1,000–2,000) from oil revenues received by private companies. The payment is called the Permanent Fund Dividend (PFD) and is paid to Alaska residents that have lived within the state for a full calendar year (January 1–December 31), and intend to remain an Alaska resident indefinitely. An individual is not eligible for a PFD for a dividend year in the following cases: during the qualifying year the individual was sentenced as a result of conviction in this state of a felony; during all or part of the qualifying year the individual was incarcerated as a result of the conviction in this state. The amount of each payment is based upon a five-year average of the Alaska Permanent Fund's performance and varies widely depending on the stock market and other factors [21].

In this social experiment the principles of universality and unconditionality are observed. In their research *Jones and Marinescu* showed that the Alaska Permanent Fund dividend increased part-time work by 1.8% points (17%); the results, that were received by the above mentioned authors suggest that a universal and permanent cash transfer does not significantly decrease aggregate employment [22]. There are some other effects, that are analyzed in connection with the UBI based on the Alaska Permanent Fund experiment. For example, in the study performed by *Watson B., Guettabi M., Reimer M.* it was revealed that there is 14% increase in substance-abuse incidents the day after the Alaska Permanent Fund payment and a 10% increase over the following four weeks. This is partially offset by a 8% decrease in property crime, with no changes in violent crimes. On an annual basis, however, changes in criminal activity from the payment are small. Estimated costs comprise a very small portion of the total payment, suggesting that crime-related concerns of a universal cash transfer program may be unwarranted [16].

One of the other experiments in UBI sphere in the USA started in 1996 and takes place in North California. There are 8000 participants in it and the condition of inclusion to the program is belonging to the Cherokee tribe. The program budget is formed by gambling revenues.

Finland is another example of advanced economy with the experiment in the sphere of UBI. The program was held in 2017–2018 with 2000 participants and 560 euro as payable amount per month. Conditions of inclusion into the program were absence of work and age from 25 to 58 years. The project has not been extended. According to preliminary results, at the end of the experiment the recipients of a basic income perceived their wellbeing as being better than did those in the control group. The recipients of a basic income had less stress symptoms as well as less difficulties to concentrate and less health problems than the control group. They were also more confident in their future and in their ability to influence societal issues [10]. Survey respondents who received a basic income described their financial situation more positively than respondents in the control group. They also experienced less stress and fewer financial worries. The basic income experiment in Finland did not increase the employment level of the participants in the first year of the experiment [10].

As for other developed countries, in 2017–2018 years there were experiments with UBI-like programs in Canada and Italy. In Canada in Ontario region there were 4000 participants of the program with a control group of 2000 people as comparison. Each single person could receive an annual sum of $16,989, or less than 50% of their employment income if applicable. Couples received an annual sum of $24,027 or less than 50% of their employment income. Some of the key areas of interest for the study were food security, employment participation, education, housing security and overall effect on health. All of the data was gathered and measured by a third party. In April 2018 the Ontario government officially concluded the first phase of enrollment for a 3-year basic income payment experiment [23].

In Italy (Livorno) in 100 poorest families received 460 euro per family per month. In 2018 the experiment was extended to another 100 families. Grants within experiment in Livorno are conditional. Recipients must complete community service and provide proof that they are actively seeking employment if they are currently unemployed. If they reject a total of three job offers, they are disqualified from the scheme [23]. Conditionality makes this project different from the true idea of a universal basic income.

There was a UBI-like experiment in India. It was held in 2010–2012 with 6000 participants. The project was backed by UNICEF, this organization provided the equivalent of $1 million to the experiment. Conditions of inclusion into the program were as follows: living in a certain village in Madhya Pradesh, that is a state in central India. Payable amount was initially set at the equivalent of US$4.40 for each adult and US$2.20 per child per month, afterwards the value was raised by 50% to adjust for inflation. The experiment involved 9 villages; performance evaluation was conducted on a wide range of factors: living conditions, nutrition quality, morbidity, etc. As the result the following was observed: improvements in health, productivity and financial stability. In terms of impacts to health, the unconditional cash transfers were associated with better food security and lower rates of malnutrition, especially in female children. Recipient villages had lower rates of illness, more consistent medical treatment and more consistent medicine intake. As a result they had higher expenditures on schooling and agricultural inputs, promoting better education and higher agricultural yields. At the same time some problems with UBI were revealed in India: they include the fact that within households women's ability to manage income will be limited in favor of men. There is also an argument about the potential overload of the banking system based on India's experience.

As for experience of low-income countries with the UBI-like programs we can name Uganda. It was the project launched by Eight, that is Belgium-based charitable organization. It is a 2-year basic income trial in rural Uganda, introduced in 2017. The participants are provided with an unconditional payment of $18.52 monthly for adults and $9.13 monthly for children. The amounts were chosen as they represent 30% of the average income for lower-income families in Uganda. The key areas of study for the project are the education of women and girls, health care, local economic development and participation in democratic institutions [23].

One of the programs with the most significant amount of participants – 21000 people- exists in Kenya. It has been operating from 2017 and its a 12-year experiment. The condition for participation is living in a certain village. 200 villages were divided into 3 groups with different duration of payments to compare effects of short- and long-term projects. In 40 villages people receive $0.75 a day for 12 years. In 80 villages they receive the same amount, but over a 2 year period. In 70 villages the amount as a one-time-only lump sum. A further 100 villages are being studied as a control group. The program is called «GiveDirectly», it differs from other schemes in that anyone can donate towards the fund, allowing it to then be distributed among the test groups. This project is considered by the specialists to be one of the most ambitious universal basic income trials currently being conducted [23].

3   Results of SWOT Analysis of the UBI

Results of SWOT analysis of some UBI-like experiments are presented in Table 1.

Table 1. SWOT analysis of UBI.

Strengths	Weaknesses
1. Poverty reduction 2. Insurance against financial disasters in light of the threat of a decline in employment due to the growing automation of processes 3. Saving on administrative costs as it replaces many other social payments 4. Reduction of the state apparatus and corruption opportunities 5. Decrease in the level of migration from the project area	1. Discouraging work and encouraging social dependency 2. Unnecessary leakage of benefits to higher-income groups 3. No effect on inequality 4. Extra income can be squandered 5. Inflation will lead to a decrease in real income, so an indexing mechanism is required 6. Limitation of women's ability to manage income within households will be limited in favor of men 7. In view of critics it can be populist and paternalistic measure
Opportunities	Threats
1. Opportunity to lead a decent life, no matter what happens 2. Avoidance of social stigma 3. Absence of information constraints 4. Possibility for people to spend more time with their families or volunteer activities 5. Increase in perceived wellbeing 6. Less stress symptoms as well as less difficulties to concentrate and less health problems 7. Confidence in the future and in ability to influence societal issues 8. Higher employment in some cases and not its decrease, as some skeptics fear 9. More opportunities for children to have schooling (education) 10. Increase in women's business activity 11. Reduction of state interference in the life of citizens 12. Contribution to economic growth, if a person invests in himself	1. burden on public finances 2. possible tax increases or growth of public debt 3. reduction of other social programs, that can be fraught with public discontent 4. risk of crowding out other high-priority governmental spending, that promotes inclusive growth 5. potential overload of the banking system if it is necessary to open and maintain accounts for people who have not previously used the services of banks 6. dismantling of labor and socially useful activities 7. polarization of society 8. loss of skills and the inability to return to the labor market

The results of the experiments are contradictory. To sum up we can write that main possible advantages of UBI are as follows: poverty reduction; possibility to lead a decent life, no matter what happens; insurance against financial disasters in light of the threat of a decline in employment due to the growing automation of processes; possibility for people to spend more time with their families or volunteer activities; reduction of corruption opportunities. Financial safety net is important for both developed and developing countries.

Some above mentioned advantages were really achieved. In developing countries, for example, in India there were significant improvements in poverty eradication.

In some experiments the results did not meet expectations in some respects. For example in some cases payouts led to higher employment and not to its decrease, as some skeptics feared.

One of the most controversial issues with UBI is the issue of eliminating of inequality. Proponents argue that UBI can reduce inequality, since income will be the same for all. Opponents claim the opposite, and their arguments are stronger. It is obvious that different people may need different income to maintain the same standard of living for various reasons: different health conditions; different climate and place of residence, etc. Thus, people with problems of any kind will be put in less favorable conditions. This fundamentally undermines the idea of a social state, which supporters of UBI advocate.

The most serious drawbacks of UBI are as follows: burden on public finances; possible tax increases; reduction of other social programs, fraught with public discontent; dismantling of labor and socially useful activities; polarization of society; loss of skills and the inability to return to the labor market. There are some other problems with implementation of UBI, for example, inflation will lead to a decrease in real income, so an indexing mechanism is required within UBI approach.

There are several aspects on which the results of the available experiments with UBI are not enough. For example, we can assume that under the conditions of UBI crime should be reduced, but we do not have enough empirical evidence for this.

4   Limitations and Potential for Future Research

As for limitations of research, we can divide them into two groups.

The first group of limitations is connected with the essence of the UBI as it is. We should mention the problem of proper defining UBI. There are difficulties with correct definition of the concept and evaluation of the results of experiments taking into consideration different factors. As it was stressed in [15], in practice, there was no full implementation of the concept of UBI in any country of the world. What is given as examples of implementation of UBI can be better defined as different forms of social benefits, which are devoid of either unconditionality, inclusiveness or regularity. The concept of a minimum guaranteed income - a social support system that provides all citizens of the country with enough income for life - is close to the idea of UBI, but is not the same. The minimum guaranteed income is fundamentally different from UBI by the application of some criteria by which the right to receive additional government benefits is assessed. As for minimum guaranteed income, various forms of it now exist in many countries.

It is also necessary to pay attention to difficulties in performing experiments of UBI. The UBI-like programs were performed on a limited scale and cannot be recognized as national scale programs. Speaking in common, we should understand that experiments in social sphere in different countries have a high degree of uniqueness and specificity, their evaluation is characterized by uncertainty and variety of factors that should be taken into consideration.

The second group of limitations is tied with the method of research – SWOT analysis. We should be aware of some shortcomings of SWOT analysis. One of them is that SWOT analysis shows only general factors. Specific measures to achieve the goals must be developed separately. SWOT analysis is typically a listing of factors, without identifying the main and secondary factors, without detailed analysis of the relationships between them. In case of SWOT analysis we have a more static picture than a vision of development in dynamics.

The results of a SWOT analysis are usually presented in the form of a qualitative description, while quantitative parameters are often required to assess a situation. Critics of SWOT analysis say, that it is quite subjective and extremely dependent on the position and knowledge of the person who conducts such type of the research. For high-quality SWOT analysis it is necessary to attract large amounts of information from various fields, which requires considerable effort and cost.

The conclusions made on the basis of SWOT analysis are descriptive in nature without recommendations and prioritization. For a more complete return on the SWOT method, the construction of options for actions based on the intersection of fields (within SWOT framework) should be also used. To this end, various combinations of environmental factors and the internal properties of the system with UBI should be further considered.

5 Conclusion

In this paper we have revealed the essence of UBI. Several experiments in the sphere of UBI in different types of countries (both developed and developing) were analyzed. We have discussed several experiments with UBI-like programs and have shown that the results of such experiments are controversial.

Despite some above mentioned limitations, SWOT analysis of UBI can be used as a base for suggestions how to increase efficiency of UBI approach as an innovation in the sphere of social policy and public finance and finally reduce poverty and inequality. The threats and weaknesses of UBI highlighted in the research are to be noted for public policy considerations so this approach will be more truly responsive to current economic and social conditions and concerns. Suggestions for future research are connected with practical recommendations for implementation of UBI.

References

1. World Bank: Poverty and Shared Prosperity (2018). https://www.worldbank.org/en/publication/poverty-and-shared-prosperity. Accessed 01 Sept 2019

2. The Global Goals for Sustainable Development. https://www.globalgoals.org/1-no-poverty. Accessed 01 Sept 2019
3. Ackerman, B., Alstott, A., Van Parij P.: Redesigning Distribution: Basic Income and Stakeholder Grants as Alternative Cornerstones for a More Egalitarian Capitalism (2003). https://www.ssc.wisc.edu/~wright/Redesigning%20Distribution%20v1.pdf
4. Carrasco, M., Mills, S., Whybrew, A., Jura A.: The Citizen's Perspective on the Use of AI in Government. BCG Digital Government Benchmarking (2019). https://www.bcg.com/ru-ru/publications/2019/citizen-perspective-use-artificial-intelligence-government-digital-benchmarking.aspx
5. Frey, C., Osborne, M.: The Future of Employment: How Susceptible are Jobs to Computerisation? (2013). https://www.oxfordmartin.ox.ac.uk/downloads/academic/The_Future_of_Employment.pdf
6. Andersson, J.O.: Will Finland be first to introduce a basic income? https://abo.academia.edu/JanOttoAndersson. Accessed 01 Sept 2019
7. Coady, D., Prady, D.: Universal Basic Income in Developing Countries: Options, and Illustration for India. IMF Working Paper (2018). https://www.imf.org/en/Publications/WP/Issues/2018/07/31/Universal-Basic-Income-in-Developing-Countries-Issues-Options-and-Illustration-for-India-46079?cid=em-COM-123-37543
8. Francese, M., Prady, D.: Univeral Basic Income: Debate and Impact Assessment. IMF Working Paper (2018). https://www.imf.org/en/Publications/WP/Issues/2018/12/10/Universal-Basic-Income-Debate-and-Impact-Assessment-46441
9. Widerquist, K.: A Critical Analysis of Basic Income Experiments for Researchers, Policymakers, and Citizens – 2018. Palgrave Pivot. Springer Nature Switzerland AG, part of Springer Nature (2018)
10. Kela: Experimental Study on a Universal Basic Income (2019). https://www.kela.fi/web/en/basic-income-experiment
11. Gontmakher, E.Sh.: Universal basic income: the political economic aspect. J. Econ. Policy **14**(3), 70–79 (2019). (in Russian)
12. Istratov, V., Belyanov, A.: Believed in a fairy tale. J. Expert **35**(1086), 36–39 (2018). (in Russian)
13. Kuznetsov, V.A.: Global challenges in payments industry: "technological unemployment" problem and universal basic approach as a possible scenario of its solution. J. Money Credit **12**, 104–107 (2017). (in Russian)
14. Kuznetsov, Y.V.: Universal basic income and the problem of information asymmetry. J. Econ. Policy **14**(3), 80–95 (2019). (in Russian)
15. Tsvirko, S.: Universal basic income: comparative analysis of experiments. In: Ahram, T., et al. (eds.) Proceedings of the 1st International Conference on Human Interaction and Emerging Technologies (IHIET 2019). Advances in Intelligent Systems and Computing, 22–24 August 2019, Nice, France, vol. 1018, pp. 719–724. Springer, Switzerland (2020)
16. Watson, B., Guettabi, M., Reimer, M.: Universal cash and crime. Rev. Econ. Stat. 1–45 (2019)
17. OECD: Basic Income as a Policy Option: Can it Add Up (2017). http://www.oecd.org/employment/emp/Basic-Income-Policy-Option-2017.pdf
18. IMF Fiscal Monitor: Tackling Inequality, October 2017 (2017). https://www.imf.org/en/Publications/FM/Issues/2017/10/05/fiscal-monitor-october-2017
19. IMF not Endorsing Universal Basic Income Over PDS in India: Official (2017). https://www.business-standard.com/article/economy-policy/imf-not-endorsing-universal-basic-income-over-pds-in-india-official-117101500053_1.html
20. Basic Income Earth Network https://basicincome.org. Accessed 01 Sept 2019
21. Permanent Fund Dividend Division. https://pfd.alaska.gov. Accessed 01 Sept 2019

22. Jones, D., Marinescu, I.: The Labor Market Impacts of Universal and Permanent Cash Transfers: Evidence from the Alaska Permanent Fund. NBER Working Paper №. 24312 (2018)
23. The-15-most-promising-universal-basic-income-trials.   https://interestingengineering.com/the-15-most-promising-universal-basic-income-trials. Accessed 01 Sept 2019

Development of Methodology of Valuation Reserves Information Disclosure of Russian Companies in Terms of Digital Economy

Ruslan Tkhagapso[(✉)] [iD], Anastasiya Trukhina [iD],
and Susanna Khatkhokhu

Kuban State University, Krasnodar 149, Stavropolskaya Street,
350040, Russian Federation
rusjath@mail.ru, nastyusha_tr@mail.ru,
s.khatkhokhu@mail.ru

Abstract. Reserves perform a set of functions related to risk insurance, regulation of financial results, evaluation of objects of accounting. On the one hand, this fact points to the exceptional importance of reserves accumulation as an instrument for covering risks ex ante inherent in the financial and economic life of an organization. On the other hand, it points to the need to draw a clear line between reserves which directly strengthen the financial state of the organization, and regulations that perform this function indirectly, by clarifying the evaluation of objects of accounting. Moreover, in the context of a variety of valuation methods inherent in modern accounting, reserves act as a link between historical value and market valuations, the crown of which is the fair value.

The authors examine the impact of reserves and regulations on the formation of financial reporting indicators in the context of digital economy. The authors' own view on approaches to reporting in Russian companies in the context of the active use of reserves accumulation with an assessment of the impact on the financial situation of the companies is proposed.

Keywords: Digital economy · Financial statements · Reserves · Financial standing

1 Introduction

The concept of a *reserve* is used in scientific economic literature and practice in relation to various economic events and processes. At the same time, the term *reserve* stands for "saving", "stock". The concept of a *reserve* is an integral one for the entire economic system, but it can have different meanings depending on a specific sphere of appliance. Reserves are considered from different points of view: state, region, industry, individual economic entity.

Historically the word "reserve" comes from the French language and has the Latin root "reservo" which means "save", "preserve".

In the economic literature information about the reflection of reserves in companies of the late XIV century – beginning of the XV century is present. In the works of R. de Roover, information about the accumulation of reserves for future expenses is mentioned. Even before the XIV century large Italian merchant and banking companies

T. Antipova and Á. Rocha (Eds.): DSIC 2019, AISC 1114, pp. 65–74, 2020.
https://doi.org/10.1007/978-3-030-37737-3_6

formed "reserves for unforeseen circumstances or subsequent settlements [1, 2]". It is also known for certain that "the statements of the company Francesco Datini in Barcelona, compiled on January 31, 1399, the balance sheet item of reserves for unpaid taxes and unforeseen expenses was indicated [3, 4]."

Many authors [5–7] make emphasis on using reserves as a deductible value from the corresponding balance sheet items when determining them. For a deeper study of the reserve category, it is necessary to pay attention to its evolution in various countries. Historically, the development process of the reserve system can be divided into three stages.

The *first stage* is associated with the emergence of reserves and is dated to the XIV–XVIII centuries, which leads to a direct understanding of the economic substance of reserves and their practical reflection in accounting and reporting. During this period there was no strict regulation of economic activity. Merchants were accumulating sources of financing from profit in order to prevent bankruptcy and pay tax payments. Thus, over the course of three centuries, reserves for risks and costs have been widely used.

The reserves development at the first stage can be characterized by the emergence of the following categories of reserves:

- reserves to compensate for possible losses on receivables;
- accumulation of reserves of the financial result by deliberately reducing profits to compensate for possible losses and risks;
- accumulation of reserves of the financial result to regroup profit in accordance with different reporting data;
- accumulation of funds for repayment of possible losses in the future.

In the *second stage*, the formation and development of reserves of economic entities was influenced by the emergence of joint-stock companies. In the period of the XIX century and the first half of the XX century joint-stock companies had their own capital, which consisted of funds received from the issue and sale of shares, and reserve capital accumulated from profit.

The development of accounting during this period is associated with the emergence of large-scale industry, the development of communication lines, an increase in the world trade, the emergence of a securities market, which increased the number of participants in market relations dramatically. Depreciation accounts, del credere, reserve accounts, provisioning, which not only acted as objects of accounting, but also had methodological accounting techniques, received widespread use in accounting organizations. The development and practical use of accounting records has contributed to the effective application of the reservation system.

J.B. Dumarchey, the 20th century French scientist [8] in his scientific works made a distinction between the concepts of reserve and reserve accounts. The reserve is always represented by the real value of the asset, while the reserve accounts have only a fictitious one. Considering the account "Trade Discount", we must classify it as a reserve, as it is regarded to a real asset – goods sold. Depreciation of fixed assets is meanwhile a reserve account, since the asset is represented by the assessment of deterioration of fixed assets. This notion made it possible to draw a clear distinction between the concept of deterioration of fixed assets (reserve account) and depreciation (reserve). In the time of J. B. Dumarchey the application of reserves in accounting was necessary in order to accumulate funds intended to prevent possible losses.

The *third stage* in the formation of the reserve system of economic entities began to develop in the middle of the XX century. The main characteristic of the third stage is the use of reserves with social and economic assignments. In accordance with the national legislation, companies could create the following reserve groups:

- additional and special pension funds;
- reserves affecting profit.

In addition, during this period a clear definition of the reserve system was proposed, which inherently affects the essence and content of the concept of "reserve".

Professor Ya.V. Sokolov describes the reserve system as "a mechanism for regulating the internal activity of an economic entity in order to maintain financial stability and ensure sustainable activity of the organization with respect to obligations to third parties [9]".

J. Richard defines reserves as "private means in case of increases in price and fluctuations in exchange rates or accumulated profits [10]". The scientist does not support the opinion that reserves are intended to cover losses from specific risks, but does not deny the priority in their formation.

J. Baetge [11], when defining the reserve system, relies on the need to implement the principles based on the accounting object belonging to the reserve system: economic burden, an undisputed obligation to a third party, quantitative certainty.

The statistics show that as of today the total number of public joint-stock companies in the Russian Federation which accumulate reserves for assets impairment is 15.2%, and only 8.3% of non-public organizations create this type of reserve. In the total number of limited liability companies, about 24% of companies form the valuation reserves.

A weak reserve system should push the management of organizations to a serious approach when making management decisions, that is, each business transaction should be considered both from the point of view of its final financial result, and from the point of view of its impact on the financial standing. One of the ways of preventing negative consequences, which can affect the financial result, is provisioning of company's own funds.

In the current conditions of digitalization of the economy, reserves become of a key importance – it becomes possible to predict risk situations more accurately and, correspondingly, to create more adequate reserves.

2 Research Results

The formation and use of reserves, like any object of accounting, directly affect the balance sheet and statement of financial performance of the organization. In practice the use of valuation reserves and asset depreciation reserve accounts has two options.

The *first option* is to reflect the valuation reserves in the asset in a special column or lines. That being the case, two sub-options occur. The first sub-option is that the reserves are deducted from each item of the asset separately: raw materials, materials, in-process inventory; goods, advance payments; receivables; financial investments; deferred expenses; intangible assets, securities. This methodology for reflecting reserves in the reporting is typical for Germany, France, Spain. With this option, it is easier to conduct financial analysis and audit control, and if the changes in assets are significant, analysts evaluate the assets at the selling price. The second sub-option requires the reflection of reserves and reserve accounts after each subsection of the asset balance or the total. This option is used by such countries as the Netherlands, Switzerland, Japan, Italy, the United Kingdom, Canada, the USA, Belgium.

The *second option* can be applied only when preparing financial reporting in the Russian Federation. In Russian accounting legislation, reserves are treated as expenses. In particular, in the Russian Accounting Standards (RAS 10/99) "Organization expenses" it is indicated that the formation of valuation reserves created in accordance with the provisions of accounting, as well as reserves associated with the occurrence of contingency of economic activity, lead to an increase in other expenses of an organization.

Figure 1 outlines the effect of three types of reserves on the financial standing and results of a company in the context of a digitalized economy.

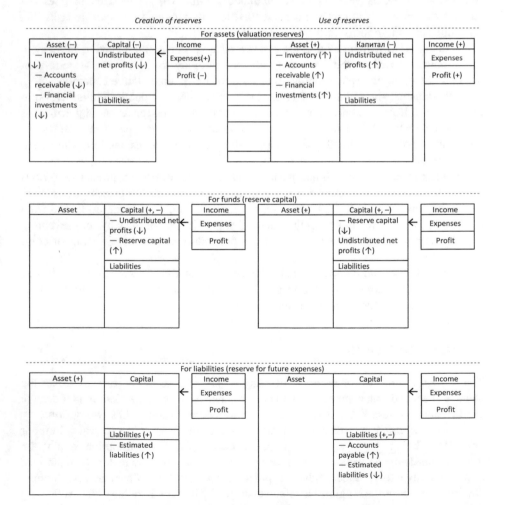

Fig. 1. Model of the effect of reserves on balance sheet ratios in the context of a digital economy

The creation of valuation reserves has a negative impact on the financial standing and financial performance of an economic entity. This is represented by a reduction in the amount of assets for which the reserve was formed, as well as by an increase in expenses, which leads to a decrease in profit and equity. As a result, the economic potential of the company decreases and its financial dependence increases.

However, the need for creation of valuation reserves makes it possible to present the most realistic picture of the current situation in the very short term prospects. Thus, the organization protects itself from financial losses in the future. In addition, digital reporting can make it possible to provide the necessary calculations to determine the forecast of company's economic performance based on financial reporting information.

Each type of valuation reserves in the balance sheet is reflected as a decreasing value of current assets. This way the provision for an impairment of inventory leads to a decrease in the balance sheet item "Inventories". For a clearer picture of the indicator "Accounts receivable" in the reporting, it is necessary to subtract the provision for doubtful debts from the amount of receivables. The provision for impairment of financial investments is also deducted from the balance sheet asset item "Financial investments".

The use of valuation reserves happens when negative forecasts are confirmed. In this case, the organization does not bear any financial losses, since the negative influence of factors is compensated by the reserve.

The provisioning for future expenses is an advance of the organization's own financial resources for the needs of current activities. The formation of the reserve involves the recognition of future highly probable circumstances that will lead to the diversion of financial flows which promise economic benefits in the future. Along with the reflection of liabilities, an expense is recognized which adversely affects the financial result, and as a consequence, the amount of capital.

Thus, the creation of a reserve for future expenses affects the balance sheet by reducing the size of retained net profit and increasing liabilities. While neither the volume nor the structure of assets are subject to change.

However, should a risky event occur, i.e. a turn of a contingent liability into a real liability does not manifest itself in a destabilizing change in the organization, since only the pattern of the nature of the liability is reflected in the balance sheet. The financial position of the organization is also preserved, since expenses are not increasing and have already ensured their influence on the financial result.

Traditionally, a reserve fund is created by the means of the amount of retained earnings of the organization. As a result of the creation of such a reserve, a change in the capital structure occurs: a decrease in the total fund leads to the creation of a special purpose reserve. In other words, within the process of the circulation of capital only the state of the capital changes whereas its value remains the same.

Fund creation can also be considered in a broader sense, not only as a redistribution of retained net profit, but also as a source of its formation. Since the reserve capital is most often allocated to covering losses of past years. Thus, we can say that the reserve capital acts as a safety bumper, affecting the negative indicators of the organization.

With reference to the foregoing, we can conclude that the valuation reserves and reserves for future expenses affect the financial performance of the economic entity, increasing other expenses or production costs; reserves created by the means of capital adjusts capitalized profit.

Figure 2 shows the effect of reserves on the formation of a financial result of an organization. We can see that reserves for future expenses contribute to a uniform change in the cost of production, reserves for doubtful debts reduce risks for doubtful accounts receivable, reserves for impairment of inventories – the value of inventories, reserves for the impairment of financial investments – the cost of financial investments, reserve capital reflects a part of retained earnings.

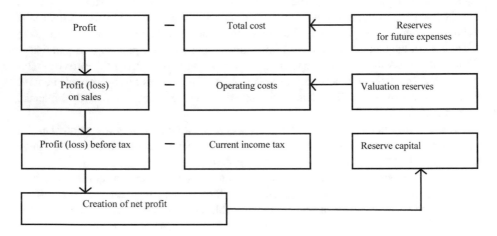

Fig. 2. The influence of reserves on a financial result of a company

In the current assets section of the Russian balance sheet, the amount of receivables under "Accounts receivable" reflects the value with the allowance for doubtful debts being considered. For an accurate reflection of the value of real receivables, it is necessary to subtract the reserve for doubtful debts from the total amount of receivables. The reserves for impairment of financial investments in the balance sheet is recognized under "Financial Investments" as a deductible amount. The amount of the reserve for the impairment of material assets reduces their value.

In the liabilities side of the balance sheet, the amount of the reserve created is not reflected individually, however, directly affects it. Since the valuation reserves affect the amount of other expenses, and as a consequence, affect the amount of profit. In the profit or loss statement, the amount of the valuation reserves is reflected as a balance with the amount of other income received from recovering the unused amount of the reserve. Thus, the size of the indicator for the item "Retained net profit" directly depends on the valuation reserves in the organization, which proves the influence valuation reserves have on total capital.

It is also important to note the need to reflect the deductible amount of valuation reserves in the Statement of changes in equity in the section "Changes in Equity" under the item "Expenses directly charged to the capital decrease" for the reporting period in which it has an impact on retained net income.

It would also be advisable to reflect the valuation reserves in the "Net Assets" section of the Statement of changes in equity for each particular asset, to determine the amount of which a reserve was created and used. These changes would allow to provide a more detailed reflection of both the total size of the valuation reserves created in the organization, and each asset subject to impairment in the reporting.

It is necessary to reflect the indicators of the movement for each valuation reserve presented, as well as their total value. The column "Balance as of the beginning of the reporting year" shows the amount of reserves formed at the end of the previous year. The line "Received in the reporting year" reflects the amount of reserves created during the reporting year. The column "Spent in the reporting year" discloses the data on the use of unused reserves during the year, unused balances of reserves liquidated before the creation of a new reserve.

Reserves for future expenses are recognized for a case of an obligation or expense which has not yet arisen, but may arise with a fairly high degree of probability. For the purposes of reflecting contingency in accounting reports, it is necessary to divide it into two groups:

- possible facts, information about which is to be disclosed in the financial statements (in the balance sheet in particular);
- possible obligations, information about which is to be disclosed in the explanatory note.

The provisioning is attributed to expenses and, depending on the type of obligation, is charged to expenses of normal activities or other expenses. The reserve is recognized at those expenses, which would be recognized in the event a liability, under which the reserve is created, occurs.

It should be noted that in the Russian accounting reporting, long-term and short-term liabilities should be presented separately. In this regard, in the analytical accounting record 96, separate accounting should be organized for long-term (maturity date of which exceeds 12 months) and short-term (maturity date of which does not exceed 12 months) valuation obligations.

Thus, it is necessary to initially open second-order accounts: "Reserves for long-term expenses" and "Reserves for short-term expenses" to the account "Reserves for future expenses". For analytical accounting of each reserve, it is necessary to open accounts of the third and subsequent orders to the specified second-order accounts, depending on the purpose of the reserve.

Some experts have put forward a proposal to permanently control the ratio of actual expenses incurred in cases of contingency of economic activity of the organization, and the amount of the reserve formed to pay the expenses on these events. If the actual costs exceed the size of the reserve formed by the time, the difference in this excess should be charged to the account "Deferred expenses".

However, the company is not always able to accumulate the necessary reserve during one reporting year. Sometimes the provisioning aimed at to covering large costs takes several years. In case if a repair of fixed assets lasts for a long period and demands a substantial amount of work, and is carried out in the next year after the reporting one, then the balance of a reserve on a case of such repairs, at the end of the year, may not be reversed. The balance at the time of preparation of the annual balance sheet will be recorded on the account "Reserves for future expenses" and reflected in the long-term liabilities of the balance sheet under the item "Estimated liabilities".

For each contingency case of economic activity (both affecting the amount of the reserve and not reflected in the accounting), the following information is disclosed in the explanatory note to the financial statements in accordance with Russian national standards:

- a brief description of the nature of the obligation and the expected term for its fulfillment;
- a brief description of the uncertainties that exist with respect to the time period for fulfillment and the amount of the obligation.

In respect of each reserve created in connection with the occurrence of a contingency, the following is additionally disclosed:

- the amount of the reserve recovered in the reporting period in connection with the recognition of an obligation previously recognized as contingent;
- unused reserve amount allocated in the reporting period to other income of the organization.

Reserves are usually accumulated and used during one reporting year and are not reflected in the balance sheet as the balance brought forward. This fact is due to the need for control over the creation and use of reserves, as well as the fact that the carry-over reserve residuals underestimates the size of the net profit of the reporting year to be distributed among the owners of the organization.

In our opinion, conducting an inventory as a control factor of accumulation and use of reserves is necessary not only at the end of the reporting period, but also at the dates of the interim reporting, which would increase its reliability and comparability with the data of annual reporting. The amount of reserves obtained in the results of the inventory must be compared with those values that are actually reserved and recorded. Based on the foregoing, we consider it possible and necessary to conduct an inventory of reserves before compiling annual and interim accounting reports, which will help increase their reliability.

Current computer and information technology provide ample opportunities for the presentation and processing of reporting information on an organization's activities, which may be useful for information users in the process of making effective management decisions. For example, software products used to compile financial statements may include additional functions of financial analysis, thus computing relevant ratios as output information.

One of the main criteria, influencing the reliability of the financial statements indicators of an economic entity is the use of a reserve system. Reflecting the reserves in the financial statements makes it possible to achieve a more accurate assessment of the assets, capital and liabilities of the organization, and as a result, its financial standing.

In the process of analysis of a company's financial standing, it is necessary to solve the main problem of the effectiveness of the formation and use of financial resources and estimating the potential for stable development of the organization. To solve this problem, it is necessary to assess the reserves of financial results growth without any additional financial flows.

The reserve system includes several tools: planning, financial analysis, internal control, accounting. Interaction and use of financial analysis and internal control will provide reliable financial stability indicators in the organization.

Financial analysis acts as an instrument of reserve policy and allows to identify the effectiveness of its application in an organization. The function, which is performed by a financial analysis on the accumulation and use of reserves provides an assessment of the appropriateness of creating reserves, and is also a means of evaluating the financial results and a financial standing of an organization, in which information about reserves plays an important role.

At present, the attention paid to the information base of financial analysis, and in particular, financial statements, is insufficient. The financial statements data serves as a source for conducting a comprehensive financial analysis in the organization. It is believed that, provided the financial statements are prepared in accordance with the requirements of the legislation, approved by the executive body of the organization, and contain an audit report, they reflect the reliable situation in the company: financial standing and the result of its business activities. Therefore, it is believed to be applicable for the purposes of financial analysis.

In our opinion, this is not always the case. The main reason for this is the relativity of accounting data, which are subject to change depending on the accounting concept and the methodology. Accounting rules are directly dependent on the goals and objectives of the organization, in connection with which they can have tremendous differences. This fact determines the variable nature of the indicators in the reporting, and allows us to conclude that they are multivariable. Consequently, the analytical indicators obtained as a result of the financial analysis are also multivariable.

The works of various authors indicate a different calculation of the liquidity ratios of the organization. The numerator and denominator may or may not include certain lines of the balance sheet and statement of financial performance, various regulatory restrictions are also proposed, but the general approach to calculating liquidity indicators is the same.

The upper limit on the value of the liquidity ratio will be the value calculated on the basis of the nominal value of the assets and liabilities of the balance sheet. That is, assets and liabilities are accepted for calculation at their initial cost: inventories – at the actual cost of acquisition, accounts receivable and payable – in the amount to be reimbursed, etc.

The lower limit of the ratio is determined by the amount calculated with the consideration of the prudence of assets and liabilities concept, according to which the value of assets should not be overestimated, therefore, they are reflected in the accounts excluding the reserve created for these assets, as for liabilities – they should not be underestimated, and therefore they are reflected with allowance made for contingent liabilities. The analysis is based on the most pessimistic forecast, i.e. indicating the maximum risk values and reserves for them. This will determine the critical level of liquidity, which will vary depending on the risk assessment.

3 Conclusion

Unlike reserves created in most areas of human activity, accounting reserves do not affect the presence and condition of material values. They have a financial nature. Creating one reserve or another, accountants follow the principle of caution in valuation (accounting conservatism concept), insuring against possible costs, which can go beyond the usual expenses. They recognize the expense (create a reserve) in advance in case of presence of a certain reasonable certainty of its occurrence in the near future.

The conducted study demonstrated that the impact of the development of the digital economy on the process of preparation and use of financial statements is difficult to overestimate. The process of expanding the scope of the digital economy offers great opportunities for improvement the quality of financial reporting and, accordingly, the effectiveness of management decision-making process.

References

1. De Roover, R.: The development of accounting prior to Luca Pacioli according to the account-books of medieval merchants. In: Littleton, A.C., Yamey, B.S. (eds.) Studies in the History of Accounting, London, pp. 114–174 (1956)
2. De Roover, R.: The story of the Alberti company of Florence, 1302–1348, as revealed in its account books. Bus. Hist. Rev. **32**(1), 14–59 (1958)
3. Kuter, M.I.: Introduction to Accounting: A Textbook, 512 p. Prosveshchenie-Yug, Krasnodar (2012)
4. Kuter, M., Gurskaya, M., Andreenkova, A., Bagdasaryan, R.: The early practices of financial statements formation in medieval Italy. Account. Hist. J. **44**(2), 17–25 (2017). https://doi.org/10.2308/aahj-10543
5. Efimova, O., Rozhnova, O.: The corporate reporting development in the digital economy. In: Antipova, T., Rocha, A. (eds.) Digital Science, DSIC18 2018. AISC, vol. 850, pp. 71–80. Springer, Cham (2019)
6. Antipova, T.: Digital view on the financial statements' consolidation in Russian public sector. In: Antipova, T., Rocha, Á. (eds.) Information Technology Science, MOSITS 2017. AISC, vol. 724, pp. 125–136. Springer, Cham (2018)
7. Khoruzhy, L.I., Gupalova, T.N., Katkov, Yu.N.: Putting in place a system of integrated reporting in organizations. Int. J. Innov. Technol. Explor. Eng. (IJITEE) **8**(7), 748–755 (2019)
8. Sokolov, Ya.V.: Accounting: ab origine till present days, 638 p. Audit, UNITY, Moscow (1996)
9. Sokolov, Ya.V.: Contingency – Unconditional Achievement of Accounting Thought. BUH.1C. 5 (2006)
10. Richard, J.: Accounting: Theory and Practice/Translated from French by Sokolov Ya.V., Finance and Statistics, Moscow, 160 p. (2000)
11. Baetge, J.: Bilanzen, überarbeitete Auflage. IDW-Verlag GBH, Düsseldorf (2000)

Transformation of Fundamental Accounting Principles in the Context of a Company Insolvency from the Perspective of the Digital Economy

Ruslan Tkhagapso(✉)

Kuban State University, Krasnodar 149, Stavropolskaya Street,
350040, Russian Federation
rusjath@mail.ru

Abstract. Bankruptcy is an integral part of a market economy, the main purpose which is the revival of the economy, business, production. The modern organization management mechanism is distinguished by a complex information system based on an abundance of external and internal information flows, as well as a variety of types of information circulating in the management system. With the growing role of information systems and digital technologies in society, the accounting information system and the professional activities of an accountant, a financial analyst and an auditor are of particular importance. The largest share in the total aggregate of economic information is information provided by accounting, which plays a dominant role in the management system as a whole.

Crisis events in a company' life stimulate the need for accounting information in order to recreate the model of the financial and economic activities of the organization in the pre-crisis period of its life cycle. Undoubtedly, the risk of termination of activity deforms the system of principles on which the organization's activity is based under normal operating conditions, which inevitably leads to the transformation of accounting methods in bankrupt organizations.

The paper examines the fundamental accounting principles in the context of a company's insolvency and their impact on the transformation of the accounting and reporting system of bankrupt organizations.

Keywords: Bankruptcy · Accounting principles · Digital economy · Financial standing

1 Introduction

According to general notion, insolvency is considered to be a definite negative state of an economic entity within the system of economic relations, which is characterized by its inability to fulfil its obligations. The liquidation of a legal entity procedure always leads to the emergence of various legal and economic issues, which are related to the process itself and closely interconnected.

© Springer Nature Switzerland AG 2020
T. Antipova and Á. Rocha (Eds.): DSIC 2019, AISC 1114, pp. 75–85, 2020.
https://doi.org/10.1007/978-3-030-37737-3_7

Professor J. Richard argues that «the use of accounting depends on the regulatory methods that this or that capitalist system uses» [1]. Discussing the four ways of regulation, J. Richard assigns the fundamental role to regulation «through bankruptcy», which represents «a specific type of accounting which focuses on the problem of liquidation of an enterprise». A textbook example of this method of regulation is the Venetian model of accounting for merchant-seafarers who distributed the financial result at the end of a commercial venture – sailing. For each voyage, the seafarers opened a separate account, which reflected the costs of purchasing goods and a vessel. At the end of the voyage, revenue from the sale of goods and income from the resale of the vessel were recorded on the account. If the merchant kept the vehicle for subsequent expeditions, it was estimated at a possible market value.

A special place in accounting is given to the development of fundamental principles, «without which it would be necessary to study new rules for each organization and it would be impossible to compare information about two organizations» [2]. In the system of the accounting principles, *the going concern assumption* is given pride of place.

It should be noted that the accounting principles are the major focus of interest of leading foreign and Russian scientists. For example, Ya.V. Sokolov [3] has examined the 24 principles proposed by eleven foreign authors and two American public accountant organizations. At the same time, Ya.V. Sokolov identifies 13 principles [4], M.I. Kuter – 19 principles [5]. The underlying accounting principles have also been studied by other Russian and foreign scientists [6–10].

In his article, Ya.V. Sokolov considers the continuity assumption «an analogue of the I. Newton's first law» [11]. M.I. Kuter writes that an enterprise, having arisen once, can work for an indefinitely long period, while avoiding a significant reduction in production and sales and, most importantly, fulfilling its payment obligations on time [5]. *In the event of insolvency, the organization cannot comply with these assumptions and therefore is obligated to state this fact in the accounting policies, which is being formed for the upcoming accounting period and in the explanatory note to the financial statements for the reporting period.*

Ya.V. Sokolov and S.M. Bychkova believe that the importance of the principle lies in the necessity and possibility for the one applying this principle to take into account six of its components:

- independence of accounting from a change of ownership;
- accounting of an organization's property at historical cost;
- use of the capitalization of cost mechanism;
- provisioning of future financial results;
- matching income and expenses of an economic entity;
- reporting for calendar periods [11].

For a business entity being liquidated, compliance with these points is impossible, first of all, due to the fact that its period of existence is now limited to the liquidation period, established by the owners or the decision-making unit which has made the decision to liquidate the company. Failure to comply with the principle of ongoing concern leads to the need to review the practice of the following principles: the periodicity assumption, historical cost, substance over form, etc. Let us consider in more detail the causes and consequences of a situation when the principles are not being applied.

2 Research Results

The going concern assumption is aimed at choosing *a method for assessing the property* of an economic entity. Under normal conditions, the property of an economic entity is valued at the initial (historical) value. The property of an organization which is on the verge of bankruptcy is included in the assessment, according to which each asset can be sold separately at the time of reporting. In 1675, J. Savary [12], proposed the lower of cost or market approach to determine the value of a possible sale in bankruptcy. It should be noted that valuation of property at a realizable value is a complex process in a market economy, when one object can have different assessments.

It should also be noted that the principle of historical cost has quite a narrow range of possible practical usage permitted by regulatory acts, although it is stated in the national standards. In the process of liquidation, the recognition of property and liabilities at cost value is all the more unacceptable due to the biased picture of the financial position of the organization being liquidated, which arises in this case due to the discrepancy between accounting indicators and the possible selling price of the property intended for these purposes (thereby, the interests of creditors will be infringed) and incomplete reflection of obligations (here the interests of both creditors and owners of the organization will be infringed). The issue about at what cost to reflect accounting objects in liquidation balance sheets is currently one of the most debated.

After the arbitral tribunal decides to declare a debtor bankrupt, the bankruptcy proceedings are approved, which is the final stage in the insolvency proceedings. This procedure is aimed at the liquidation of an economic unit by means of its sale as a property complex or its individual components to proportionately settle the claims of creditors. According to Russian law, two months from the date of publication of information on declaring bankruptcy and the opening of bankruptcy proceedings are allotted to the debtor organization to accept the claims of creditors.

A court-appointed trustee makes settlements with creditors according to the register of creditors' claims in the order of priority stated in Article 134 of the Bankruptcy law [13]. The claims of the creditors of each priority rank are satisfied after the full repayment of the requirements of the previous priority rank, with the exception for «secured» creditors. If there is insufficient amount of money to fulfill payment obligations to creditors of one category, these funds are distributed among the creditors in proportion to the amounts of their claims included in the register.

Claims of creditors within the insolvency proceedings may be classified into claims *filed* (including recognized and unrecognized) and *unasserted*. Recognized claims in accordance with the provisions of the Bankruptcy Law include: statutory (undisputable) claims; other requirements recognized by the court, in respect of which there is a valid court decision on their satisfaction. Structurally, debt to the creditor consists of: the principal debt amount; assessed penalty for late payments; claims for state duty and reimbursement of legal expenses.

The accounting procedure for these process should be reflected in the accounting policies from the date of recognition of the organization as a bankrupt (without waiting for the start of the new reporting year) due to a significant change in the conditions of activity, as well as in the work plan of accounts, analytical ledger, document flow rules, which will ensure meeting the requirements for completeness of accounting, timely reflection of business transactions in accounting registers.

The accounting methodology in bankrupt organizations should be based on compiling all accounting data in accordance with the actual availability of property identified as a result of the inventory, and accounts payable – with the results of consideration of creditors' claims. Changing the conditions and the very essence of the economic activity of a bankrupt organization necessitates a revision of the set of accounts and accounting methods, since different requirements to the quality, reliability and objectivity of accounting information are set. At the same time, an efficiently developed accounting system contributes to the systematization and accumulation of information, has a direct impact on the reliable reflection and disclosure of data in the interim liquidation balance sheet, thereby satisfying the changing requirements of information users.

In the condition of insolvency, priority is given to regrouping analytical data on accounts in order to obtain the data from the perspective of the order of repayment, which is necessary for subsequent payments to creditors and reporting. Sub-ledger accounts are opened for each creditor of a bankrupt organization. To control the procedure of making payments to the budget arising in the course of the bankruptcy proceedings, analytical accounts for each tax and charge should be kept.

The recording of an organization' liabilities should be kept in the order of repayment of claims, not based on the economic content of relations with counterparties, as it is the case with normally functioning enterprises. The data on the accounts are recorded in accordance with the register of creditors' claims. *The allocation of separate second-order accounts for each priority rank with analytical account for each creditor is of high practical importance, as it increases the efficiency of payments to creditors and facilitates the analysis of the dynamics of debt repayment.* If a creditor has claims of different priorities, they should be reflected in different accounts (subaccounts), even if they arise from one contract or obligation. Debts to other creditors for received products, services and various sanctions technically relate to other debts, but since sanctions are not repaid earlier than the primary debt, they are recoded on different accounts to keep the order of repayment under control. In order to find out which requirement should be reflected on which account, one needs to refer to the register of creditors' claims. According to most experts, the analytical accounting of obligations under liquidation should be organized in the context of groups of creditors.

The business entity concept, the pioneer of which was J.P. Savari, assumes that the property and liabilities of the enterprise are recorded separately from the property and liabilities of the owners, that is to say, a change of ownership of the organization should not affect the accounting of this enterprise.

H. Vannier set out a principle, which later became the basis of this concept: *accounting is carried out on behalf of the company, not the owner, and, therefore, the owner is responsible for the debts within the share of his contribution* [14]. Since the owners of an organizations on the verge of bankruptcy are trying by all means to save the property from forcible withdrawal for repayment of debts, the role of this principle is increasing, along with the first historical task of accounting – *to ensure the safety of property* of the owner. Under such conditions, the main task of an auditor is to verify the presence and safeguard of assets of an economic entity.

To organize synthetic accounting by the order of repayment of claims of creditors, separate accounting of costs and settlements arising within the process of liquidation of an organization, it is necessary to re-group the data between analytical and synthetic accounts. The introduction of various accounting and analytical registers in the process of accepting creditors' claims will help to avoid inconsistencies or losses when transferring data from one account to another. Table 1 demonstrates the system of accounts developed by the author, which helps to improve the quality of control during the process of accumulation of information regarding the obligations of the debtor and their repayment from the perspective of their priority.

Table 1. Indicative list of accounts for obligation recording of a bankrupt organization

Account title	Characterization of the account
Trade accounts payable	Summarizes the information regarding the current liabilities of the debtor to its counterparties (security and pre-sales preparation of property services, notification of creditors services, etc.), occurred within the procedures of bankruptcy administration
Compulsory budget payables	Reflects the information regarding the taxes and charges debts to the budgets of different administrative levels. Book records should consider the order of creation of obligations
Extra-budgetary funds payables	Includes the information regarding the compulsory payments to extra-budgetary funds of different administrative levels grouped in accordance with the priority of their settlements
Current payroll obligations	Summarizes the information regarding the payroll payments, arisen after the initiation of insolvency proceedings
Current claims	Summarizes the information regarding the extraordinary payments: (1) «Litigation costs» excluding the expenses on a court-appointed trustee; (2) «Payments for a court-appointed trustee services»; (3) «Payments for other services appointed by the court-appointed trustee to participate in the bankruptcy proceedings» – an independent assessor service fee, a registrar service fee, etc.; (4) «Other extraordinary payments» – utility services payments, etc.

(continued)

Table 1. (*continued*)

Account title	Characterization of the account
Settlement of claims related to compensation for loss of health or life	Reflects the information regarding the settlements of the first priority rank of claims for loss of health or life
Settlement of claims of employees for wages	Summarizes the information regarding the payroll payments, accrued before the initiation of insolvency proceedings
Secured creditors payables	Includes the information regarding the settlements with creditors secured by a property pledge of a bankrupt organization
Other creditors payables	Reflects the settlements with creditors not secured by a property pledge, as well as other accrued financial sanctions, divided in accordance with types of obligations
Unasserted claims	Considers the obligations to counterparties for which claims from creditors have not been received
Settlements of claims	Reflects the information on contracts, voided by a court-appointed trustee

Recording the data on new analytical positions of accounting registers is only possible upon completion of acceptance of creditors' claims.

To keep the records of the liabilities, arising in the course of current settlements, separate accounts «Trade accounts payable» and «Property disposal account», which are used to record information about the assignment of claims. At the same time, settlements of compulsory payments are reflected in the accounts «Taxes and duties payable» and «Payable to state extra-budgetary funds» divided in accordance with the priority and the direction of payment. In accordance with Russian legislation, the compulsory payments claimed after closing of the register, including those arisen after the opening of bankruptcy proceedings, are paid to the budget only after satisfying the claims of creditors asserted on time, that is after the registry is closed, the organization has an obligation to prepare and file a tax declaration in case it performs taxable activities, but these taxes are paid to the budget after all registry claims are paid.

Payroll obligations arising within the bankruptcy administration procedures are recorded in the «Current payroll obligations» account, and the obligatory payments arising from this are recorded in the account «Extra-budgetary funds payables».

For the convenience of settlements and monitoring by the liquidation commission, all subsequent obligations of the bankrupt organization are reflected in the account «Settlements with other debtors and creditors» with the opening of accounts of the second and third order. This way on the account of the second order Settlement of claims related to compensation for loss of health or life» the obligations of the bankrupt organization to citizens to compensate for harm to life or health, as well as moral harm is recorded.

The debts to workers, which were formed at the time of the application for declaring the debtor bankrupt and were not repaid during the administrative receivership, if carried out, are paid in the second order, according to its priority rank. These claims of creditors are recorded in the second-order account «Settlement of claims of employees for wages».

The procedure of settlements with creditors *of the third priority rank* is of particular interest. Within this category, a kind of sub-ranks emerges, which is due to the fact that secured creditors' claims are of the priority with the exception of those creditors in the second and the priority ranks, claims for which arose prior to the conclusion of a relevant pledge agreement. This means that when deciding on the priority, the bankruptcy trustee must pay attention to the date of conclusion of the pledge agreement. This operation is of necessity in order to satisfy the claims of creditors of the first and second priority rank which arose prior to this date, then make settlements with secured creditors and only after that return to satisfying the remaining claims of the first and second priority rank. If, at the expense of the value of the subject of the pledge, the claim of the pledge holder is not fully satisfied or not settled, this requirement shall be fulfilled within the frames of the third category without any advantages.

Regardless of the source, penalties paid last after the repayment of the principal debt are reflected in the second-order account «Settlements with other creditors» with a separate analytical position «Settlements of financial sanctions».

Amounts of creditors' claims, not submitted within the established time, but revealed in the accounting data, are recorded on a separate second-order account «Unasserted claims». The need for this account is determined by the possibility of incurrence of the claims after the deadline set by the bankruptcy trustee. These claims can be satisfied in case there are funds available, after the settlements with creditors who have made claims within the established time are completed.

Claims of unsecured creditors and authorized bodies made after the closing of the register (except for the claims of creditors of the first and second priority ranks) and recognized in the established manner, as well as obligations of mandatory payments which arose after the initiating of bankruptcy proceedings, regardless of the time of the claims are satisfied at the expense of the property of the debtor remaining after satisfaction of the claims of creditors included in the register. Thus, in fact, *the fourth category* of creditors is formed, which include, among other obligations, taxes and fees accrued during the period of bankruptcy proceedings, settlements on which are carried out only after satisfying the requirements of the creditors of the previous priority ranks.

The Russian liquidating practice shows that insolvent organizations usually only have enough funds to pay off debts of the first and second priority ranks and claims secured by a pledge. Therefore, taxes and fees arising in the course of bankruptcy proceedings are only accrued, but not paid due to the insufficiency of the property of the debtor.

The going concern assumption also implies *the accrual principle assumption*. Here it is appropriate to give an interpretation of M.I. Kuter, which states that selling expenses (symmetrical to income from sales) are recognized in that accounting period in which income, possible because of these expenses, is generated. Non-operating expenses and non-operating income are recognized not at the time of payment or receipt of money, but at the time of their occurrence [5]. When conducting bankruptcy administration, on the contrary, the time of receipt or payment of cash and repayment of the corresponding debt obligations becomes of the key importance, and «boiler» method of cost accounting is used, when income and expenses are recognized in accounting not at the moment of their occurrence, but at the moment of cash outflow or inflow.

According to the conservatism concept, which emerged one of the first, an enterprise can use profit (or distribute losses) only in case of liquidation, that is to say, the result should not be calculated before the moment of closure of the enterprise and ensuring the coverage of its debts. In accounting, conservatism concept in valuation usually means a greater willingness to recognize expenses and liabilities than possible income and property. The liquidation balance sheet should contain no items of deferred income or expenses, or reserves. The implementation of the conservatism principle is limited in bankruptcy, as it can lead to an overestimation of the losses of an economic unit. The principle plays an important role at the bankruptcy prevention stage though, ensuring the organization's continuous and uninterrupted operation.

In the liquidation of an organization process, special attention is paid to the creation and movement of capital of the bankrupt organization. At this stage, the division of the organization's capital in accounting into individual components does not have any sense as it does not carry any information load for the users of the reports. It is also worth noting that the latter are more interested in the real data on capital (or its deficit) as a source of funds guaranteeing the ensuring of satisfying creditors and owners of the organization's interests. Consequently, the indicator of the capital of a bankrupt organization in the liquidation balance sheet should be uniform. At this stage, many authors suggest closing the accounting of the elements of a company's own capital considering the account of the authorized capital, which seems somewhat controversial and not lacking some disadvantages.

In the capital structure of an economic entity, several components can be distinguished: initially invested capital, which includes the authorized capital of the organization and funds contributed by the owners in excess of the amount of registered capital (in Russian accounting it is reflected in the second-order account opened to the account «Additional capital»); reinvested capital, which includes the reserve capital and the capitalized part of retained net profit («Retained earnings» account); the property revaluation fund formed during the pre-assessment (reflected in a separate second-order account opened for the «Additional capital» account), which acts as a reserve in case of a decrease in the value of the revaluation of property.

The substance over form principle was originally earmarked in the Anglo-American accounting system, where the company accepted for accounting and recognized in the reporting not only their own possessions, but resources involved in the economic turnover and generating revenue for this company, even though they are owned by another economic entity. Later other countries – adherents of the continental accounting model, began to concede this interpretation in their own regulations. Under the conditions of insolvency, the limitation of this principle lies in the impossibility to recognize income from the use of property not belonging to the organization in the foreseeable future and, as a result, in the arising need to reflect only those objects that belong to the organization in the balance sheet asset on the right of ownership, in order to avoid an overestimation of the property mass.

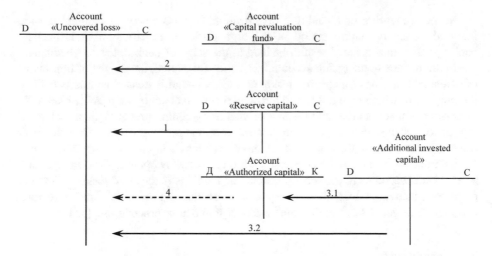

Fig. 1. Model of accounting records of capital for a company being liquidated

At the time of liquidation, in the accounting of a bankrupt organization, it is necessary to close the accounts of elements of equity (Fig. 1). According to the decision of the meeting of founding members (shareholders), the reserve capital account is closed first (record 1) with the allocation of funds from the reserve fund to cover losses incurred before the introduction of the bankruptcy administration procedure.

Recognition of past revaluation amounts in the accounting of a bankrupt organization is a subject of particular interest. Upon disposal of a long-term asset, *the amount of its revaluation is transferred from the additional capital of the organization to the retained earnings of the organization* (record 2).

The most controversial issue in accounting of an organization's capital is the procedure of closing an account of additionally invested capital. Many scholars in their studies suggest that these amounts be used to increase the authorized capital of the organization (entry 3.1), but any change in the authorized capital requires a re-registration of the company's charter, which in the face of insolvency seems to devoid of any practical sense. In this regard, the author considers it more correct *to allocate these funds to repayment the amounts of uncovered losses that occurred before the organization was declared bankrupt* (record 3.2) within the insolvency proceedings with the consent of the members of the liquidation commission and the participants of the business entity (without any infringement of the owners' interests).

Upon completion of the bankruptcy proceedings, in case of insufficiency of property and other liquid assets of the liquidated legal entity, which could be used to cover the occurred losses, it is possible to use the authorized capital of the bankrupt organization (record 4).

In the accounting of a bankrupt organization, it is necessary to separate the loss incurred during the entire period of the organization's activity until it is declared bankrupt from the financial result obtained at the stage of bankruptcy proceedings.

In the process of analyzing accounting records for capital, the degree of importance of «Retained earnings (uncovered loss)» and «Profits and losses» accounts is defined, the purpose of which is to reflect the financial result obtained by an organization at the final stage of its operation with the aim of increasing control and analytical features of liquidation balance sheets. To ensure control, it seems necessary to separate the profit (loss) generated before the date the decision on company's liquidation was made from a similar indicator related to the liquidation period directly, by opening the corresponding second-order accounts to the account «Retained earnings (uncovered loss)»: «Retained earnings (uncovered loss) before liquidation of the organization» and «Retained earnings (uncovered loss) during the period of liquidation procedures» [15].

3   Conclusion

Despite the apparent autonomy, the principles have a definite interrelationship, since the set of organizational, technical, methodological and methodical elements of the accounting procedure is determined by the means of a specific set of initial assumptions, a change in one of which entails the need to revise the whole set. This, consequently, can lead to inability to follow one or more principles.

Under the current conditions, the issues of accounting for settlements with creditors in the course of repayment of obligations of an economic entity are particularly relevant. We propose to perform the accounting of obligations based on the priority rank of claims of creditors established by the Bankruptcy Law, with the allocation of separate synthetic accounts and analytical positions for each line. The main advantage of this approach is the formation in the accounts system of data on the amount of debt for each category, which is necessary for arrangement of payments to these creditors. Moreover, the proposed system of accounts allows one to divide the accounting obligations arising before and after the date of declaration of bankruptcy, while considering the factor of differences in the legal regime for their satisfaction. Such construction of accounting requires a regrouping of the data of analytical accounts on the synthetic accounts in comparison with the accounting procedure before the declaration of the business entity a bankrupt.

References

1. Richard, J.: Accounting: Theory and Practice/Translated from French by Sokolov Ya.V., Finance and Statistics, Moscow, 160 p. (2000)
2. Anthony, R., Reece, J., Merchant, R., Hawkins, D.: Accounting: Text and Cases, 11th edn., 1008 p. McGraw-Hill Education, New York (2003)
3. Sokolov, Ya.V.: Accounting: ab origine till present days, 638 p. Audit, UNITY, Moscow (1996)

4. Sokolov, Ya.V.: Theoretic Framework of Accounting: A Textbook. Finance and Statistics, Moscow, pp. 31–41 (2000)
5. Kuter, M.I.: Introduction to Accounting: A Textbook, pp. 94–103. Prosveshchenie-Yug, Krasnodar (2012)
6. Baetge, J.: Bilanzen, überarbeitete Auflage. IDW-Verlag GBH, Düsseldorf (2000)
7. Chiwanza, M.R., Dumbu, E.: Principles of Accounting, 272 p. LAP Lambert Academic Publishing, Saarbrucken (2013)
8. Needles, B.E., Powers, M., Crosson, S.V.: Financial and Managerial Accounting, 10th edn., 1280 p. Cengage Learning, Boston (2013)
9. Needles, B.E., Powers, M., Crosson, S.V.: Principles of Accounting (Financial Accounting), 11th edn., 1328 p. Cengage Learning, Boston (2011)
10. Wild, J.J., Shaw, K., Chiappetta, B.: Fundamental Accounting Principles, 23rd edn., 1248 p. McGraw-Hill Education, New York (2016)
11. Sokolov, Ya.V., Bychkova, S.M.: The going concern assumption. Accounting 4, 56–59 (2001)
12. Savary, J.: Le parfait négociant ou instruction générale pour ce qui regarde le commerce … et l'application des ordonnances chez Louis Billaire…; avec le privilège du ROY. (Reproduction en fac similé de la 1ère édition par Klassiker der Nationalökonomie, Allemagne) (1993)
13. On Insolvency (Bankruptcy): Federal law of Russian Federation dated 26 October 2002 No127-FZ with changes and amendments
14. Oberbrinkmann, F.: Statische und dynamische Interpretation der Handelsbilanz: Eine Untersuchung der historischen Entwicklung, insbesondere der Bilanzrechtsaufgabe und der Bilanzrechtskonzeption (German Edition), 348 p. IDW-Verlag, Düsseldorf (1990)
15. Kuter, M.I., Tkhagapso, R.A.: Accounting in Insolvency: A Textbook, 204 p. Kuban State University, Krasnodar (2005)

The Equilibrium Model for the Price of Digital Cellular Services

Ilona Tregub[✉], Nataliya Drobysheva, and Andew Tregub

Financial University under the Government of the Russian Federation,
49, Leningradsky Prospect, 125993 Moscow, Russia
ITregub@fa.ru

Abstract. Nowadays the development of mobile communications is proceeding at an accelerating pace. Cellular operators are no longer limited to providing only basic voice services and short text or multimedia messages. So-called digital value-added services have become more popular among cellular subscribers. A subscriber who has a phone with two SIM cards of different operators can choose which provider to use at the current moment of time. At the same time, in the conditions of a uniform network coverage area and the same quality of cellular communication services, only the service price has a significant effect on the consumer's choice.

Keywords: Demand function · Market equilibrium · Mobile services

Introduction

In the modern context, any economic entity performance is primarily aimed at making a profit from its business. The work of mobile operators is no exception. The struggle for market share forces equipment manufacturers to invent new and more functional mobile devices. Operators, in turn, competing for subscribers, develop and implement new mobile services in the market. These services are supposed to provide information to a mobile phone for an additional fee. These include services such as locating a mobile terminal (LBS), media-integrated interactive services (voting, polls, etc.), business applications (diaries, translators, mobile office, etc.), transaction services (management of a bank account via a mobile phone, payment by a phone account), data transmission via voice channels using a subscriber-operator-service (IVR) scheme, mobile television, mobile Internet.

The appearance of mobile phones supporting two subscriber identification modules (SIM cards), as well as aggressive promotion of cellular operators' services by various marketing ploys on the consumer market (the most striking example of which being to grant a certain operator SIM card when buying a phone) has led to a sharp increase in Cellular density. It has long exceeded 100%, which means that today almost every subscriber has two or more SIM cards.

The number of users who are subscribers of two mobile operators is also growing. All this, on the one hand, results in the increased competition in the mobile services market. On the other hand, if there is an effective contract with two mobile operators, subscribers tend to choose services only from those provided by these operators. The choice by the subscriber who is a two SIM cards phone owner of one or the other

T. Antipova and Á. Rocha (Eds.): DSIC 2019, AISC 1114, pp. 86–93, 2020.
https://doi.org/10.1007/978-3-030-37737-3_8

operator as a service provider out of the two is made, primarily, due to the price. Such a mechanism for providing and consuming a service fits into the concept of a price duopoly model, which allows one to conclude that the subject-matter of the research is relevant.

1 Russian Mobile Market

Four large mobile operators, whose share exceeds 90%, represent the cellular market in Russia. The operators make profit by providing their consumers not only with communication services, but also with value-added services (VAS), which are particularly popular among consumers.

Despite the growth of the national economy as a whole, the dynamics of mobile service revenues and operators' OIBDA in 2018 turned out to be negative. It was caused by a number of factors such as the negative dynamics of real wages, the unstable exchange rate of the ruble, increased competition, etc.

1.1 The Dynamics of Real Wages

In the conditions of a long and significant decline in real wages, Russian households have shown a previously uncharacteristic tendency to save on communication services. According to the Russian Federal State Statistics Service, from 2013 to 2018, the share of household expenditures decreased by 2% and amounted to 3.3% of total expenditures. Consumers switched to cheaper tariffs and refused additional SIM-cards and additional services.

1.2 The Ruble Exchange Rate

A large share of operating expenses of mobile operators is historically tied to the dollar rate. In addition, the weakening of the national currency reduces revenues from overseas roaming amid falling foreign tourist traffic, as well as other categories of revenue from VAS services.

1.3 Increased Competition

The entry of Tele2 into the key market for the Russian operators in the Moscow region had a negative impact on pricing. In fact, the oligopolistic model of competition of "the Three" has made a big step towards the monopolistic competition. Combined with the fall in real wages that had intensified by then and the weakening of the ruble, the companies in the sector showed a record simultaneous drop in profits.

1.4 Attempts to Diversify

The decline in profits and the strengthening of state regulation naturally led to a decrease in the profitability of investments. In the mobile market, a trend to diversify activities has emerged. The "Megafon" company bought a controlling stake in Mail.ru

Group. The "MTS" operator made investments in the "Ozon" company. In some cases, mergers and acquisitions greatly unbalanced the cash flow and overloaded companies with debts. In particular, against this background, the "Megafon" company refused to pay dividends.

1.5 Non-market Pricing Attempts

After a record decline in the mobile revenue and a decrease in the profits in the second half of 2016, market participants began to synchronously change the pricing strategy on the principle of "more services at a higher price", including archived tariffs. The first step was the abolition of unlimited data tariffs in early 2017.

The second step was the increase in prices for VAS services, which since that time have mostly become package. The actions of the operators coincided with the local recovery of real wages of the population and led to a noticeable acceleration of the mobile revenue growth rate in Russia, as well as a natural increase in profits. This led to the fact that the actions of mobile operators made antitrust authorities interested. The introduction of new regulatory documents by the state structures again resulted in the operators' profits reduction.

1.6 Abandoning Low Profitable Retailers

In 2017, the operators reached an agreement to reduce the excess number of cellular phone stores. By the end of 2017, the aggregate number of outlets decreased by 10–15%. This had a positive effect on the profits of companies and on the profitability of OIBDA [1] (see Fig. 1).

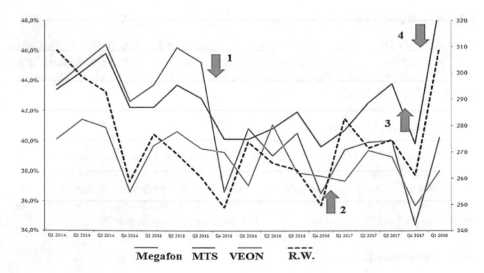

Fig. 1. OIBDA profitability of the Russian segment of Megafon, MTS, VEON and the real wage index. (**1** - launch of Tele2 in Moscow; **2** - reduction of cellular phone stores, price increases, abandoning unlimited tariffs; **3** - transition to new financial reporting standards; **4** - the law by Yarovaya.)

However, by mid-2018, the number of aggressive marketing campaigns for the sale of contracts began to increase.

In general, 2018 was the year of the Russian mobile market recovery after falling over the 2014–2017 period. Nevertheless, in the medium term, the market situation retains a negative potential [2].

Currently, the mobile market has reached saturation, and the potential for organic growth in the number of subscribers has been exhausted. According to the Accounts Chamber, telecommunication services in Russia have begun to increase at rates that exceed the average incremental price and tariffs for services in various industries. Revenues of telecom operators are declining, while the Tele2 operator is steadily increasing its subscriber base, and the company's revenue continues to grow. The rest of the Russian operators have the indicator of the subscriber base "net outflow" at 50–60% per year.

In 2018, cellular operators, including Tele2, continued to form a system of new mobile services. Tele2 has secured the status of a company that knows and anticipates the development trend of the Russian telecom market. At the same time, regardless of the market strategy of telecom operators and the dynamics of real wages of the population, the pricing peculiarities in the market with four federal participants (MTS, Megafon, Beeline, and Tele2) has led to a decrease in the profits of each individual cellular operator compared to the previously existing oligopoly of the "Three". The first consequences of Tele2's entry into the Moscow and Moscow region markets have sharpened the competition, which caused the greatest damage to the profits of Megafon and Beeline.

New services launched by mobile operators force subscribers, especially active users of different operators' services, to make a choice in favor of one of them, thereby increasing the revenue of the selected operator. All this aggravates the competition in the cellular market and encourages operators to engage in price wars. In conditions when, due to the design features of a smartphone, a consumer chooses the cheapest services of one of the two operators, the pricing process in the market can be described by a price duopoly model. It is the model of price duopoly that we will explore using the example of IVR services in the Russian cellular market.

2 Development of the Optimizing Model

In our work, we use the classic models of the duopoly of Cournot [3] and Bertrand [4] as well as its modifications [5, 6]. The demand of the regional and federal mobile operators' subscribers for additional services is considered from the perspective of a potential and actual one.

The potential demand is determined by the number of services in quantitative terms which is desirable for an individual to consume at the stated service price level.

The amount of the potential demand was determined on the basis of the survey data collected by analytical agencies. It was also based on the data from billing systems (Telecommunications billing) of operators and service providers on the results of targeted interaction with consumers of various groups via mobile polling tools [7]. The awareness and demand for various mobile services, the general and current structure of demand for VAS services were studied on the basis of data from the ROMIR Moni-toring research holding.

Actual demand is equal to the factual volume of services rendered. The value of the actual demand for IVR services is equal to the volume of services provided. This value was determined on the basis of commercial performance reports of cellular operators and service providers.

Potential industry demand for IVR services can be represented by a hyperbolic function of price p

$$D(p) = \frac{a}{b+p} \tag{1}$$

Parameter b in the denominator is introduced to limit the demand function at zero price ($p = 0$). The demand level at the zero price of the service is set by parameters b and a. As mentioned above, we will assume that a consumer's choice is limited to the services of only two operators. The actual demand function $D_1(p)$ is the number of services rendered by the first operator if subscribers who are able to use the services of both the first and second operators consume IVR services of only the first operator during the given period. The demand value depends on the number of subscribers (N_1) and the average number of services per subscriber for a unit of time, for example, for one month.

$$D_i(p) = N_i \bar{q}_i(p); \quad i = 1, 2. \tag{2}$$

$\bar{q}_i$ is the averaged demand function of the i company. We assume that initially the prices for services of both operators are the same and equal to the minimum (p_{min}). The potential demand of each operator in this case can be described by an empirical function of the actual demand, the analytical form of which coincides with the function of industry demand and takes the form of a hyperbola. Estimates of coefficients b and a can be obtained by processing statistical data using the least squares technique [8–10].

When making a decision to establish new levels of their prices for services, operators may be guided by competitors. In this case the demand function of operator 1 that is the first to increase the price from the p_{min} level to p_1 will be

$$D_1(p_1) = D_1(p_{min}) - \Delta D(\Delta p) \tag{3}$$

Where

$$\Delta D(\Delta p) = N_{12}\bar{q}_1 = N_0(p_1 - p_{min})\frac{a_1}{b_1 + p_{min}} \tag{4}$$

Formula (3) takes into account the fact that the first operator knows part of its subscribers N_{12} will switch on to the second operator.

Provided that at the first step, the second operator does not increase its price simultaneously with the first operator ($p_2 = p_{min}$), the problem of the first operator's profit maximization will be

$$\pi_1 = \left(\frac{a_1}{b_1 + p_2} - \frac{a_1 N_0 (p_1 - p_2)}{b_1 + p_2} \right) p_1 - p_1 c_1 \rightarrow \max_{p_1} \tag{5}$$

The problem (5) solution is the function of

$$p_1 = \frac{a_1 - b_1 c_1}{2 a_1 N_0} + \frac{a_1 N_0 - c_1}{2 a_1 N_0} p_2 \tag{6}$$

3 Results

Let us consider the response curves of operators in model 1 and model 2 in more detail. The empirical functions of the operators' demand are presented in Fig. 2.

The estimation of the demand functions parameters was carried out by the least squares method in the econometric package of GRETL. It was based on the previously linearized statistical data of the federal cellular operators' billing systems.

We assumed that the number of subscribers who switched from one operator to another is $N_0 = 20$. Costs for the content production and the provision of one service by the operators are the same and equal to 1.5. In the future, this assumption can be replaced by a detailed theoretical or empirical analysis of the disloyal subscribers' number effect on operator profits and market equilibrium.

We take into account different parameter values of the empirical demand functions of operator 1 and operator 2 as well as the number of subscribers who switched from the first operator to the second one (N_0). Table 1 gives the values of the equilibrium prices.

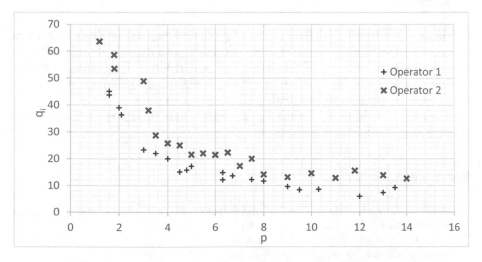

Fig. 2. Empirical functions of operators' demand for IVR services.

Table 1. Equilibrium price levels for operator's response model. Case $N_0 = 20;\ c_1 = c_2 = 1,5$

Model		
Price	(a) Operator 1 starts first	(b) Operator 2 starts first
p_1'	2.697	–
p_2'	1.370	–
p_1''	7.126	–
p_2''	3.583	–

Operator 1 is the first to start increasing the price of digital services. The parameters of its estimated demand function are $a_1 = 90;\ b_1 = 5$.

Following the increase in price by operator 1, the second operator can also change the price for its services. In case the first operator increases its price with a focus on the actions of the second operator, we receive a very interesting result: there are two equilibrium points in the market (p' and p'').

Equilibrium at point p' can be achieved if operator 2 reduces the price of services below the level of costs. This situation seems unlikely. When the first operator raises the price, the second operator's gains increase due to the influx of new subscribers. The equilibrium at point p'' is achieved at price levels lower than those existing under the conditions where the first operator does not respond to the second operator's actions.

In the other situation, we will suppose that the second operator starts to increase the price first. Parameters of its demand function are following $a_2 = 70;\ b_2 = 2$.

Market equilibrium is of a response from the first operator to the price increase made by the second operator. There is no price market equilibrium, in this case.

4 Conclusion

In this paper, we have solved the problem of profit maximization of a cellular operator for interactive voice services on the assumption of a price duopoly model. We found explicit empirical functions of the demand for the services of two leading cellular operators. In addition, the conditions for the existence of an equilibrium in the value-added services market were analyzed.

The results can be useful for mobile operators to develop strategies of behavior in setting prices for digital cellular services.

References

1. Russian Internet Forum. https://journal.open-broker.ru/investments/investicii-v-akcii-rossijskih-telekommunikacionnyh-kompanij
2. Foh, K.-L.: Introducing DIAL's Mobile Market Model, 20 July 2018. https://digitalimpactalliance.org/introducing-dials-mobile-market-model
3. Cournot, A.: Recherches sur les principles mathematique de la theorie des richesses, ch. VII, Paris (1938)

4. Bertrand, J.: Theorie Mathematique de la Richesse Sociale. Journal des Savants, pp. 499–508, September 1883
5. Michalakelis, C., Sphicopoulos, T., Varoutas, D.: Modeling competition in the telecommunications market based on concepts of population biology. IEEE Trans. Syst. Man Cybern. Part CAppl. Rev. **41**(2), 200–210 (2011)
6. Tregub, I., Drobysheva, N., Tregub, A.: Digital economy: model for optimizing the industry profit of the cross-platform mobile applications market. In: Antipova, T., Rocha, A. (eds.) International Conference on Digital Science, DSIC 2018. AISC, vol. 850, pp. 21–29. Springer, Cham (2019). https://doi.org/10.1007/978-3-030-02351-5
7. Tregub, I.V.: Forecasting demographic indicators of the regions of Russia. In: Proceedings of 2018 11th International Conference; Management of Large-Scale System Development, MLSD 2018, Moscow, pp. 1–5, IEEE (2018)
8. Tregub, I.V.: On the applicability of the random walk model with stable steps for forecasting the dynamics of prices of financial tools in the Russian market. J. Math. Sci. **5**(216), 716–721 (2016)
9. Tregub, I.V.: International diversification. In: PSTM, Moscow (2015)
10. Tregub, I.V.: Econometric analysis of influence of monetary policy on macroeconomic aggregates in Indian economy. IOP Conf. Ser. J. Phys. Conf. Ser. **1039**, 012025 (2018). https://doi.org/10.1088/1742-6596/1039/1/012025

The Formation of Data Bases
at the Technogenic Risk Management System

Kirill Litvinsky[1]([✉]) [iD] and Elena Aretova[2] [iD]

[1] Kuban State University, 350040 Krasnodar, Russia
litvinsky@econ.kubsu.ru
[2] Kuban State University, Stavropol'skaya st., 149, 350040 Krasnodar, Russia

Abstract. The article reviewed the issues of economic and mathematical modeling of the process of knowledge acquisition in the knowledge bases of the technological risk management system, taking into account the need to determine methods and approaches to their replenishment. An attempt was made to describe and solve the problem of acquiring knowledge that arises when solving such tasks within the framework of the functioning of the technological risk management system. The author's interpretation of the concept of "obtaining knowledge in intellectual system" and, on this basis, strategies for acquiring knowledge are formulated, based on a specific set of rules. Formation Strategies allowed to describe and visualize the process of knowledge transfer from an expert to the knowledge base, as well as the domain formation algorithm in the form of interactive dialogue "knowledge base - expert" with the aim of forming the domain structure. As a result of the implementation of this algorithm a global object is created in the knowledge base which is based on the attribute name and sets of its meanings. During algorithm operations the initial knowledge base is filled with terms about specific subject area. The semantic relations are determined in the direct acquisition of knowledge in the knowledge base of the system for managing the technological risks of business entities with the aim of constructing a heterogeneous semantic network.

Keywords: Knowledge management · Knowledge base · Interactive knowledge interpreter · Semantic network · Technological risk management system

1 Introduction

When researching the issues of economic and mathematical modeling of the process of knowledge acquisition in the knowledge bases of the technology-related risk management system, it is necessary to determine methods and approaches to replenish them. In our opinion, the replenishment of knowledge is carried out by the system itself on the basis of the existing knowledge base and various programs for inferring of new knowledge in order to solve the problem. However, there is a problem of acquiring knowledge that arises when solving such problems within the framework of the functioning of the system of management of technology-related risks.

T. Antipova and Á. Rocha (Eds.): DSIC 2019, AISC 1114, pp. 94–103, 2020.
https://doi.org/10.1007/978-3-030-37737-3_9

2 First Section

2.1 Review of Prior Literature

The acquisition of knowledge, in terms of replenishment of the knowledge base (KB) [1–3], is carried out by various methods: while communicating with the external environment, while setting new tasks and achieving new goals. Currently, the main method is direct dialogue with an expert [2]. Such a dialogue allows to solve the main problems arising in extraction of knowledge: to overcome cognitive protection, to some extent help the expert to structure his knowledge in a certain area.

In addition, knowledge replenishment is carried out by the system itself based on the existing knowledge base and machines (programs) of the logical derivation of new knowledge, intended for the direct use of knowledge from the knowledge base in order to solve the problem. This leads to the elimination of the difficulties associated with limited knowledge. In the course of solving a specific problem, the main functions of the inference machine are:

- determination of a specific set of facts, rules and relations required to solve the problem among the selected facts and rules;
- alignment of selected rules, relationships, algorithms and procedures into logical decision-making chains and the implementation of these chains.

The problem of acquiring knowledge arose while solving such problems as: understanding the natural language, finding answers to questions to the knowledge base, analyzing situations, etc. in the framework of the technogenic risk management system (TRMS) at the enterprise of the real sector of the economy (ERSE). All approaches to some extent use the idea of production rules [4, 5].

The analysis of various sources showed that the concept of "obtaining" knowledge in an intellectual system has a lot of variations [2–4, 6]. As synonyms other terms such as: "replenishment", "extraction", "acquisition", "production", "formation" are quite often used. In foreign literature, two terms are most often used: acquisition and extraction.

In order to uniquely formulate the term of "obtaining" knowledge, we consider three strategies based on three rules [2, 7].

Rule 1. Extraction of knowledge is the procedure of obtaining knowledge in which transfer of knowledge is carried out through direct interaction between the knowledge engineer (cognitologist) and the source of knowledge.

Rule 2. Acquisition of knowledge is a procedure of obtaining knowledge in which the transfer of knowledge from an expert to a cognitive engineer is carried out using computer software.

Rule 3. The formation of knowledge is a procedure of obtaining knowledge in which training programs are used.

The most widely interpreted term is "Acquisition of knowledge". In [2] it is noted that the process of acquiring knowledge can be characterized by the following stages and features:

- the acquisition of knowledge;
- modeling the knowledge acquisition;
- nomenclature of knowledge acquisition;
- level of knowledge;
- mechanisms for debugging knowledge.

The first stage is mostly technological in nature, and consists of phases of knowledge acquisition. In practice there are three phases which reflect the variation of functions of the cognitive engineer and expert at the first stage [2]:

- the preparatory phase is the phase of the direct extraction of knowledge from any source of knowledge at the stages without the use of computer technology;
- the initial phase is the phase at which the knowledge base is filled with knowledge about a specific knowledge domain;
- the filling phase is a phase that is relevant at the stages of implementation and testing and is associated with solving of certain problems such as: detection of incomplete knowledge, its inconsistency or inaccuracy; the extraction of new knowledge that can eliminate the detected inaccuracies and errors; transformation of the acquired knowledge into a form that is understandable to the knowledge base.

Therefore, generally, the procedure of acquisition of knowledge implies the presence of all the above phases, and in the narrow sense, this is the initial phase (accumulation of knowledge) in which knowledge is directly transferred to the system.

However, the most difficult remains the phase of knowledge extraction in which formalization of processes is a very difficult task.

2.2   Statement of Basic Materials

Extraction of knowledge in KB TRMS at the ERSE is based on the procedure of transferring the knowledge from each expert to the knowledge base. This process is represented in Fig. 1.

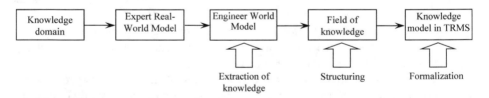

Fig. 1. The process of transferring knowledge from an expert to a knowledge base

Most often, the formation of a field of knowledge in the form of a certain mathematical and physical knowledge representation received from an expert in a less formalized form, in fact, is the final step of the procedure of obtaining knowledge from an expert (structuring) [2, 5, 8, 9].

In the process of extracting knowledge during the creation of the TRMS ERSE it is rational to apply the textological method [5], which involves the extraction of knowledge from paper media such as textbooks, instructions, technological descriptions, etc. The structure of the process of extracting knowledge in this case has the form (1).

$$MA_1 \rightarrow Verbal\, Analysis \rightarrow Perception\, of\, the\, Text \rightarrow Understanding \rightarrow MA_2, \quad (1)$$

where MA_1 is the author's model of the world;

MA_2 is a model that is formed after reading a text.

Models MA_1 and MA_2 most often do not coincide due to the distortion of meaning in the verbal analysis of MA_1 and interpretation in MA_2. The analysis shows that the scientific text contains the following main components [2, 5]:

- presentation of objective information α;
- systems of scientific terms β;
- scientific views, author's experience γ;
- the common text δ;
- borrowings from other sources of information ε.

So the author's model should be presented as a tuple (2):

$$MA_1 = \, <\alpha, \beta, \gamma, \delta, \varepsilon>. \quad (2)$$

The MA_2 model is formed from a tuple $<\alpha, \beta, \gamma, \delta, \varepsilon>$ by synthesizing the perception of the text and reflecting the individual qualities of a person (engineer), characterized by the following elements:

- experience φ;
- erudition ξ;
- knowledge of the knowledge domain η.

Then, the MA_2 model will take the form (3):

$$MA_2 = [<\alpha, \beta, \gamma, \delta, \varepsilon>, <\varphi, \xi, \eta>]. \quad (3)$$

Since there is a difference between the MA_1 and MA_2 models, we can conclude that there is insufficient correspondence between the received and initial information. To fill the knowledge base of TRMS ERSE a specific algorithm for working with text materials shall be proposed [5]:

- compiling a list of basic literature for a more detailed knowledge of the knowledge domain (by specific ERSE);
- selection of a specific text;
- preliminary acquaintance with the text;
- thorough study of the text, the highlighting of keywords and persistent expressions;
- formation of links between keywords, the formation of a generalized graph of text;
- formation of a knowledge model.

The type and content of knowledge sources affects the understanding of the text. The easiest way to work with textbooks in which knowledge is well structured while minimizing subjective factors. Working out technical descriptions is much more difficult. It is even more difficult to analyze scientific articles. Therefore the presence of an experienced expert is a prerequisite for the effective extraction of knowledge from scientific publications.

The identification of the structure of terms in the direct acquisition of knowledge in TRMS knowledge base at the ERSE is based on the fact that in the direct acquisition of knowledge, the computer system is the intermediary between the knowledge base and the knowledge source [2]. This approach can be implemented using a model in the form of a heterogeneous semantic network (HSN) in the knowledge base. Such a network in TRMS ERSE is described by (4):

$$M = <F, R_c, R_f, R_\in>,\tag{4}$$

where F is the frame forming a specific term; R_c is the relation of connectivity (incidence) between frames; R_f is a function characterizing the properties of frames; $R_\in$ is the membership relation between frames and functions.

In TRMS ERSE using an on-line interpreter of knowledge (OIK) an heterogeneous semantic network is formed. The objects of work of OIK are the names of objects, their properties, processes and procedures, the range of values of properties and relationships on a variety of objects and processes. OIK is designed to convert information from an expert into mathematical formulas and compile them into the HSN [5].

The strategy for the direct acquisition of knowledge in STRM is a sequential breakdown into steps [2]. This strategy is designed to identify the structure of terms of a specific knowledge domain and implements the scenario "Name - Property". For the formation of the structure of the knowledge domain using OIK, an algorithm is designed that takes the form of a dialogue "System - Expert" (see Fig. 2).

An example implementation of the dialogue has the following form:

1. Enter the name of the term - a system for monitoring and recording information.
2. Enter the name of term attribute - objective control system (state of the system).
3. Are there several values of the entered characteristic? (Yes/No).
4. If the answer is "No" (for a term attribute of an objective control system), then the name of term attribute is stored in the form of the name of the term. In this case, a pair of event names is formed (a system for monitoring and recording information, objective control systems). If the name of the 2nd event is new for the knowledge model, then go to step 2.
5. If the answer in paragraph 3 is "Yes" (for the attribute of system status), then the question arises: Determine the type of set (Continuous/Discrete) - for example, "Discrete".
6. If the type is Continuous, then the following question: Set the range limits.
7. If the type is Discrete, then the next question: List the elements of the discrete set (for example, on, off).
8. Set the unit of measure for the characteristic — dimensionless quantity.

9. Set a subset of the property values of the event in question - (enabled).
10. The presence and determination of all the attributes of the terms is checked.

As a result of steps 2–10, a global object is created in the database, consisting of: the name of the attribute and the set of its values. During the operation of the algorithm, the knowledge base is initially filled with terms about a specific knowledge domain.

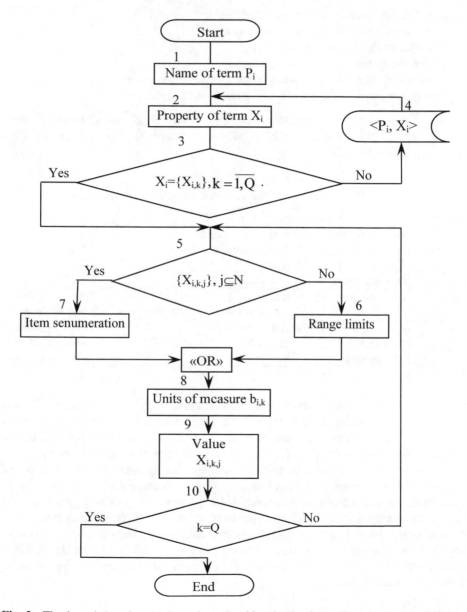

Fig. 2. The knowledge domain formation algorithm in the form of an interactive dialogue "Knowledge Base - Expert"

The definition of semantic relationships in the direct acquisition of knowledge in the knowledge base of the technogenic risk management system at the ERSE is based on the statements of the expert. The process of identifying semantic relationships is involved in the construction of the HSN. The main types of connections used in the HSN are shown in Table 1. The criteria used to determine the type of connection that two event names (network nodes) have can be formulated as follows:

- transformation (establishment of transitivity);
- permutation (establishment of symmetry);
- substitution (establishing reflexivity);
- circulation (establishment of asymmetry);
- modality (distinguishing between relationships by modality).
- time of occurrence of events (establishment of momentum).

We denote: *Aref* - antireflexivity, *Asi* - asymmetry, *Ansi* - antisymmetry, *Nref* - non-reflexivity, *Nsi* - non-symmetry, *Ntrn* - non-transitivity, *Ref* - reflexivity, *Si* - symmetry, *Trn* - transitivity.

Table 1. Relationships and criteria in a heterogeneous semantic network.

Class №.	Type of relationship	Form of relationship	Attribute
1	*Gn* – generative	X is an element of Y	*Aref, Nsi, Ntrn*
	St – situational	X is in situation Y	*Aref, Asi, Trn*
	Ng – negative	X denies Y	*Aref, Si, Ntrn*
2	*In* – instrumental	X is a means of Y	*Nref, Nsi, Ntrn*
3	*Cm* – commutative	X is followed by Y	*Ref, Ansi, Trn*
	Cr – correlative	X sometimes increases the possibility of Y	*Ref, Si, Ntrn*
4	*Fn* – target	X is the target of Y	*Aref, Nsi, Ntrn*
	Cs – causal	X causes Y	*Nref, Nsi, Trn*
	Pt – potential	X may cause Y	*Nref, Nsi, Ntrn*

This table show types of relationships which are divided into four classes. Classes 1–3 define one-stage dependences (turning on the switch - lighting operation), class 4 - the multi-stage dependence between elements or procedures in the technological system (turning on the device power - lighting and information screen operation).

In classes 1–3, the types of relationships are distinguished by the property of reflection. Within groups, the types of relationships are distinguished by the presence of symmetry. Since in class 4 all relationships have the property of non-symmetry, the properties of transitivity and reflection are used to distinguish the relationships.

Therefore, such properties as symmetry, reflexivity, simultaneity, transitivity should be used as criteria for identifying types of relationships [2, 10]. In Fig. 3, a comparison of the criteria and branches of the tree for determining the type of relationships is carried out.

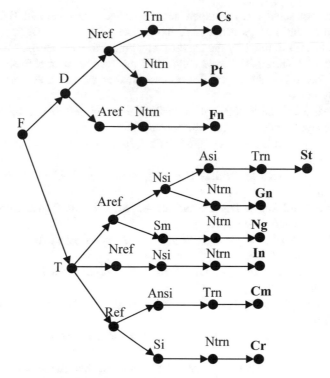

Fig. 3. Tree for determining the type of relationship

It is necessary to establish two-way relationships between events A and B in the knowledge base of the STRM ERSE. They are as follows [5]:

- in the presence of event A, there may be event B (positive relationship);
- in the presence of event A there is always B (strong positive relationship);
- in the presence of event A, usually B does not happen (negative relationship);
- in the presence of event A, there is always no B (exclusive connection).

3 Results

Using the OIK, the type of relationship between the event names is determined by the following algorithm [5]:

1. The expert is offered a list of event names and is invited to identify pairs of related events.
2. If the expert does not determine such a pair, the dialogue ends and control is transferred to the OIK. If an expert selects a pair of concepts (A, B), then it is substituted into all forms of communication in a certain order: first $XR_iY: = AR_iB$ (first part of the list), and then $XR_iY: = BR_iA$ (second part of the list).

3. From the generated list, the expert needs to select the statement *Wj*, which most reflects the relationships between A and B. Let's say this is *AR$_i$B*.
4. If the statement defined by the expert corresponds to the first part of the list, then the attribute *F* (First) is entered (Fig. 3), if the second, then *S* (Second).
5. For the selected *L*, the expert determines whether events A and B appear simultaneously or not.
6. If the answer is simultaneous, then attribute *T* is entered; if not, attribute *D*.
7. The statement *L* is checked by the substitution criterion.
8. If attributes *T* and *Ref* or *Aref* are selected, then *L* is checked by the criterion of permutation.
9. If attribute *T* is selected and *Si* is absent, then *L* is checked by the criterion of circulation.
10. If attribute *T* and *Ref* or *Aref* are selected and *Si* is absent, then *L* is checked by the criterion of transformation.
11. If there are attributes *D* and *Nref*, then *L* is checked by the criterion of transformation.
12. If there are attributes of *F*, *T*, *Aref*, *Ntrn* and no *Si*, then for *L* the relationship *Gn* (A, B) is determined.
13. If there are attributes *F*, *T*, *Aref*, *Asi*, *Trn*, then for *L* the relationship *St* (A, B) is determined.
14. If there are attributes of *F*, *T*, *Aref*, *Si*, then for *L* the relationship *Ng* (A, B) is determined.
15. If there are attributes of *F, T, Nref* and *Si* is absent, then the relationship *In* (A, B) is determined for *L*.
16. If there are attributes of *F*, *T*, *Ref*, *Trn* and *Sm* is absent, then the relationship *Cm* (A, B) is determined for *L*.
17. If there are attributes of *F*, *T*, *Ref*, Si, then for *L* the relationship *Cr* (A, B) is determined.
18. If there are attributes *F*, *D*, *Aref*, *Ntrn*, then for *L* the relation *Fn* (A, B) is determined.
19. If there are attributes of *F*, *D*, *Nref*, *Trn*, then for *L* the connection *Cs* (A, B) is determined.
20. If there are attributes *F*, *D*, *Nref*, *Ntrn*, then for *L* the connection *Pt* (A, B) is determined.

4 Conclusion

The semantic relations are determined in the direct acquisition of knowledge in the knowledge base of technological risks management system of business entities with the aim of constructing a heterogeneous semantic network. In the knowledge base of technological risk management system at an enterprise in the real sector of the economy, two-way relationships between events are established, and then, using a dialogue knowledge interpreter, the type of connection between the names of events is determined by a certain algorithm. Thus, in the knowledge base of TRMS ERSE it is supposed to use two options for acquiring knowledge: direct acquisition of knowledge on the basis of sequential division into steps with the help of OIK (in the presence of low qualification of an expert) and direct extraction of knowledge.

References

1. Arsenyev, Y.N., Davydova, T.Y., Minaev, V.S.: Knowledge management based on the tools of synergetics and cognitive science. Bull. Tula State Univ. Econ. Leg. Sci. **2-1**, 60–69 (2015)
2. Litvinsky, K.O., Malyshev, V.A., Nikitenko, Y.V.: The model of the decision support subsystem in the management system of technogenic risk enterprises of the fuel and energy complex. Econ. Sustain. Dev. **1**(21), 91–100 (2015)
3. Pasmurnov, S.M., Firtych, O.A.: The formation of the knowledge base of the project management system with predictable risks. Bull. Voronezh State Tech. Univ. **11–3**, 82–85 (2015)
4. Malyshev, V.A., Litvinsky, K.O., Nikitenko, Y.V.: Economic and mathematical modeling of basic operations at enterprises of the real sector of the economy. Sustain. Dev. Economics. **2**(22), 184–189 (2015)
5. Malyshev, V.A., Nikitenko, Y.V.: Theoretical Foundations of Building A Technological Risk Management System at Industrial Enterprises. Scientific Book, Voronezh (2015)
6. Rybina G.V., Danyakin I.D.: Combined method of automated temporal information acquisition for development of knowledge bases of intelligent systems. In: Proceedings of the 2017 2nd International Conference on Knowledge Engineering and Applications, pp. 117–123. IEEE (2017)
7. Tzacheva, A.A., Bagavathi, A., Ganesan, P.D.: MR – random forest algorithm for distributed action rules discovery. Int. J. Data Min. Knowl. Manag. Process. (IJDKP) **6**(5), 15–30 (2016)
8. Arsen'yev, Y.N., Davydova, T.YU., Minayev, V.S.: Knowledge Management Based on Synergetics and Cognitive Science. Izve. Tula State Univ. Econ. Leg. Sci. (2-1), 60–69 (2015)
9. Barkalov, S.A., Dushkin, A.V., Kolodyazhny, S.A., Sumin, V.I.: Introduction to systems design of intelligent knowledge bases. In: Novoseltseva, V.I. (ed.) Goryachaya liniya - Telekom, Moscow, 108 p. (2017)
10. Litvinsky, K.O.: Methodology of construction management models of actors of nature. In: Advances in Intelligent Systems and Computing, vol. 850 (2019)

Intellectual Capital of a Company: Presentation and Disclosure of Information

Yana Ustinova[✉] [iD]

Novosibirsk State University of Economics and Management,
630099 Novosibirsk, Russia
ustinova_pr@mail.ru

Abstract. The paper substantiates the need to disclose information on the intellectual capital of the company in addition to traditional financial statements in order to provide investors and creditors with the opportunity to adequately assess the financial position of the company and its future development prospects. The paper contains an analysis of the work of scientists studying the problems to be solved in the report on intellectual capital, its composition and structure, their dependence on the industry sector of the company, its scope and specifics of the activity, as well as the impact of the report on statement users' perceptions of the company. The content and structure of information on intellectual capital presented in the public financial statements of the largest Russian companies, as well as the information needs of statement users, are examined. Recommendations on overcoming the identified information deficit are formulated.

Keywords: Intellectual capital · Information limitations of traditional financial statements · Information needs of users of financial statements · Intellectual capital report

1 Introduction

To date, an analysis of the multiple informational limitations of financial reporting in relation to intangible assets forces us to recognize the inadequacy of traditional financial reporting to comprehensively reflect the intangible, intellectual resources of a company. As a result, users of financial statements, including investors and creditors, are unable to adequately assess the financial performance of the company, its solvency, the profitability of its assets, investment attractiveness and prospects, which affects the company's capitalization.

At the same time, the growth of intellectual capital is closely related to the growth of capitalization of companies. In particular, according to the results of V. Dzenopoljac (2018), obtained on the basis of analysis of reporting data of oil and gas companies included in the 2016 Forbs global 2000 list for the period from 2000 to 2015, an increase in the estimated value of intellectual capital by 1% point leads to an increase in capitalization companies for 113–157 million dollars. At the same time, 35% of the formed sample of companies did not disclose information on intellectual capital in general, and 65% of companies disclosed it only partially [1, pp. 14–15].

© Springer Nature Switzerland AG 2020
T. Antipova and Á. Rocha (Eds.): DSIC 2019, AISC 1114, pp. 104–113, 2020.
https://doi.org/10.1007/978-3-030-37737-3_10

Accordingly, in a competitive environment, this state of affairs with the disclosure of the intellectual resources of the company in its financial statements may adversely affect the possibility of sustainable development of the company, especially in conditions of unstable economic environment. This issue is particularly relevant in the context of intra-industry competition in an industry that has a fairly long history, when key players in the market have approximately the same material and technical level, and intellectual capital is the decisive factor in differentiation.

At the same time, the insufficiency of information disclosed in traditional financial statements, recognized by both foreign and Russian researchers, requires the submission of additional information on intellectual capital to users of financial statements. Its volume, structure, presentation form are the subject of serious scientific discussions.

The work aimed on a comprehensive study of theoretical and practical aspects of presenting and disclosing information on the intellectual capital of the company, corresponding to the information needs of financial statements users and determining their favorable perception of the company's prospects in general.

2 Literature Review

An analysis of the work of foreign and Russian researchers on the study of overcoming the information gap in terms of disclosing intellectual capital in financial statements made it possible to perform a brief review of the recommendations contained therein (see Table 1).

Table 1. Recommendations of researchers regarding the development of the concept of accounting and of information disclosing on intellectual capital in financial statements

Key recommendations	Authors
Supplementing traditional financial statements (while preserving the accepted concept of its preparation) with a report on intellectual capital (in order to expand the content of intangible assets and/or disclose their value, worth for the company, unlike existing accounting standards)	Kaplan and Norton [2, p. 57], Edvinsson and Malone [3, pp. 32–39], Wyatt and Abernethy [4, pp. 15–16, 22–25], Oliveras and Kasperskaya [5, pp. 4, 12], Marr, Gray and Neely [6, p. 447], Hunter, Webster and Wyatt [7, pp. 11–12], Starovic and Marr [8, pp. 24–26], Abeysekera [9, pp. 3–4], Cormier and Ledoux [10, pp. 3–6, 18], Artsberg and Mehtiyeva [11, pp. 20–22], Krstic and Dordevic [12, pp. 336, 346], Heinrich [13, p. 35], Chander and Mehra [14, pp. 2–5, 24], Rea and Davis[15], Bulyga [16, pp. 173–175], Wu and Lin [17, p. 346], Kuzubov and Yevdokimova [18, p. 29]
Modification (expansion, revision of the preparation concept) of traditional financial statements	Sullivan Jr. and Sullivan [19, pp. 330–335], Lev [20, pp. 125–130], Lev and Daum [21, pp. 8–12], Lev and Raygopal [22]
Addition or modification of traditional financial statements (the choice is up to the standard developers or, according to their decision, to the companies)	Stewart [23, p. 23], Sveiby [24, pp. 67–75], Holland [25, p. 513], Lopes [26, pp. 103–104]

As can be seen from Table 1, foreign and Russian researches offer a solution to the issue of incomplete reporting information on intellectual capital through the addition of traditional financial statements, or through a review of the concept of its preparation. Moreover, in practice, the first option was most widely used. Estimated by R. P. Bulyga (2012), the formation of a special external reporting unit containing information on intellectual capital, is the main direction of modernization of the information coverage system of the company's activities in the UK and Continental Europe [16, pp. 175–176]. Moreover, it is the lack of traditional financial statements regarding intellectual capital, according to F. Castilla-Polo and M. C. Ruiz-Rodriguez (2018), explains the strategic advantage of voluntary disclosure in additional reports as long as the transparency achieved by such disclosure helps to strengthen the reputation of internal and external stakeholders, support the strategy of industry differentiation, reveal the company's competitive advantages, increase the assessment of its economic potential and attractiveness in the capital market [27, pp. 4–5].

If the need to provide additional information on intellectual capital is not called into question, then the volume, structure, and form of its presentation are subject to serious scientific discussions. Numerous works by researchers are devoted to the reviews of solutions to this problem (see Table 2).

Table 2. The main directions of research of the information content of intellectual capital reports in the practice of companies

The main areas of research	Authors
Study of the content of intellectual capital report and its impact on the perception of the company financial performance by financial statements users	Brennan [28], Lev and Daum [21], Guthrie, Petty and Ricceri [29], Chander and Mehra [14], Abhayawans and Guthrie [30], Singh and Narwal [31]
Study of the structure of intellectual capital report	Castelo, Sá, Delgado and Sousa[32], Abhayawans and Guthrie[30]
Study of the dependence of intellectual capital report content on industry specifics, company size, and other similar factors	Guthrie, Petty and Ricceri [29], Whiting and Woodcock[33], Santos and Venancio [34]

As can be seen from Table 2, the research interest is not only the structure and content of intellectual capital reports, their dependence on industry specifics, but also their relationship with the perception of information by financial statements users.

At the same time, the researchers noted the lack of uniformity in the disclosure of information even on basic issues (structure, presentation and disclosure rules, etc.), as well as a clear deficit of disclosed information, complicates the perception of information on intellectual capital by users of financial statements, which requires the development of appropriate recommendations.

3   Methodology

The study was based on the results of the work of foreign and Russian researchers aimed at developing a concept and tools for reflecting information on intellectual capital in financial statements that would allow users of financial statements to form an adequate idea of the impact of this object on the financial performance of the company in the future.

In addition, an analysis of the data of the consolidated financial statements for 2015–2017 was carried out (compiled in accordance with IFRS, including explanations), of the largest Russian companies (included in the rating 2000 Global from the journal website http://www.forbes.com), in terms of the presentation and disclosure of information on intellectual capital.

As part of the study, we used materials from surveys of the company's administration, financial analysts, lenders and investors regarding information on intellectual capital that they would like to see in the reporting data (traditional financial statements (explanations) or additional reports).

The article analyzes the information gap between the data of the consolidated public financial statements and the information needs of the main groups of financial statements users in terms of information on intellectual capital, which made it possible to formulate practical recommendations for overcoming it.

4   Results

According to the consolidated financial statements for 2015–2017 (compiled in accordance with IFRS) of the largest Russian companies (included in the rating 2000 Global from the journal website http://www.forbes.com in May 2015), information on intellectual capital is disclosed as follows (see Table 3).

As can be seen from Table 3, the following trends are observed in the structure of disclosure of information about intellectual capital:

- 29.6–33.3% of cases of disclosure - information on the intellectual property of the company;
- 37.0–40.7% of cases of disclosure - information about goodwill;
- 51.9–55.6% of cases of disclosure - information about other elements of intellectual capital, including brand, reputation, staff qualifications, corporate culture and management policy;
- 74.1–81.5% of cases of information disclosure in the discursive form, in addition to numerical indicators;
- 18.5–25.9% of companies did not disclose relevant information at all.

At the same time, numerical information is disclosed only for aggregated indicators (without further elaboration). The disclosed discursive information contains only brief explanations of the data included in the aggregated indicators. The procedure for the formation of these indicators, including the uncertainties associated with them, plans for their further use and prospects for the development of companies do not find adequate disclosure.

Table 3. Information on the disclosure of intellectual capital in public financial statements

Company	2015				2016				2017			
	Numerical			Discursive	Numerical			Discursive	Numerical			Discursive
	Intellectual property	Goodwill	Other		Intellectual property	Goodwill	Other		Intellectual property	Goodwill	Other	
1. Gazprom	+	+	+	+	+	+	+	+	+	+	+	+
2. Rosneft			+	+			+	+			+	+
3. LukOil		+		+		+		+		+		+
4. Sberbank			+	+			+	+			+	+
5. Surgutneftegas		+	+	+		+		+		+		+
6. Transneft		+		+		+		+		+		+
7. Norilsk Nickel			+	+			+	+			+	+
8. VTB Bank			+	+			+	+			+	+
9. Magnit	+	+	+	+	+	+	+	+	+	+	+	+
10. Tatneft									+		+	+
11. Novatek			+	+			+	+			+	+
12. IDGC Holding												
13. Novolipetsk Steel												
14. MegaFon	+	+	+	+	+	+	+	+	+	+	+	+
15. Sistema												
16. RusHydro												
17. UC Rusal	+		+	+	+			+	+		+	+
18. Rostelecom	+		+	+	+		+	+	+		+	+
19. Severstal	+	+	+	+	+	+	+	+	+	+	+	+
20. X5 Retail Group		+		+		+		+		+		+
21. Inter Rao									+			+
22. Moscow Exchange												
23. United Aircraft Corporation		+	+	+		+	+	+		+	+	+

(continued)

Table 3. (*continued*)

Company	2015				2016				2017			
	Numerical			Discursive	Numerical			Discursive	Numerical			Discursive
	Intellectual property	Goodwill	Other		Intellectual property	Goodwill	Other		Intellectual property	Goodwill	Other	
24. Nomos Bank			+	+			+	+			+	+
25. Mechel	+		+	+	+		+	+	+	+	+	+
26. Mail.ru Group Ltd. Sponsored GDR RegS	+	+	+	+	+	+	+	+	+	+	+	+
27. Alrosa			+	+			+	+			+	+
Total, % of the total number of companies	29,6	37,0	51,9	74,1	29,6	37,0	51,9	74,1	33,3	40,7	55,6	81,5

Thus, the analysis of the structure and content of the information on intellectual capital disclosed by the largest Russian companies in the consolidated public financial statements allows us to conclude that it is insufficient for users to make adequate economic decisions.

Of course, along with traditional financial statements, companies use other types of reporting to disclose information about their intellectual resources: intellectual capital report, integrated reporting, etc. However, the focus of this study is on financial statements as public available reporting, the preparation procedure for which is established in accounting standards, which provides a single basis for the perception of accounting information by company stakeholders.

The author's analysis of the main trends in the formation of intellectual capital reports in practice, along with an analysis of the information needs of users, made it possible to establish that the disclosure of following items would be optimal from the position of financial statements users:

- information about the goals and strategies of the business, about target indicators of financial performance and forecasts for their change;
- information on the key components of intellectual capital that ensure the development of the strategy, on available and necessary resources for their effective use, on potential risks associated with their use;
- analysis of the sources of formation of intellectual capital, assessment of the independence of the organization in this aspect;
- classification of components of intellectual capital;
- the field of application of intellectual capital (in terms of actual and potential application, ongoing and planned projects, stages of formation of company value, etc.);
- the effectiveness of the use of intellectual capital: the actual and expected;
- assessment of the components of intellectual capital (taking into account various options for its use), potential opportunities for their alienation and alternative use;
- analysis of market advantages provided by the organization's intellectual property (as a component of intellectual capital);
- assessing the estimated duration of these benefits, the risks that threaten them, and safety measures;
- assessment of the necessary additional investments in intellectual capital and analysis of their effectiveness, etc.

Disclosure of this information seems to be a necessary condition for effective communication with investors and creditors of the company, a key parameter of its intra-industry differentiation, its competitive advantage in the capital market. The widespread use of modern digital technologies in the context of the general digital transformation of society and the economy can help in the preparation of this information. At the same time, cluttering up with excessive information, which users do not feel the need for, entails a decrease in the efficiency and quality of the generated financial statements.

5   Conclusion

In general, it should be noted that overcoming the information deficit of traditional financial statements in terms of intellectual capital becomes a fundamentally important issue when designing a company's development strategy, maintaining and enhancing its competitive advantages, including within the industry, in assessing its profitability and investment attractiveness. At the same time, the study shows the actual gap between the information on intellectual capital disclosed in public traditional financial statements and the information needs of financial statements users. In particular, financial statements users to form their own models of economic decision-making need not only a valuation of aggregate indicators of intellectual capital, but also the disclosure of the order of its formation and the uncertainties associated with these indicators, plans and forecasts, estimates of efficiency, strategic and competitive advantages.

An analysis of theoretical and practical aspects of disclosing information on the intellectual capital of a company provides the ability to create a report that is most relevant to the information needs of reporting users and determines their favorable perception of the company's prospects in general.

References

1. Dzenopoljac, V., Muhammed, Sh., Janosevic, S.: Intangibles and performance in oil and gas industry. Manag. Decis. (2018). https://doi.org/10.1108/MD-11-2017-1139. Accessed 18 May 2019
2. Kaplan, R.S., Norton, D.P.: The balanced scorecard: measures that drive performance. Harv. Bus. Rev. **70**(1), 71–79 (1992)
3. Edvinsson, L., Malone, M.S.: Intellectual Capital: The Proven Way to Establish Your Company's Real Value by Measuring Its Hidden Brainpower. Platkus, London (1997)
4. Wyatt, A., Abernethy, M.A.: Framework for measurement and reporting on intangible assets. Working Paper no. 12/03. Intellectual Property Research Institute of Australia, Melbourne (2003)
5. Oliveras, E., Kasperskaya, Y.: Reporting intellectual capital in Spain. Working Paper. Universitat Pompue Fabra (2003)
6. Marr, B., Gray, D., Neely, A.: Why do firms measure their intellectual capital? J. Intellect. Cap. **4**, 441–464 (2003)
7. Hunter, L.C., Webster, E., Wyatt, A.: Measuring intangible capital: a review of current practice. Intellectual property Institute of Australia Working Paper, no. 16/04 (2005)
8. Starovic, D., Marr, B.: Understanding Corporate Value: Managing and Reporting Intellectual Capital. Chartered Institute of Management Accountants. Cranfield University. School of Management (2005)
9. Abeysekera, I.: Intellectual Capital Accounting: Practices in a Developing Country. Routledge, New York (2007)
10. Cormier, D., Ledoux, M.-J.: The Influence of Voluntary Disclosure about Intangible Assets Reported in French Financial Statement: The Role Played by IFRS. Corporate Reporting Chair (2010)
11. Artsberg, K., Mehtiyeva, N.: A literature review on Intangible assets. Critical questions for Standard setters. Working Paper. School of Economics and Management (2010)

12. Krstic, J., Dordevic, M.: Financial reporting on intangible assets – scope and limitations. Facta Univ. Ser. Econ. Organ. **7**(3), 335–348 (2010)
13. Heinrich, R.: Valuation in Intellectual Property Accounting. UNECE Team of Specialists on Intellectual Property, Bishkek (2011)
14. Chander, S., Mehra, V.: A study on intangible assets disclosure: an evidence from indian companies. Intang. Cap. **7**(1), 1–30 (2011)
15. Rea, N., Davis, A.: Intangible assets: what are they worth and how should that value be communicated (2012). www.buildingipvalue.com. Accessed 21 May 2017
16. Bulyga, R.P.: Kontseptsiya intellektual'nogo kapitala kak osnova povysheniya prozrachnosti i dostovernosti vneshney otchetnosti organizatsii [The concept of intellectual capital as a basis for improving the information transparency and reliability of the organization's external reporting]. Innovatsionnoye razvitiye ekonomiki [Innovative development of the economy] **2**(8), 170–176 (2012). (in Russ.)
17. Wu, J.-Ch., Lin, Ch.: A balance sheet for knowledge evaluation and reporting. In: International Conference on Management, Knowledge and Learning, Zadav, Croatia, pp. 341–348 (2013)
18. Kuzubov, S.A., Yevdokimova, M.S.: Povysheniye stoimosti publikatsii publikatsiy nefinansovykh otchetov po standartam GRI? (na primere strannykh BRIKS) [Does the cost of a company increase the publication of non-financial reports on GRI standards? (by the example of the BRICS countries)]. Uchet. Analiz. Audit [Accounting. Analysis. Audit] **2**, 28–36 (2017). (in Russ.)
19. Sullivan Jr., P.H., Sullivan Sr., P.H.: Valuing intangible companies. An intellectual capital approach. J. Intellect. Cap. **1**(4), 328–340 (2000)
20. Lev, B.: Intangibles: Management, Measurement and Reporting, Brookings. Institute Press, Washington, D.C. (2001)
21. Lev, B., Daum, J.H.: The dominance of intangible assets: consequences for enterprise management and corporate reporting. Meas. Bus. Excel. **8**(1), 6–17 (2004)
22. Lev, B., Raygopal, Sh.: FASB prizval pereyti na printsipy [FASB called for a switch to principles] (2016). https://gaap.ru/news/149381. Accessed 20 May 2018. (in Russ.)
23. Stewart, T.A.: Intellectual Capital: The New Wealth of Organizations. Doubleday/Currency, New York (1997)
24. Sveiby, K.E.: The New Organizational Wealth: Managing and Measuring Knowledge-Based Assets, pp. 123–148. Berrett-Koehler, San Francsico (1997)
25. Holland, J.: Financial institutions, intangibles and corporate governance. Account. Audit. Account. J. **14**(4), 497–529 (2001)
26. Lopes, I.T.: The boundaries of intellectual property valuation: cost, market, income based approaches and innovation turnover. Intellect. Econ. **1**(9), 99–116 (2011)
27. Castilla-Polo, F., Ruiz-Rodriguez, M.C.: Intangible Assets Disclosures in the Olive Oil Differentiation Strategy: A Theoretical Review. Agric. Res. Technol. Open Access J. **14**(1), 1–8 (2018)
28. Brennan, N.: Reporting intellectual capital in annual reports: evidence from Ireland. Account. Audit. Account. J. **14**(4), 423–436 (2001)
29. Guthrie, J., Petty, R., Ricceri, F.: The voluntary reporting of intellectual capital: comparing evidence from Hong Kong and Australia. J. Intellect. Cap. **7**(2), 254–271 (2006)
30. Abhayawans, S., Guthrie, J.: Importance of intellectual capital information: a study of australian analyst reports. Aust. Account. Rev. **24**(1), 66–83 (2014)
31. Singh, R.D., Narwal, K.P.: Intellectual capital and its consequences on company performance: a study of Indian sectors. Int. J. Learn. Intellect. Cap. **12**(3), 300–322 (2015)
32. Castelo, M., Delgado, C., Sá, M., Sousa, C.: An analysis of intellectual capital disclosure by Portuguese companies. EuroMed J. Bus. **5**(3), 258–278 (2010)

33. Whiting, R.H., Woodcock, J.: Firm characteristics and intellectual capital disclosure by Australian companies. J. Hum. Resour. Costing Account. **15**(2), 102–126 (2011)
34. Santos, J.C., Venancio, M.T.: Intellectual capital: information disclosure practices by Portuguese companies. Revista Universo Contábil **9**(2), 174–194 (2013)

Digital Economy in Competitiveness of Modern Companies

Elena Alexandrova[(⊠)] [iD]

Kuban State University, Stavropolskaya St., 149, 350040 Krasnodar,
Russian Federation
al-helen@mail.ru

Abstract. The paper considers the main problems of introduction of digital economy technologies to activities of companies in order to create their sustainable competitive advantage. Technologies and tools of the digital economy compel companies to reconceive customary standards of doing business and traditional business processes. Digital ecosystems and innovative business models are coming to the forefront. In a number of industries and scopes of activities, including the company's retail network, modern consumers value convenience and environmental friendliness. They demand personalized approach and seek to save their time. The need for being aware of the latest technological trends and using every opportunity to make their businesses more efficient and competitive forces retailers to develop their own competence in the field of designing and introduction of the digital economy solutions. The choice of digital economy technologies in network retail is defined by financial capabilities of companies, the expected effect of the solution introduced, the level of technological development in the industry (country), the skill level of personnel, the process of organizing of introduction. The scientific article discusses the main changes and characteristics of the competitive environment of modern companies in the context of the digital economy. The paper provides the main solutions and objectives of the introduction of digital economy technologies in the company of network retail, as well as recommendations for their development in order to improve competitiveness and meet the needs of consumers of network companies.

Keywords: Digital economy · Digital economy technologies · Competitive advantage · The company's retail network

1 Introduction

Increasing competitiveness of modern companies depends on various price and non-price factors, including improving business processes and the quality of goods and services produced. Nevertheless, cloud solutions, artificial intelligence, Internet technologies, automated machine learning and other digital economy technologies have a significant impact. Introduction of digital economy technologies sets modernization of traditional sectors of the economy and creates new industries, which becomes the basis for economic growth. Introduction of new technologies increases labour productivity and accessibility of information, decreases business expenses, reduces barriers to entry

© Springer Nature Switzerland AG 2020
T. Antipova and Á. Rocha (Eds.): DSIC 2019, AISC 1114, pp. 114–125, 2020.
https://doi.org/10.1007/978-3-030-37737-3_11

new markets and ultimately has multiplier effect on economic development in general. Accelerated development of the digital economy leaves no doubt as to the importance and necessity of its technologies for companies, but their equally fast obsolescence, need for financial and human resources, ease of replication require a more responsible approach to building sustainable competitive advantages of companies.

Changes occurring in business environment under the influence of the digital economy in scientific literature has been termed Industry 4.0 (the fourth industrial revolution), which is considered to be symbiosis between the material traditional world and the virtual one [1, 2].

In modern literature various interpretations of the concept of the digital economy are presented. The digital economy is also sometimes called the Internet Economy, the New Economy, or Web Economy. The term "Digital Economy" was coined Don Tapscott to show how the Internet would change the way we did business [3]. Based on the "technological shifts" Tapscott identifies some "key areas in the economy on which these advances have and will have a major impact upon". This is a number of areas, including: retailing and distribution, manufacturing, tourism and other. Author reveals how "the new technology and business strategies are transforming not only business processes but also the way products and services are created and marketed, the structure and goals of the enterprise, the dynamics of competition, and all the rules for business success" [3].

The digital economy is defined as "the integration of complex physical machinery and devices with networked sensors and software, used to predict, control and plan for better business and societal outcomes" [4]. Other definition of the digital economy is "the economic activity that results from billions of everyday online connections among people, businesses, devices, data, and processes. The backbone of the digital economy is hyperconnectivity which means growing interconnectedness of people, organizations, and machines that results from the Internet, mobile technology and the Internet of things (IoT)" [5].

In the digital economy "importance for businesses to both optimize their internal processes as well as to understand the digital users in order to be successful on the market" [6].

The digital economy transformed of industrial manufacturing based on digitization and exploitation of new technologies, enables and conducts the trade of goods and services through electronic commerce on the Internet. Digital economy technologies are transforming business models, facilitating new products and services, creating new processes, generating greater utility. According to Bernard and Jensen, technological innovations are an important driving force in allocation of resources seeking to lead not to long-term changes in growth rates, but to one-time level changes [7]. Their research furthermore suggests that actions that contribute technological innovations in order to reduce trade costs can be more efficient government policies in comparison with actions, that are aimed at export promotion.

The digital economy is based on following pillars: supporting infrastructure (hardware, software, telecoms, networks, etc.), e-business (processes that an organization conducts over computer-mediated networks) and e-commerce (transfer of goods online). There are four fundamental areas of digital transformation central to business success in the digital economy: the formation of remote ecosystem of talent and enable digital business processes that prove to be effective, even when distributed across various places and time zones; customers (regardless of model – B2B, B2C) interact with company way that is most convenient for them, when and where they want; digital supply networks (creating new intelligent digital networks of networks); the Internet of Things (IoT) [5].

Information economy technologies include numerous tools and theories that justify scopes of the optimal use of a certain software architecture. Among the most defining digital economy technologies for a number of industries and areas the following ones should be noted: information and communication technology (ICT); network communications; big data and cloud computing, modelling, virtualization and simulation, distributed computing; Internet of things; blockchain; digital twins; augmented reality (AR); additive manufacturing; robots and cognitive technologies, etc.

Digitalization of business processes (processes of production, distribution, exchange, consumption and utilization of goods and services) is profitable both for large and small businesses. Active introduction of digital tools occurs in all of the fields and scopes of activities of a company, including marketing, logistics, finances and accounting [8, 9]. Nowadays development and implementation of digitalization strategies are priorities not only for most of companies regardless of industry sector, business specifics or legislative regulation, but also for states. Contribution of digital economy in developed countries is estimated on average at 6–7% of GDP: France – 5,7%, Germany – 6,3%, the UK – 7,1%, the USA – 7,4%, Sweden – 8,6%. Contribution of advanced technologies to GDP in Russia is 3% [10]. In the USA, where investments of companies in development and production of information technologies significantly exceed the level of other countries, annual growth rate of hourly production in non-agricultural industries, for instance in 1995–2000, were 2,5%, which is significantly higher than the rate of previous two decades.

On the one hand, information economy technologies catalyse other factors of productivity increasing in business sector, such as innovations and dynamism of business activities. On the other hand, it would be improperly to rely only on the capacity of the digital economy to solve a number of problems, for example in management area or transport infrastructure. For many countries, the "old" development problems, such as institutes, infrastructure and qualifications of the staff are the key reasons of slow growth of introduction digital economy technologies to activities of national companies. In order for technology leaps to open a new path to development for low-income countries, these issues cannot be ignored [11]. It is obvious that companies from such countries also will not be able to use all of the advantages of the digitalization under conditions of, for example, inefficient institutional environment and lack of funding.

Competitive forces encourage companies to search for new growth areas based on either expansion of portfolio or penetration and expansion in new markets or combination of both. Under modern innovation conditions, information economy technologies and strategies of the product differentiation serve as engines of competitiveness of companies in a number of industries and fields of activities.

In the first part of the study, the issues of the influence of the digital economy technologies on companies' competitiveness, including ones based on network retail, were considered, the main problems of introduction of them into business processes of retail chains are analyzed and possible courses of solutions based on modern trends and scopes of digitalization in this area are defined. In the second part of the study, the main conclusions and results are presented, and areas for further work in this area are identified.

2 Trends and Problems of Digitalization of Retail Companies as a Factor of Their Competitiveness

Companies willing to obtain a competitive advantage have to focus relatively equally on such activities as production, product turnover, supply and customer service on the one hand, and on the other hand human resources management, information technologies [12]. It seems that M. Porter's approach to the definition of competition as struggle of competing companies, taking into account their external environment, most severely corresponds to the issue about the influence of the information economy on the competitiveness of companies [13]. Any activities of a company in this area take into consideration both behavior of rivals when choosing information economy technologies and their external environment of possible technological solutions. In this regard, adoption of information technologies may be considered not only as a conscious and aimed aspect of firm's activities, but also as an element of its general competitive strategy. Referring to the possibility of applying this approach to the study of the role of information technologies in the increasing competitiveness of companies, M. Porter proves that efficiently organized systems of information management can make a significant contribution in the enhancing companies' competitiveness. For example, augmented reality (AR) "change how enterprises serve customers, train employees, design and create products, and manage their value chains, and, ultimately, how they compete" [14].

If one considers certain examples of information economy technologies in the competitiveness of modern companies, using the Internet should be noted primarily (for example, social networks) for promotion of reward and coupons for consumers, searching for alternative sources of supply on the B2B Marketplace Internet service, formation unique characteristics of the product or service. For instance, the augmented reality application IKEA Place allows you to see how exactly its furniture will look like in the interior of a certain room. Web-platforms like trendhunter.com track emerging trends (from dynamics of smartphone size to changes in chocolate packaging colours) and allows companies to form relevant competitive advantages in the industry or in regional markets. PwC identifies three increasingly popular information technologies – robotization of processes, transactional analytics and intellectual process. At the end of 2019, 40% of all initiatives connected to IT will be supported by cognitive technologies and artificial intelligence. According to Gartner, analytical agency, in the short term artificial intelligence and the Internet of things along with blockchain technologies will become the main areas of the strategic investment [15]. In 2018 the volume of these segments globally amounted to about $3.7 trillion.

Information technologies lead to radical changes in the competitive environment of national and global companies [16]: online-shopping is becoming more efficient than shopping in usual store; digital marketing is more efficient than high cost of print, television and radio advertisements; social networks are more efficient than visiting clubs; cloud computing is more efficient than private computer network.

Modern companies forming their competitive advantages according to information economy solutions generally have the following characteristics: they have more accurate business planning, form an efficient marketing system, demonstrate high volumes and rates of sales, carry out real-time monitoring; management is character-ized by systematic and complex tasks, constant and prompt customer support is pro-vided non-stop (through several communication channels, including email, social media platforms, webinars).

Long-term business growth is very complicated to achieve without impetus of information technologies. Information technologies in business help a company monitor costs and profit promptly, which allows leadership to act more flexibly towards reducing costs or refocusing sales team when it is necessary. They give a competitive advantage to companies, allowing them to access larger markets, expand products or service lines more efficiently and monitor their rivals as well. Information economy technologies allow an enterprise to make more efficient decisions by attracting teams through videoconferences, analyzing public mood in social networks and industry forums, and also through online surveys for obtaining customers' opinions. Nowadays internet marketing with the use of methods of online advertising (SEO, PPC, Facebook Ads) is a tool, that is more accurate than traditional marketing of searching for target audiences, identification of their needs and making marketing campaigns. Cloud computings allow company employees to use any device to access enterprise software anywhere.

Digital economy technologies significantly change economic and organizational processes, ways of communication between suppliers and consumers of goods and services in a number of scopes of activities by affecting the competitiveness of modern companies. One of the areas most susceptible to various technologies and solutions of the information economy is retail, which is largely due to the high level of competition in this industry.

According to BrandZ "Top 100 Most Valuable Global Brands 2018", it is the second year in a row when retail is considered to be the most fast-growing category [17]: in 2018 the growth was 35%. Such networking companies as Amazon, Alibaba, The Home Depot and others are at the top of the list of the most expensive brands in retail area. It is presented in Table 1.

Table 1. Retail Top 20 Most Valuable Global Brands 2018.

Rank	Brand	Brand Value 2018, million US dollars	Brand Value, % Change 2018 vs. 2017
1	Amazon (USA)	207,594	+49%
2	Alibaba (China)	113,401	+92%
3	The Home Depot (USA)	47,229	+17%
4	Walmart (USA)	34,002	+22%
5	JD.com (China)	20,933	+94%

(*continued*)

Table 1. (*continued*)

Rank	Brand	Brand Value 2018, million US dollars	Brand Value, % Change 2018 vs. 2017
6	Costco (USA)	18,265	+12%
7	IKEA (Sweden)	17,481	−8%
8	Ebay (USA)	14,829	+20%
9	ALDI (Germany)	13,785	+12%
10	Lowe's (USA)	13,111	−2%
11	7-eleven (Japan)	9,227	+1%
12	Tesco (Great Britain)	9,079	+13%
13	Walgreens (USA)	8,842	−13%
14	CVS (USA)	8,450	−13%
15	Lidl (Germany)	8,219	+14%
16	Target (USA)	7,620	−12%
17	Whole Foods (USA)	7,088	New
18	Woolworths (Australia)	6,880	+5%
19	Carrefour (France)	6,607	−3%
20	Falabella (Chile)	5,373	New

Source: BrandZ Top 100 Most Valuable Global Brands 2018.

The Table 1 shows the retail category includes physical and digital distribution channels in grocery and department stores and specialists in drug, electrical, DIY and home furnishings. Amazon appears within retail because it achieves approximately 90% of its sales from online retailing.

The most competitive global network retailers have a number of key characteristics defining their success in the market, allowing them to conquer new markets successfully and form sustainable consumer preferences. Such characteristics include focusing on customers at every point of contact with them; using artificial intelligence, augmented reality technologies (AR) for client interface; changing business model under the influence of the external factors; integration of online and offline channels for consumer interaction. The competitive advantages listed are mostly formed under the influence of advanced technologies and innovative digital economy solutions. Due to such technologies retail turns into complex networks that are able to recognize and satisfy consumer's needs and desires in a wider range of goods and services.

Under the pressure of constantly changing consumers' preferences and information technologies networking retail is developing along the path of simplicity and convenience for the consumers. Modern consumers make extensive use of digital technologies; hence in order to meet their preferences and compete successfully in the market, introduction and application of modern technological solutions are required.

Advanced digital economy technologies make it possible to automate the operation of industrial complexes, optimize information exchange between the company and customers, organize decentralized storage of information without the possibility of unauthorized changes, etc. (see Fig. 1). Such solutions as Big Data, artificial intelligence, augmented reality, blockchain and others help us not only cut costs, but also increase profits in the only possible way – by retaining existing clients and attracting new ones. Automation and introduction of digital technologies for increasing efficiency of retail companies are the key tasks of their development. IoT, autonomus vehicles, drones, robotics and artificial intelligence are projected to be extremely important for retail due to their widespread use, ability to manage the efficiency and influence on labour. Some of the technologies at Picture 1 are applied at every stage of companies' supply chains, but this doesn't necessarily correspond to the greater general benefit for business. Autonomus vehicles, for instance, have relatively small application range, but undoubtedly they are going to change the way of transporting goods. The specific level of commercial benefits from the considered tools will be different for each particular network retail company, but in order for any of perspective technology to realize its potential, it has to become an integral part of the overall business and competitive strategy of the company.

A special place in development of modern retail takes electronic commerce. The penetration of e-commerce into the area of network retail, according to various estimates, in the medium term will grow from about 10% today to more than 40% in 2026 [18]. The prices of e-commerce brands have recently grown significantly, and companies of this category have adapted to failures and crises in the global and regional markets. Traditional retail chains are forced to spend huge sums on the development of Internet services and other technologies of the digital economy in order to maintain the advantage over growing Internet retail companies, such as Amazon.

For a number of companies, formation their competitive advantages is based on coordination of online and offline sales, development of new consumer experience, information integration of products and logistics (from search to delivery). Transactions on the aquision of stationary retail chains by online stores are fairly significant. Amazon purchased Whole Foods, the network of stationary grocery stores in the USA. Chinese leaders of e-commerce, Alibaba and JD.com, are connected to physical retail chains as well. Chinese corporation Alibaba is a successful example of merger of online and offline shopping. In 2016 they opened Hema store (in 2018 there were 65 of them). Customers are to install the app attached to Taobao or Alipay account. You can order food from Hema and it will be cooked by store's own cooks. Then couriers deliver the order to the customer in 30 min. In offline store people can order takeaway food made from ingredients they bought in the very store and then eat in special zone. Every good has a QR-code on it, which contain the information about the product: its origin, history, the exact date of delivery and cost. Customers pay for their purchases themselves, with Alipay. In Hema partially automated process is applied. Special robots transfer chosen goods from shelves to the kitchen and then deliver prepared food to the table chosen by customers [19].

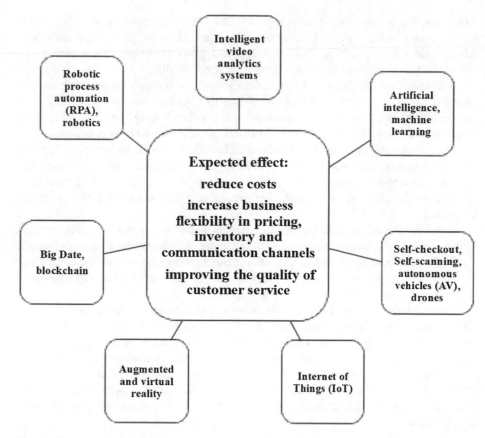

Fig. 1. The main solutions and aims of introduction of digital economy technologies into network retail companies.

In network retail the following areas of the introduction of digital economy technologies for increasing the Competitiveness and maximum satisfaction of consumers' needs have the priority [20]: working with consumers, including all stages from emergence of interest in trademark (brand) to the moment of purchasing; ensuring operational efficiency, working with products, price tags, merchandising, etc.; logistics and supply control; controlling work of IT-infrastructure and security system. Retailers have been experimenting with robots that help customers find goods they need. In 2016 Lowe company launched concierge-robot LoweBot in stores in San Francisco. Customers can ask Lowebot a question or use touchscreen for searching. The robot also tracks the availability of goods in stock in real time when passing the shelvings by.

Preferences of customers all over the world and in countries' markets, achievements in innovation sphere, informatization and digitalization largely influence on the pace and directions of the introduction of modern technologies in network retail, which allows companies to expand their borders in the market significantly, reduce costs, increase quality of goods by changing the character of competition in the industry. The future of retail is already seen in the case of China, which is, according to a number of indicators, the leader in the world. It partly shows the susceptibility of Chinese consumer to new technologies. Thus, up to 90% of purchases on Alibaba are made with mobiles on AliPay platform. Chinese consumers pay their rent, buy goods for their houses and order taxis at the same platform in their mobiles.

Network retail companies will successfully compete through modern technologies only in case of introduction of tools according to objective assessment of their costs and expected benefits [21]. Particularly, when using information technologies in development of competitive advantage, retailers may face the following problems [20].

First, high cost and certain difficulties of introduction of information technology tools to company's business processes. The company must have significant capital and capacities essential for introduction of new technologies. In practice, firms often have outdated technologies or lack qualified personnel that limit the introduction of digital solutions.

Second, pace of the digital economy changes is ahead of firms' capacity to move with the times. The adoption of a new business model by a company requires the new level of operational flexibility, which affects the structure, opportunities, culture and decision-making process. It is obvious that not all firms are ready for radical changes of their operating activities and business models.

It should also be noted that the effectiveness of modern technologies in network retail depends not only on a particular technology, but also on the process of organizing its introduction in the company's business model. The key role is played by personnel. If a company does not have the experience and personnel for introduction of information technology, it is more efficient to outsource business process/function/operation. Thus in X5 Retail Group 70% of IT tasks are solved on their own and 30% are outsourced. External contractors are partially transferred the tasks of testing, support, platform development.

3 Results

As network retail companies develop their business models based on modern digital economy technologies, it is crucial for them to have the appropriate basic capabilities to service new solutions. These solutions are to require companies to be prepared for innovation and cooperation. So, as network companies move into the online space, they will face the need for reviewing the role of retail space in order to optimize the profitability of assets and retaining customers' loyalty. Retail companies should think of reprofiling their offline stores and cooperation with companies of other industries and scopes of activities (for example, delivery service) in order to increase customer's satisfaction with purchase process, and also for revaluation of the value of a physical asset in terms of the income of a network retail company.

For the successful implementation of modern technologies in the business model of a network company, the strategic fit of offline online models is important. Companies should define the offline-online model that is the most appropriate for their strategic priorities and will help them maintain their competitive advantage in their industry. Even within the same product categories, different models may be better for different companies. It depends on the individual organizational capabilities that exist or can be developed in the short and medium term.

As the study showed, the tasks of building deep understanding and communication with their customers, the rapid introduction of modern digital economy technologies, the transformation of business models in both online and offline space and creating the key competitive advantages on this basis are in the first place for modern network retail companies. Modern information technologies used in business models of network retail companies form an added value of retailers' offer for final consumer through prices, range, convenience and experience. As e-commerce plays an important role in the future of retail, industry players should consider maintaining priorities and setting up e-commerce businesses to take an advantage of this growth path. This transformation gives rise to a number of problems that will need to be solved by companies' active preparation for changes and introducing the necessary technologies to solve issues related to the closure of "physical" stores and training (searching) for specialists who are able to effectively apply digital economy technologies.

4 Conclusion

The aim of the study was to identify the main areas and problems of using digital economy technologies to increase the competitiveness of companies. When considering and evaluating the specific effects of the introduction of technologies and solutions of digital economy in activities of companies, their influence on the competitiveness of the business, industry specifics must be taken into account. One of the industries in which the maximum effect of the application of the digital economy can be achieved is network retail. In this regard, the main results of the work are related to the formulation and systematization of the main trends and problems of the impact of the digital economy on the development of network retail companies.

In the sphere of retail digitalization, the following interrelated trends are highlighted: the active use of modern technologies, the desire for in-depth analysis of consumers, the coverage of transformational business models in both online and offline space, the creation of distinctive competitive advantages. The availability of a digital economy technology for network retail companies is a matter of sustainable demand for it. Thus in the future the active distribution of mobile devices will allow to collect more and more data about users, their habits and preferences, which will directly affect the number of various services that develop the direction of big data and their availability for more retailers. Augmented reality technologies (AR) will also help in promotion and sales by reducing the cost of devices and increasing audience.

In the study, the main problems of digitalization of business processes in network retail are identified. Among these problems are high cost and certain difficulties in introduction of information economy tools into the company's business processes, lack of operational flexibility affecting the structure, capabilities, culture and decision-making process in the digital economy; companies are not ready for radical changes in their operating activities and business models; lack of qualified personnel with practical skills in introduction and working with specific tools and solutions.

The solution of these problems in most cases requires companies to develop basic capabilities for servicing new solutions of the digital economy, which form the basis of their competitive advantage. First of all, such opportunities require companies to be ready for innovation and cooperation. Another area should be an integrated approach to the strategic fit of offline-online models. Companies should adopt an offline-online business model that best suits their strategic priorities and is able to maintain a competitive advantage in the industry. It is also necessary to ensure consistency in the tasks of building long-term relationships with consumers, customers, rivals, transforming business models in both autonomus and online spaces, depending on the technologies introduced in the digital economy and creating key competitive advantage on this basis.

The study does not pretend to be a complex description of all the problems of the digital economy's impact on the competitiveness of companies, taking into account their industry-specific activities. However, consistent work on monitoring and solving the existing limitations of digitalization of business processes will significantly advance in this direction. A promising area for further research is the development of business models for the introduction of selected digital technologies considered in the study in accordance with the stages of the value chain in the company.

References

1. Maresova, P., Soukal, I., Svobodova, L., Hedvicakova, M., Javanmardi, E., Selamat, A., Krejcar, O.: Consequences of industry 4.0 in business and economics. Economies 6(3), 46 (2018). http://dx.doi.org.aucklandlibraries.idm.oclc.org/10.3390/economies6030046
2. Melnik, M., Antipova, T.: Organizational aspects of digital economics management. In: Antipova, T. (ed.) Integrated Science in Digital Age. ICIS 2019, LNNS, vol. 78, pp. 148–162. Springer, Cham (2020). https://doi.org/10.1007/978-3-030-22493-6
3. Tapscott, D.: The Digital Economy: Promise and Peril In The Age of Networked Intelligence: A Textbook, p. 368. McGraw-Hill, New York (1994)
4. Industrial Internet Consortium: A Global Industry First: Industrial Internet Consortium and Plattform Industrie 4.0 to Host Joint IIoT Security Demonstration at Hannover Messe (2017). https://www.iiconsortium.org/press-room/04-20-17.htm. Accessed 23 July 2019
5. What is digital economy? Unicorns, transformation and the Internet of things. Deloitte. https://www2.deloitte.com/mt/en/pages/technology/articles/mt-what-is-digital-economy.html. Accessed 23 July 2019
6. Leimeister, J.M., Winter, R., Brenner, W., Reinhard J.: Research program 'digital business & transformation IWI-HSG'. University of St. Gallen's Institute of Information Management Working Paper No. 1, 25 September 2014. http://dx.doi.org/10.2139/ssrn.2501345
7. Bernard, A., Jensen, J.B.: Exceptional exporters' performance: cause, effect or both? J. Int. Econ. 47, 1–25 (1999)
8. Kuter, M., Lugovsky, D., Mamedov, R.: Depreciation policy is an accounting policy element to ensure the financial strategy of the owner. Econ. Anal. Theory Pract. 29(158), 17–23 (2009)
9. Lugovsky, D., Kuter, M.: Accounting policies, accounting estimates and its role in the preparation of fair financial statements in digital economy. In: Antipova, T. (ed.) Integrated Science in Digital Age. ICIS 2019, LNNS, vol. 78, pp. 165–176. Springer, Cham (2020). https://doi.org/10.1007/978-3-030-22493-6_15

10. Titov, B.: Russia: from digitalization to digital economy. http://stolypin.institute/institute/rossiya-ot-tsifrovizatsii-k-tsifrovoy-ekonomike/. Accessed 23 July 2019
11. The Global Competitiveness Report 2018. The World Economic Forum. https://www.weforum.org/reports/the-global-15competitveness-report-2018. Accessed 23 July 2019
12. Porter, M.E.: Competitive Advantage. Creating and Sustaining Superior Performance. The Free Press, New York (1998)
13. Porter, M.E.: The five competitive forces that shape strategy. Harv. Bus. Rev. **86**(1), 78–93 (2008). Special Issue on HBS Centennial
14. Porter, M.F., Heppelmann, J.E.: Why every organization needs an augmented reality strategy. Harv. Bus. Rev. **95**(6), 46–57 (2017)
15. Master Today's Technology Trends. Gartner Insights. https://www.gartner.com/en/information-technology/insights/trends-predictions. Accessed 28 July 2019
16. Sisk, A.: Importance of Information Technology in the Business Sector. https://bizfluent.com/about-6744256-importance-information-technology-business-sector.html. Accessed 25 July 2019
17. BrandZ: Top 100 Most Valuable Global Brands 2018. https://brandz.com/Global. Accessed 26 July 2019
18. Digital Russia. New reality. Research by McKinsey Global Inc., July 2017. http://www.tadviser.ru/images/c/c2/Digital-Russia-report.pdf. Accessed 26 July 2019
19. Autonomous stores, digital shelves, drones and augmented reality: what technologies are implemented in retail. https://vc.ru/future/56618-avtonomnye-magaziny-cifrovye-polki-drony-i-dopolnennaya-realnost-kakie-tehnologii-vnedryayut-v-riteyle. Accessed 26 July 2019
20. Shaping the Future of Retail for Consumer Industries. World Economic Forum, January 2017. http://www3.weforum.org/docs/IP/2016/CO/WEF_AM17_FutureofRetailInsightReport.pdf. Accessed 26 July 2019
21. The Digital Enterprise Moving from experimentation to transformation. World Economic Forum, September 2018. http://www3.weforum.org/docs/Media/47538_Digital%20Enterprise_Moving_Experimentation_Transformation_report_2018%20-%20final%20(2).pdf. Accessed 26 July 2019

Automation and Algorithmization of "Smart" Benchmarking of Territories Based on Data Parsing

Daniil Kurushin⑩, Julia Dubrovskaya⁽✉⁾⑩, Maria Rusinova⑩,
and Elena Kozonogova⑩

Perm National Research Polytechnic University, Perm 614990, Russia
uliadubrov@mail.ru

Abstract. Benchmarking is one of the modern tools for regional development. The tasks of regional benchmarking include determining the leading territory, analyzing the main factors of its success and adapting the identified benefits to the analyzed region. At the same time, a feature of "smart" benchmarking is a preliminary selection of structurally similar territories and further comparison of the object of study only with identical regions.

In the article the algorithm of automation of the procedure of regional benchmarking based on parsing sites for its further use in the software environment for statistical data processing. The algorithm is tested on the example of the subjects of the Russian Federation. Calculation of indicators is carried out by a flexible algorithm, which can be changed at the request of the researcher. The parser receives the calculation formula as input in Python or LaTeX notation and looks for indicators in the collected data suitable for substitution into the formula. When all indicators are collected, the database is calculated and updated. The visualization of the "smart" benchmarking procedure was carried out for the Perm Territory in the form of a geographical map of identical regions.

Keywords: "Smart" benchmarking · Regional development · Algorithmization · Visualization · Parser

1 Introduction

Issues of the need to improve the activities of economic entities in various fields, including through automation of processes, are widely considered by scientists. This is partly due to the global trend of increasing interest in digitalization issues both from representatives of the scientific sphere, and from corporate managers and government executives.

Indeed, without automation it is impossible to imagine a solution to operational management tasks, or the implementation of long-term projects, or the implementation of subsequent control. With algorithms and software, process automation allows managing efficiently at all stages of both commercial and government activities.

The Digitalization Program until 2020 adopted in Russia in 2017 implies that the successful functioning of the digital economy is possible with electronic interaction in

© Springer Nature Switzerland AG 2020
T. Antipova and Á. Rocha (Eds.): DSIC 2019, AISC 1114, pp. 126–134, 2020.
https://doi.org/10.1007/978-3-030-37737-3_12

public administration (public services), the formation of relevant information databases, and the expansion of electronic commerce [1]. Thus, automation and algorithmization issues become paramount for a wide range of economic entities: commercial enterprises [2–5], government [6–9], financial institutions [10–13].

This article is devoted to the problem of algorithmization and automation of regional benchmarking based on data parsing. Like benchmarking of companies, benchmarking of the territory is carried out on the basis of continuous monitoring of best practices, the search for new ideas for forming development priorities based on identifying the unique opportunities of the territories (Groenendijk [14], Iurcovich [15], Koellreuter [16]). Regional benchmarking is an inter-regional comparison of processes, practices, activities, policies and further use of the information obtained to improve the regional development system.

In recent years, the concept of "smart" benchmarking has been widely developed [17, 18]. According to it, the comparison of territories should be carried out on regions with similar institutional conditions.

The methodology developed by the Basque Institute for Competitiveness [19] and used by European regions and countries to create effective innovative strategies for territorial development was used as the methodological basis for the benchmark procedure. As the basic factors to compare regions, we selected the ones that show the strengths and weaknesses of the territory in the best way, and do not tend to change in the short term.

The authors identified 7 criteria for regional development: natural features of the region, education, innovation, sectoral structure, investment climate, transparency of the region, social values. The criteria include 12 factors, which are a system of statistical indicators (34 indicators in total), on the basis of which the territories are compared.

More information about the author's methodology for benchmarking procedure, the main stages of its implementation, and rationale for the selection criteria is in the paper [20]. Note that the methodology was finalized by changing the content component of the indicators according to the criterion "industry structure". As a statistical indicator characterizing this criterion, the indicator "average annual labor productivity by type of economic activity" was chosen instead of "distribution of the average annual number of employed population by type of economic activity".

This paper presents an algorithm for the automated collection of unstructured information, its transformation and issuance in a structured form for further use in a software environment for statistical data processing.

2   Algorithmization of the Procedure for Conducting "Smart" Benchmarking of the Territory

The process of "smart" benchmarking takes place in two main stages: data collection and finding identical regions. In this case data collection is based on the parsing of relevant sites; determination of identical regions – by calculating the indexes of the structural distance. Let us dwell on them in more detail.

2.1 General Algorithm for Parsing Necessary Data

Data parsing is performed according to the algorithm presented in Fig. 1.

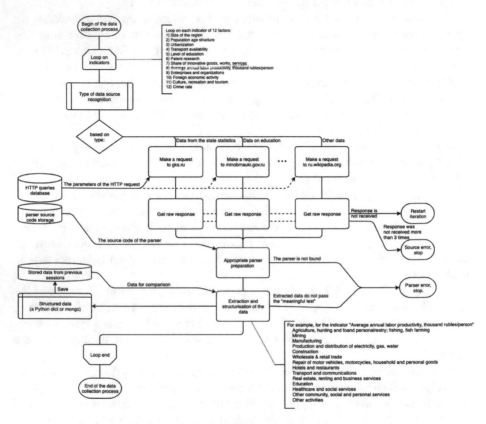

Fig. 1. Data parsing.

For each statistical indicator, on the basis of which the territories are compared, the following actions are performed from the section "source data" (the list may vary depending on the further calculations) the following steps are performed:

(1) Determining the preferred data source (Sites of the Federal State Statistics Service, the Central Bank, the Ministry of Education, etc.), in this case, if the source is not identified, the website ru.wikipedia.org is used.
(2) Program has access to the database of the HTTP query rules for the different sites. Depending of the chosen one the appropriate rule is loaded.
(3) HTTP request formed and sent to the data source (gks.ru, cbr.ru etc.).
(4) The un- or semistructured response comes as an answer. It can be in the form of an html page, a JSON structure, office document, depending on the internal logic of the source. If there is no reply from a source, the iteration is repeated. In case of multiple failures, the software stops gathering data and informs the administrator.

(5) Suitable parser is extracted from parser code storage.
(6) The parser is being executed. It transforms the un- or semistructured data into a predetermined structure. The data are compared with previously extracted and pass the "meaningful test". The obtained values are compared with known and if there is a statistical outlier manual checking must be performed.
(7) Data is stored in structured format in the RAM and in a file for future use.

At the next stages of data processing in order to obtain identical regions, some of the indicators that are used to compare territories are calculated, while others are used in their original form. Based on the collected statistical data, indicators such as "Number of people with higher education", "Average annual labor productivity" are calculated. At the first and second stage, an algorithm for calculating these indicators is given, the principal feature of which is the implementation of the so-called "data-driven" algorithm [21].

Stage 1. Calculation of the indicator "Number of people with higher education"

(1) In the loop the indicators to be calculated, are requested from the database generated during the previous step.
(2) If indicator is not received, an error is generated. Administrator attention is required. This means that it was not collected during the previous step, but the data collection stage has not fixed this problem. Probably a bug in the data collection stage.
(3) The extracted indicator is added to the formula.
(4) If the calculation is successful, the data is stored into the database and saved for further work.
(4.1) If not, the cycle repeats.

A detailed algorithm for the implementation of stage 1 is shown in Fig. 2.

Stage 2. Calculation of the indicator "Average annual labor productivity"

(1) In the loop the indicators to be calculated, namely, the GRP indicator by type of economic activity and the average annual number of employees by type of economic activity, they are requested from the database formed on the basis of the general algorithm for benchmarking.
(2) If indicator is not received, an error is generated. Administrator attention is required. This means that it was not collected during the previous step, but the data collection stage has not fixed this problem. Probably a bug in the data collection stage.
(3) The extracted indicator is added to the formula.
(4) If the calculation is successful, the data is stored into the database and saved for further work.
(4.1) If not, the cycle repeats.

A detailed algorithm for the implementation of stage 2 is shown in Fig. 3.

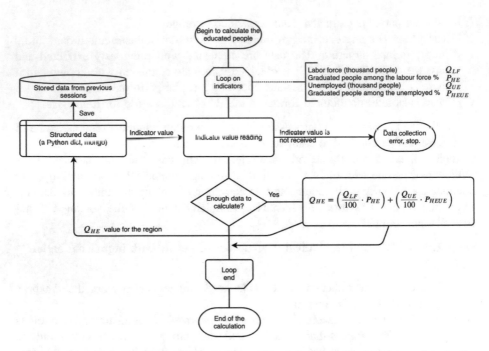

Fig. 2. Algorithm for calculating the indicator "Number of people with higher education".

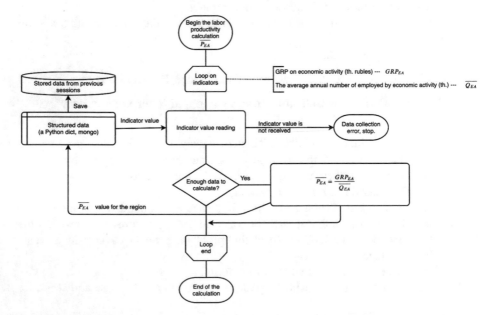

Fig. 3. Algorithm for calculating the indicator "Average annual labor productivity".

Also, note that a flexible algorithm, which can be changed at the request of the researcher, performs the calculation of indicators by stages. The parser receives the calculation formula as input in Python or LaTeX notation and looks for indicators in the collected data suitable for substitution into the formula. When all indicators are collected, the database is calculated and updated.

2.2 Determination of Identical Regions by Calculating the Indexes of the Structural Distance

To complete the benchmarking procedure, the asymmetry of the distribution of indicators is assessed. The asymmetry value characterizes the degree of distribution dissymmetry of a statistical indicator relative to the average indicator value in the country. If the asymmetry value is over 0.5 in absolute value, then to smooth out "outliers" (extreme values), each value of the indicator is transformed. Also, values are reduced to a single scale.

Thereafter, the structural distance index calculated according to formula (1):

$$d(i, i') = \sum\nolimits_{j=1}^{k} m_j \left(\overline{x_{ij}} - \overline{x_{i'j}} \right)^2 \tag{1}$$

where $d(i, i')$ – structural distance index of the i-th region;

$\overline{x_{ij}}$ – value of j-index of i-initial region;
$\overline{x_{i'j}}$ – value of j-index of i-"another" region;
m_j – weighting factor

Weighting factor is calculated according to formula (2)

$$m_j = \frac{1}{a} / j \tag{2}$$

Where a – number of criteria for comparing regions;

j – number of statistical indicators characterizing the criterion

Sequentially calculating the structural distance index between the analyzed region and other regions within each criterion, the distance matrix is constructed. In this case, the smaller the value of structural distance index, the most similar regions. The regions with the lowest values of structural distances indices are identical regions.

3 Results

Testing of the procedure of "smart" benchmarking was carried out in accordance with the steps described by us. As an experimental base, we used the statistical data of the Russian regions for 2015 published by the Federal State Statistics Service. As a result, a complete distance matrix was constructed for the regions of Russia. We have chosen the Perm region as the analyzed region.

On order to determine, the regions identical to Perm Krai used the values of the global distance matrix for Russian regions. Table 1 shows a fragment of the matrix containing regions identical to the Perm Region.

Table 1. Distance matrix of regions identical to Perm Krai

Region	Perm Krai	Republic of Tatarstan	Krasnoyarsk Krai	Saratov Oblast	Volgograd Oblast	Primorsky Krai	Arkhangelsk Oblast	Kaliningrad Oblast	Kaluga Oblast
Perm Krai		0,112	0,168	0,184	0,200	0,220	0,279	0,291	0,343
Republic of Tatarstan	0,112		0,235	0,352	0,268	0,415	0,410	0,421	0,545
Krasnoyarsk Krai	0,168	0,235		0,541	0,420	0,184	0,316	0,292	0,689
Saratov Oblast	0,184	0,352	0,541		0,088	0,375	0,280	0,382	0,104
Volgograd Oblast	0,200	0,268	0,420	0,088		0,291	0,149	0,223	0,109
Primorsky Krai	0,220	0,415	0,184	0,375	0,291		0,165	0,099	0,381
Arkhangelsk Oblast	0,279	0,410	0,316	0,280	0,149	0,165		0,111	0,204
Kaliningrad Oblast	0,291	0,421	0,292	0,382	0,223	0,099	0,111		0,293
Kaluga Oblast	0,343	0,545	0,689	0,104	0,109	0,381	0,204	0,293	

According to the Table 1, the Republic of Tatarstan has the lowest index of structural distance along with the Perm region.

Next, the regions identical to the Perm region was visualized (Fig. 4).

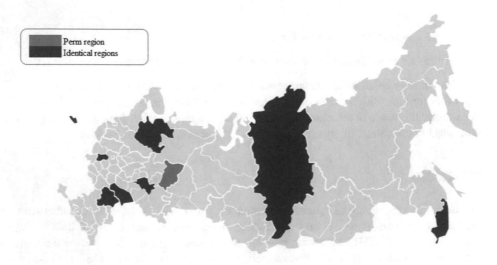

Perm region
Identical regions

Fig. 4. Map of regions identical for Perm Region

4 Conclusion

The presented algorithm for automating the process of conducting the benchmarking procedure can make a significant contribution to the process of monitoring, developing and implementing innovative strategies for the development of territories. This is due to the possibility of a holistic visual presentation of the relevant statistics. We believe that the improvement and development of a regional benchmarking procedure for territorial entities requires further research and it is objectively of scientific interest.

The described algorithm for automating the benchmarking procedure is universal and can be applied to any territorial units of all levels, including municipal and regional.

In the future, we plan to develop and register a software tool that synthesizes regional statistics on the basis of site parsing and allows us to identify structurally similar subjects of the Russian Federation in order to model the optimal development option for a specific territory.

Acknowledgment. The reported study was funded by RFBR according to the research project № 19-010-00449.

References

1. http://government.ru/docs/28653/. Accessed 03 Aug 2019
2. Anikanova, M.A., Morgunov, A.F.: Criterial evaluation of the possibility of small businesses business process automation on public cloud platform. Busi. Inform. **3**(33), 55–64 (2015)
3. Kek-Mandzhieva, Z.V.: Automation of accounting and analytical management system as a necessary component of effective enterprise management. Sci. Modernity **35**, 183–191 (2015)
4. Nitsenko, V.S., Vysochinskaya, L.N., Marojko, M.N.: Automated forms of accounting as an instrument of increasing intensity information processing. J. Econ. Bus. **4**, 120–122 (2016)
5. Rabinovich, L.A., Libman, M.L.: The main provisions of the feasibility study of new means of automation of production processes. News of Volgograd State Technical University, no. 1, pp. 7–9 (2004)
6. Lastochkina, M.A.: Development of methodology and tools for assessing the degree of modernization in Russia's regions. Probl. Dev. Territory **4**(78), 69–79 (2015)
7. Nemets, K.A.: Automation of regional management processes and the formation of "Electronic Government". Econ. Sci. **4**(127), 85–89 (2011). Scientific and technical statements of the St. Petersburg State Polytechnic University
8. Nemtsev, A.N., Shtifanov, A.I., Belenko, V.A., Zagorodnyuk, R.A., Nemtsev, S.N., Galtsev, O.V.: Designing the automated information system for monitoring of educational departments and provision the opportunity to use "electronic" services in education sector. Comput. Sci. **13–1**(108), 65–76 (2011). Scientific reports of Belgorod State University. Series: Economics
9. Robozov, S.A., Erunova, M.G., Maltsev, K.V.: Quality assessment of regional and municipal management in an automated information system for monitoring municipalities. J. Sci. Technol. **1**, 176–181 (2010)
10. Vasilieva, E.E.: Modeling language credit risk assessment of banking activity in the Russian regions, based on fuzzy sets methods. Internet J. Sci. Sci. **8**(6), 1–18 (2016)

11. Ryzhkov, O.Y., Bobrov, L.K.: Complex automation of actuarial work of the insurance company. Comput. Sci. Inform. **2**, 98–108 (2014). Vestnik of Astrakhan State Technical University. Series: Management
12. http://www.standardandpoors.com/prot/ratings/articles/ru/ru?articleType=HTML&assetID= 1245322330515. Accessed 03 Aug 2019
13. Solntsev, O.G., Mamonov, M.E., Pestova, A.A., Magomedova, Z.M.: Experience in developing early warning system for financial crises and the forecast of Russia banking sector dynamic in 2012. J. New Econ. Assoc. **4**(12), 41–76 (2011)
14. Groenendijk, N.: EU and OECD benchmarking and peer review compared. In: Laursen, F. (ed.) The EU and Federalism: Polities and Policies Compared, pp. 181–202. Ashgate Publishing Company, Farnham (2010)
15. Iurcovich, L., Komninos, N., Reid, A.: Mutual Learning Platform: Regional Benchmarking Report: Blueprint for Regional Innovation Benchmarking. European Commission, Brussels (2006)
16. Koellreuter, C.: Regional Benchmarking as a tool to improve regional foresight. European Commission-Research DG-Directorate K (2002)
17. Navarro, J., Smart, J.P.: Specialisation benchmarking and assessment: pilot study on wind energy. Publications Office of the European Union, Luxembourg (2017)
18. http://www.academia.edu/24150494/Policy_learning_through_benchmarking_national_ systems_of_competence_building_and_innovation-learning_by_comparing. Accessed 03 Aug 2019
19. http://s3platform.jrc.ec.europa.eu/regional-benchmarking. Accessed 03 Aug 2019
20. Dubrovskaya, Y.V., Kudryavtseva, M.R., Kozonogova, E.V.: "Smart" benchmarking as a basis for strategic planning in regional development. Econ. Soc. Changes Facts Trends Forecast **11**(3), 110–116 (2018)
21. https://homepage.cs.uri.edu/~thenry/resources/unix_art/ch09s01.html. Accessed 03 Aug 2019

XBRL as a Tool for Integrating Financial and Non-financial Reporting

Olga Efimova[✉][iD], Olga Rozhnova[✉][iD],
and Olga Gorodetskaya[✉][iD]

Financial University Under the Government of the Russian Federation,
Moscow 125993, Russia
{Oefimova, ORozhnova, OGorodetskaya}@fa.ru

Abstract. The paper examined the ways that XBRL technology may potentially improve company reporting system to fully meet stakeholder information requests. The study investigated the existing system of financial and non-financial information disclosure and the ways that XBRL may influence or transform that process. As a part of the study we used companies' interviews and investor surveys. We also used a systematic literature review method to accumulate reporting development research results. The study results help to reveal the benefits of XBRL for the formation of various types of reporting and formulated two main goals of XBRL application - ensuring effective management of companies and their business processes, and high-quality meeting of stakeholder information requests. The achievement of these goals will increase the image and value of the company, contributes to its sustainable development, creates an opportunity for the sustainable development of society.

Based on a detailed analysis of outlined issues we conclude that to solve them, it is necessary to focus on two areas - improving the training of accounting professionals in the field of digital technologies and harmonizing principles and standards of financial and non-financial reporting, that is made possible through the use of XBRL technologies.

Keywords: XBRL application · Sustainability reporting · Reporting integration

1 Introduction

Financial reporting has been developing during the entire period of its preparation [15]. In recent decades, new accounting objects connected with "intellectual economy" have emerged, that leads to the growth and significance of intangible assets, which are not provided for in traditional financial reporting. New non-GAAP indicators (not formally defined in the standards) are actively using by investors and other capital providers. They also show increasing interest in corporate reporting, especially in the area of long-term sustainability.

The significance of long-term development aspects and the need to analyze environmental risks in the process of making investment and financial decisions led to the emergence of a new type of accounting - accounting for sustainability with its new

© Springer Nature Switzerland AG 2020
T. Antipova and Á. Rocha (Eds.): DSIC 2019, AISC 1114, pp. 135–147, 2020.
https://doi.org/10.1007/978-3-030-37737-3_13

accounting objects [1]. As a result, more and more public companies began to disclose not only financial, but also non-financial (ESG) indicators material for interested parties.

The influence of technology is growing throughout the cycle of formation, promotion, and use of financial statements in the decision-making process.

All these very significant trends cannot affect the financial statements. Nevertheless, financial statements remain the most important source of information for investment decisions that are based on expected estimation of future cash flows. In turn, the likelihood of obtaining expected cash flows from business or individual assets is highly dependent on the influence of interested parties and the associated non-financial risks and opportunities (ESG risks and opportunities). In this regard, interested users of financial statements need a system of interconnected financial and non-financial information, which comprehensively characterizes both current financial results and cash flows, as well as their forecast. The probabilistic nature of future cash flows necessitates the development of mechanisms for the disclosure of significant risks and opportunities and professional judgments in close connection with the disclosed financial reporting information [12].

According to research results [13], investors do not seek to reduce the information disclosed because they have the technology to analyze them. Investors require a better use of technology capabilities for the formation of integrated reporting information and its disclosure [4, 14]. At the same time, determining how to efficiently collect, aggregate and disaggregate data, taking into account user information requests, presenting and analyzing interconnected financial and non-financial data, is a matter that investors view as one of the most important [3, 6]. Thus, the problem should be focused on how technology can be used to reform the entire process of generating financial and related non-financial reporting as a single process from beginning to end, and not just submitting it to the regulatory authorities. The solution of the problem is seen in the use of the international reporting format - XBRL (eXtensible Business Reporting Language), which provides issuers and users with a standard reporting algorithm and allows to present and to use the information in various ways depending on requests from regulators, capital providers and other interested parties.

2   Literature Review and Current State of the Issue Assessment

From the position of financial capital providers, a number of issues connected with providing financial reports to interested parties remain unresolved or insufficiently developed. These matters include emerging trends in technology and connectivity, the inability of the existing accounting model to provide investors with sufficient decision-useful information in a new economy, and the lack of a measurement framework that can inform the disclosures necessary to make such measurements meaningful [3].

Among the most significant financial drivers of value creation, investors tend to include non-GAAP indicators: EBIT, EDITDA, FCF, and their modifications. These indicators are the basis for calculating the most important financial multipliers, such as EV/EBITDA, Net Debt/EBITDA. The numerous techniques used by companies to calculate them, on the one hand, make it impossible their comparative analysis and, on the other hand, the manipulation of these indicators for the purpose to increase investment attractiveness. The lack of transparency in the disclosure of the methodology for calculating these key indicators represents one of the most important problems of investment analysis [4].

Stock exchanges are actively involved in the process of integrating ESG factors into the disclosure requirements of listed companies. In 2018, the World Federation of Exchanges (WFE) issued an updated version of Exchange Guidance and Recommendations on Sustainability [23], confirming that investors are the target audience for listed company ESG disclosures. In turn, global rating agencies (Fitch Ratings, Moody's Investor Services, S & P Global Rating) published information on how ESG-factors are taken into account in the framework of investment or credit analysis. A well-known practice is when the ESG aspects have changed the initial rating or forecast regarding companies' rating. Thus, it becomes generally accepted that ESG-factors should be integrated into issuer reporting and investment decision-making system [9].

Despite the growth of disclosed information, the research conducted on the satisfaction of the investment community with its quality [4, 14] indicates serious problems and frustration in the usefulness of the data provided.

Investors and other interested parties confirm the growing volume of information that companies publish in various reports. At the same time, users note that companies use different methods of collecting and disclosing non-GAAP and ESG data due to it is difficult to quickly navigate and use in the analysis process.

The complaints from the investment community are caused primarily by the lack of comparability of disclosed non-GAAP indicators and, moreover, non-financial information, differences in approaches to the formation of financial and non-financial reporting in terms of principles, requirements, periodicity and other aspects [4]. Thus, improving the practice of both the request from the investment community and the system disclosure of ESG data on a comparable basis from reporting companies remains a highly relevant task [8]. Experts see the solution in the more active use of IT technologies from both the investment community and companies.

To close this gap, Lubin and Esty [16] described "new approach to sustainability reporting". Seele [19] argued that XBRL in particular can help to close this "sustainability gap". This unified and interlinked reporting concept will also allow companies to integrate practices within "operations and measuring and reporting on those integrated practices in an aggregated, machine-readable, XBRL format" [22].

This approach is close to Eccles and Armbrester' concept, who categorize the transformation brought about by integrated reporting and computing as "disruptive ideas" that "enable companies to make much more informed decisions about how they are using financial, natural and human resources to meet both financial and nonfinancial performance objectives" [7].

A first application field in which to apply XBRL to sustainability reporting was reporting of the energy performance of buildings. Gräning and Kienegger [17], inspired by Basel II and financial reporting, suggested that XBRL can be used "as a mean of standardization for the reporting concerning the energy performance of buildings" and discuss the generalization of XBRL and the possibility of applying it to other domains. Another aspect has been described as "inter-organizational sustainability reporting" [20]. Seele [19] proposed a new concept called "digitally unified reporting" that combines digital data management of sustainability performance with digitally standardized sustainability reporting. Finally, GRI in collaboration with Deloitte Netherlands had developed one of the first XBRL taxonomies for sustainability reporting. These XBRL taxonomies had intended to help investors, auditors and analysts to access information in sustainability reports faster, and more simply. By tagging the data once, the GRI says organizations can gain better control over the quality and integrity of their sustainability performance data. This will help investors, auditors and other report users to access and compare GRI data without the need for excessive manual work[1].

3   The Main Body

3.1   Approach

The paper examined the ways that XBRL technology may potentially improve company reporting system to fully meet stakeholder information requests. We investigated the existing system of separate preparation and presentation of financial and non-financial information and the ways that XBRL technology may potentially improve or when necessary to transform that process.

As a part of the study we used the companies' interviews and investor surveys. We also used a systematic literature review method to accumulate reporting development research results.

At last we assessed how XBRL may potentially transform the financial and non-financial reporting production, delivery and consumption processes to make them more effective and transparent for investors. For this purpose, we analyzed the financial statements of major Russian companies as well as their non-financial reports, registered in the National Register of corporate reports. We focused on the financial and non-financial reporting relationship and interactions to assess their usefulness for capital providers and other interested parties.

3.2   Statement of Basic Materials

The study made it possible to identify the following basic properties of the XBRL format, which provide the advantages of its use and are significant for the integration of financial and non-financial reporting. The following Table 1 gives a summary of the identified positions.

---

[1] https://www.environmentalleader.com/2013/08/gri-digitizes-g4-uses-open-source-tagging-language/.

Table 1. The advantages of XBRL.

Properties	Benefits
Offers a standard algorithm for report preparation, which can later be presented in various ways; automates the process of comparing financial information	Computer programs can efficiently and reliably extract information according to users' requirements
Taxonomy available for IFRS, Basel II, III, Solvency II	Provides the basis for the development of the required functionality and provides components for decision making Compatibility with financial reporting systems and requirements is possible Data exchange among different regulators and departments is possible
Multidimensional data model. Companies collect and transmit not individual reports, but a multidimensional data file, formed according to a special architecture	Faster data collection, aggregation and sorting analysis [11]
Global standard with embedded semantic rules	Data exchange between company business systems is simplified
Has a flexible format and the ability to expand	Business processes of collecting, analyzing and exchanging data among the information within the organization are optimized
Proceeds from a form-centric approach to data-centric approach	The data is formed as multidimensional structures built on the basis of the registers of accounts, individual accounts and indicators that are necessary for advanced data analysis
Covers all stages of producing and consuming business information	The quality of business information is improved, its collection, processing and analysis is optimized, the costs of all participants involved in the process are reduced The risk of human error is eliminated
Has extensibility, i.e. allows to: add one or more concepts; prohibiting the use of existing resources and relationships; add one or more resources and relationships	Has the property of extensibility, i.e. allows the addition of one or more concepts or resources; the prohibition of existing resources' usage

As follows from Table 1, the above properties of XBRL allow to solve the problems that the last decades have been especially important for accounting:

– elimination of redundancy and duplication of reporting data through the creation of a unified system for collecting and processing of reports;
– improving the reliability and quality of reporting data through the unification and automation of processes;
– increasing the transparency and openness of financial information;

– improving the accuracy of decision making due to either identifying business problems or identifying ways to solve them and create mechanisms to prevent their occurrence in the future.

Using semantic tools XBRL allows to submit a system of financial and non-financial indicators in accordance with generally accepted requirements of international standards in electronic form. Software products with XBRL technology contain the process of collecting, selecting and pre-processing information to further provide the results to different users. Sources of XBRL data can serve as data of the accounting system, as well as other company's information systems (Fig. 1).

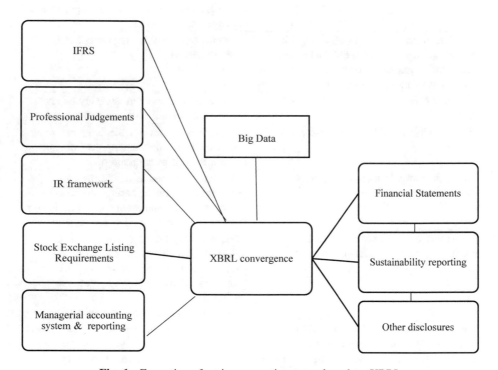

Fig. 1. Formation of various reporting types based on XBRL

According to Fig. 1 XBRL provides enhanced ability to customize reporting to meet the specific needs of information users such as investors and analysts. Since the goal of the XBRL format is the preparation of analytical materials, a single data warehouse allows to process large amounts of data with the necessary analytics.

XBRL reports are dimensional and for that they can be transformed and automated in the Multidimensional Data Model (MDM) so they can be analyzed by business users in the most effective ways [18].

On-Line Analytical Processing (OLAP) - multidimensional data analysis technologies that provide interactive work with information in real time and allow to aggregate, drill down data, change analytical sections, calculate derived indicators, and perform advanced graphical analysis [13]. OLAP tools give capacity to the user to analyze multidimensional data from multiple perspectives. All the OLAP tools are built upon three basic analytical operations:

- **Consolidation:** performs data aggregation that can be computed in many dimensions.
- **Drill down:** a contrasting technique to consolidation that allows users to navigate through data details in a reverse approach to consolidation.
- **Slicing and dicing:** a technique in which users take out (slice) a set of data called OLAP cube and then further dice the data cube (slice) from different viewpoints.

The OLAP analysis process is a set of analytical operations with multidimensional data: their detailing, consolidation, formation of slice and rotation. Consolidation operations represent a transition from a separate presentation of information to an aggregated one or more dimensions, and in the case of drill down, a large data set is divided into separate parts. A cube slice is formed by fixing the value (or values) of a particular dimension. The rotation operation consists in changing the position of the measurements - the axes of the cube. The "point of view" on the data changes upon completion of the rotation.

OLAP provides the user with an intuitive multi-dimensional data structure by collecting them into an OLAP cube or hypercube [13].

$$G = <D, F> \tag{1}$$

- a model of the logical multidimensional representation of data, characterized by two sets of parameters: indicators and measurements.

$$F = <f1, f2, \ldots, fn> \tag{2}$$

- indicators (measures) of the hypercube: each indicator has many values that quantitatively characterize the analyzed process.

$$D = <d1, d2, \ldots, dm> \tag{3}$$

- hypercube dimensions: each dimension is an ordered set of values of a certain type. Dimensions can be organized as an ordered hierarchical structure. Many dimensions form the faces of a hypercube.

An example of a cube based on which further analysis can be carried out is shown in Fig. 2.

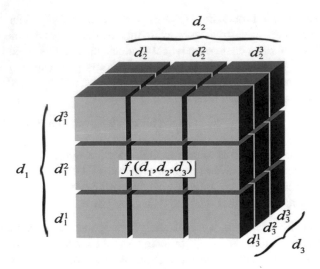

Fig. 2. OLAP cube example

Figure 2 shows the analytical capabilities of the technology. For stakeholders, the use of OLAP technology means high flexibility of solutions, as it allows to present the necessary information to the extent of aggregation or detail, as necessary for interested parties. This makes it possible to generate various types of reporting (financial, non-financial, managerial) on the basis of a single massive data, which ensures their compliance and mutual complementation. Thus, XBRL in combination with MDM technologies creates an opportunity to solve the most relevant and important problem at the present stage of development of accounting and reporting - the creation of a single enterprise accounting and reporting and a global network of reporting data that can easily be transformed into any reporting format depending on the needs of external and internal users.

Issues surrounding XBRL technology proliferation in Russia.

Example of Russia demonstrates that transition of companies' reports to XBRL is an important trend affecting not only the accounting functions, but also whole financial markets of a country. Commencing with January 1, 2017, following decision of the Bank of Russia, non-credit financial institutions began transition to a Unified chart of accounts (UCA) and industry accounting standards, accompanied by the introduction of XBRL.

It was expected that the introduction of digital technologies in the form of XBRL would result in – mutual understanding between organizations and the regulator, transition to a unified business reporting. Moreover, the preparation was expected to be "quick and easy". However, problems were encountered on the way that resulted in contraction of the financial market. An examination of these issues is important taking into account that in Russia insurance brokers, microfinance institutions should introduce XBRL in 2019. Then it will be the turn of pawnshops, credit consumer cooperatives, leasing companies. The study [interviews] of IT companies' specialists identified the following issues:

(a) the difficulty of acquiring the right IT solutions;
(b) lack of qualified professionals;
(c) absence of a methodology for transition to XBRL;
(d) continuous changes in IFRS.

Analysis of these issues demonstrated:

(a) Many companies postpone acquiring XBRL software, continue to input data manually, or acquire cheap software. Lacking necessary functionality, inexpensive software often does not work well. Due to absence of appropriate IT-specialists in some companies choosing proper IT-solutions is difficult. In addition, there are not many well performing and universal software products on the offer. Usually companies acquire an off-the-shelf product that requires significant adjustment to the company's business. Such tuning is expensive, but own experts are non-existent.
(b) Operating of XBRL requires not only IT specialists, but also specialists in IFRS, since industry accounting standards closely follow IFRS. The Unified Chart of accounts for non-credit financial institutions (Russia) is largely similar the chart of accounts for credit institutions. It is designed for detailed accounting of various operations and objects. The Unified Chart of accounts systematizes and prepares information for the use of XBRL. Accounting and reporting under IFRS requires significant volumes of raw data. This is especially true for first time reporting (preparation of initial report). The costs of information gathering, specialists training for IFRS, or outsourcing to third party experts, can be significant.
Usually there is a lack of expertise in IFRS in non-credit organizations implementing XBRL due to inadequate accounting education in Russia.
(c) Many companies do not develop the concept and methodology of their accounting system. The introduction of XBRL demands a new approach to the accounting system, - requires development of its own concept and methodology.
(d) Financial reporting rules are constantly changing, which require re-configuration of XBRL. Recent changes include significant transformation of accounting for financial instruments, revenue, leases that was connected with the adoption of IFRS 9, IFRS 15, IFRS 16. An accountant needs now more and more additional information to make a professional judgment in various situations. It is becoming increasingly difficult to decide on accounting treatment of a transaction or an object.
(e) various non-financial reporting standards are currently used worldwide. In Russia, as in the world, most public companies use GRI standards or the Conceptual framework of integrated reporting. The existence of some significant differences between these standards, as well as differences in the conceptual framework of financial and non-financial reporting make it necessary their integration on the basis of XBRL.

These issues need to be addressed. Moreover, XBRL capabilities already include some solutions to these problems. Previously, software products and technologies were chosen for accounting. Rapid development of information technology, the era of digitalization present a new stage of relations between accounting and information technology. Nowadays technologies, such as XBRL, set the trend for the development of accounting, pushing in the direction of fusion of different types of accounting in a unified reporting system [9].

Decision on Training
IFRS education pays little attention to digital technologies. Although certain IT courses are present in universities, they are meant for IT students. As a result, many companies that need to implement XBRL, neither have specialists who master both IFRS and IT (including digital technologies), nor have specialists of either of these groups.

Decision on Concept and Methodology
The concept of unified accounting based on XBRL involves the use of a multidimensional description of each operation and object. The methodology should include the possibility of adjusting disclosures to the requirements of key stakeholders, primarily regulators and investors. New requests should be processed and taken into account in a timely manner.

Decision on Change of Accounting and Direction of Its Future Development
Harmonization of financial and non-financial reporting standards is required. In addition, non-GAAP disclosure requirements are needed for measurements used by the organization in management accounting. XBRL already allows to do this, i.e. organize, process such information, get it in the form of the necessary reports. It is critical to ensure effective cooperation between development of accounting methodology (both financial and nonfinancial) and digital technologies.

4   Results

The results of our study are important because XBRL is been widely using for financial information be reported in a standardized and digital manner to regulators such as the SEC, the Committee of European Banking Supervisors (CEBS), The Central Bank of the Russian Federation [10]. At the same time, the analysis of the possibilities of its application for the formation of non-financial reporting and their integration has been studied to a much lesser extent.

The study shows that XBRL:

- brings improvements to sustainability reporting;
- plays a crucial role in facilitating progress toward more comprehensive sustainability reporting;
- gives opportunities to develop different fields of application in sustainability reporting, benefiting from the opportunities in data management based on standardized XBRL taxonomies [19];
- affects governance, transparency, data management (extension, mutual complement and linking data) and cost effectiveness in business reporting (XBRL, being already

established in mandatory financial reporting, offers opportunities to develop rigorous sustainability metrics that increase comparability and cost reduction).
- reduces information asymmetry [2]. Next to the cost benefits of not obtaining third-party assurance on the disclosure document;
- works as a concept incentive to enhance the financial reporting supply chain and to involve major stakeholders in the use of XBRL-based reporting both financial and non-financial [5].

We have shown that the composition of the information that fills the XBRL is determined by the rules: reporting under IFRS (or another system of financial accounting and reporting); additional requirements of the regulator, non-financial reporting (according to certain standards); management accounting and reporting rules (according to the standards established by the company).

We have identified the possibilities of XBRL for the formation of various types of reporting and formulated two main goals of filling XBRL with information - ensuring effective management of the company, its business processes, and high-quality satisfaction of information requests of stakeholders [21]. The joint achievement of these goals increases the image and value of the company, contributes to its sustainable development, creates an opportunity for the sustainable development of civilization.

We identified (formulated) the actual problems of XBRL technology distribution in Russia, summarizing the experience of its use in the financial sector under the guidance of the Central Bank of Russia (the issue of acquiring the right IT solutions; lack of specialists with the necessary qualifications; lack of a methodology for the transition to the implementation of XBRL; constant change of IFRS standards).

Based on a detailed analysis of these issues, we found that to solve them, it is necessary to focus on two areas - improving the training of accounting professionals in the field of digital technologies and harmonizing the concept and standards of financial and non-financial reporting, which is made possible through the use of XBRL technologies.

5   Conclusion

Our study, of course, does not cover all issues of using XBRL in the accounting area. We investigated only some problems of this topic and their solutions. In addition, our study has a number of limitations, which should include insufficient coverage of respondents, limited list of questions, lack of periodic representative sample surveys in the context of various groups involved in the field of accounting and reporting. However, we believe that, despite these limitations, our results will be useful for other studies, and will also contribute to the development of accounting in the context of digitalization.

We consider it necessary to develop XBRL application its existing advantages, as well as other aspects that will help in improving company's reporting system, and a deeper penetration of digital technologies in the accounting area, as well as the further development of the accounting sphere under the influence of digital technology. In addition to the described advantages of XBRL-based business reporting, other aspects can be identified that will help in developing company's reporting to become more holistic with regard to both content and managerial application. Such advantages arise from the combination of XBRL with Big Data and Business Intelligence technologies, which can significantly extend the benefits of structured integrated reporting.

References

1. Barker, R., Eccles, R.: Should FASB and IASB be responsible for setting standards for nonfinancial information? Green Paper. University of Oxford, Sand Business School (2018)
2. Blankespoor, E., Miller, B.P., White, H.D.: Initial evidence on the market impact of the XBRL mandate. Rev. Account. Stud. **19**, 1468–1503 (2014). https://pennstate.pure.elsevier.com/en/publications/initial-evidence-on-the-market-impact-of-the-xbrl-mandate
3. CFA Institute: Financial reporting disclosure. Investor Perspectives on Transparency, Trust, and Volume (2013)
4. CFA Institute: ESG_Survey_Report July_2017 (2017). https://www.cfainstitute.org/learning/future/Documents/ESG_Survey_Report_July_2017.pdf
5. Doolin, B., Troshani, I.: XBRL: a research note. Qual. Res. Account. Manag. **1**(2), 93–104 (2004)
6. Eccles, R.G., Krzus, M.P.: One report: integrated reporting for a sustainable strategy. Financ. Exec. **26**(2), 28–32 (2010)
7. Eccles, R.G., Armbrester, K.: Integrated reporting in the cloud. IESEinsight (2011). http://www.people.hbs.edu/reccles/Insight_Article_2011.pdf
8. Eccles, R.G., Kastrapeli M.: The Investing Enlightenment: How Principle and Pragmatism Can Create Sustainable Value through ESG, Environmental, Social and Governance (ESG) (2017)
9. Efimova, O., Rozhnova, O.: The corporate development in digital economy. In: Antipova, T., Rocha, A. (eds.) Advances in Intelligent Systems and Computing, vol. 850, pp. 71–81. Springer Nature, Switzerland AG (2019)
10. Getting started for regulators/XBRL International Inc. https://www.xbrl.org/the-standard/how/getting-started-for-regulators/. Accessed 29 May 2019
11. Gomaa, M.I., Markelevich, A., Shaw, L.: Introducing XBRL through a financial statement analysis project. J. Account. Educ. **29**(2–3), 153–173 (2011)
12. Girella, L.: The Boundaries in Financial and Non-Financial Reporting. A Comparative Analysis of their Constitutive Role. Routledge, London (2018). https://doi.org/10.4324/9780429504341
13. Gorodetskaya, O.Y.: The technologies for effective project management. Khronoe'konomika **4**(6), 19–23 (2017). Available in Russian
14. Kahn, R.N.: The future of investment management. The CFA Institute Research Foundation (2018). https://www.cfainstitute.org/en/about/press-releases/2018/future-of-investment-management
15. Kuter, M., Gurskaya, M., Andreenkova, A., Bagdasaryan, R.: The early practices of financial statements formation in Medieval Italy. Account. Historians J. **44**(2), 17–25 (2017)
16. Lubin, D., Esty, D.: Bridging the sustainability gap. MIT Sloan Manag. Rev. (2014). http://sloanreview.mit.edu/article/bridging-the-sustainability-gap/. Accessed 29 May 2019
17. Piechocki, M., Gräning, A., Kienegger, H.: XBRL as eXtensible Reporting Language for EU Reporting. In: New Dimensions of Business Reporting and XBRL. http://www.bookmetrix.com/detail/chapter/6c2a1a0a-b0a0-46c2-b913-5c5595e7975. Accessed 21 Mar 2019

18. Santos, I., Castro, E.: XBRL and the multidimensional data model. In: WEBIST 2011 - 7th International Conference on Web Information Systems and Technologies. https://www. scitepress.org/papers/2011/33990/33990.pdf. Accessed 05 Feb 2019
19. Seele, P.: Digitally unified reporting: how XBRL-based real-time transparency helps in combining integrated sustainability reporting and performance control. J. Cleaner Prod. **136**, 66–77 (2016). https://doi.org/10.1016/j.jclepro.2016.01.102
20. Solsbach, A., Isenmann, R., Gómez, J., Teuteberg, F.: Inter-organizational sustainability reporting – a harmonized XRBL approach based on GRI G4 XBRL and further guidelines. In: EnviroInfo 2014 – ICT for Energy Efficiency, Proceedings of the 28th International Conference on Informatics for Environmental, University of Oldenburg (2014). Accessed 21 Apr 2019
21. Zhu, H., Wu, H.: Quality of data standards: framework and illustration using XBRL taxonomy and instances. Electron. Mark. **21**(2), 129–139 (2011)
22. Watson, L.A., Monterio, B.J.: The next stage in the evolution of business reporting – the journey towards an interlinked, integrated report. Chartered Accountant, 75–78 (2011). http://220.227.161.86/23478july2011journal_75.pdf. Accessed 25 June 2019
23. WFE ESG Guidance and Metrics Revised June 2018. https://www.world-exchanges.org/ news/articles/world-federation-exchanges-publishes-revised-esg-guidance-metrics. Accessed 05 July 2019

Determinants of the Foreign Direct Investments in Terms of Digital Transformation of the Ukrainian Economy

Svitlana Tkalenko[1] , Natalya Sukurova[2](✉) ,
and Anastasiia Honcharova[1]

[1] Department of European Economics and Business, Kyiv National Economic University named after Vadym Hetman, pr. Peremogy 54/1, Kiev 03057, Ukraine
sv.tkalenko@gmail.com, goncharova.a.o@gmail.com
[2] Department for International Economic Relations, «KROK» University, Tabirna Street, 30-32, Kiev 03113, Ukraine
natalill@ukr.net

Abstract. This article is a research study of the role and necessity of digital transformation of Ukrainian economy in the context of digitalization of the world economy, which will promote economic growth and increase the international competitiveness of national economy, which in its turn will contribute to the further formation of an attractive investment environment. The modern tendencies of information economy development are disclosed. Changes of the determinants of FDI in the Ukrainian economy in terms of its digital transformation are revealed. It is proved that the increase of investment attractiveness will ensure a constant inflow of foreign capital, creation of new jobs, new technologies and products. Much attention in the article is paid to the role of information, development of information sphere, digitalization, formation of a new information economy, foreign direct investments, which act as a driving force in the development of a competitive national economy. The necessity of this study is due to the modern trends of development in the world economy and the digital transformation of the national economy. The role and importance of foreign direct investments in the digital transformation of the national economy has been proved on the basis of a broad analysis of research works of foreign and national specialists, analysis of economic situation in Ukraine and the main macroeconomic indicators identified the role and necessity of transformation processes, the world market of foreign direct investments have been analyzed, the necessity and requirements for the formation of an attractive investment environment, favorable investment climate and business environment have been identified. Using the econometric model, the indicators of GDP, exports and imports of goods and services, inflation have been analyzed in the context of their impact on the volume of FDI attraction, their forecasting have been made, and their influence on the formation of economic digitization have been proved.

Keywords: Digital transformation · Globalization · Information economy · Information technology · Foreign direct investments

© Springer Nature Switzerland AG 2020
T. Antipova and Á. Rocha (Eds.): DSIC 2019, AISC 1114, pp. 148–164, 2020.
https://doi.org/10.1007/978-3-030-37737-3_14

1 Introduction

The process of integration of Ukraine into the world community and the European integration vector of development of the national economy presupposes the formation of an attractive investment environment for attracting foreign direct investments (FDI). FDI stimulate economic growth, promote job creation (i.e. reduce unemployment), promote the development and implementation of innovative, digital technologies and provide further economic growth. Innovative and digital technologies will contribute to the formation of national advantages and enable to keep competitive positions on the global arena. The dynamic development of information and communication technologies, the scientific and technological revolution and the processes of globalization determine the role and the necessity of digital transformation of the Ukrainian economy.

The purpose of our research is to identify changes of the determinants of FDI in the national economy in the context of digital transformation as a general trend of further development of the world economy. Based on the goal, the objectives are to research the basic principles of the formation of digital economy and to identify its impact on the process of attracting of FDI into the national economy.

Theoretical and applied aspects of issues of digital transformation of economy and digitization were considered in the research works of famous Ukrainian scholars, such as, Dulska [1], Dzhusov [2], Kolyadenko [3], Meshko [4], et al. The analysis of research works prepared by these authors have shown that they have paid considerable attention to the formation of high-tech development in the context of globalization, to the role and importance of the digital economy at the current stage of development, to the necessity of the introduction of digital technologies, which will promote the realization of synergistic effect and economic growth.

Among foreign scholars, the foundations and development of the digital economy, digital society and global changes in the world are considered in the research works of Castells [5], Grimes [6], Niebel [7], Webster [8], et al.

Consequently, the digital world economy is of great importance, promoting the creation of new information and communication technologies, new products and shaping new market needs. Investment flows are mostly directed into new digital technologies. For Ukraine, this segment is not sufficiently developed, but it has huge potential and prospects for future development.

Therefore, the authors of this article prove the actuality and necessity of digital transformation of the Ukrainian economy with the purpose to increase rates of economic growth and national competitiveness on the world level. The role of foreign direct investments in the digital economy and the determinants of attracting foreign direct investments into the national economy are determined. Based on econometric modeling using E-Views program, the impact of macroeconomic factors on FDI attraction is determined.

2 Materials and Methods

XXI century is characterized by new qualitative features of globalization as a new phenomenon in the development of humanity, its economic, civil and political structures. Globalization represents by itself the emergence of a single economic and informational space in a planetary scale. Just globalization today reflects the new reality of the growth of global interdependence, which is mostly due to the new communication technologies. Consequently, the current stage of development of the economies of the countries of the world is characterized by an increase of the role and importance of information and knowledge that has become the dominant factors in the process of formation and use of intellectual capital, which in turn contributes to the creation of new high-tech and scientific products that will ensure the appropriate level of countries competitiveness on the world markets. The modern era is associated with an era of constant transformation and the transition of society from the outdated entropy-market system to a new highly organized economic system [9]. Such system will require the attraction of additional financing and foreign direct investments, which will inflow to the country's economy in case of creation of attractive and competitive environment.

In order to achieve the setted goal it is necessary to research the current stage of digitization; to reveal the modern trends of information economy development; identify and disclose factors of formation of a new information economy; to prove the role and necessity of FDI in terms of formation of information economy, in particular in Ukraine; using the econometric model and forecasting methods to show the influence of the most significant factors on the attraction of FDI into the economy of Ukraine, and to prove their influence on the formation of digitization of the Ukrainian economy.

Transformation of national economy assumes the transition·to an information society, which means that the level of socio-economic development is directly related to access to information resources. Thus, information becomes a means by which corporations prove that consumption is an essential and inevitable element of life [10]. This means that the growth of the role of information and the development of the information sphere has become an important condition for the spread of consumerism (the mass consumption of material goods and the influence of consumers on the producers of products). The modern process of production, distribution and use of knowledge forms is the basis of a new information economy, and the global information network is its infrastructure. The development of communications, the commercialization of knowledge, electronization, and the transformation of knowledge into the product is the basis of the modern stage of digital transformation. So, the digital economy is a new paradigm of world economic development, and in particular in the context of the digital transformation of the national economy.

It should be noted that in developed countries, digitalization takes a leading position in the policy of the governments. Just such an active policy of countries helps to stimulate business and their engagement in the processes of digitalization; business expansion will require the attraction of private investors, as well as their involvement in the creation of digital platforms. Thus, the digital economy promotes the development of world trade that displays through the speed of turnover and increase of efficiency.

The term "knowledge-based economy" by itself is used to refer to two concepts: first, the information economy is a modern stage in the development of civilization characterized by the primary role of creative labor force and information products; and secondly, the information economy is an economic theory of the information society [11]. The information economy focuses on the following aspects: the study of information asymmetry; economics of information products; economics of information technologies.

The development of information products and information technologies, while ensuring the rights and freedoms of citizens to use these technologies and access to information, are key factors in the effective transition to a new information economy. It should be noted that some countries can turn information technology into an engine of their development, move from an agrarian or industrial base to a new information economy, others may be far behind. Such a process of transition to a new economy is objective and requires the consciousness of society.

Bases for the formation of a new information economy are the following factors, which explain the rapid distribution of information, namely:

- factors directly related with the scale of multinational companies activity: by placing their brunches around the world, these companies cannot operate without communication and information infrastructure which provide their activities. In addition, information networks are vital not only for a particular company, they link to a single whole all the market agents, without which the global market cannot function. Therefore, the international financial networks in the information environment take a leading place;
- the results of information revolution, which were first used by transnational corporations (all the more so, because long-term judgments about the need of information networks contributed to this).

The formation of the information economy takes place against the backdrop of the main challenges of a rapidly changing world. The unique international challenges faced recently by the economies of the world, including the national economy, are significantly different from the problems that were inherent two or three decades ago. The scale and innovation of changes have led to widespread use of the terms "globalization" and "informatization" of society and the economy. Crisis phenomena at the turn of the XX-XXI century indicate on the need to revise the principles of consumption and production from the point of view of sustainable development of society and certain spheres of economy. Thus, the current crisis is a crisis of industrial-market paradigm, ideology, concept, model of development of the world economy, which should be replaced by a new innovation-synergistic information paradigm, on the model of socio-economic development and formation of knowledge-based economy [9]. In such a model of development, the main role is played by highly skilled professionals who are the bearers of intellectual capital, creative and innovative abilities, and are the main drivers in ensuring business competitiveness.

Economic theory has revised its own preconditions and tasks of theoretical knowledge, which is due to the achievements of humanity in the creation and implementation of information and telecommunication technologies. The question about the entrance of world civilization in the post-industrial era was the subject of discussion just in the early 1960's. This was due to the achievements of humanity in the creation and implementation of information and telecommunication technologies. Bell [12], Beck [13], Castells and Himanen [14], Schiller [15], Toffler [16], Wallerstein [17] and others stood at the roots of the theory of information and communication technologies. They describe in different ways in their research works the information economy and society, its economic principles, which are based on information and telecommunication technologies. In particular, American Daniel Bell, believed that just now the process of formation of information society is going on [18]. American Herbert Schiller recognizes the fact that today information technologies have a significant impact on economic development, but does not see a fundamentally new phenomenon. It is the phenomenon that reveals a new paradigm of the information economy, which is based on knowledge and information. British Peter Golding and Graham Murdoch have recognized the influence of information on society development and growth in the XXth century, and agree with its axial purpose for events in the world, but argue that information and communication are only the main components of the long-established and well-known capitalism formation [19]. So, such issues in the global information economy are also relevant today.

Consequently, today information is the driving force in the development of society, economy and business. Information affects human activity, the economy of the countries of the world, the surrounding world, becomes a giant, technical, socio-economic, political and cultural power. At the current stage of business development, the most competitive is that business entity which can produce more information of the best quality, introducing it into all spheres of life, using modern information technologies. In its turn, for Ukraine, the use of up-to-date information, modern technologies will mean attraction of foreign direct investments.

Undoubtedly the leading places in the modern information economy take big companies - transnational corporations. Just these corporations offer innovative products and services that are characterized by global innovation; and are the major source of FDI. The rapid development of such corporations and their expansion today has a significant impact on the economic development of other countries of the world, including Ukraine.

From the second half of the XXth century the movement of capital is taking place at a high rate. During the last fifty years, capital inflows have increased in 133 times (to $1.42 trillion in 2017), outflows of foreign direct investment have increased in 104 times (to $1.4 trillion in 2017) [20]. The FDI in stock, possible potential, namely the cost of capital share and reserves (including retained earnings) related to the parent company, as well as the net indebtedness of branches of parent companies, also significantly increased over the past thirty-five years: stock inflows have increased almost in 29 times and the stock outflows increased almost in 45 times.

However, it should be noted that from the middle of XXth century, the main volumes of inflow-outflows of foreign capital have just developed countries: they account for about 55% of capital imports, and 72% of capital exports. On the one hand, this is due to the development of the scientific and technological revolution, later with the formation of Western European integration, the liberalization of foreign economic relations, the activities of transnational corporations, the development of international financial markets; on the other hand, the main donors of capital are precisely economically developed countries, exactly in these countries there is the largest number of parent companies, which in its turn are owners of capital.

For Ukraine, the driving forces in ensuring the international competitiveness of the national economy are investments and technologies. The dynamics of FDI attraction is the basis of long-term development of the economy, the basis of the reproduction process, which will contribute to the expansion of high-tech forms of reproduction of fixed capital and the accumulation of highly intellectual capital. One of the driving forces of such fundamental changes in the Ukrainian economy is foreign direct investments. Figure 1 shows the total volume of attracted FDI in the economy of the country. Statistics shows that the growth in 2008 was due to the deployment of privatization processes; since 2014 there has been a slowdown in economic development and the decrease in volumes of FDI, due to the political and economic situation in the eastern part of the country.

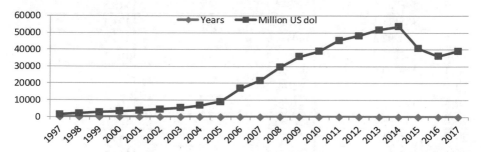

Fig. 1. Dynamics of foreign direct investments inflow into the Ukrainian economy from 1997 to 2017, million US doll.

Overall, Ukrainian economy is characterized by a low level of attraction of foreign direct investments, which is primarily due to the underdeveloped financial and credit system, unfavorable investment environment, constant changes in the tax code, and in the context of the crisis, due to the low level of savings of population in the investments. According to the data of European Business Association, the investment attractiveness index of Ukraine at the beginning of 2019 fell to the level of 2016 and amounted 2.85 (out of 5), and is in the negative plane (i.e. below 3.0 points).

At the same time, positive factors should include the stability of the national currency, the liberalization of currency legislation, the reduction of inflation since 2015, and other political factors. The main macroeconomic indicators, GDP, exports and imports, and inflation, which have the greatest impact on FDI attraction, are given in Table 1 [21].

Table 1. Indicators of economic development of Ukraine

Period	GDP, million US doll	EI (export and import), million US doll	I (inflation), million US doll
1997	50150.4	37516.6	106.4
1998	41883.2	32563.7	110.0
1999	31581.0	28154.8	122.1
2000	32375.3	33334.6	127.8
2001	39309.6	36919.3	112.3
2002	43956.4	40432.6	101.0
2003	51331.4	52077.1	105.0
2004	67217.6	69338.5	108.9
2005	89239.4	79749.7	113.6
2006	111884.8	94929.0	108.2
2007	148733.9	124344.7	110.3
2008	188111.1	17124.2	121.4
2009	122992.5	100444.9	114.9
2010	141209.9	129505.3	109.1
2011	169333.0	169926.7	104.6
2012	182592.5	174378.4	99.8
2013	190498.8	161926.7	100.5
2014	133503.4	128936.7	124.9
2015	91031.0	92687.3	143.3
2016	93356.0	90674.6	112.4
2017	112190.4	108216.0	113.7

This table presents the statistical data about basic macroeconomic indicators of economic development of Ukraine, which have the greatest influence on attracting FDI.

Presented economic indicators from 2014 due to the political and economic situation in Ukraine deteriorated.

3 Results

Ukraine in the period of the digital transformation of the national economy needs to attract significant volumes of foreign direct investments. At the same time, it is necessary to create the most attractive environment for their attraction and effective use. Our research, based on many of the analyzed macroeconomic indicators for the period 1997–2017, made it possible to highlight those that have the greatest impact on the investments movement, in particular GDP, foreign trade turnover (exports and imports) and inflation rate (Table 1). Let's examine their impact on FDI, as well as our hypothesis regarding their significance and important role for FDI. The results of our study made it possible to explain the relationship between the selected variables using the E-Views-7 package.

The statistical data for the analysis in the model include 21 observations (1997–2017). A general view of the FDI model from the main variables chosen by us can be described by the following equation:

$$FDI = f\,(GDP,\,EI,I),$$

GDP - GDP of Ukraine, millions US doll; EI - amount of export and import, millions US doll; I – inflation, %.

To explain the relationship between selected variables we construct the correlation matrix (see Table 2).

The indicated matrix of correlation coefficients shows a strong correlation between FDI and GDP, FDI and the volume of exports and imports - more than 80%. An acceptable correlation result between variables actually confirms that our model is successful. Results of multi-factor regression are given in Table 3.

R2 shows on how much selected variables and their quantities explain the volume of foreign direct investments in Ukraine, i.e. the selected variables explain on 86.8% the attracted volumes of FDI. The presence of a sufficiently strong correlation indicates a correlation coefficient, which is 0.84; F-statistic = 0.000000, the probability of accepting the null hypothesis, confirms the need to take an alternative hypothesis, which certifies the significance of the equation as a whole.

Table 2. Correlation matrix of selected variables

	FDI	GDP	EI	I
FDI	1	0.83229	0.82496	0.060595
GDP	0.83229	1	0.75414	−0.186899
EI	0.82496	0.75414	1	−0.27980
I	0.06059	−0.18689	−0.27980	1

Source: authors development
This table presents a correlation matrix that explains the relationship between selected variables and shows their impact on foreign direct investments. Matrix constructed by us confirms the success of the model.

The model is verified on explanatory capability, since it should accurately reflect FDI inflows using available independent variables (see Fig. 2). As can be seen from the graph, the fitted values (Fitted) have the same trends with actual values (Actual), so the model is fully acceptable under this criterion.

At the same time, we consider that more precise results allows us to obtain the vector model of autoregression (VAR-model), which we disclose through a set of dynamic time series in which current values depend on these values of time series in past periods and which are constructed according to a stationary time series.

Table 3. The results of multi-factor regression

Dependent variable: FDI. Method: least squares				
Included observations: 21				
Variable	Coefficient	Std. Error	t-Statistic	Prob.
GDP	0.165302	0.047207	3.501621	0.0027
EI	0.215163	0.053247	4.040849	0.0008
I	558.8983	168.9684	3.307709	0.0042
C	−74670.82	20204.14	−3.695818	0.0018
R-squared	0.867951	Mean dependent var		23671.19
Adjusted R-squared	0.844649	S.D. dependent var		19118.55
S.E. of regression	7535.510	Akaike info criterion		20.86228
Sum squared resid	9.65E + 08	Schwarz criterion		21.06124
Log likelihood	−215.0540	Hannan-Quinn criter.		20.90546
F-statistic	37.24677	Durbin-Watson stat		1.031405
Prob (F-statistic)	0.000000			

Source: authors' development
This table shows the results of multi-factor regression using the least squares method and different coefficients, which as a whole prove the significance of the equation.

To research the cause-effect dependence between selected variables and FDI, Granger's test was used. With the help of the Granger test, the zero hypothesis is tested A is not the reason of changes B. The criterion for accepting the hypothesis is probability; if the value of probability (Prob.) is less than 0.05, then the null hypothesis is not accepted.

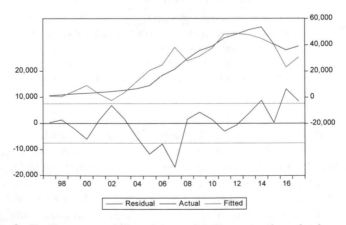

Fig. 2. Explanatory capability of the model (*Source: authors development*)

Tests were conducted for lags 2, 3, 4. The results of the test are given in Table 4.

Table 4. The results of Pairwise Granger Causality Test on cause-effect dependence for the period 1997–2017

	Null hypothesis:	Pairwise Granger Causality Tests						
		Lags: 2			Lags: 3		Lags: 4	
		F-Statistic	Prob.	Conclusion	F-Statistic	Prob.	F-Statistic	Prob.
1	GDP does not Granger Cause FDI	6.76318	0.0088	Reject	5.37611	0.0160	2.11596	0.1704
	FDI does not Granger Cause GDP	0.13880	0.8716	Accept	0.28524	0.8351	2.90721	0.0930
2	EI does not Granger Cause FDI	0.49116	0.6221	Accept	0.61124	0.6216	1.87727	0.2079
	FDI does not Granger Cause EI	2.40086	0.1269	Accept	3.01012	0.0763	2.07168	0.1767
3	I does not Granger Cause FDI	2.13300	0.1554	Accept	1.86684	0.1936	2.33051	0.1434
	FDI does not Granger Cause I	0.11594	0.8914	Accept	1.78669	0.2078	1.63345	0.2569
4	EI does not Granger Cause GDP	12.5795	0.0007	Reject	8.11319	0.0039	4.80195	0.0286
	GDP does not Granger Cause EI	1.23916	0.3195	Accept	1.31604	0.3185	1.25114	0.3637
5	I does not Granger Cause GDP	1.23260	0.3213	Accept	0.74259	0.5487	0.38899	0.8111
	GDP does not Granger Cause I	0.74241	0.4938	Accept	0.33563	0.8000	0.52855	0.7187
6	I does not Granger Cause EI	0.62307	0.5505	Accept	0.47612	0.7053	0.65277	0.6411
	EI does not Granger Cause	0.17098	0.8446	Accept	0.94843	0.4507	1.94775	0.1959

Source: authors development
This table shows the results of Pairwise Granger Causality Test on cause-effect dependence with 2, 3 and 4 lags, which show the effect of one factor on another in our model.

Thus, according to the results of the Granger test, we may do a number of general conclusions: the greatest influence on FDI has GDP, that is, the more on the national market products are produced, which confirm the growth the more FDI will be obtained. Cause-effect dependence from other variables has a manifestation with delay – lag 3, 4.

The mean, standard deviation, skewness and kurtosis, Jarque-Bera statistics, and the probability of hypothesis acceptance were examined (see Table 5).

The largest standard deviations have all indicators except inflation, which confirms the general tendency for their variability, changes, both from external and internal factors, which influence in general on the economic development of the country. The asymmetry coefficients in the model are significant. For symmetric distributions,

Table 5. Descriptive analysis of variables

	FDI	GDP	EI	I
Mean	23671.19	101546.7	85865.78	112.8667
Median	21607.30	93356.00	90674.60	110.3000
Maximum	53704.00	190498.8	174378.4	143.3000
Minimum	1438.200	31581.00	17124.20	99.80000
Std. Dev.	19118.55	54392.50	49343.90	10.39473
Skewness	0.165640	0.249685	0.333510	1.245516
Kurtosis	1.418033	1.751547	1.950905	4.608193
Jarque-Bera	2.285821	1.582006	1.352325	7.692590
Probability	0.318890	0.453390	0.508565	0.021359
Sum	497094.9	2132482.	1803181.	2370.200
Sum Sq. Dev.	7.31E + 09	5.92E + 10	4.87E + 10	2161.007
Observations	21	21	21	21

Source: authors development
This table shows the parameters that characterize the quantitative criteria of the chosen indicators and prove the success of this model.

kurtosis indicator is calculated that shows how sharp our variables are. In our model, kurtosis indicates a sharp vertex distribution and the jump is considered to be significant. The null hypothesis of all variables except inflation is normally distributed at 5%. The standard deviation, compared to the mean, is low, indicating a small coefficient of variation. Graphical representation of actual variable data in VAR model is shown in Fig. 3.

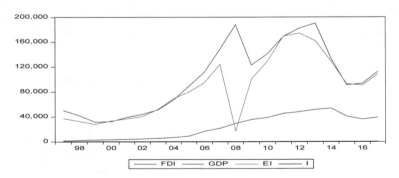

Fig. 3. Graphical representation of actual variable data in VAR model for the period 1997–2017 (*Source: authors development*)

Further, in VAR model, we firstly check the stationarity of all the variables we examine, and secondly, the temporality of shocks, how they are scattered and returned to the medium or long-term time series (see Table 6). In our case, we accept the null hypothesis, and the series is stationary.

Table 6. Verification of variables on stationarity by Dickie-Fuller test

Null hypothesis: unit root (individual unit root process)							
Series: FDI, GDP, EI, I. Sample: 1997 2017							
Automatic lag length selection based on AIC: 0 to 1							
Method				Statistic		Prob.**	
Im, Pesaran and Shin W-stat				5.52194		0.0000	
** Probabilities are computed assuming asymptotic normality							
Intermediate ADF test results							
Series	t-Stat	Prob.	E(t)	E(Var)	Lag	Max Lag	Obs
D(FDI)	−2.5636	0.1174	−1.520	0.865	0	4	19
D(GDP)	−3.4957	0.0199	−1.520	0.865	0	4	19
D(EI)	−5.2208	0.0005	−1.520	0.865	0	4	19
D(I)	−5.1955	0.0007	−1.511	0.953	1	4	18
Average	−4.1189		−1.518	0.887			

Source: authors development
In this table, authors tested variables on stationarity by using the Dickie-Fuller test based on the null hypothesis. Verification is made for all studied variables.

Next step is to test VAR-model on autoregression, which means the assessment of series value with the past values of the same series. The autoregression results are shown in Table 7.

Table 7. Results of coefficients estimation in VAR model for the period 1997–2017

Vector autoregression estimates				
R-squared	0.994938	0.949351	0.721166	0.522552
Adj. R-squared	0.990888	0.908831	0.498098	0.140593
Sum sq. resids	31627320	2.65E + 09	1.20E + 10	1005.691
S.E. equation	1778.407	16278.82	34615.86	10.02841
F-statistic	245.6895	23.42944	3.232947	1.368084
Log likelihood	−163.0482	−205.1170	−219.4515	−64.66525
Akaike AIC	18.11034	22.53863	24.04753	7.754237
Schwarz SC	18.55770	22.98600	24.49489	8.201602
Mean dependent	25978.58	107392.0	91215.85	113.3579
S.D. dependent	18630.96	53913.71	48861.38	10.81765
Determinant resid covariance (dof adj.)	2.58E + 25			
Determinant resid covariance	1.98E + 24			
Log likelihood	−639.3074			
Akaike information criterion	71.08499			
Schwarz criterion	72.87445			

Source: authors development
This table shows verification of VAR model on autoregression, which means the assessment of series value with the past values of the same series, i.e., all coefficients are evaluated - they have a high value, proving the success of the model.

Further, we analyze endogenous variables (see Table 8). Since VAR model describes the evolution of endogenous variables, it should be noted that in our model with lag 2, there is a delay of all selected criteria, and such analysis once again confirms the stationarity of all variables.

Table 8. Analysis of endogenous variables in VAR model for the period 1997–2017

VAR lag order selection criteria					
Endogenous variables: FDI GDP EI I. Included observations: 19					
Lag	Log L	LR	FPE	AIC	SC
0	−723.9506	NA	2.23e + 28	76.62638	76.82521
1	−669.0658	80.88287	3.92e + 26	72.53324	73.52739
2	−639.3074	31.32464*	1.22e + 26*	71.08499*	72.87445*
* indicates lag order selected by the criterion					
LR: sequential modified LR test statistic (each test at 5% level)					
FPE: Final prediction error					
AIC: Akaike information criterion					
SC: Schwarz information criterion					
HQ: Hannan-Quinn information criterion					

Source: authors development
This table shows the analysis of endogenous variables, which confirms the stationarity of all variables.

Further, we identify endogehous and exogenous variables in VAR - model. The test results are shown in Table 9. Test shows the probable exogeneity of such variables as foreign trade turnover and inflation, and the probable endogeneity of variables FDI and GDP. This means that GDP and FDI are most influenced by internal factors, such as, for example, creation of an attractive internal environment with the favorable tax code for business that will lead to increase of national production and, in general, the growth of national economy.

Table 9. Identification of endogenous and exogenous variables in VAR model

VAR Granger Causality/block exogeneity wald tests Sample: 1997 2017. Included observations: 19	Prob.
Dependent variable: FDI, All	0,0000
Dependent variable: GDP, All	0,0000
Dependent variable: EI, All	0,2991
Dependent variable: I, All	0,6979

Source: authors development
This table presents the results of identification of endogenous and exogenous variables in VAR model.

Next step in the VAR - model was to determine the impulse response function, where the impulse is a one-time indignation which is given to one of the parameters. This function describes the reaction of dynamic series in response to some external shocks. The quantity of 10, 20 and 50 periods was studied. Data analysis showed a big long-term remoteness - the response of selected variables, including GDP, on the positive shock of attracting investments indicates that there is no long-term relationship. Exports and imports, inflation, on the contrary, testify the availability of long-term connection with FDI attraction. The shocks of these variables directly affect FDI attraction: the higher the inflation, the less investments inflow in the country; and the smaller the foreign trade turnover (i.e. in the context of export and import decline), the less foreign investments will be obtained. Results of autocorrelation residuals indicate their absence (see Table 10). The probability for almost all lags (12) is zero.

Table 10. Results of autocorrelation residuals in VAR-model

VAR residual portmanteau tests for autocorrelations					
Null hypothesis: no residual autocorrelations up to lag h					
Lags	Q-Stat	Prob.	Adj Q-Stat	Prob.	df
1	11.68836	NA*	12.33771	NA*	NA*
2	25.92512	NA*	28.24939	NA*	NA*
3	44.52869	0.0002	50.34112	0.0000	16
4	57.50755	0.0037	66.78102	0.0003	32
5	66.37912	0.0405	78.82100	0.0033	48
6	83.58480	0.0507	103.9678	0.0012	64
7	95.12020	0.1191	122.2322	0.0017	80
8	104.1265	0.2681	137.7884	0.0034	96
9	113.0753	0.4538	154.7911	0.0046	112
10	118.7919	0.7081	166.8596	0.0119	128
11	123.3074	0.8931	177.5838	0.0299	144
12	129.0498	0.9655	193.1705	0.0378	160
*The test is valid only for lags larger than the VAR lag order.					
df is degrees of freedom for (approximate) chi-square distribution					

Source: authors development
This table presents the results of autocorrelation residuals in VAR-model, which confirms their absence and success of constructed model.

The estimation of dispersion decomposition for VAR - model variables over 10 periods in graphical representation is shown on Fig. 4. The analysis of graphs shows the dependence of FDI from the selected variables. It should be noted that GDP has a greater impact on FDI than vice versa.

Based on estimated VAR model, we make forecast of researched variables till 2022. The graphical representation of forecasting is shown on Fig. 5.

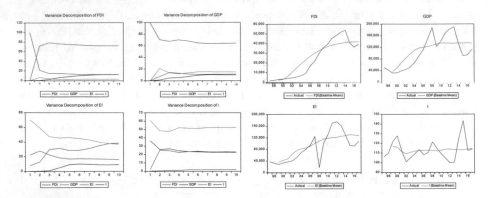

Fig. 4. The estimation of dispersion decomposition for VAR - model variables over 10 periods (*Source: authors development*)

Fig. 5. Forecasted values of indicators included in VAR – model

Therefore, this VAR - model has the following form shown in Table 11.

Table 11. VAR - model - substituted coefficients

FDI =	0.299319096824 * FDI(−1) + 0.475096727832 * FDI (−2) + 0.213637448867*GDP(−1) − 0.0646555045327 * GDP (−2) + 0.0275668085137 * EI(−1) − 0.0959427637655 * EI (−2) − 43.6456676686 * I(−1) + 76.423802117 * I(−2) − 5344.26578357
GDP =	−2.28000492827 * FDI(−1) + 1.20486913908 * FDI (−2) + 1.37523015198 * GDP(−1) − 0.307449226152 * GDP (−2) + 0.823663601355 * EI(−1) − 0.53943156713 * EI (−2) + 716.887988411 * I(−1) + 596.07916614 * I(−2) − 150688.521397
EI =	6.16311879572 * FDI(−1) − 1.17622248497 * FDI (−2) + 0.543526083388 * GDP(−1) − 0.99973960888 * GDP (−2) + 0.0099229252134 * EI(−1) − 0.722383301616 * EI (−2) − 249.031343009 * I(−1) − 623.168589389 * I(−2) + 172912.644517
I =	0.000880116065046 * FDI(−1) − 0.00107400128189 * FDI (−2) − 0.000311105290488 * GDP(−1) + 0.000214597892738 * GDP (−2) − 5.12834128857e − 05 * EI(−1) + 0.000227237642828 * EI (−2) + 0.231604688775*I(−1) − 0.185065455398*I(−2) + 106.51056203

Source: authors development
This table presents equations of this constructed VAR-model (for all selected variables).

The construction of this model is solved using the stochastic method. Thus, within existing factors of economic development, FDI have a slow upward trend. GDP and foreign trade turnover have also grown very slowly over the last ten years and, moreover, there is a probability of a fall in their value under the influence of unfavorable factors. Inflation has more equivalent probability of both increase and decrease.

4 Conclusion

The research of this problem made it possible to disclose the essence and irreversibility of the process of formation of information economy. There is a transformation of national economies, which characterizes the current stage of civilization development, the scales of TNCs activity are increasing, that lead to the growth of FDI movement, and the information acts as a driving force for the development of countries economies. For the Ukrainian economy, digitalization has advantages; in particular with the help of the main subjects of TNCs, the domestic economy will have technologies, innovations and will lead to economic growth.

Directly in this study, authors analyzed the determinants of foreign direct investments that are necessary for the digital transformation of Ukrainian economy. Attraction of foreign investments was considered in the context of improving the investment climate and investment environment, which as a rule are directed on innovative technologies and on the process of creating new products. The formation of an investment attractive national economy will show new opportunities for economic growth.

The authors paid special attention to forecasting of FDI inflows in the national economy, which showed slow growth trends that could be accelerated by the digital transformation of the country's economy. This will mean more efficient use of intellectual capital; promoting the creation of new high-tech, innovative and scientific products; the creation of digital platforms that will lead to the increase of volumes of trade due to the speed of turnovers The study confirms that foreign direct investments, as a rule, come from transnational companies, which are leading players on the capital market. Their rapid development and expansion at the end of the last century had a significant impact on the economies of other countries as well as on the economy of Ukraine. Over the last twenty years FDI in Ukraine's economy have increased almost in three times which have inflow mainly through transnational corporations.

In order to increase the rates of economic growth, it is necessary first of all to assist the digital transformation of the national economy and the creation of an attractive investment environment, which will create the basis for further increase of FDI volumes.

Research prospects should focus on support of innovative companies and on the readiness to implement innovate, high-tech, competitive products, create and promote digital platforms, primarily for business. At the same time, further steps are studies that disclose such important questions as: how should be supported companies that invest in innovative, high-tech products and what incentives should be provided to them? What legislative support will be realized in the context of the digital transformation of the national economy? What steps should be taken first to implement digital transformation reforms?

References

1. Dulska, I.V.: Digital technologies as a catalyst for economic growth. Econ. Forecast. **2**, 119–133 (2015). [in Ukrainian]
2. Dzhusov, A.O.: The digital economy: structural shifts in the international capital market, no. 9 (2016). [in Ukrainian]. http://journals.iir.kiev.ua/index.php/ec_n/article/view/3058/2746
3. Kolyadenko, S.V.: Digital economy: preconditions and stages of formation in Ukraine and in the world. Manag. Topical Issues Sci. Pract. **6**, 105–110 (2016). [in Ukrainian]
4. Mesko, N.P.: Strategies for high-tech development in the context of globalization: national and corporate aspects: Monograph. South-East, Donetsk (2012). [in Ukrainian]
5. Castells, M.: End of Millennium, The Information Age: Economy, Society and Culture, vol. III. Wiley-Blackwell, Cambridge (2010)
6. Grimes, A., Ren, C., Stevens, P.: The need for speed: impacts of internet connectivity on firm productivity. J. Prod. Anal. **37**(2), 187–201 (2012)
7. Niebel, T.: ICT and Economic Growth: Comparing Developing, Emerging and Developed Countries. ZEW Discussion Paper 14–117, Germany (2014)
8. Webster, F.: Theories of the Information Society, 3rd edn, USA and Canada (2006)
9. The state and the market: mechanisms and methods of regulation in the transition to innovative development: a collective monograph, St. Petersburg (2010). [in Russian]
10. Tkalenko, S.I.: Necessity of the trunk of the new international economic order for information economy. International Business in Information Suspension: Collective Monograph, Kyiv (2012). [in Ukrainian]
11. Korneychuk, B.V.: Information Economy, St. Petersburg (2010). [in Russian]
12. Bell, D., Inozemtsev, V.L.: The Age of Disunity: Reflections on the World of the 21st Century, Moscow (2007). [in Russian]
13. Beck, W.: Cosmopolitan worldview, Moscow (2008). [in Russian]
14. Castells, M., Himanen, P.: The Information Society and the Welfare State: The Finnish Model. Oxford University Press, Oxford (2004)
15. Gerbert, I., Schiller, G.: Information Inequality: The Deepening Social Crisis in America. Routledge, New York (1996)
16. Toffler, E., Toffler, X.: Revolutionary Wealth, Moscow (2008). [in Russian]
17. Wallerstein, I.: The end of the familiar world. Sociology of the XXI century, Moscow (2003). [in Russian]
18. Bell, D.: The coming post-industrial society. Experience of social forecasting, Moscow (2004). [in Russian]
19. Webster, F.: Theories of the Information Society, Moscow (2004). [in Russian]
20. Website Unctad Stat. http://unctadstat.unctad.org/wds/TableViewer/tableView.aspx
21. Website State Statistics Committee of Ukraine. www.ukrstat.gov.ua/

What Does ICT Mean for Tourism Export Development?

Aleksandr Gudkov$^{(\boxtimes)}$ ⓘ and Elena Dedkova ⓘ

Orel State University, Orel 302020, Russia
sashaworld777@gmail.com

Abstract. Tourism is the industry that currently has the widest boundaries, bringing together participants from different countries and destinations to provide tourist services that best meet the needs of tourists. All this would be practically impossible without the use of information and communication technologies (ICT). In the modern digital era, ICTs play an important role in the tourism, travel and hospitality industry. ICT makes it easier for enterprises and people to access information about tourism products from anywhere in the world at any time. The purpose of this article is to consider the challenges and opportunities of the ICT in the tourism industry from the standpoint of increasing its export potential in digital age and ensuring a high level of attractiveness of tourist destinations. For the purposes of the study, we used scientific analysis, systematization and literature review. We primarily analyzed the role of ICT in tourism industry for: (1) the service providers (airlines, hotels, catering, etc.), the travel firms, and other intermediaries, (2) the final consumers, and (3) the countries.

Keywords: Tourism · ICT · Export

1 Introduction

The existence of international tourism is impossible without the export of various types of services. Export allows industries to develop in conditions of internal limitations and degradation, forming a competitive environment based on the interaction of international market participants [5, 9, 11].

The Russian government is making a huge effort to improve the country's tourism infrastructure and to help the Russian regions in their movement toward sustainable tourism. The Concept of the Federal Target Program "Development of domestic and incoming tourism in the Russian Federation (2019–2025)" addresses the still remaining shortages and gaps in infrastructure, technologies and quality of services. It is planned to give priority to the cluster approach in tourism and to boost five priority types of tourism, namely, cultural, health, active, cruise and ecotourism. Other key objectives set in the Concept for the next few years are as follows: ICT infrastructure development, advanced training programs for the tourism industry, the promotion of Russia as a tourist destination and plans for subsidizing entrepreneurial and public initiatives in regional tourism [23].

© Springer Nature Switzerland AG 2020
T. Antipova and Á. Rocha (Eds.): DSIC 2019, AISC 1114, pp. 165–174, 2020.
https://doi.org/10.1007/978-3-030-37737-3_15

One of five major objectives stated in the concept of tourism development in Russia is "Development of the ICT infrastructure for tourism industry management". Because digitalization and use of ICT became an important integral part of contemporary tourism, a number of projects have been initiated, among them:

- development of a unified information system "Electronic tour";
- creation of an information system that provides, on a reciprocal basis, visa-free group tours;
- creation of a pan-Russian database of tourist facilities, tourism industry organizations and tourism specialists; and
- creation of a database of activities related to the use of foreign tourist vessels [24].

In this regard, there is an objective need to consider tourism as the most promising direction for the development of Russian exports, taking into account the possibilities of increasing its competitiveness in world markets due to the formation of an effective modern system of information support for tourism processes.

2   Analysis of the Recent Research and Publications

Given the importance of tourism as a major industry and its significance for a wide range of other activity sectors P. Gouveia analyzes and compares the tourism exports cycles from all regions of the world [7]. In times of crisis tourism becomes of prominent interest in public and political debates; however, it decreases drastically during years of economic growth [16].

ICT is helping us develop sharing economy in recent years which is becoming a new paradigm of the global economy [20]. V. Katsoni and M. Sheresheva on example of Greece discussed the opportunities and threats for new business models of sharing economy in hospitality and tourism, as well as the need for existing market players to adapt for new conditions and to improve legal framework and mechanisms for their implementation [18]. The contemporary sharing economy is enabled by ICT-based platforms, making sharing heavily dependent on ICT [17]. By providing a technological infrastructure for sharing, ICT has shaped new forms of social connections that can be upscaled to social relationships reliable enough to perform economic activities [22].

Smartphones have become an integral part of everyday life, and more people are acquiring the latest technology. With the mounting ownership of multiple technology devices, and the contribution of mobile travel bookings to the overall travel market, smartphones and smart tourism are now powerful tools for tourists, owing to the mobile applications that make travelling easier [1, 25]. Nowadays creations and promotion of mobile applications for tourist attractions in various cities significantly increases the awareness of tourists and improves perception. J. Im and M. Hancer in their study enhances the understanding of travel mobile application usage behavior by investigating interrelationship of utilitarian/hedonic motivation and self-identity on attitude toward using travel mobile application [12, 13]. The tourism mobile applications currently can be said as the most proficient applications that can facilitate the explorers' movement of travelers. The tourism mobile application deals not only the descriptive text of the information provided, but shows pictures of hotspots. In addition to that, tourists will be able to check for the facilities available surrounding certain places like entertainment and restaurants, hotels and others [10].

S. Ivanov et al. provided a comprehensive review of research on robotics in travel, tourism and hospitality. The incorporation of robotics came relatively late to the industries involved in travel, tourism and hospitality, probably since many of the services provided require sophisticated reactions to the needs of the customer. However, modern ICT technologies make it possible to adapt robots even for such industries, and intelligent software can cover a significant number of different operations [14]. Of course, not all service processes can and have to be automated or performed by robots – at the end of the day it is the economic efficiency, customer experience, company's competitiveness and other factors that will determine whether to automate and robotize [15].

The role of social media in tourism is particularly significant and the impacts of social media use by tourists, destinations and tourism providers are manifold. The high involvement, high risk and often international aspects of tourism consumption that leave tourists with very little opportunity to legally or politically voice their opinions makes social media activism a particularly important topic in tourism, yet systematic studies on its forms, reach, emergence, longevity, success rates, or results are currently lacking from the literature [8]. B. Dedeoğlu examines the impact of social media sharing on the recognition of a travel brand and the quality of a destination and the quality of service, and explores the role of a moderator of a country of origin as a destination and quality of service in Alanya, Turkey [4]. Interaction with interested parties in social networks is also a form of corporate social responsibility of companies. CSR practices are considered one of the key success factors influencing firm performance. According to the results of the research of M. R. González-Rodríguez, organizational culture influences different dimensions of CSR. Through improved reputation, CSR practices positively influence firm performance [6].

3 Goal of the Research

To reduce the dependence of the economy on domestic consumption, it is necessary to significantly increase its external component – export. Tourism is one of the leading and steadily developing industries in the modern world. However, the development of its export potential is impossible without the formation of ICT infrastructure. Thus, in the framework of this study, three hypotheses are formed: ICT forms tourism export and provides participants access to any information; ICT expands the boundaries of tourism and makes it maximally accessible to the majority of the world's population; ICT allows for the implementation of various processes and services without the participation of human resources, which affects responsiveness, consistency and quality.

4 Methodology

The authors used such methods as literature review, synthesis, generalization, theoretical modeling, analysis, graphical construction, formalization, forecasting. The positions of the research are argued also with the help of private scientific methods of research (formal, comparative, functional, concretization, etc.).

5 Results

Tourism involves the interaction of various governments, businesses and consumers from around the world. There can be thousands of kilometers between them, but ICTs act as a connecting link, allowing not only to ensure interconnection, but also to make it convenient, cheap and often without human support [2].

Figure 1 shows the basic relationships between governments, businesses and consumers in the sphere of tourism and travel.

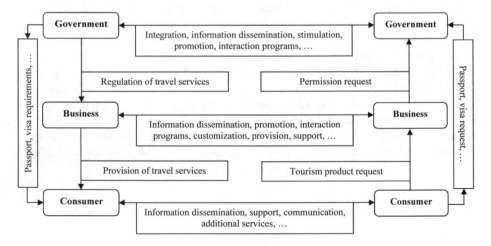

Fig. 1. Basic relationships between governments, businesses and consumers in tourism.

Relying on the fact that tourism covers participants from different countries and the provision of tourist services often requires the interaction of various businesses and governments it becomes obvious that without ICT the existence of such a complex system is impossible. Nowadays Internet is the most used means of finding the necessary information. And information search is one of the most important activities in consumer's consumption of products [2]. In tourism, we observe a large number of sites that help tourists learn about the visa policies, purchase tickets, book hotels, plan a trip, rent a vehicle, read and watch reviews and blogs of other tourists, and so on.

In Russia the share of group trips is still high [9]. Tourist product – a set of transportation and accommodation services provided for the total price (regardless of the cost of excursion services and (or) other services included in the total price) under the agreement on the sale of a tourist product. The interaction of the main participants of the tourist services market is presented in Fig. 2.

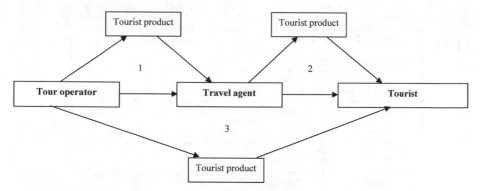

1 - contractual relationship between the tour operator and the travel agent (usually through the use of ICT);
2 - contractual relationship between the travel agent and the tourist (sometimes through the use of ICT);
3 - contractual relationship between the tour operator and the tourist (usually through the use of ICT).

Fig. 2. Scheme of interaction of market participants in tourism services.

The tour operator is the most important subject of the market of tourist services, which, usually through interaction with travel agents, sells the tourism product to the final consumer. At the same time, the tourist product is formed by the tour operator from the services provided by various objects of the tourism industry (Fig. 3). After the tourist product has been formed, the tour operator proceeds to the following steps that relate to the issues of bringing the created tourist product to the final consumer: promotion of the tourist product (a set of measures aimed at selling the tourist product – advertising, including participation in specialized exhibitions, fairs, the organization of tourist information centers, the organization of seminars, study tours, the publication of catalogs, booklets, etc.), the sale of a tourist product (activities related to the conclusion of an agreement on the implementation of the tourism product and activities of a tour operator or a third party for the provision of tourist services in accordance with this agreement).

As we see the tour operator and travel agent are mutually complementary subjects of the market, while the activities of the travel agent are impossible without interaction with the tour operator, but not vice versa. The interaction of these elements almost always occurs in online reservation and support systems. This allows to quickly monitor the processes of acquiring tours, form tourist groups by directions, control free places, change pricing, maintain reporting and so on [2].

ICT has extensively altered the role of each player in the value-creation process on the tourism and hospitality market. Evidence designates that an effective application of information technology has turned out critical for the attractiveness and prosperity of tourism enterprises, since it has persuaded their ability to distinguish their offerings, as well as their manufacture and transport costs [19].

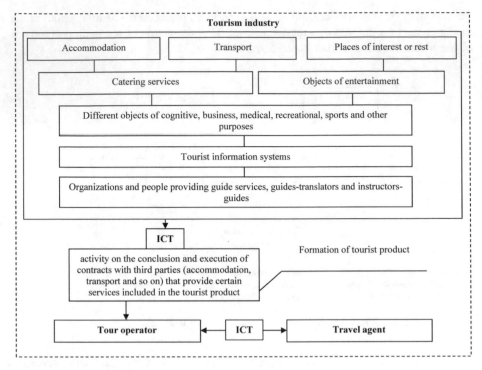

Fig. 3. Formation and promotion of a tourist product by a tour operator.

Since suppliers of tourism services are in most cases distant from consumers and travel agencies, interaction with them is also carried out using ICT. In Table 1 you can see the matrix of ICT tools used by participants in the tourism market.

Table 1. Matrix of main ICT tools used by participants in the tourism market.

	Governments	Travel intermediaries	Travel service providers	Travel firms	Consumers
Governments	Websites with information about visas and passports, travel advisories, statistic and reports (Russia tourism, Russia. Travel). Embassy websites (U.S. Embassy) et al.				
Travel intermediaries		Travel blogs, forums, social media (TripAdvisor, TopHotels). Websites with on-line reservation and support systems (Booking, Airbnb, Expedia, AutoEurope) et al.			
Travel service providers			Websites with on-line reservation and support systems (Alamo, Turkish Airlines, Hilton), social media et al.		
Travel firms				Websites with on-line reservation and support systems (Thomas Cook Group, TUI), social media et al.	
Consumers					Travel blogs, forums, social media (YouTube, Instagram, Facebook) et al.

E-tourism and the Internet in particular support the interactivity between tourism enterprises and consumers and as a result they re-engineer the entire process of developing, managing and marketing tourism products and destinations [3].

Social media are now the most important and influential ICT resources in tourism [8]. With their arrival that consumers became freer from travel agencies in obtaining timely and reliable information about the state visa policy, accommodation facilities, transport, catering, entertainment, etc. The high reliability of such a tool for obtaining information is ensured by the fact that in addition to textual information, social networks use photos and sound, and recently video blogging has become widespread. Tourism industry could also benefit further from developing more video content on YouTube and creating more graphic and moving images.

Table 2 shows the general statements and interpretations regarding the role of ICT in tourism export development (proof of three hypotheses).

Table 2. General statements and interpretations regarding the role of ICT in tourism export development (proof of three hypotheses)

Statements about ICT in tourism	Interpretations
ICTs widen tourism	ICT contributes to the development of tourism exports: - access to any travel information; - on-line booking and registration; - communication with travel firms, travel service providers and consumers; - et al.
ICTs foster tourism information	ICTs provide unique access to information: - interactive maps with panoramas; - travel videos; - audio guides; - mobile applications for routing; - mobile translators; - travel portals of countries; - web sites of attractions, places of visit, travel agencies; - aggregators and booking systems of vehicles; - et al.
ICT is better than human sources	ICT is better in tourism for its: - fastest exchange of information; - relative indestructibility; - clear sequence; - constancy of quality; - minimal cost; - ease of control; - et al.

According to Table 2, all three hypotheses identified in the goal of the research section received positive confirmation. ICT really allows us to develop tourism exports and ensure the availability of information for the majority of the world's population, allows us to expand tourism through multimedia opportunities and to increasingly involve robots in the process of providing services for tourists.

6 Conclusion

Until the advent and spread of the Internet, tourism enterprises relied on analogue (telephone, fax), costly, time-consuming processes of communication with their consumers. Now, with the advent of the Internet and the digital boom, each client has online access to various tourist information resources. To a large extent, such access to information, the possibility of comparing destinations, an unlimited choice of services and the simplicity of their reservation have made tourism the fastest growing export industry in the world.

In the 2000s social media revolutionized the Internet. Using social media applications consumers have equal opportunities to create and spread information not only to hospitality and tourism firms, but also to other consumers. This has largely shifted the power from firms towards customers and forced tourism and hospitality organizations to rethink their engagement with consumers [8, 21, 26].

After remains a large scope for tourism industry to develop and online presence after making significant improvements in recent years [19]. From the point of view of promoting the country and increasing the export potential of tourism, it is necessary:

- formation and development of the national tourism portal with all relevant information with a description of the regions, including tourist facilities, organization of the tourism industry;
- creation of an information system that provides on a reciprocal basis visa-free group tours, the provision of visa-free transit entry from 24 to 72 h;
- development of mobile applications for transport and other systems with the possibility of online payment;
- use of automated computer passport control at the state borders.

For the service providers (airlines, hotels, catering, etc.), the travel firms, and other intermediaries:

- paying more attention to social media (especially YouTube), providing unique content about destinations and travel services;
- development online travel portals with combine a range of services, including booking tools, social media, feature articles and other content to offer travelers a compelling mix of trip planning tools;
- creation of new international searches and aggregators for comparing prices for travel services;
- distribution of the sharing economy and the use of robots in hotels, catering facilities and on routes;
- development of smart mobile applications for tourists.

References

1. Adeola, O., Evans, O.: Digital tourism: mobile phones, internet and tourism in Africa. Tourism Recreation Res. **44**(2), 190–202 (2019)
2. Benckendorff, P.J., Xiang, Z., Sheldon, P.J.: Tourism Inf. Technol., 3rd edn. CABI, Oxfordshire (2019)
3. Buhalis, D., O'Connor, P.: Information communication technology revolutionizing tourism. Tourism Recreation Res. **30**(3), 7–16 (2005)
4. Dedeoğlu, B.B., van Niekerk, M., Küçükergin, K.G., De Martino, M., Okumuş, F.: Effect of social media sharing on destination brand awareness and destination quality. J. Vacation Market. (2019)
5. Dedkova, E., Gudkov, A.: Tourism Export Potential: Problems of Competitiveness and Financial Support. Lecture Notes in Networks and Systems, vol. 78, pp. 187–202 (2020)
6. González-Rodríguez, M.R., Martín-Samper, R.C., Köseoglu, M.A., Okumus, F.: Hotels' corporate social responsibility practices, organizational culture, firm reputation, and performance. J. Sustain. Tourism **27**(3), 398–419 (2019)
7. Gouveia, P., Guerreiro, R., Rodrigues, P.: The world tourism exports cycle. Economic Bulletin and Financial Stability Report Articles and Banco de Portugal Economic Studies, Banco de Portugal, Economics and Research Department, pp. 69–91 (2013)
8. Gretzel, U.: Social media activism in tourism. J. Hospitality Tourism **15**(2), 1–14 (2017)
9. Gudkov, A., Dedkova, E., Dudina, K.: The main trends in the Russian tourism and hospitality market from the point of view of Russian travel agencies. Worldwide Hospitality Tourism Themes **10**(4), 412–420 (2018)
10. Hashim, N.L., Isse, A.J.: Usability Evaluation Metrics of Tourism Mobile Applications. J. Softw. Eng. Appl. **12**, 267–277 (2019)
11. Hjerpe, E.E.: Outdoor recreation as a sustainable export industry: a case study of the boundary waters wilderness. Ecol. Econ. **146**, 60–68 (2018)
12. Im, J., Hancer, M.: What fosters favorable attitude toward using travel mobile application? J. Hospitality Market. Manage. **26**(4), 361–377 (2016)
13. Im, J.Y., Hancer, M.: Shaping travelers' attitude toward travel mobile applications. J. Hospitality Tourism Technol. **5**(2), 177–193 (2014)
14. Ivanov, S., Gretzel, U., Berezina, K., Sigala, M., Webster, C.: Progress on robotics in hospitality and tourism: a review of the literature. Journal of Hospitality and Tourism Technology forthcoming (2019)
15. Ivanov, S.H., Webster, C., Berezina, K.: Adoption of robots and service automation by tourism and hospitality companies. Revista Turismo Desenvolvimento **27**(28), 1501–1517 (2017)
16. Jóhannesson, G., Huijbens, E.: Tourism in times of crisis: exploring the discourse of tourism development in Iceland. Curr. Iss. Tourism **13**(5), 419–434 (2010)
17. Kathan, W., Matzler, K., Veider, V.: The sharing economy: your business model's friend or foe? Bus. Horiz. **59**, 663–672 (2016)
18. Katsoni, V., Sheresheva, M.Y.: Sharing economy in hospitality and tourism. Moscow Univ. Econ. Bullet. **1**, 71–89 (2019)
19. Khan, M.Y.H., Hossain, A.: The effect of ICT application on the tourism and hospitality industries in London. SocioEcon. Challenges **4**(2), 60–68 (2018)
20. Ko, Y.S.: A study on sharing economy of the ICT development. e-Bus. Stud. **15**(6), 77–100 (2014)
21. Leung, D., Law, R., Van Hoof, H., Buhalis, D.: Social media in tourism and hospitality: a literature review. J. Travel Tourism Market. **30**(1–2), 3–22 (2013)

22. Pouri, M.J., Hilty, L.M.: Conceptualizing the digital sharing economy in the context of sustainability. Sustainability **10**, 4453 (2018)
23. Sheresheva, M.Y.: The Russian hospitality and tourism market: what factors affect diversity and new destination development? Worldwide Hospitality Tourism Themes **10**(4), 510–522 (2018)
24. Sheresheva, M.Y.: The Russian tourism and hospitality market: new challenges and destinations. Worldwide Hospitality Tourism Themes **10**(4), 400–411 (2018)
25. Tan, G.W.-H., Lee, V.H., Lin, B., Ooi, K.-B.: Mobile applications in tourism: the future of the tourism industry? Ind. Manage. Data Syst. **117**(3), 560–581 (2017)
26. Xiang, Z., Gretzel, U.: Role of social media in online travel information search. Tourism Manage. **31**(2), 179–188 (2010)

Accounting and Macroeconomic Issues of Evaluation of Financial Instruments in Preparing Financial Statements

Svetlana Grishkina⬤, Vera Sidneva$^{(\boxtimes)}$⬤, Vladimir Skalkin⬤,
Yulia Shcherbinina⬤, and Elena Astakhova⬤

Financial University under the Government of the Russian Federation,
Moscow 125167, Russia
v.sidneva@gmail.com

Abstract. The article is devoted to the analysis of modern problems of valuation in reporting of financial assets and liabilities based on the fair value
concept presented in IFRS 13. The development trends of valuation methods
and requirements for financial assets and liabilities accounting in national and
international financial reporting standards are considered. The application of the
International Financial Reporting Standard (IFRS) 9 (mandatory introduction
from January 1, 2018), which requires the use of a new classification of financial
instruments and a model for recognizing expected credit losses, is analyzed. The
comparative characteristic of accounting rules for financial instruments of
international and Russian standards is given. The practice of applying the new
rules in the reporting of Vnesheconombank Corporation and the group of
companies in the oil and gas sector is examined and their influence on the
financial performance of companies is revealed. Based on the results of the
study, the main directions of activity in the field of solving problems arising in
the process of evaluating financial instruments and the requirements for the
formation and disclosure of an effective basis for determining an estimate at fair
value are formulated.

Keywords: International financial reporting standards · Financial instruments ·
Financial investments · Fair value · Amortized cost · Impairment

1 Introduction

In different national and international accounting systems, there is a search going on for
the most correct ways to evaluate the elements of financial reporting to ensure the
usefulness of accounting information, including its attribute as forecasted value. Thus,
to achieve this characteristic, in our opinion, during the long period of time they were
directed the efforts of supporters of fair value concept. The use of fair value in the
preparation of financial statements makes it possible to objectively reflect the financial
position of the company and support the company stakeholders with true and fair view
on company value prospects.

To evaluate accounting elements at fair value, market indicators are required, and in
their absence, appropriate methods for its calculation are used [1, 2]. To create a single

© Springer Nature Switzerland AG 2020
T. Antipova and Á. Rocha (Eds.): DSIC 2019, AISC 1114, pp. 175–184, 2020.
https://doi.org/10.1007/978-3-030-37737-3_16

basis for measurements at fair value, IFRS 13 "Fair value measurement" is used, which sets out the requirements for determining the fair value of accounting elements in various situations.

In international practice, fair value accounting has become widely used, first of all, for evaluating financial instruments, which is quite justified, since information on their market quotations is more often available for financial assets and liabilities than other accounting elements. At the same time, due to the volatility of financial markets, many experts point out the negative aspects of applying fair value in accounting for financial instruments.

It is widely believed that the overestimation of the value of derivative financial instruments with a tie to their valuation at fair value has caused the global financial crisis. Some researchers point out that accounting for financial assets and liabilities at fair value and cost violates the principle of prudence. The authors of this article believe that the conservatism principle violates not very application of the fair value but the use of insufficiently justified methods of its calculation, as well as the desire of many compilers to "window dress" financial performance.

2 Trends in the Development of Accounting Rules and Valuation of Financial Instruments

Basic approaches to the valuation of assets and liabilities in Russia are formulated in the Concept of Book Accounting in the Market Economy of Russia which was approved by the Ministry of Finance and the IPA (Institute of Professional Accountants) of Russia, December 29, 1997, the "Regulation on Accounting and financial reporting in the Russian Federation", specified in the relevant accounting standards (FSBU- Federal Standards of Accounting). In accordance with the Concept, the following values can be used for valuing assets and liabilities: actual value, current (replacement) value and current market value (realized value). The present discounted value can be used in the absence of another measurement base for valuation.

The most common methods for evaluation of assets and liabilities in Russian accounting practice are valuation at the actual cost of acquisition (historical cost). It is recommended that certain assets be reported at market value; however, the market value is mainly applied when the carrying amount of the asset exceeds its market value. For example, the RAS instruction "Accounting for Inventories" recommends creating provisions for impairment of tangible assets. Accordingly, inventories are recorded in the financial statements at purchase cost minus reserves.

An exception to the previous instruction is the reporting of financial investments in the financial statements at market value in accordance with RAS 19/02 "Accounting for Financial Investments". In addition, in accordance with this regulation, for debt securities and loans granted, an organization may calculate their valuation at a discounted value. However, they are reported in the balance sheet at historical cost, and information on discounted value is disclosed only in the Notes to Financial Statements.

The discounted value method is not applied for the measurement of liabilities, except for long-term estimated liabilities (RAS 8/2010). The revenue approach is provided for in IFRS 13 "Fair Value Measurement" which is based on the calculation of discounted cash flows.

RAS is different from IFRS by describing only accounting rules for financial assets (investments) but does not contain requirements for financial liabilities accounting. The authors have devoted some publications where the difference between RAS and IFRS is addressed in detail [3–5]. The main complexities are related to evaluation for assets and liabilities, accounting for their impairment. It makes development and application of specific accounting techniques relevant for reporting financial position of an entity in true and fair view. For that the Notes to Financial Statements shall include information about accounting techniques used.

The scientific literature it is widely represented whole host of loopholes in the rules of accounting and evaluation of financial instruments. This is one reason why developers of national and international standards are constantly making changes to them. Meanwhile, the new rules are constantly getting more complicated and require compilers to take into account internal and external factors and apply professional judgment in each case, when evaluating financial assets and liabilities. Particular attention is paid to risks, since recent changes require mandatory disclosure of information about them in the financial statements.

So, IFRS 7 "Financial Instruments: Disclosures" requires an entity to disclose "information that enables users of its financial statements to evaluate the nature and extent of risks associated with financial instruments that the entity is exposed to at the end of the reporting period" [6]. First of all, we mean disclosing information about the risks arising in connection with the evaluation of financial instruments at fair value (Table 1).

Table 1. Risks of the organization related to the assessment of the financials tools in the financial statements

Types of risks associated with financial instruments	Nature of risk
Credit risk	The risk that one of the parties to a financial instrument will incur financial loss due to non-fulfillment of obligations by the other party
Currency risk	The risk that the fair value of a financial instrument or future cash flows will fluctuate due to changes in foreign exchange rates
Interest rate risk	The risk that the fair value of a financial instrument or future cash flows will fluctuate due to changes in market interest rates
Liquidity risk	The risk that the organization will encounter difficulties in fulfilling obligations associated with financial obligations involving cash or another financial asset

(continued)

Table 1. (*continued*)

Types of risks associated with financial instruments	Nature of risk
Market risk	The risk that the fair value of a financial instrument or future cash flows will fluctuate due to changes in market prices. Market risk includes three types of risks: currency risk, interest rate risk and another price risk
Other price risks	The risk that the fair value of a financial instrument or its future cash flows will fluctuate due to changes in market prices (excluding those related to interest rate risk or currency risk), regardless of whether these changes are caused by factors that are characteristic only for a specific financial instrument or its issuer, or factors affecting all similar financial instruments exposure to the market

Source: Compiled by the authors on the basis of IFRS 7 "Financial Instruments: Disclosures"

An even greater consideration of external and internal factors affecting the real value of financial assets and liabilities is required by IFRS 9 "Financial Instruments", which has recently been phased in and supersedes of IAS 39 "Financial Instruments: Recognition and Measurement".

The current edition of IFRS 9 is effective as of 2018 [7]. The most common innovations related to the full application of the requirements of IFRS 9 [8] is as follows. Firstly, the classification of financial assets has been changed. Instead of four categories, financial assets must now be classified into three categories due to methods of accounting (amortized cost, fair value through other comprehensive income and fair value through profit or loss). Secondly, the classification of debt instruments depends on the financial asset management business model. Third, IFRS 9 completely changes the accounting for impairment losses on loans, which are calculated on the basis of a new approach based on the forecasted expected credit loss model instead of the incurred loss model provided for in IAS 39.

3   Methodological Aspects of the Assessment of Provisions for Expected Credit Losses

The most significant innovation of IFRS 9 "Financial Instruments", in our opinion, is the transition from the model of losses incurred to the model of expected credit losses. The new model introduces fundamental changes to the accounting and reporting system, since information on future events largely prevails over information on past transactions, which are traditionally the main objects of accounting and reporting. The new model, on the one hand, enforces the predictive value of accounting information, and on the other, applies the principle of prudence.

However, it is obvious that forecasts are always associated with subjectivity in making professional judgments. In this regard, the new rule will lead to an increase in the quality of reporting only if the forecast estimates are sufficiently substantiated. In addition, since the procedure for assessing future credit losses is established by the company on its own to ensure the relevance of accounting information by its users and its usefulness for making decisions regarding the future, sufficiently clear and understandable disclosures are needed on the methods used to estimate the allowance for expected credit losses.

The provision for ECLs (Expected Credit Loss) is estimated at the amount of credit losses that are expected to arise over the life of the asset (expected credit losses for the entire term) if the credit risk for the financial asset has increased significantly since initial recognition. Otherwise, the allowance for losses will be estimated at 12 months of expected credit losses. 12-month ECLs are part of the full-time ECL, which is the ECL arising from defaults on a financial instrument expected within 12 months after the reporting date.

New impairment loss recognition model [9] requires the development of new approaches to credit risk assessment, since it is credit risk that is fundamental in determining the amount of recognized impairment loss. The new model for recognizing impairment losses in effect in connection with the application of IFRS requires the development of new approaches to assessing credit risks, since it is credit risk that is fundamental in determining the amount of recognized impairment loss.

Analysis of financial statements of entities applying IFRS 9, allows us to conclude that the recognition of the provision for expected credit losses companies use different procedures to identify credit risks [10]. For example, they apply testing, distribute financial instruments into various qualitative stages, use statistical data and expert estimates to determine the probability of default, the level of losses during default, etc. For example, the Notes to the Consolidated Financial Statements of the State Corporation Vnesheconombank Group disclose a mechanism for transition from the model of actually incurred losses to the model of expected losses [11] (as of January 1, 2018, the company switched to the application of requirements (IFRS) 9 "Financial Instruments" instead of IAS 39 "Financial Instruments: Recognition and Measurement").

The new expected credit loss assessment model (ECL) involves the recognition of the allowance for losses, and its assessment depends on the degree of deterioration in the credit quality of the financial instrument after its initial recognition. In accordance with the new approach, for debt financial instruments carried at amortized cost and fair value through other comprehensive income, a different time horizon for assessing ECLs is used, based on the distribution of financial instruments into 3 qualitative stages:

1. Upon initial recognition of a loan, an allowance for impairment is recognized in an amount equal to 12-month ECL. Stage 1 also includes financial instruments for which the credit risk has decreased to such an extent that they have been transferred from stages 2 and 3.
2. If the credit risk of a financial instrument has increased significantly since initial recognition, an allowance for impairment is recognized in the amount equal to ECL for the entire life of the instrument. Stage 2 also includes assets for which the credit risk has decreased to such an extent that they have been transferred from stage 3.
3. For the financial instruments that are credit-impaired the Company recognizes an allowance for impairment in the amount equal to the full-time asset assessment. Acquired or created credit-impaired financial assets (ACCIs) are assets for which

there was a credit impairment at the time of initial recognition. At initial recognition, they are carried at fair value, and subsequently, interest income on them is recognized based on the effective interest rate adjusted for credit risk. The provision for ECL is recognized or ceases to be recognized only to the extent that there has been a change in the amount of expected credit losses.

If a company does not have reasonable expectations regarding the full or partial recovery of a financial asset, then it reduces the gross carrying amount of the financial asset. Such a reduction is considered to be a partial derecognition of a financial asset. At each reporting date, an assessment is performed to identify a significant increase in credit risk and evidence of impairment since the initial recognition of the financial instrument.

Financial instruments go from stage 1 to stage 2 if the following factors are present: presence as of the reporting date of overdue debt on the principal, debt, and or interest, as well as other payments stipulated by the contract, for a period of 31 to 90 days; decrease in the internal rating by three or more notches or a decrease in the rating by one or several international rating agencies by two notches; initiating or carrying out restructuring with a significant change in the terms of the transaction in favor of the borrower in connection with the deterioration of its financial situation.

The third stage includes financial instruments for which one of the following events has occurred:

- the delay in any material loan obligation exceeds 90 days;
- the counterparty, financial instrument, project is assigned a rating from the notch of the master scale corresponding to default;
- downgrade by one or several international rating agencies by three notches;
- events indicating a negligible probability of repayment of obligations by the debtor from the available resources. For example, bankruptcy proceedings have been instituted or bankruptcy proceedings have been introduced; the borrower has the license for the main type of activity revoked; the process of liquidation of the borrower has been initiated, or the exclusion of the borrower from the USRLE (Unified State Register of Legal Entities in the Russian Federation) by decision of the tax authority; appeal of the borrower with a proposal to accept its property to the company-lender in repayment of debt.

The forecast credit loss estimate (ECL) calculation algorithm consists of the following elements:

1. The assessment of the probability of default is carried out over a given time period. A default can occur only at a certain point in time during the period under review, if the asset has not been derecognized, and it is still part of the portfolio.
2. The value at risk of default is determined. It is a calculated estimate of the amount that is subject to default at some future date, taking into account the expected changes in this amount after the reporting date, including repayment of the principal amount of the debt principal and interest stipulated by the contract or otherwise, repayment of loans issued and interest accrued as a result late payments.
3. The level of losses due to default is calculated, that is, a calculated assessment of losses arising in the event of default at a certain point in time is carried out. The

fundamental principles for calculating credit losses recognized as an allowance reserve are compatible with the provisions of the Basel Committee on Banking Supervision and comprise the application of risk components, including indicators of the probability of default, the level of losses during default, the amount exposed to default risk. To assess the likelihood of default, internal and external statistics are used, namely, quarterly samples of migration of ratings of counterparties (issuers). The depth of historical data, which are used to determine default probabilities is at least five years.

For impaired assets classified in stage 3, expected losses and allowance for losses are estimated based on at least two scenarios of expected future cash flows discounted at the effective interest rate at the time the asset is recognized or materially modified. In addition, in order to switch from the model of actually incurred losses to the model of expected losses, the probability of default is adjusted from the calculated taking into account past economic cycles into the respective ratio at the time of calculation.

At least two macroeconomic scenarios are used to calculate expected losses: optimistic and conservative, taking into account the weights of the probability of each scenario. The probability of default is reassessed on a quarterly basis upon updating statistics and revaluating the forecast of macroeconomic scenarios. The selection and justification of the elements for macroeconomic scenario analysis is carried out on an annual basis.

In some cases, when it is necessary to exclude the influence of factors that are not relevant for estimating expected losses, take into account current observable information or expected events, assumptions or adjustments based on professional judgment can be applied to the calculated values of allowances for impairment. The decision on the application of adjustments is made in accordance with the established distribution of powers in identifying significant economic and other factors that affect the amount of reserves. Adjustments based on professional judgment that result in a significant reassessment of expected losses should be disclosed as part of information about the assessment of impairment losses.

4 Effect of IFRS 9 on Financial Performance

The new procedure for evaluating financial instruments leads to a change of lines in the financial statements. So, the analysis of financial ratios of State Corporation "Vnesheconombank" Group [11], shows the following. Since, the loans were previously measured at amortized cost, they are classified as financial assets at FVPL (Fair Value through Profit and Loss). In addition, part of the portfolio of assets previously classified as available for sale was classified as debt instruments measured at amortized cost.

As of January 1, 2018, a part of the financial instruments available for sale was reclassified: financial assets that are valued at FVPL, since the instruments did not meet the definition of equity. In addition, since a number of instruments are held in the framework of the business model, the purpose of which is to obtain the cash flows stipulated by the contract and the sale of financial assets, they were classified as debt instruments measured under the FVCI (Fair Value through Comprehensive Income).

Similarly, a study of the financial reports of the largest companies in the oil and gas sector also shows that innovations in accounting for financial instruments lead to a decrease in the value of their net assets (Table 2).

Table 2. The impact of new accounting rules for financial instruments on the value of net assets of oil and gas companies

Indicator				Mln rub.
	PJSC «NK Rosneft»	Group «Tatneft»	PJSC «Lukoil»	PJSC «Novatek»
Net asset value when applying IAS 39 as at 1 January 2018	4 183 000	718 729	3 490 399	775 659
Recognition of CMOs in accordance with IFRS 9 for assets measured at amortized cost	(26 000)	(9 012)	–	–
Recognition of CMOs in accordance with IFRS 9 for assets at fair value through other comprehensive income	(1 000)	(193)	–	–
Recognition of CMOs on credit related commitments	–	(710)	–	–
Revaluation of loans and advances to customers at fair value through profit or loss	(5 000)	(717)	–	–
Other revaluation	–	(144)	–	–
General effect on retained earnings before income tax	(33 000)	(10 776)		
Deferred tax in relation to the above	5 000	1 769	–	–
Total effect of applying IFRS 9 with deduction of tax	(28 000)	(9 007)	(6 831)	0
Including non-controlling interest	(1 000)	(2 048)	–	–
Net asset value when applying IFRS 9 as at January 1, 2018	4 155 000	709 722	3 483 568	775 659

Source: Compiled by the authors based on the consolidated financial statements of oil and gas companies [12–15].

As the data in the table shows, in three of the four companies studied, the application of IFRS 9 "Financial Instruments" led to a decrease in the value of net assets. In the company PJSC "NK Rosneft" this decrease amounted to 0.7%; in the Group "Tatneft" - 1.3%; in PJSC "Lukoil" - 0.2%.

The deterioration in financial performance is primarily due to the recognition of expected credit losses. So, in PJSC "NK Rosneft", the overall effect of the adoption of IFRS 9 net of tax amounted to RUB 28,000 million, of which 23,758 (28,000 × 28,000/33,000) from recognition of expected credit losses, in the Group "Tatneft" - 9,007 million rubles; at PJSC "Lukoil" - 6,831 million rubles.

Analysis of the structure of the notes to the consolidated financial statements allows us to conclude that they are "overloaded" with a description of the standards themselves, which, in our opinion, is not entirely justified. It would be better to disclose the business model used, the external and internal factors of the company that affect the classification of financial instruments, and therefore the cost at which they are reported in the financial statements.

5 Conclusion

Valuation of financial assets and liabilities is the main type of accounting work when recognizing financial instruments in financial statements and disclosing information about them in notes. Recent changes in international financial reporting standards for the accounting of financial instruments make this process even more relevant, which is associated primarily with the introduction of the new version of IFRS 9 "Financial Instruments".

Meanwhile, the system of Russian Federal Accounting Standards (FSBU) provides a simplified procedure for assessing and accounting for financial assets (RAS 19/02 "Accounting for Financial Investments"), there are no requirements for accounting for derivative financial assets and financial liabilities, which leads to some distortions in accounting information, and also to the incomparability of reporting data presented in RAS and IFRS.

The expected credit loss model envisaged by IFRS 9 "Financial Instruments" makes fundamental changes to the accounting system, enforces the forecast value of accounting information, and also applies the principle of prudence. Analysis of the financial statements of credit organizations and oil and gas companies showed that due to the application of the new rules, the financial indicators of the companies under study are deteriorating, which is associated with the recognition of expected credit losses.

At the same time, the new IFRS rules do not eliminate all the problematic issues associated with the valuation and accounting for financial instruments. On the contrary, providing autonomy to companies in making professional judgments about future events requires the use of reasonably sound methods and the disclosure of understandable information about them in the Notes to Financial Statements. In addition, in our opinion, factors affecting the valuation of individual objects should be sufficiently clearly and concisely disclosed. A study of the Notes to the Financial Statements of companies in the financial, oil, and gas sector shows that they pay excessive attention to the description of standards, instead of presenting specific techniques and algorithms used in the process of evaluating future events that affect the value of accounting elements.

References

1. Fawzi, A., Al, S.: Fair value accounting: a controversial but promising system. Account. Financ. Res. **1**(5), 88–98 (2016)
2. Magnan, M., Menini, A., Parbonetti, A.: Fair value accounting: information or confusion for financial markets? Rev. Account. Stud. **1**(20), 559–591 (2015)
3. Grishkina, S.N., Sidneva, V.P.: Accounting for financial instruments according to Russian and international standards. Econ. Bus. Banks **4**(25), 96–108 (2018)
4. Sidneva, V.P.: Actual issues of fair value measurement of liabilities and equity instruments. Econ. Manage.: Prob. Solutions **3**, 2–8 (2014)
5. Grishkina, S.N.: Modern approaches to the formation and disclosure of information about the financial assets of the organization. Accounting, analysis and audit: realities and development prospects. In: Proceedings of the II All-Russian Scientific and Practical Conference. Kerch, Publishing House: Federal State-Funded Educational Institution of Higher Education "KSMTU", pp. 14–20 (2018)
6. International Financial Reporting Standard (IFRS) 7 "Financial Instruments: Disclosures"
7. International Financial Reporting Standard (IFRS) 9 Financial Instruments
8. Gryazeva, V.: Accounting for financial instruments in accordance with IFRS 9: what has changed since January 2019? Taxation, Acc. Report. Commercial Bank **9**, 27–32 (2018)
9. Druzhilovskaya, TYu., Dobrolyubov, N.A.: Modern problems of accounting for financial instruments of organizations. Acc. Budget Non-profit Organ. **13**, 2–12 (2017)
10. Kuvaldina, T.B., Lobachev, E.V.: IFRS 9: assessment of expected credit losses and their reflection in statements. Int. Acc. **4**, 364–378 (2019)
11. Interim Condensed Consolidated Financial Statements of the Group of the State Corporation Bank for Development and Foreign Economic Affairs (Vnesheconombank) as of June 30, 2018. https://web.rf/files/?File=3f212ed2cc61641e3a77dbd0d0e8996b.pdf. Accessed 21 July 2019
12. Consolidated financial statements of Rosneft December 31, 2018. https://www.rosneft.ru/upload/site1/document_cons_report/Rosneft_FS_12m2018_RUS.pdf. Accessed 21 July 2019
13. Tatneft Group Consolidated Financial Statements in accordance with International Financial Reporting Standards as of and for the year ended December 31, 2018. https://www.tatneft.ru/storage/block_editor/files/ff2ba6f00bf89e868155f7af7d8ae007084ce2d0.pdf. Accessed 21 July 2019
14. PJSC LUKOIL Consolidated Financial Statements December 31, 2018. http://www.lukoil.ru/FileSystem/9/328349.pdf. Accessed 21 July 2019
15. PJSC NOVATEK Consolidated Financial Statements Prepared in accordance with IFRS for the Year Ended December 31, 2018 and Auditor's Report. http://www.novatek.ru/ru/investors/disclosure/ifrsreporting. Accessed 21 July 2019

Information Risks and Threats of the Digital Economy of the XXI Century: Objective Prerequisites and Management Mechanisms

Lyudmila Popova⑩, Irina Korostelkina⁽⊠⁾ ⑩, Elena Dedkova⑩,
and Mikhail Korostelkin⑩

Orel State University, Naugorskoe Highway, 40, Orel 302020,
Russian Federation
cakyra_04@mail.ru

Abstract. The development of national economy, as shown by numerous studies and practices, depends not only on the quality of the entrepreneurial climate, human capital and high safety standards, but also on the fundamental prerequisites, among which is the level of development of the digital economy. The widespread digitalization of the economy, in addition to the obvious positive aspects, is accompanied by various risks and threats. This is especially relevant with regard to information risks, because information is the platform for digital economy. The article provides a theoretical analysis of the categorical apparatus and elements of the digital economy, information security and risks, examines the indicators of the development of information security in the conditions of digitalization of the economy, as well as a mechanism for managing information risks in the digital economy. The implementation of measures for the further spread of digital technologies in Russia and the regulation of information security risks help reduce uncertainty and increase economic and information stability.

Keywords: Digital economy · Information risk · Threat · Control · Challenges · Information security

1 Introduction

XXI century is accompanied not only by globalization, the growth and competitiveness of national economies, but also by scientific and technological progress and the innovativeness of all spheres of social development. Risks of intentional or unintentional distortion of information, its inaccuracies, subjectivity and irrelevance lead to a slowdown in the digitalization of the economic space, reduce its effectiveness. In this regard, it seems relevant to consider the types of information risks and threats and the possibilities for leveling them, to study the global practice of regulation and management of information risks in the digital economy.

The foundation of modern society is the knowledge digitalization economy, which includes information resources, intellectual capital and innovation. The innovation system and knowledge economy (digital economy) of developed countries are the basis

© Springer Nature Switzerland AG 2020
T. Antipova and Á. Rocha (Eds.): DSIC 2019, AISC 1114, pp. 185–194, 2020.
https://doi.org/10.1007/978-3-030-37737-3_17

for the formation of innovative model for developing countries. At present, the digital economy in the Russian Federation is among the priority areas of the Strategy for Scientific and Technological Development and the Strategy for the Development of the Information Society.

Theories and concepts of the knowledge economy, information security and risks are explored in the modern scientific space, taking into account the historical aspect and the application of the best world and domestic practices. The digital economy is part of the innovation.

The digital economy as a virtual business system in theory began to develop in 1994 with the release of the book of Don Tapskotta «Digital Economy». Further, in his publications, the author substantiates the need for a transition to new business models [1] and proves that the growth of digital technologies provides a transition to the next economic cycle [2].

In the scientific world, the issues of economic digitalization and virtual technologies are paid attention by Dobrynin, Chernykh, Kupriyanovsky, Kupriyanovsky, Sinyagov [3], Baller, Dutta, Lanvin [4], Brynjolfsson, Kahin [5], Dosi [6] and others. In particular, Brynjolfsson and Kahin defined digital economy as «... the transformation of all sectors of the economy using computerized digitization of information» [7].

At the seminar of the World Bank in Russia, which took place on December 20, 2016, the essence of the digital economy was characterized as a set of installations to enhance economic development through the use of digital technologies. According to Dosi [6], the procedures and the nature of «technologies» are suggested to be broadly similar to those which characterize «science». In particular, there appear to be «technological paradigms» (or research programmes) performing a similar role to «scientific paradigms» (or research programmes).

In developed countries, due to its innovativeness, the digital economy is growing rapidly. However, there are technological and organizational gaps between the economies of different countries, due to the fact that the share of countries on the sectoral market is determined by technological factors, while cost advantages (disadvantages) do not play a significant role [8]. Such country gaps are estimated using the international index of the digital economy and society, which takes into account the parameters of connectedness, human capital, use of the Internet, the integration of digital technologies, the development of an electronic state. The average value of this indicator across EU countries is 0.54, in the USA – 0.62, in Russia – 0.47 [9].

The digital economy in Russia is still emerging. Its formation leads to global changes in all spheres of human life. The study of world practice, features and trends in the development of the digital economy makes it possible to most effectively exploit and develop the existing innovative potential of the country, create conditions for economic growth and promote the improvement of national well-being.

Information security, risks and threats to innovation, including the digital economy are studied by Lebedeva [10], Baskakov, Ostapenko, Shcherbakov [11], Kalashnikov [12], Galushkin [13] and others. Scientists are developing both theoretical and methodological aspects of security, as well as specific areas of information security policy making, information risk management mechanisms. In particular, Teoh and Mahmood [14] investigated cyber threats and cyber risks for different countries in the form of malware: online commerce, mobile and cloud technologies. The authors believe that «In order for digital economy to thrive, the digital confidence of the stakeholders should be high». Pouri and Hilty [15] consider the conceptualization of the impact of digitalization and information and communication technologies on the sustainability of the modern economy, the positive and negative potentials of digital exchange. Bubnova, Efimova, Sokolov and Akopova explore digital technology as a platform for risk management and competitiveness [16].

The aim of the article is to study the global and domestic experience in the development of the digital economy, on the basis of which information risks and threats are analyzed and the mechanism for managing components of the digital economy is proposed. In the future, the comprehensive digitalization of the Russian economy will create a platform for high-quality transformation and the realization of potential opportunities. In order to work for the good of society, innovative technologies must be safe and cost-effective. In this regard, information security, risks and threats to the national economy and methods of management are important for further innovative development.

2 Materials and Methods

The methodological basis of the study was a set of general scientific (analysis, synthesis, induction, deduction, analogy) methods, system-structural and functional approaches, modern analytical tools and technologies. The provisions of the study are argued by using particular scientific research methods (formal, comparative and functional methods, specification, etc.). In the process of work, the tools of graphic interpretation, comparative analysis, method of attendant changes were used.

3 Discussion and Results

The historical perspective of the global economy indicates the high efficiency of innovation. The conditions for realizing new opportunities for economic growth in European countries have historically been provided by an optimal combination of scientific knowledge, commercial calculation and reasonable organization of labor. This contributed to the most important technological breakthroughs, scientific discoveries and their widespread application.

The second half of the 20th century is characterized by a significant accumulation of new knowledge, while the speed of information dissemination has significantly increased, the diffusion of innovations and multidisciplinary technologies have been actively pursued. In the 21st century, scientific and technological progress, technological breakthroughs, global innovation have given a powerful impetus to the wide spread of the digital economy, which structurally combines the market, technological and infrastructure components. Guo, S., Ding, W. and Lanshina, T. (2017) determine that «international society has to govern the digital economy properly to eliminate distortions between developed and developing countries, ensure cyber security and achieve a higher quality of life for all» [17].

Russia takes 45th place in the international ratings of the development of the digital economy with an ICT (Information and Communication Technology) index value of 7.07, ahead of Slovakia and Italy. This indicator characterizes the level of development of ICT infrastructure, its demand for the population and is used in international statistics to estimate the scale of the «digital divide» between developed and developing countries. According to this indicator, the most developed countries in 2017 were Iceland, Korea, Switzerland [9].

The «Digital Economy of the Russian Federation» program implemented in Russia includes a number of infrastructure and institutional areas, among which information security occupies an important place [18]. Information security is the stability of the information system and the degree of security of information resources of the state. Information security should ensure that the state of the information system in which it is least susceptible to interference and damage from third parties (state of protection) [19].

Information security is strengthened through the implementation of data integrity and confidentiality requirements in combination with its availability to all users. In the digital economy, characterized by the widespread introduction of digital technologies in all areas of activity of participants in economic relations, the state of protection of the information environment is regularly exposed to risks and threats. The threat in this context is intent for the purpose of deliberately harming the subject of the information sphere. In the current conditions of global development of information and telecommunication technologies and resources, the role and importance of information for subjects of the digital economy is increasing. Information threats can have a significant negative impact on information security in particular, and national security in general. In this case, the main problem of ensuring information security is precisely the risks of implementing information threats, that is, the likelihood of an adverse event occurring as a result of the use of information technologies.

Information security threats in the digital economy are in violation of its integrity, completeness and accessibility [20]. Such threats are associated with the leakage of information through technical means and unauthorized access to protected information, which is widespread in the digital economy due to cyber crime and hacking. To prevent these threats, special protective computer programs are created that require constant improvement in connection with the increasing incidence of their hacking. In addition, scientists propose various decision-making models in the context of informational threats, for example, Grubor, G., Barac, I., Simeunovic, N. and Ristic, N. have proposed the use of the CCIDFI make optimal management decisions based on multi-criteria and expert approaches [21].

In Russia, mass digitalization is hampered not only by the existing gaps in legal regulation, but also by the problems of ensuring digital identity, safety of information data, virtual threats, lack of powerful security systems and well-established information infrastructure, low level of innovation and staffing in the field of information security.

The global cyber security index is used to characterize the level of information security in the country which defines five main aspects of information security: legal, technical, organizational, innovative, international (Table 1, Fig. 1).

Table 1. Components of the global cyber security index by country for 2017.

Country	Legal	Technical	Organizational	Capacity building	Cooperation
USA	1	0,96	0,92	1	0,73
Australia	0,94	0,96	0,86	0,94	0,44
France	0,94	0,96	0,6	1	0,61
Russia	0,82	0,67	0,85	0,91	0,7
Singapore	0,95	0,96	0,88	0,97	0,87
Malaysia	0,87	0,96	0,77	1	0,87
Oman	0,98	0,82	0,85	0,95	0,75

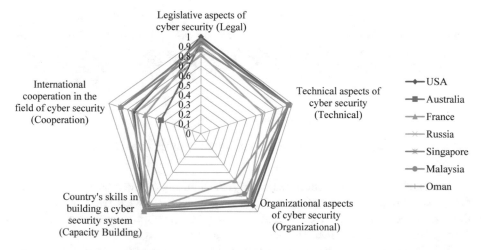

Fig. 1. Indicators of the development of information security, which are part of the global cyber security index (compiled from [9])

According to the estimates of the International Telecommunication Union in terms of this indicator, Singapore, United States, Malaysia, and Oman have the greatest development of legal, technical and organizational measures, experience and international cooperation in the field of cyber security. Russia (with an index value of 0.788) ranks 10th, ahead of Japan, the United Kingdom, and Norway. The top 20 closes Israel in 2017 (Fig. 2) [22].

Information risks and threats can be global (intercountry), national and local [23]. All of them are classified for various reasons (external and internal, intentional and unintentional, direct and indirect, etc.). With the development of the information environment of the digital economy and the emergence of new (previously unknown) threats to information security, the signs of classification can expand, and the types of risks and threats increase and transform [24]. Therefore, information risks require timely identification, systematic causal analysis and effective management (Fig. 3), which is possible within the framework of the implementation of a special mechanism based on the information risk management strategy and including a number of successive stages.

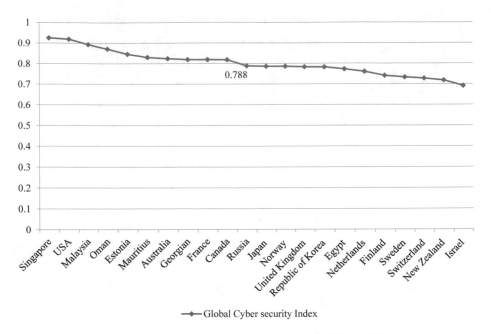

Fig. 2. Global cyber security index (compiled from [22])

Development of information risk management strategy is necessary to determine the goals and objectives, including reducing the probability of occurrence of threats and their negative consequences, forecasting risks and threats, identifying the most appropriate and effective methods and techniques of leveling.

It is important to ensure the possibility of adjusting the information risk management strategy based on the results of monitoring and control over its implementation. The flexibility and variability of the strategy ensures its adaptation to the current situation in the digital economy and possible changes in the future.

The implementation of the stages of the information risk management mechanism is influenced by what form of management is used (traditional, flexible or innovative) and what kind of control action. Depending on this, some stages may not be fully implemented (skipped), or extended with additional activities. For example, at some stages it is possible to quantify risks or involve experts in the analysis of the consequences of possible information risks and the severity of their impact on the processes taking place in the digital economy.

Information security of the state is assessed by indicators of virus infection, spam, unauthorized access to ICT, the use of mobile Internet, etc. In 2017, the total amount of information risks decreased by 5.4% compared with 2015 and amounted to 28.8%. The greatest risk was spam (unauthorized sending) - 18.5%, and also virus infection - 11.4% [9].

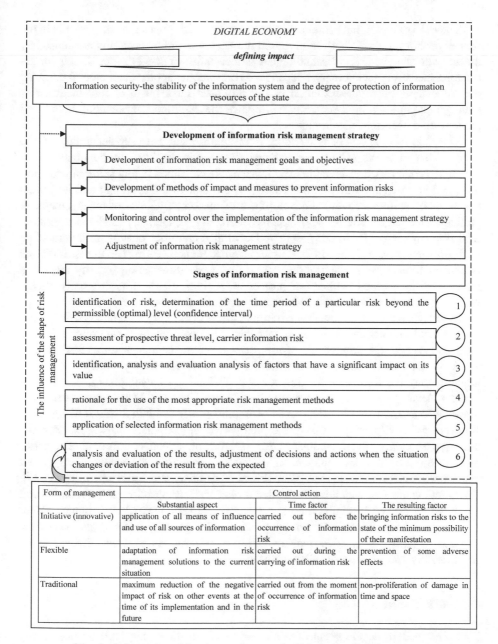

Fig. 3. Information risk management mechanism in the digital economy

Rogers, R., Apeh, E. and Richardson, C.J. (2016) in their study detailed the technical component of e-Commerce and information security, while offering dramatic changes in the construction of technological platforms. Scientists believe that «… Information Assurance is paramount to the success of IoT, specifically its resilience and dependability to continue its safe support for our digital economy» [25].

The risks of the digital economy, in fact, are potential threats that are possible in the future (near or not so distant). Information threats to the digital economy are caused by the growth of cybercrime, technological vulnerability of the existing digital infrastructure, increasing technological dependence on foreign suppliers and other factors. At the same time, the implementation of the proposed mechanism of information risk management in the digital economy will reduce the likelihood of potential threats and prevent their occurrence in the future.

The ratio of methods of influencing information risks can vary depending on the state of the information environment and the results expected to be obtained. In developed countries, considerable experience has been gained in managing information risks and ensuring the information security of the innovation economy and its infrastructure, taking into account the development of market conditions of management. In particular, the largest number of people in Croatia (41%), Hungary (36%) face viruses. In Russia, this indicator in 2017 is 11%. Developed countries occupy an intermediate position: Germany, Austria - 14%, Belgium - 20%, France - 29%. The lowest indicator is observed in the Netherlands - 6% [9]. The Russian information security market is represented by various methods of protection, including anti-virus programs (48%), virtual private networks (29%), cyber-attacks detection tools (10%).

In today's Russian information environment, risks are very high and are due to the widespread introduction of digital technologies. The participant in economic relations must choose appropriate methods of information risk management for each situation.

At present, each country implements its own national security strategy. Russia is implementing the Strategies for Science and Technology Development, Information Society Development, National Security, the Digital Economy Development Program, however, our country is in the process of implementing the initial stages of the transition to mass digitalization and the innovative development paradigm. For the domestic economy, the prerequisites for creating an effective information security mechanism are the development of high-tech breakthrough technologies, rapid response to economic sanctions, geopolitical instability, and crisis phenomena [26]. Analytical platforms, which are integrated tools for ensuring information security and the basis of society's digitalization, are being actively created and implemented. Creating a solid foundation for ensuring information security in the digital economy will contribute to the effective implementation of government programs that shape the sustainability and security of information infrastructure.

4   Conclusion

Thus, the modern development of world economic relations, the globalization of the economy determine the need for digitalization of the national economy, the trends and development prospects of which suggest that now it is necessary to actively enter the information and technological flows of innovatization and apply them everywhere with maximum use of potential.

World space is currently making the transition to a post-industrial society, in which information is the main resource, while information security is a major factor of economic growth. Russia is not among the countries leading in the digitalization of the economy, but the development of digital technologies in our country is proceeding at a significant pace.

Ensuring the information security of the national economy should be accompanied by the timely identification of risks and threats, the maximum use of the state's resource potential in combination with the use of the world's best practices for the sustainable development of society. The purpose of this article is to examine the information risks and threats of the digital economy with the application of the world's best management practices and the identification of measures to strengthen information security. The novelty of the research is the proposed mechanism for managing information risks in the digital economy.

The general and particular research methods, methods and tools of graphical interpretation, comparative analysis, and accompanying changes act as methodological tools of the present study. The implementation of information risk management methods and tools at various levels of regulation of economic relations will reduce the uncertainty of the business environment and increase the validity of the implemented policy to ensure the national security of the state.

The digitalization of the economy actualizes the need for innovation, the creation of value which should be considered in the context of infrastructure concepts (systems evaluation, accounting, analysis and control of value added innovative products). This will contribute to the formation of timely and reliable information and will allow to predict the level of value added and adjust the proportions of its distribution for the purposes of effective public administration in the innovative sphere of the digital economy.

Acknowledgments. The article was prepared within the framework of the project part of the state task in the field of scientific activity in accordance with task No. 26.2758.2017/ 4.6 for 2017-2019 «System of analysis of the formation and distribution of the cost of innovative products based on the infrastructure concept».

References

1. Tapscott, D.: Strategy in the new economy. Strategy Leadersh. **25**, 8–14 (1997)
2. Tapscott, D.: Growing up digital: the rise of the net generation. NASSP Bullet. **83**, 86–88 (1999)
3. Dobrynin, A.P., Chernykh, K.Y., Kupriyanovsky, V.P., Kupriyanovsky, P.V., Sinyagov, S. A.: Digital economy - different ways to effectively apply technology (BIM, PLM, CAD, IOT, SmartCity, BIGDATA and other). Int. J. Open Inf. Technol. **1**, 4–11 (2016)
4. Baller, S., Dutta, S., Lanvin, B.: The global information technology report 2016. World Economic Forum, pp. 1–307 (2016)
5. Brynjolfsson, E., Kahin, B.: Understanding the Digital Economy: Data, Tools, and Research, p. 408. MIT Press, Cambridge (2000)
6. Dosi, G.: Technological paradigms and technological trajectories: a suggested interpretation of the determinants and directions of technical change. Res. Policy **3**, 147–162 (1982)
7. Brynjolfsson, E., McAfee, A., Jurvetson, S., O'Reilly, T., Manyika, J., Tyson, L., et al.: Open letter on the digital economy. Technol. Rev. **118**, 11–12 (2015)

8. Dosi, G., Grazzi, M., Moschella, D.: Technology and costs in international competitiveness: From countries and sectors to firms. Res. Policy **44**, 1795–1814 (2015)
9. Abdrakhmanova, G.I., Vishnevsky, K.O., Volkova, G.L., Gokhberg, L.M.: Indicators of the digital economy: 2018: statistical collection/60 National research. University «Higher school of Economics». – Moscow: HSE, p. 268 (2018)
10. Lebedeva, T.V.: Risks Information risk assessment and management. Bullet. Russian New Univ. Ser.: Complex Syst.: Models Anal. Manage. **3**, 64–66 (2010)
11. Baskakov, A.V.: Information security policy as the main document of an organization in creating an information security system. Inf. Secur. **2**, 43–47 (2006)
12. Kalashnikov, A.O.: Information risk management in interacting organizational systems. Innov. Develop. Inf. Commun. Technol. **1**, 123–124 (2008)
13. Galushkin, A.A.: On the question of the meaning of the concepts «national security», «information security», «national information security». Hum. Rights Advocate **2**, 8–10 (2015)
14. Teoh, C.S., Mahmood, A.K.: National cyber security strategies for digital economy. J. Theor. Appl. Inf. Technol. **95**(23), 6510–6522 (2017)
15. Pouri, M.J., Hilty, L.M.: Conceptualizing the digital sharing economy in the context of sustainability. Sustainability **10**(12), 1–19 (2018)
16. Bubnova, G.V., Efimova, O.V., Sokolov, Y.I., Akopova, E.S.: Management of risks and economic processes in Russian railways OJSC in digital economy. Adv. Intell. Syst. Comput. **726**, 320–325 (2019)
17. Guo, S., Ding, W., Lanshina, T.: Global governance and the role of the G20 in the emerging digital economy. Int. Organ. Res. J. **12**, 169–184 (2017)
18. Order of the Government of the Russian Federation of July 28, 2017. p. 1632 «On approval of the program «Digital Economy of the Russian Federation» available at: http://www.consultant.ru/document/cons_doc_LAW_221756/
19. The decree of the President of the Russian Federation of may 9, 2017, p. 203 «On the Strategy of information society development in Russian Federation to 2017–2030». http://www.garant.ru/products/ipo/prime/doc/71570570/
20. Markov, A.A.: The concept and characterization of informational risks dangers and threats. Bullet. Volgograd State Univ. Ser. 7 Philos. **1**(11), 123–129 (2010)
21. Grubor, G., Barac, I., Simeunovic, N., Ristic, N.: Achieving business excellence by optimizing corporate forensic readiness. Amfiteatru Econ. **19**(44), 197–214 (2017)
22. Global Cybersecurity Index – 2017. https://www.itu.int/dms_pub/itu-d/opb/str/d-str-gci.01-2017-pdf-e.pdf
23. Kiseleva, I.A., Iskadjian, S.O.: Information risks: evaluation and analysis methods, IT portal, 2(14) (2017)
24. Vasilyeva, N.F.: Information risk as a component of economic risk in crisis management». Ind. Develop. Strategy Regulatory Mech. **3**, 37–49 (2011)
25. Rogers, R., Apeh, E., Richardson, C.J.: Resilience of the Internet of Things (IoT) from an Information Assurance (IA) perspective. In: 10th International Conference on Software, Knowledge, Information, Management and Applications, pp. 110–115 (2016)
26. Korostelkina, I.A., Popova, L.V., Dedkova, E.G.: Research of modern paradigm of innovative development of national economy. Bullet. Omsk Univ. Ser.: Econ. **4**, 44–56 (2018)

Digital Education

E-Learning Efficiency: Linguistic Subject Taught via Electronic Educational Resources

Aigul Sushkova⬛, Albina Bilyalova(✉)⬛, Dinara Khairullina⬛,
and Chulpan Ziganshina

Kazan Federal University, Kazan, Russia420008
abil171@mail.ru

Abstract. The globalization processes occurring in the world as well as constant scientific and technical progress are reflected in the field of education. Requirements for the quality of education along with teachers' qualification are constantly growing, the role of e-learning education is constantly increasing, and, as a result, boundaries between classroom/full-time and distance education via electronic technologies are blurred. The number of universities using e-learning technologies is growing year by year. The current study presents a research on results of e-learning in language instruction among students of Elabuga Institute of Kazan Federal University. The study aims at determining efficiency of implementing electronic educational resources education in linguistic subjects. The results of the study show that the majority of students approve of studying linguistic subjects via electronic educational resources. The study discusses various aspects of language instruction by means of electronic educational resources.

Keywords: Digitalization · Digital education · E-learning · Electronic educational resources · Language instruction · Linguistic subjects

1 Introduction

The era of digitalization, designed to optimize and increase productivity of business processes, remains impacting educational technologies since sharp changes in the labor market require new approaches to human resources. These requirements are dictated by the need to provide the development of scientific and technological progress by competent professionals and the need to multiply the existing human asset where each person is given the opportunity to work comfortably, getting satisfaction from his or her occupation.

The days when students were supposed to just sit quietly at desks, listening to their teachers and producing all they have heard and written are gone, thus, transmission of ready knowledge and skills from teacher to students is no more first consideration. Instructors should take advantage of new communication and information technologies [8]. The analysis of the literature on the investigated problem allows us to speak about the complex of knowledge, the ways and means to work with information resources, and about technical resources, which scientists have linked with the technological chain, ensuring the accumulation, storage, classification, output and distribution of

© Springer Nature Switzerland AG 2020
T. Antipova and Á. Rocha (Eds.): DSIC 2019, AISC 1114, pp. 197–206, 2020.
https://doi.org/10.1007/978-3-030-37737-3_18

information [1, 2, 6, 11, 14–16]. A lot of works deal with the question of using technology in teaching English [6, 12, 18]. Some scholars have noted that the widespread use of identified technologies can significantly increase the effectiveness of active learning methods to all forms of organization of educational process in the study of a foreign language, namely: practical, individual lessons, during independent work [8]. Other scientists have considered the use of ICT as a means of increasing motivation, commitment to the systematic study of a foreign language, which allows to obtain quick results in learning a foreign language [13]. The use of ICT in the educational process directed to the full immersion of students in the language environment that contributes to the formation and development of their communicative competence, the part of which is the sociocultural competence. It equips learners with digital age literacy, inventive thinking, creative thinking, higher-order thinking, effective communication, and high productivity.

The developing of information culture is one of the most important tasks of the education system nowadays. Requirements for students' skills have changed, since it is necessary not only to read, write and count, but also to be able to organize data resources, cooperate productively, collect, evaluate and use information [3–5]. We are living and working in times where new innovations are being developed exceptionally quickly. These in turn are driving faster growth and efficiency for businesses and society but there is a lack of digital skills. It is widely accepted that there is no single definition of the term 'digital skills' but most definitions make reference to the ability to use computers and digital devices to access the internet, the ability to code or create software, with higher level digital skills including the ability to critically evaluate media, and to navigate knowingly through the negative and positive elements of online activity. Digital skills are defined as a range of abilities to use digital devices, communication applications, and networks to access and manage information [12]. They enable people to create and share digital content, communicate and collaborate, and solve problems for effective and creative self-fulfillment in life, learning, work, and social activities at large.

In this regard, adaptation to changing conditions and requirements is necessary, which entails the digitizing education. This type of education would not be therefore possible without rapid development of computers and the Internet [13]. The prevalence of computers and broadband Internet has given a very strong impulse to use them in the educational activity.

Modern higher education is aimed at creating specialists striving for continuous self-education and self-development, ready for steady innovations and possessing character traits like mobility, creativity and competence. Student should develop skills for independent search and use the information obtained. This independent work is a necessary foundation of the educational process. Both e-learning and use of various digital technologies serve that purpose. Therefore, they have become a real must for present-day education worldwide.

Chris Rothwell, Director of Education, Microsoft UK says: "The role of the teacher has never been more vital than it is today. Within a rapidly changing world, the next generation must be prepared with the confidence, skills and lifelong learning mindset needed to succeed. Teachers have a key role to play in instilling this. Our research shows that teachers, as always, are eager to go above and beyond to nurture future-ready skills and innovate in the classroom. "What's important is that they have the support that they need to help them get here: access to great learning environments, opportunities for strong professional development and the chance to work in evolving, transformational environments that support our future leaders" [17].

In addition, time has grown into a most valuable aspect of life, thus, time management tips (proper planning and controlling one's time) are to be swallowed. E-learning, digital learning technologies provide students with immense access to education in comparison to traditional methods of teaching since students can study whenever and wherever it is convenient to them, at the same time they are given the option to study part-time or full-time. E-learning has modified the educational sector by enabling students to get, share and implement knowledge in a comparatively easy way. It should be noted that students who study one and the same subject by means of e-learning tools score considerably higher than the lecture condition [10].

According to the data from the Statistics Portal, a considerable ratio of faculty worldwide has expressed their desire to support education models in favour of digital rather than traditional. As a matter of fact, about more than a half of faculty approves of the use of open educational resources in teaching and gives preference to the competency-based education system. Students worldwide have also shown eagerness to apply to digital learning technologies and practices alongside with a desire to use various devices in the learning process [8].

The US is the leader in online education, and together with Canada it occupies more than 50% of the entire e-learning education market. It should be noted though that in the future the US and European online education markets are expected to reduce their shares due to the high growth rates of online education in Asia, with China taking the first place in the world in terms of growth of e-learning education services. Such dynamics is due to the fact that the markets of e-learning education in North America and Western Europe have already developed and passed the stage of avalanche growth while the markets in other countries are just beginning to introduce e-learning education, which can probably lead to such results. At the same time, Russian e-learning education market is increasing year by year [2].

Russian economist, public figure, founder and rector of the National Research University "Higher School of Economics", Y. Kuzminov gave an interview to a famous Russian newspaper informing that in 5 years' time the university will refuse from traditional forms of lectures, they will be replaced by electronic educational resources. To proves the necessity of such dramatic changes, he gives statistics that only 15–17% of students of Russian higher educational institutions attend lectures. At the same time, due to a huge classroom academic load teachers do not have enough time to get engaged in scientific work. Electronic educational resources, to his mind, are aimed to encourage students to get more involved in the education process and to increase the quality of Russian higher education. Last but not least, electronic educational resources will help teachers get rid of some academic load thus making them more involved in scientific work without changes in their salaries [9].

Digital education is sure to be more effective when students have motivation: they have a clear view of what they want to get from education and they should be able for self-organization and self-education [7]. Of course, this is easier said than done, but it is obvious that taking such responsibility for the radical changes associated with digitalization, students should not be left alone to go through all the difficulties of life. In this case, teachers should be mentors and guide students until they learn how to properly use electronic educational resources and use them to their own benefit. Besides electronic educational resources help students adequately assess themselves, which is so relevant among students who are apt to think teachers are subjective while grading them.

Digital education develops skills necessary for digital communication by future specialists. They will be ready to work in the digital educational environment and will be aimed at ongoing professional self-development. As a result, their personal digital profiles of competencies will include those that will allow them to use their knowledge not only for self-education, but also to make use of digital elements necessary for their occupation in the digital world thus optimizing their work.

Keeping up with time and influence of digital era, Kazan Federal University has accepted a new corporate educational programme – KPI (Key Performance Indicators) which gives a clear view how effectively the university is achieving key business objectives. Along with that, the university faculty is succeeding in conducting a part of classes with the help of electronic educational resources.

Information and communication technologies (ICT) are of key importance at all levels of the educational system. At each stage of cognitive activity, research and practical applications in all branches of knowledge ICT perform both the functions of tools and objects of knowledge. Consequently, ICT innovations not only provide a revolutionary development in this branch of knowledge, but also have a direct impact on the scientific and technological progress in all areas of society. Thus, information and communication technologies are a class of innovative technologies for the rapid accumulation of intellectual and economic potential of strategic resources, ensuring sustainable development of society [3].

The current study intends to find out how effectively e-learning education (classes conducted with the help of electronic educational resources) is implemented in studying linguistic subjects.

2   Materials and Methods

2.1   Methodology

The study is based on the theory of foreign and domestic scientists. To achieve the aims of the study the authors strove to obtain data by employing questionnaire survey in order to determine willingness for e-learning and effectiveness of electronic educational resources in linguistic subjects while learning foreign languages among the students of Elabuga Institute of Kazan Federal University. Besides, the study used literature analysis, direct and indirect observations.

2.2   Participants

The primary focus of the research was on the students of the faculty of foreign languages. The participants were 80 students with a year of study ranging from 2 (60%), 3 (20%) to 4/5 (20%). 56% of the respondents had 2–3 subjects taught by means of electronic educational resources.

2.3   Measure

The study used anonymous questionnaire developed by the authors in order to get a complete understanding of the issue mentioned and make the process of interpreting objective. The questionnaire consisted of eleven questions aimed to identify effectiveness of electronic educational resources while learning foreign languages as long as students' willingness to increase further number of subjects conducted by means of them.

3   Results

We first present the collected data concerning respondents' willingness to take classes using electronic educational resources. When students were asked whether they are eager to increase number of language courses carried out via electronic educational resources, the high percentage, over 45%, answered that they would like to have more linguistic subjects taught via electronic educational resources while the others did not show eagerness at all or were puzzled and uncertain about the answer, 30% and 25% correspondingly.

Figure 1 given below shows the range of positive responses to study by means of electronic educational resources among students of different years of study.

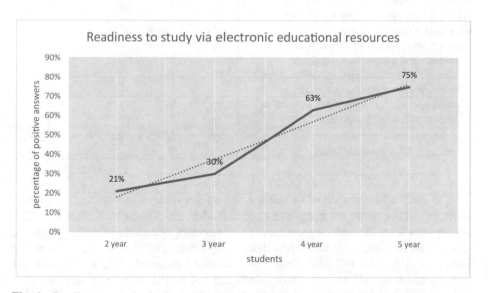

Fig. 1. Readiness to study via electronic educational resources among students of different years of study.

As it is seen, it should be taken into account that the majority of positive answers were given by senior students (4/5th year of study), yet sophomores (2^d year of study) and junior students (3^d year of study) were either apt to give negative responses or not sure about the answer at all.

4 Discussion

The analysis of the responses given by the students in regard to general presentation of subjects by means of electronic educational resources is seen from the table below (Table 1).

Table 1 Results of the questionnaire

Statement	Excellent	Good	Reasonable	Poor
organization of the educational process through electronic educational resources (schedule, weekly academic load, use of modern learning technologies, opportunity to learn from the comfort of home, etc.)	20%	65%	15%	0%
informative value of educational material (how fully topics are covered, new and relevant language teaching approaches	12.5%	70%	17,5%	0%
clear and reasonably structured curriculum, use of illustrative means (information coherence and consistency, clarity of presentation of educational material, use of visual aids such as schemes, models, drawings, examples, animations, interactive schemes, audio/video aids, etc.	22.5%	70%	5%	25%
learning via electronic educational resources – all aspects included	7.5%	67.5%	22,5%	2,5%

The study of the responses in Table 1 shows the students' assessment of general structure and content of the disciplines taught via electronic educational resources. As it can be seen from the table, a great number of students, approximately 67,5%, gave a good mark to electronic educational resources, about 17,5% assessed them at the highest level, for about 11,25% of students considered electronic educational resources quite satisfactory, and the remaining 1,25% did not find them instructive at all. It is obvious from the statistics in the table that the average mean value was "good", which indicates that the majority of the students tend to agree that there are still some drawbacks which withhold students from adopting electronic educational resources to full advantage in their learning process.

We should mention here that to know the technological side of electronic resources is only 1 step. Digital higher education is supposed to give some freedom to arrange the educational route, timing, pace and order of its implementation, to provide freedom of choice of educational content, including means of control and self-assessment, to provide digital devices and tutor support. For this purpose, classrooms should be re-equipped, curricula – reviewed, guidelines need to be prepared, assessing criteria and schemes – compiled [7].

Organization always plays almost the most important role in any activity. Hence in such a complicated and significant sphere as education, which influences all the other spheres of community life, organisation should be developed and considered thoroughly. Teachers should collaborate with professional peers before launching electronic educational resources. Along with this, as it has already been mentioned before, time management rules undoubtedly should be incorporated in the organization of electronic resources, i.e. timing should be comfortable for all students so that they can choose the most suitable period for studying themselves. Likewise, all human beings have different psychological peculiarities and temperaments, at the same time, our education has become inclusive. These facts prove the idea that everyone works at his or her own pace, e.g. completing one and the same task will take phlegmatic students and sanguine students different period of time due to objective psychological ground, the same can be said about average students and students with disabilities, especially if these disabilities concern mind's work or physical ones interfering performance of the task. That is why we should take into account all these aspects to make the process of digital education comfortable for each student.

Dwelling on the results of the third pattern in Table 1 we have to apply to psychology again. All people know that people have different ways of perception: visual – people learn the world and are affected with the help of their eyesight; auditory – they are forced and motivated by listening to others or audio records; kinesthetic – people explore the world, some new information through their experience and feelings, this type presupposes personal involvement, the use of gestures, touches; and the last type – digitals – they learn new things using logical chains, they structure and classify information in the right order. Reviewing all these types of perception one cannot but come to conclusion that electronic educational resources should be developed meeting the needs of each type, i.e. tasks set by the teacher should combine all types of such tasks not to make some students feel ill at ease.

According to the data regarding the students' perceptions concerning their comprehension level, the students were asked the following question: "Was the explanation of the educational material via electronic educational resources clear?" and the answers turned out like this (Table 2):

Table 2 Comprehension level

Answer	Percentage
Yes, absolutely	17%
Yes, yet there were some difficulties	75%
No, I would prefer teacher's explanation in the classroom	7,5%

The results in Table 2 show that the majority of students (75% per cent) admit that comprehension of disciplines by means of adopting electronic educational resources is quite possible, though sometimes deals with certain difficulties. The reasons for it could be quite diverse: either the technical errors or teachers' miss. In both cases developing of tasks/curricula should be developed carefully and review and tested several times by several peers, since the things obvious for a teacher can be not clear or comprehensible for students, thus every detail missing can play even a fatal role in understanding the material.

At the same time, the learners do not treat the conservative way of teaching as the only possible one (17%), it ceased to be their first consideration because they keep up with the times: the way of thinking is changing, it is rational and constructive, students are becoming independent.

According to the data obtained, the majority of students (87,5%) are satisfied with the present system of assessment via electronic educational resources: "Yes, completely; it does not imply subjectivity of a teacher" (32,5%), "Yes, quite; still there are some technical debts" (55%). On the contrary, only 12,5% of the students still tend to the conservative way of getting assessed: "Not really; I consider teachers' evaluation during classes to be more effective than any other because active work during classes and individual characteristics as well as many other aspects should be taken into account in the process of assessment".

The questionnaire contained the following question: "Did you get enough knowledge for further professional activity via electronic educational resources?" Table 3 contains answers to this question.

Table 3 The results of the responses

Answer	Percentage
Yes, enough	52%
Yes, but I would prefer to combine it with classroom work	40%
No, they lack theoretical knowledge	2%
No, they lack practical skills and abilities	6%

The results of the responses shown in Table 3 suggest that not all students are yet ready for getting professional activity by means of electronic educational resources. They strongly believe that it is impossible to obtain practical skills necessary for further professional life through the computer. Such conservatism of thinking can be caused by the conservatism of teachers in school most of which are opponents of changes inspired by the time. At the same time, it should be noted that overall digitalization cannot but influence the majority of students, they are aware of time-management principles and ready to adjust for rapidly changing rules of modern digital world.

5 Conclusions

The research shows that students of Elabuga Institute of Kazan Federal university are contented with learning languages via e-learning tools (in particular, electronic educational resources). The results of the questionnaire suggest that the majority of students believe they have acquired good knowledge of linguistic subjects in the process of e-learning. Meanwhile, as it is seen from the data obtained, it should be acknowledged e-learning via electronic educational resources has a number of advantages. Firstly, electronic educational resources facilitate the process of studying due to reasonably structured curriculum and clarity of presentation, with its range of illustrative means (visual aids such as schemes, models, drawings, examples, animations, interactive schemes, audio/video aids, etc.). Secondly, they economize time spent on way to and from the university, as a practical aspect of time management. Thirdly, electronic educational resources make the process of getting knowledge easier for people with disabilities. Fourthly, they are characterized with flexibility: students can choose the time and the place of studying themselves. Finally, they provide objectivity of assessing process.

We cannot but pay attention to the fact that still there is much work to be done, since the data obtained revealed some difficulties emerging in the process of studying via electronic educational resources. All these difficulties can be grouped in one: technical debts. While creating electronic educational resources, instructors should pay much attention and think over all the problems which can further arise. Some minor shortcomings are difficult for the author of the course to find, thus, before launching courses, several instructors should test them and give reviews in order to avoid technical debts.

All in all, we should admit that the state-of-the-art way of teaching in the digital era provides students with effective instruction, enabling them to develop relevant and necessary competencies to meet the needs of the modern market.

Finally, we would like to emphasize the following: digital tools are rapidly changing the educational landscape, and the future-ready teacher is truly a lifelong learner, ready to adapt to the frequent changes to teaching and learning. The future-ready teachers don't fear what they don't yet know, because they are confident in their ability to learn. They know that they can use their digital skills to support their own learning, to find new digital resources, and to connect to the online community of educators. As lifelong learners, the future-ready teacher knows that whatever new technologies enter their classroom, they have the digital skills to make sense of them and to seek out professional development opportunities that will enable them to learn and grow as an educator. But most importantly, these digital skills will ensure that their students have the skills to be future ready as well.

References

1. Aoki, K.: The use of ICT and e-learning in higher education in Japan. Int. J. Soc. Behav. Educ. Econ. Bus. Ind. Eng. **4**(6), 986–991 (2010)
2. Bataev, A.: Analysis of the world market of distance education, Young scientist, №20, 205 (2015)
3. Bilyalova, A.: ICT in teaching a foreign language in high school. Educ. Health ICT Transcultural World **237**, 175–181 (2017)

4. Bilyalova, A., Vasilyeva, A., Islamova A., Akhmetshina A.: Teaching academic writing in a foreign language with information and communication technologies. In: International Conference « Topical Problems of Philology and Didactics: Interdisciplinary Approach in Humanities and Social Sciences « (TPHD 2019). Advances in Social Science, Education and Humanities Research (ASSEHR), vol. 312, pp. 67–71 (2019)
5. Bilylova, A., Salimova, D., Zelenina, T.: Digital trnsformation in Education. In: Antipova, T. (ed.) Itergrated Science in Digital Age. ISIC 2019. Lecture Notesin in Networks and Systems, vol. 78. Springer, Cham (2019)
6. Dudeney, G.: How to Teach English with Technology. Pearson, Longman (2008)
7. Dyakova, E., Sechkareva, G.: Digitalisation of education as a basis of training teachers of the XXI century: problems and solutions. Bulletin of Armavir state pedagogical Universit, No. 2, 24–35 (2019)
8. E-learning and digital education – Statistics & Facts. The Statistics portal. https://www.statista.com/topics/3115/e-learning-and-digital-education/
9. Kuzminov, Y.: Lectures are transferred into the online classroom. Yaroslav Kuzminov is going to oblige teachers to read courses in the digital format. Kommersant, 179 (2018)
10. Lee, P.Y., Hui, S.C., Fong, A.C.M.: An intelligent categorization engine for bilingual web content filtering. IEEE Trans. Multimedia 7(6), 1183–1190 (2005)
11. McKinney, D., Dyck, J.L., Luber, E.S.: iTunes University and the classroom: Can podcasts replace professors? Comput. Educ. 52(3), 617–623 (2009)
12. Rakhimova, A.: Advantages of the use of computer technology in teaching foreign language. Foreign Lang. Sch. 10(2012), 56–60 (2012)
13. Seljan, S., Berger, N., Dovedan, Z.: Computer-assisted language learning (2004)
14. Tinio, V.L.: Survey of ICT Utilization in Philippine Public High Schools (2002)
15. Pegu, U.K.: Information and communication technology in higher education in india: challenges and opportunities. Int. J. Inf. Comput. Technol. 4(5), 513–518 (2014)
16. Wheeler: Information and communication technologies and the changing role of the teacher. J. Educ. Media 26(1), 717 (2001)
17. Vrana, R.: ICT-supported communication of scientist and teaching staff at the faculty of humanities and social sciences in Zagreb. New Library World 111(9/10), 413–425 (2010)
18. Youngsters risk leaving school unprepared for workplace, Microsoft research reveals. https://news.microsoft.com/en-gb/2019/01/23/youngsters-risk-leaving-school-unprepared-for-workplace-microsoft-research-reveals
19. Zakharova, I.G.: Information Technologies In Education: Textbook for Higher Educational Institutions. Academy (2008). 338 p.

Higher Education in Digital Age

Albina Bilyalova[1]($\boxtimes$) ⓘ, Daniya Salimova[1] ⓘ, and Tamara Zelenina[2]

[1] Kazan Federal University, Kazan 420008, Russia
abill71@mail.ru
[2] Udmurt State University, Izhevsk 426034, Russia

Abstract. The article presents a theoretical understanding of the content of the stages of university development and its transformations depending on the development of society. The paper describes four university models that were specific for historical stages (University 1.0, University 2.0, University 3.0, University 4.0.). Particular attention is paid to identifying features of the University 4.0., a promising model of universities that combines physical and virtual space, developing on digital platforms. The authors conclude that at the present stage of development of society there are obvious trends in the development of universities and higher education, including a change in the status of universities due to increased risks in science and experimentation, the transition from competition to partnership interaction, access to large databases (Big Data), the transition to multi-format open educational resources (Open Online Resources), a combination of new and traditional training formats, redevelopment of classrooms in the format of open space (Open Space Education). Obviously, such technologies as mobile training (Bring Your Own Devices), transition to the "inverted classes" (Flipped Classroom), the creation of "wearable" technologies like Google Glass, the development of adaptive learning through the introduction of digital platforms and the spread of the Internet of things will be widespread in the nearest future.

Keywords: University · Higher education · University model · Digitalization, science · Business · Corporate university · Research university · Innovative university · Digital university

1 Introduction

Discussion of the university's role in society, its mission and key tasks has been and remains a subject for scientific and public debate for centuries [1, 2, 9–11, 14, 15].

Since the University is a centuries-old civilization project, changes in society, economy and culture entail changes in the understanding of the university. It is important to note that the transformation of universities is associated, on the one hand, with changes in the external to the universities of the world (economy, technology, society and man), on the other hand, with the development logic of the universities themselves as leading cognitive institutions. Historical experience shows that universities play not a simple "function" of social systems, but they are active participants and actors in shaping the future, having their own ideas about the future, priorities, tools and resources for their advancement.

T. Antipova and Á. Rocha (Eds.): DSIC 2019, AISC 1114, pp. 207–219, 2020.
https://doi.org/10.1007/978-3-030-37737-3_19

The development of new frames of university ideas can be justified from the standpoint of Bauman's "fluid modernity", which postulates the variability and openness to changes, while not denying previous eras, but emphasizing the instrumental crisis, when the available tools do not correspond to the new problem formats [3]. The creation of a unified European educational space, network models of open online resources are attempts to find new, more convenient and effective tools for the promotion, development and enrichment of higher education content.

This also implies the statement that the search for a tool from the standpoint of only one scientific discipline was doomed to failure, because for a long time the sciences distanced themselves from each other, outlining the boundaries around themselves in the hope of understanding and objectifying the subject and the results of research to the maximum.

The modern era differs from the previous ones in the development of digital technologies, globalization processes that converge ideas, meanings, cultural codes, technologies, ways and methods of working with information and materials. It is logical that changing the conditions of existence of an object affects the object itself. But at the same time, it is important to take into account that throughout the entire existence, universities have remained a mobilization resource for the development of society and culture, having a lasting importance. Consequently, universities, as a centuries-old civilization project acting as an axiological core, at the same time undergo some transformation. What is the transformation expressed in? First of all, it is expressed in changing the instrumental set and expanding the range of social order. The transformation of a university implies a change in its mission, its functions in society and in accordance with this it changes the complex of actions performed, the types of technologies used and organizational forms.

What are the prospects for the development of a modern University? What does it retain as a construct and what are subjects to change? To answer these questions, it is necessary to turn to the historical stages of development of the University and give a detailed description of each of university generation.

2   Research Methods

The research methodology of the study is based on the approach to ontological analysis of the education phenomenon and its institutions. The object is studied in three fundamental and interrelated dimensions: universum – social, economic and cultural realities (context); generative-constituting – institutionality and activity; ontogenic – models, paradigms, universals.

In the cultural and historical genesis of the University, global factors that determined advancement of the fourth mission of University were studied. Methods of cultural-historical reconstruction, social modeling, economic analysis, theory of learning cognition were used.

Basic types of the University activity were studied from the perspective of national prerequisites for its development. Methods of structural-functional analysis, theory of innovations, were applied.

The study of the social and cultural conditionality of the University models was based on the methods of anthropology, psychology of culture, theory of motivation, axiology.

3 Corporate University (University 1.0.)

The word "University" comes from latin "universitas magistrum et scholarium", which means "teachers and students corporation" [18]. Like any structural unit of language, the word "universitas", obeying the cultural and temporal dynamics of the significate, has undergone changes. Initially, this term was used in the meaning of "corporation", and later as "association", "aggregate". Why "corporation"? The emergence of the first European universities dates back to the XI-XIII centuries, the period of development of medieval cities, where the majority of teachers and students were consolidated. Actually they did not have the rights of citizens, so they were forced to unite in corporations so called workshops, which served to form a special atmosphere. The discourse of the University, which included education, upbringing, psychotherapy and internal dialogue, developed a special language spoken only by those who were involved in it. On the one hand, such language made university discourse strictly directed and closed, almost ritual. On the other hand, a lifestyle based on dialogue and the pursuit of knowledge made this structure open. "Corporate University" met the needs of society in training professionals through the corporate culture of the first European universities. The axiological basis was the "cultural model" of intelligence, the criteria components of which was the Trinity: the Culture of Thinking, the Culture of the profession, the Culture of the University Corporation. Important was not only knowledge of the profession from within, but also profile ritual practices. Actually, culture was the external referent of the university. The key category was "education", which was the most important point in the organization of the educational process at the entrance and exit, and the methodological basis was the classical paradigm of education, the ultimate goal of which was to prepare for life and work.

Mostly the first European universities were focused on the "free arts", medicine, law and theology. Mastering the "free arts", students not only received knowledge, they received the language spoken by all the educated people of that era. The medieval university carried out the reproduction of the elite, who later possessed the instruments of social control. The church needed a rational theological doctrine, the city needed artisans, doctors and lawyers - hence the dual nature of the European University of the Middle Ages. On the one hand, this is a market relationship when the teacher received payment from students for classes, on the other hand, a benefit from the church, which posits that education is a gift of God, therefore, the sale of knowledge is a sacred trade [13]. Such a dichotomy still manifests itself in the conflict between the private form of higher education organization focused on applied research and specific needs of the consumer, and the state form of organization focused on culture and fundamental research. In the medieval University there was practically no deep research, and teaching was often reduced to the reading of texts which commented on by the Professor, which was explained as a lack of books and scholastic system of knowledge.

4   Research University (University 2.0.)

The development of the "Research University" was influenced by the enlightenment, in the center of which were the products of human activity, and the logical way out was the scientific and technical revolution and the construction of industrial society. The University postulated learning through research as a possible extension of the horizons of human thinking, embodying it in the philosophy of science, which became the new religion of European society. The axiological basis was the "academic model" of intelligence, the criteria indicators of which were: the ability to hypothetical-deductive thinking and knowledge of the classics in the original. The key categories in University life were "education" and "research", organically synthesized into a single instrument of knowledge of the material and intangible world. "Research University", which became a collective image of various forms of University life ("Intellectual University", "Research University" and later "University of culture"), was a place where universal knowledge (stadium generale) was taught. Free thinking through the search for truth, objectified by the language of philosophy and cleared of all "superficial", became one of the distinctive features of the Research University model. The focus of attention is not so much corporate culture (professional culture) but life guidance and eternal values laid down by the enlightenment, which turned its attention to the origins of European culture. Knowledge of the original texts, which were the language of communication between different professional groups, became even more important, instead of the hermetic corporate culture, which, moreover, had not accumulated much of their own works at that time. The ability to think, to have their own opinion about the subject became important for the development of a free personality, who was able to rise to the cultural level of the era.

The external referent of the University 2.0. was the truth. The search of the truth was a process and can only be achieved through research and dialogue, the so-called "socratic" communication. The state began to act as the main initiator of university life, displacing the church from these positions. The need to develop a civil society and the self-justification of the national state, dictated by the conditions of the time, forced the state to pay attention to universities, to finance them and to carry out social orders. Thus, the University could not be divorced from the state, since it became an approbation platform for the realization of the plans of the latter.

5   Innovative University (University 3.0.)

The development of a "Innovative Technological University" was triggered by the restructuring of the socio-economic architecture of a "third wave" society, the so-called post-industrial society, where the basic components were: idea, technology and capital [4, 17]. States began to need the development of science to support production processes and stabilize economies. The crisis of classical science (pure science) was expressed in opposition to applied research. In the early 2000s, universities began to play a leading role in the commercial development of scientific knowledge. As a result of the interaction of universities and industry, scientific discoveries are translated into innovative products and commercialized in the framework of suitable business models.

Mature entrepreneurial universities simultaneously carry out educational, research, and commercial activities that stimulate each other [11]. University campuses turned out to be platforms for the development of innovative culture and high technology with the subsequent output by students to create their own businesses, providing them with jobs as an example of the highest educational results and providing the development of competencies, as an example of core compliance [8]. In that period of time a self-construction of the profession was observed, which entails a crisis of professional culture and a crisis of the academic model of intelligence, which is not capable of ensuring the breakthrough nature of the learning process. That's why the competency model of University 3.0., especially at the early stages, served as an axiological basis, as an attempt to reorganize the educational process. Transdisciplinarity was developed as a comprehensive tool for learning and finding opportunities for pedagogical work with implicit knowledge and understanding of the principles, including through a matrix of competencies (knowledge, skills, marstering). Nowadays the educational process is increasingly being produced through group (network) interaction, for example such forms as the "inverted class" appear [6, 7]. There is a paradigm shift towards the postclassical paradigm of education, the purpose of which is already to ensure the conditions for self-determination of the student. The functioning of the university is provided through the integration of three key categories: "education", "science" and "business". Multiple innovative breakthroughs emerge which form entire spaces: Silicon Valley, Road 128, technopolis cities, etc. Business increasingly interferes with university activities using venture investment tools, creating science parks, business incubators, etc. Science has been put on the conveyor, and production is no longer without new technologies. There is a phenomenon of "multiversity" by K. Kerr. In these conditions we can observe that the national state, which acted as an intermediary between universities and society, itself began to experience a crisis in the context of globalization convergence. The crisis is expressed in the total bureaucratization of higher education, extrapolating the principles and mechanisms of functioning of factories to universities. Universities, education and science (up to academic publications) began to turn into a profitable business, reinforcing the phenomenon of "academic capitalism" [19]. The need for narrow specialists, the possibility of self-determination multiplied the number of disciplines, blurring the disciplinary core of educational programs. The external referent of the university becomes quality, measured at different levels - from monitoring public and government organizations to international ratings (THE, QS, ARWU) and organizations (OECD, UNESCO).

In the strategy of transition to the University 3.0 model, the following main components can be distinguished: (1) socio-academic: the transformation of the university structure; changes in the academic environment, the educational process and teaching activities; advanced scientific and educational development; (2) research and innovation: the development of centers of research and technological superiority, the development of a system of open innovation, the implementation of the concept of "university in the center of the innovation-business ecosystem"; (3) economic: flexible response to labor markets (dialogue with industry), orientation to the principles of a network economy, management of intellectual property, economically promising elements of corporate and multi-campus universities models.

In its complex social role, University 3.0 not only supplies personnel or research products. To a much greater extent, its role is to educate innovative-type specialists who are competent to move from research to development with their subsequent commercialization. The social role of university 3.0 implies the creation of basic structures of the knowledge society. University 3.0 becomes the basis of the global competitiveness of national economies, and its entrepreneurial ecosystem forms new, fast-growing industries, promising technology markets, and economically leading administrative and territorial spaces.

6   Digital University (University 4.0.)

A fourth-generation university at a given time can only be described by a draft design, since its pragmatic and sociocultural aspects are still being formed. The development of a "Digital University" is a perspective due to the development of digital platforms and analytical applications and a new industrial revolution. A change in the ecosystem of a society entails a change in the social order. Already today one can observe the development of educational hubs, network communities and a number of other new forms of organization of university life activity, open educational resources with varying combinations of education are rapidly developing.

Breakthrough technologies require universities to restructure the organization and essence of education. The target unit, apparently, will be ways of "opening up" the talents of a person and "flashing" his life scenarios, through the synthesis of biology and "smart technologies", as well as the development of the noosphere. Evaluation of current processes is already outdated only taking into account industrial indicators and an academic model of intelligence. It is likely to be replaced by a model of multiple intelligence, when the evaluation criteria are concretized under given conditions, adapting to the peculiarities of a person's thinking, and not vice versa. It is obvious that the academic model of intelligence, when various ways of working with information are evaluated one-sidedly (in fact, the assessment is only one of the many characteristics of the tool), has exhausted itself, because it limits the hidden possibilities of human intelligence.

Digital applications (scripts) for example electronic educational resources will be fully adapted to the needs of the person and finally supplant the classical educational programs and the linear way of transmitting information. Students will be able to study anywhere and at any time [5]. The disciplinary core will finally be destroyed, which will be replaced by thematic education when the phenomenon is investigated, which will enhance transdisciplinarity both in science and in education. The university will stop only two-dimensional existence in the physical space, expanding its presence in virtual reality through cloud technologies, including in the format of a network partnership with distributed control. The key categories will be: "creativity" (a person acting as a creator), "ecosystem" (development and sometimes creation of which will become one of the key educational goals) and "business" (as a regulator of inter-institutional relations). Metaindividuality will be fixed in the educational process, and self-design through the tools of educational design and intelligent machines will become a form of education.

The formal assessment system will cease to exist because of its meaninglessness. A viable product designed by a student, or a relevant contribution to the development of a local ecosystem of society, will make an assessment of the effectiveness of the development of educational programs. It is the solution of specific global and local problems (resources, hunger, ecology, epidemics, viruses, etc.) that will reinforce the need for societal involvement in higher education and university science.

Digital platforms, institutionally replacing the former channels of storage, processing and receiving information, have a disruptive effect [5], becoming "destructive" for the usual educational forms. Analytical applications and educational resources, adapted to the needs and capabilities of a person, fundamentally change the ways of education, therefore, its architecture, principles, goals and essential characteristics. The phenomenon of Learning on Demand occurs when an educational product is constructed according to an individual demand.

Education, as a process and a product, goes beyond the framework of a national state and ethnic culture. Multicultural education arises when professors and students begin to think in categories combined from two levels (national and supranational), which ultimately leads to asynchronization of identity, multiplicity of cultural realities, "doubling of cultures". In the educational content, the share of theoretical knowledge is gradually reduced, compensated by specialized applications. Theoretical knowledge will change the character of education, whose bases are not knowledge of facts and material, but complex analytics, postulated on a chain of multiple cognitive operations.

Apparently, the social dichotomy between the mass "pop education" integrated into the game shell through gamification and adaptation and the elite education accumulating the expensive "engineering" knowledge about the content of advanced technologies will increase. Here it is necessary to underline that the development of gaming techniques and the Internet availability of information content of educational platforms will create the illusion of the availability of high-quality higher education. Thus, scientific and technological progress is changing the face of modern society. Waves regularly emerge of further breakthroughs in various fields: from deciphering information recorded in human genes to nanotechnology, from renewable energy to quantum computing. It is the synthesis of these technologies and their interaction in the physical, digital, and biological domains that make up the fundamental difference of digital era [16]. Sinusoidal development of technologies provokes not only the emergence of new ways of producing goods and services, but also axiologically restructures human needs and capabilities. Some professions are interchanged by others, extrapolating the social order to universities. Higher education is defragmented, turning into cloud-based packages of courses demanded by a person at various stages of his professional life. The constantly updated reality of education will be constructed in a dialogue mode, being in a process that has no framework, no end and no results. Already, there is the question of the final point of transition from formalized to person-centered practices.

"Lifelong learning" becomes an objective necessity, in this connection, the classical forms of education are to be transformed and in the future will become narrow-targeted short-term courses with a way to confirm the core competencies. The development of "smart technologies", digital applications and educational resources MOOC (Massive Open Online Courses), SPOOC (Self-Paced Online Courses) and a number of others, is a disruptive factor of the global educational space, gradually changing the ways of education and its organization. Academic institutions and universities are often viewed as the forefront of developing progressive ideas, especially in the realities of global risks. At the same time, recent studies show that professors, based on career considerations and financing conditions, prefer phased conservative research instead of bold innovative projects [12]. The antidote to the conservatism of research in the scientific community, experts say, is to stimulate the commercialization of scientific research, the active participation of governments in progressive projects, the private sector and local communities, whose needs can first be met by the latest developments.

The model of University Model 4.0. (see Fig. 1) has a beneficial effect on universities' impact on the economic and social development of society, which is implemented through the quality of education, compliance of graduates with market demands, conducting qualitative and significant research, interaction with various groups of stakeholders of the University and taking into account their interests, competitiveness in the international market of educational services, making stable income flows and less dependence on budget funding.

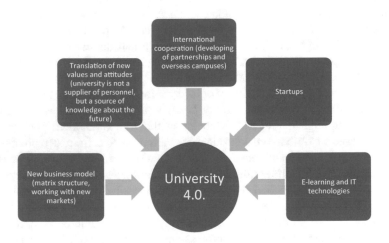

Fig. 1. The model of University 4.0

University Model 4.0 includes: (1) the development of new business model (matrix structure, working with new markets); (2) the translation of new values and attitudes; (3) international cooperation; (4) launching of startups; (5) deployment of a communications network (e-learning and IT technologies).

A comparison of the conceptual models of Universities Format 1.0, Format 2.0, Format 3.0 and Format 4.0 is presented below (see Table 1).

Table 1. Comparison of conceptual models of university generations

University Generation	Description	
University 1.0.	*Goal*	Elite reproduction
	Consumer	Church, city
	Key category	Education
	Axiological basis	Cultural model (culture of thinking, culture of the profession, culture of the university corporation)
	Features of the organization	Students and Teachers Corporation
	Features of the educational process	Scholastic teaching system
	Format of study	Monologue
University 2.0.	*Goal*	Creating a civil society; national idea (self-justification)
	Consumer	State
	Key category	Training, research
	Axiological basis	Academic model (knowledge of the classics in the original, the ability to hypothetical-deductive thinking)
	Features of the organization	A cluster of university and state
	Features of the educational process	Learning via research
	Format of study	Dialogic or Socratic Communication
University 3.0.	*Goal*	Innovation and technology development
	Consumer	State, manufacturing sector
	Key category	Education, science, business
	Axiological basis	Competency model (knowledge, skill, mastering)
	Features of the organization	A cluster of university and state
	Features of the educational process	Training through specialized competencies
	Format of study	Group (network) interaction
University 4.0.	*Goal*	The development of the noosphere and individual ecosystems of society
	Consumer	Different communities (formal and informal), transnational organizations, individual
	Key category	Creativity, ecosystem, business
	Axiological basis	Multiple intelligence
	Features of the organization	Physical and virtual (cloud) existence, network organization
	Features of the educational process	Meta-individuality
	Format of study	Self-designing

The table presents comparative analysis of conceptual models of university generations based on the following criteria: Goal, Consumer, Key category, Axiological basis, Features of the organization, Features of the educational process, Format of study.

According to the comparative analysis presented it is quite obvious that in University model 4.0. a formal assessment system will cease to exist beyond meaninglessness (as a relic of church control, then state control), and a viable product designed by a student or a relevant contribution to the development of the local ecosystem of society will evaluate the effectiveness of mastering educational programs regardless of the phenomenon being studied. In fact, it is the solution of specific global and local problems (resources, ecology, epidemics, viruses, etc.) that will strengthen the need for the involvement of societies in higher education and university science.

7 Conclusion

To date, there are no uniform methodological approaches to separate the levels of development of universities. In the study, we adhered to the following classification of universities: Corporate University (University 1.0.); Research University (University 2.0.); Innovative Technological University (University 3.0.); Digital University (University 4.0.).

University 1.0 is an institute of society, implementing the function of education. It is effective in the translation of knowledge, the development of students' talents, training (focused on traditional sectors of the economy), is able to act as a social elevator.

University 2.0 is an institute of society, implementing on a par with the educational, research function. We call it "Classical University" in the spirit of Humboldt University. It is capable of generating new knowledge through research activities, it performs research on the orders of the industry, as well as creates technologies under the order of partners, and it is not able to manage intellectual property.

University 3.0 is an institute of society, which implements the function of technology transfer and delivery to end users along with educational and research functions. In "University 3.0 » there is an effective process of commercialization of technology, in this model of universities entrepreneurial culture is developed, technological start-UPS are created, patents are registered for the University (which are further can be sold to partners, or individual rights can be transferred on the basis of license agreements). The University is able to establish an effective dialogue with representatives of the business community: quickly responds to requests for the release of new specialists that meet the expectations of the market; carries out research in areas of interest to industry representatives.

University 4.0 is a university in a cognitive society. Its mission is to ensure the production, reproduction and application of intelligence in the scales and forms characteristic of this society. The university of this generation is technologically connected with the digital-communicative revolution. As the intellectual leader of this revolution, the university itself becomes networked, geographically distributed, international; its key activities (education and awareness, research and production of innovations, expertise and consulting, design and creation of new practices) are largely transferred to the virtual space. Digital University is the University of Future.

If we speak about the landscape of modern education it is diverse. Higher education systems are developing today as institutionally complex structures that build learning based on organizations from different professional spheres of society. A socially and economically significant element of this structure is the higher education (University 3.0.). Its institutional base consists of scientific institutions, high-tech companies, innovative firms, industry consortia, and institutes of innovative growth. Entrepreneurial ecosystems are becoming a place for the development of effective mechanisms for the transfer of technology, scientific and engineering innovations. The universities that make up this sector carry out three main social missions: education, research, commercialization of knowledge. They are built on the basis of interconnected models of network, creative, innovative and entrepreneurial universities. Digitalization leads to the using of the network model that forms a cross-institutional environment for creative learning and creates cost-effective structures for scientific and educational cooperation. The creative model provides training for scientifically and economically productive workers in the intellectual sphere necessary for the university's business ecosystem in the digital era. An innovative and entrepreneurial model forms the structures and processes that ensure the competitiveness of network innovation partnerships and the socio-economic "output" of individual creativity.

The new role of universities in the knowledge society is manifested in the fact that, firstly, the contribution of modern universities to the innovative development of the country is increasingly determined by the value of created and commercialized intellectual property. Secondly, modern universities are the institutions of society that best solve the problem of transferring knowledge to intellectual capital through the use of resources of globality, openness, dynamism, a constant influx of active youth. Thirdly, universities no longer only execute research and development orders, but actively create technologies and technology companies themselves. Fourthly, universities today are becoming leaders and centers for creating new technological industries.

In the current situation, there are obvious trends in the development of universities and higher education, including changing the status of universities by increasing risks in science and experimentation, the transition from competition to partner interaction, access to large databases (Big Data), transition to multi-format open educational resources (Open Online Resources), a combination of new and traditional formats of preparation, redevelopment of classrooms in the format of open spaces (Open Space). Such technologies as mobile training (Bring Your Own Devices), transition to the "inverted classes" (Flipped Classroom), creating "space designer" (Makerspaces) - high-tech platforms using 3-D printers, creating "wearable "Technologies like Google Glass, the development of adaptive learning through the introduction of digital platforms and the spread of the Internet of things will be the main tools at university training [5].

Changes in university identity are inevitable. The tendency of reconstructing the format of its conceptual model is obvious. The upcoming digital era will require all participants to integrate activities of universities, local communities, states, international organizations, transnational companies, as the degree of interdependence of actors increases without regard to binding to the educational area and ultimate goals of education.

On the way to "University 4.0" Russian universities still need to go a long way. In our opinion, the following should be considered as the key tasks on this way:

1. To transform Russian universities in the direction of taking into account the trends of the knowledge economy.
2. To implement the interaction between universities and the business community as efficiently as possible.
3. Transformation of the institute of higher education in the process of developing talents, including by embedding training in working out in modern economic and social realities into the learning process.
4. Transformation of universities into centers of regional and sectoral ecosystems, focusing on them tools and resources of cluster and sectoral development.

References

1. Auzan, A.A.: Missiya universiteta: vzglyad ekonomista [University's Mission: Views of Economist]. Voprosy Obrazovaniya **3**, 266–286 (2013)
2. Baughn, C.C., Neupert, K.E.: Culture and national conditions facilitating entrepreneurial start-ups. J. Int. Entrepreneurship **1**, 313–330 (2003)
3. Bauman, Z.: Liquid Modernity, p. 228. Blackwell, Cambridge (2000)
4. Bell, D.: The Coming of Post-Industrial Society: A Venture in Social Forecasting. Basic Books, New York (1973). 616 p
5. Albina, B.: ICT in teaching a foreign language in high school. In: Education, Health and ICT for a Transcultural World, vol. 237, pp. 175–181 (2017)
6. Bilyalova, A., Salimova, D., Zelenina, T.: Digital Transformation in Education. In: Antipova T. (ed.) Integrated Science in Digital Age. ICIS 2019. Lecture Notes in Networks and Systems, vol 78. Springer, Cham (2020)
7. Albina, B., Nailya, S., Anifa, A.: Electronic educational resources in foreign languages teaching. In: Proceedings of the International Conference on the Theory and Practice Of Personality Formation in Modern Society (ICTPPFMS 2018), Vol. 198, pp. 173–177 (2018)
8. Clark, B.: Creating Entrepreneurial Universities: Organizational Pathways of Transformation. Emerald Group Publishing Limited, Bingley (2001). 180 p
9. Collini, S.: What Are University For?. Penguin Group, London (2012)
10. Cole, J.R.: The Great American University: Its Rise to Preeminence, Its Indispensable Nation Role, Why It Must be Protected. PublicAffairs, New York (2010)
11. Etzkowitz, H.: The Triple Helix: University – Industry – Government. Innovation in Action. Routledge, London (2008)
12. Foster, J.G., Rzhetsky, A., Evans, J.A.: Tradition and innovation in scientist's research strategies. Am. Sociol. Rev. **80**(5), 875–908 (2015)
13. Kislov, A.G., SHmurygina, O.V.: University idea: retrospective, versions and perspectives. Obrazovanie i nauka. №8. p. 96–122 (2012)
14. Marhl, M., Pausits, A.: Third mission indicators for new ranking methodologies. Eval. High. Educ. **5**(1), 43–64 (2011)
15. Readings, B.: The University in Ruins, vol. 2009, №11, pp. 50–52. Harvard University Press, Cambridge (1996)
16. Schwab, K.: The Fourth Industrial Revolution. Crown Business, New York (2017). 192 p
17. Toffler, A.: The Third Wave. Bantam, New York (1984). 560 p

18. UniversityOnline Etymology Dictionary. [Electronic resource]. - Access mode: https://www.etymonline.com/search?q=university
19. Welsh, R., Glenna, L., Lacy, W., Biscotti, D.: Close enough but not too far: assessing the effects of university-industry research relationships and the rise of academic capitalism. Res. Policy **37**(10), 1854–1864 (2008)

"Foreign Language" Potential Characteristics in Intercultural Competence Formation Process

T. V. Lyubova[1,2]([✉]) [ID]

[1] Kazan Federal University, Kazan 420008, Russia
lubovatv@bk.ru
[2] Esenin Str., 4,45, Naberezhnye Chelny 423802, Russia

Abstract. The article justifies the formation of intercultural competence need among university students in the terms of digital education. The priority role of a foreign language academic discipline in the formation of the foregoing competence is determined. The main problems of the intercultural competence development, a variety of basic pedagogical approach provisions aimed at tolerance nurturing of the younger generation and facilitating to a holistic perception of the cultural picture of the world structured by languages and the value systems of each nationality are considered.

Keywords: Digital education · Intercultural competence · Globalization · Bologna process · Pedagogical approach · Academic discipline foreign language · Tolerant consciousness · Intercultural communication · Competitiveness · Multiculturalism

1 Introduction

In the middle of the last century, the digital revolution began as a multifactor transition from analog to digital processing, storage and transmission of data, and, accordingly, the rapid development of the hardware and software serving these processes [12]. But the role of a foreign language in the formation of a full-fledged, diversified and competitive personality in a modern multicultural society is difficult to overestimate. The social order today is directly related to the rapidly developing globalization of the world community. As a result, it has changed dramatically and influenced the modernization of the goals of education in general and the teaching of a foreign language in particular. Education in foreign universities and professional internships in foreign companies become commonplace. The precedent of Russia's entry into the Bologna system and the approval of the new Federal State Educational Standards of High Professional Education is also one of the fundamental factors that changed the attitude of the state and society to the problem of learning a foreign language. The globalization of the world economy and the world community as a whole, the intensive growth in the number of companies actively cooperating with foreign partners, increase the importance of learning a foreign language so as a specialist speaking a foreign language properly becomes more competitive in a modern labor market. Career growth is

© Springer Nature Switzerland AG 2020
T. Antipova and Á. Rocha (Eds.): DSIC 2019, AISC 1114, pp. 220–228, 2020.
https://doi.org/10.1007/978-3-030-37737-3_20

inextricably linked with knowledge of a foreign language in the same way, very often we can observe the top managers of large enterprises that have barely reached the age of thirty-five, speaking in international forums masterfully and negotiating in English fluently. In the last decades teachers of foreign languages were at the epicenter of public attention: battalions of specialists from various fields of science, culture, business, technology and other areas of human activity demanded quick studying of foreign languages as an effective instrument of production and an important tool for building a successful career. Languages, and first of all, English, are required as practical applications in various areas of society, and most importantly - as a means of direct communication with people from other countries. Thus, today we can observe cardinal changes in the people minds, and the development of a completely new progressive type of thinking: this is modern man need for self-actualization and self-realization. Until recently, the purpose of training was focused only on the formation and development of skills and abilities, today time dictates its own laws, that is, the need to educate a new type of personality with different types of competencies.

The primary purpose of higher education is the formation of a student as a strong, thinking person, a specialist, ready to make independent effective decisions, able to search and dialogue in the process of research and solve fundamental, applied, most relevant problems in a science, technology, culture and society. Consequently, the state, and, therefore, the state educational system is faced with the task of a qualitatively new foreign language teaching. Today, teachers should press for such a high level of language teaching, so that the student can know a foreign language on a par with their native one. An integral part of this process of foreign language teaching should be learning the peculiarities of intercultural communication.

2 Methodology

Currently, the thesis of the interpenetration of language and the surrounding reality, and the relationship of language and culture interdependence based on it, being the object of many studies, has become an indisputable fact of pedagogical reality, which confirms Yan Komensky statement about the need to learn languages (native and foreign) and the knowledge of the material world simultaneously. This is evidenced by the position of sociolinguists who claim that so as a person lives, works, learns and orients himself in the outside world with the help of language, the world around him, to some extent, functions based on various linguistic social groups language habits. Thus, vision of the fullness of the world picture and its diversity, one way or another, is formed and supported by the factor of language. In this regard, learning a foreign language should be built as a kind of comprehension of the culture of the country of the target language. Today it becomes one of the foregoing problems of the teaching of a foreign language process reorganizing, in which questions of the goals and content of education orient students not to the generally accepted knowledge of the language, but to the foreign language culture in general. In the scientific literature, the term "intercultural education" [17] is defined as an organic and dynamic process that ensures the development of intercultural dialogue between representatives of different cultures and as a result of which a complex set of different identities of different cultures interlocutors is formed.

The overall goal of intercultural education is to prepare trainees for successful professional cooperation in the modern multicultural world by means of a foreign language. For the implementation of the tasks in modern methods language teaching, the competence approach is generally accepted, [4].

Competence is considered as a requirement for the preparation of the student, "expressed by a set of interrelated semantic orientations, knowledge, skills, and experience of the student in relation to a certain range of objects of reality, necessary for the implementation of personally and socially significant productive activities" [3, p. 9]. Based on the provisions of the competence approach, it is possible to determine its main didactic and methodological principles of teaching foreign language communication in non-linguistic universities: communication, intercultural, professional and self-education.

In this regard, on the one hand, goal-setting is determined by a specific social order, based on the increasing integration of nations, on the other, these goals themselves determine the educational system as a whole, determining both the content of this system and its structure.

This is especially true of the communicative side, which is reflected in the structuring of the goals and content of teaching a foreign language in the following sequence:

learning a foreign language → learning a foreign speech → learning a foreign language speaking → learning to communicate → forming intercultural competence.

It should be emphasized that a foreign language as an educational discipline will be able to realize its potential fully only if all aspects of the learning process are considered not autonomously, but interrelated and equivalent, especially since their combined use makes possible the qualitative formation of intercultural competence. students.

It is known that being a student is a period characterized by a favorable environment for understanding the need to possess intercultural competence: an interest in lifestyle, attitudes, values of others [10, p. 44], attention to questions of identity, defining one's own position in the sphere of human relations often through differentiating others into "like me", "mine" and "different", "others") [6, p. 74], [8, p. 17], [11, p. 125], the formation of a sense of social responsibility [5, p. 92], [9, p. 31], as well as the development of reflection and reasoning skills [4, p. 521]. In this regard, the education of the student personality with a multicultural consciousness and attitudes will become even more relevant.

As already mentioned, various academic subjects can and should become a means of educating a multicultural student's personality, and in our case, a foreign language that not only introduces and joinsits speakers culture, but also contributes to the formation and enrichment of the cultural identity of the individual.

Among many general educational and special disciplines of higher educational institutions, a foreign language occupies a special place in the development of students' intercultural competence. So, Zimnyaya [7, p. 213] indicates that "Foreign language in the process of learning implies a large proportion of the formation of speech skills and at the same time not less than for the exact science subjects, the amount of language knowledge in the form of rules, patterns, programs, solutions of various communicative tasks. However, these rules are not self-valued as in other scientific disciplines, they relate to the construction, implementation of language activities."

We believe that a foreign language, like any other academic subject, should become an essential factor that forms a personality, which is necessary for the diversified development of an individual and the full realization of his potential in independent living. In this regard, the priority goal of teaching a foreign language is the upbringing, development and education of students in the means of this subject on the basis and in the process of learning foreign language speech activity by them. Thus, the above goal is revealed in the unity of its four components: educational, developmental, educational and practical. The educational component is the formation of an active personality of a foreign language, personality, which is characterized by ideological conviction, patriotism, and with all this - the culture of interethnic communication, autonomy, hard work and tolerant attitude towards others.

In connection with the definition of the goals and content of learning a foreign language, it should be noted that as early as the 1960s, the Council of Europe [4, p. 22] identified a number of measures aimed at developing a program for the intensification of teaching foreign languages. As a result of studying the issue, scientists began to do some researches in the area under our consideration, which put forward the idea of developing specific goals of learning foreign languages, such as the idea of threshold levels (threshold levels). Based on this, the main method of teaching a foreign language is also defined - an integrated-communicative method that aims to form the following cultural competences: language material knowledge for its use in the form of speech statements (language competence); the ability to use language units in accordance with communication situations (sociolinguistic competence); degree of familiarity with the socio-cultural context of the language (socio-cultural competence), the ability to compensate gaps in language knowledge by verbal and non-verbal means (strategic competence) and intercultural competence itself, which is based on the ability for communication in the inter-lingual space. Taking into account that good knowledge of a foreign language becomes a means of communication and mutual understanding, the need to create conditions for the formation of intercultural competence is emphasized today, which means, in fact, a significant strengthening of the pragmatic aspects of language learning. Consequently, it is not only about achieving high-quality results in studying foreign language communication, but also about finding real interaction with a different culture and its carriers, in other words, about practical language skills. And this means that a foreign language, being an instrument of communication and self-realization in the outside world, has become a real demand, and therefore the status and educational potential of a foreign language as an academic discipline has increased significantly.

In short, a foreign language is able to make a huge contribution to the formation of a tolerant consciousness and the ability to move towards a dialogue as the dominant form of communication between people of different cultures. That is why one of the main tasks of the teacher is to create conditions for the formation of a respectful attitude towards Western culture without prejudice to their own national and Russian multinational culture; the formation of positive psychological attitudes towards other peoples, which are laid not only in the learning process, but also in a real social multicultural environment, where native and non-native languages often coexist.

3 Results

Based on the foregoing, it is possible to identify the main pedagogical conditions for raising the educational potential of the foreign language academic discipline in the development of the intercultural competence of technical university students, taking into account the complex conceptual pedagogical bases on which the process of preparing the future specialist for professional activity is built. Among them, the most important are: the creation of a humanitarian environment in a technical college as a factor of humanization of the educational process; cultivating an imperative among students for developing their skills and abilities for languages; the formation of values for tolerant attitude to the others and the surrounding reality; the inclusion in the content of education a system of concepts that orients future specialists to reflection and the ability to intercultural communication; the use of such active methods and forms of education that would saturate the educational process with personal meaning and thereby to "provoke" his desire for the full formation and improvement of intercultural competence.

Thereby, an increase in the efficiency of educating students tolerance, openness, respect for the culture of other nations as part of the culture of their homeland and world culture as a whole, as well as spiritual involvement in a foreign culture and comprehending the mentality of another people is achieved, which allows, in the end, to comprehend and originality of their own culture. Language and culture today are so closely interrelated that any knowledge of a foreign language implies a certain degree of acculturation (that is, students, as a result of communicating with native speakers of the language being studied, adopt certain features of the culture of a given people). And the better a person speaks a foreign language, the higher the expectations of the foreign interlocutor in the field of intercultural competence are. Ignoring one or another significant cultural specificity of the language being studied, as a rule, complicates further fruitful dialogic communication with a representative of another culture.

At the same time, the practice of higher professional education shows that the effectiveness of the educational and educational potential of teaching a foreign language in higher education conditions increases under the main condition, namely: the co-creation of two individuals - a teacher and a student. The personality of the teacher is no longer dominant, his influence on the audience becomes more intimate at the same time, not diminishing, but significantly increasing. A modern teacher is the organizer and coordinator of group interaction [11, p. 261]. A good teacher should "love a foreign language" [12, p. 137], be aware of all new trends in the field of methods of language teaching and have its extensive knowledge [13, p. 218], have an idea of the emerging trends in language dynamics [1]. Attitudes and motivations in learning a language often unconditionally and unconsciously influence both students and teachers [15, p. 151]. M. McGroarty notes that wise teachers should make every effort to identify the factors that contribute to situational motivation. Thus, the negative image of the teacher-mentor is gradually becoming a thing of the past, the teacher-observer, the teacher-organizer comes to the foreground [14, p. 263].

In this regard, based on the system of cultural, psychological and pedagogical conditions of multicultural education, the process of linguistic and cultural self-determination of an individual can be successful only when there is a cultural interaction between the subjects (teacher-student, student-student, native speaker of the language) in the multilingual space will be event-driven in nature. Integrated educational situations most effectively function, in our opinion, in cases where:

1. the teacher does not broadcast, but organizes linguistic communication on the basis of creating intercultural space, in which he carefully guides the development and self-development of the individual as a person of culture;
2. an emotionally positive attitude is created to expand the students' language space;
3. conditions are created for raising the general culture of students (moral, scientific, aesthetic, political, and in general, the worldview of the personality), designed to provide a qualitatively new level of communication culture of the personality as an integral part of intercultural competence.

Thus, the formation of the intercultural competence of the individual in the educational space of the university will take place when, firstly, in the pedagogical process the goal setting of the multicultural and personal development of a young person is achieved at a high level, due to the fact that today, here and now he can have the ability to carry out their livelihoods effectively. Secondly, if an individual's desire to understand another culture is consistent with his upbringing and self-education, then this leads to the activity manifestation of his individuality. Thirdly, the process of student personal growth in the multicultural sphere and its movement towards self-realization in a multilingual environment can be of real importance, in a case when teachers fulfill such current pedagogical tasks as the spiritual development of a young person, education of a good citizenship, upbringing of tolerant, actively self developing, aspiring to human values, personality. And this means that only in the process of humanization of education is it possible to provide effective assistance to young people in the development of the spiritual and intellectual aspect of their personality.

An important role in stimulating the process of personal growth or self-development of future professionals plays the support of the teacher, and its effectiveness is primarily due to the integrity of the pedagogical process, in which we include the indissoluble unity and autonomy of the components of this process (education, training, development); the presence of a common, inherent to all components of the educational process and the preservation of their specific features; integrity and subordination of internal and autonomous systems that are part of the pedagogical process.

In a word, education is assigned one of the leading roles in the formation of a multicultural personality, intercultural competence formation and striving for tolerant relations between people. Forming the personality of a future specialist in any field through the development of a interethnic relations culture, the university simultaneously determines the life of a future generation. Therefore, the skills of subsequent generations largely depend on the quality of training of students of higher education today, on the development and democratization of society as a whole, studying the culture of interethnic relations, which is closely related to intercultural competence, and all these problems are solved primarily through multicultural education.

One of the main components of intercultural competence is, in our opinion, political tolerance, which is the normative basis, personal conviction in the process of interethnic interaction. In this regard, higher education institutions should prepare not only highly qualified specialists, but also cultural, intellectually developed, spiritually rich, solving problems professionally and emotionally stable; independent, responsible, self-confident individuals with flexible and critical thinking and an adequate perception of a multicultural environment. The formation of intercultural competence contributes to achieve this goal. In addition, we include such parameters as: adequate, effective, most rational ways of behavior in real-time conditions in this multilingual public system; regulation of their own lives in accordance with the requirements of others and the choice of the best way to combine social and individual interests.

It should be added that intercultural competence contributes to the process of self-realization of the individual, being based on the existing potential socium. In essence, the university forms this competence and as a young person's personal education, which combines a valuable understanding of social reality and specific language skills, which act both in professional activities and in social activity as a guide to action.

4 Conclusions

Continuing to study the problem of developing intercultural competence, we note that it is inextricably linked with sociocultural and regional geographic knowledge (so-called "secondary socialization"), and not knowing the socio-cultural stratum of the country of the language being studied, it is not possible to form the aforementioned competence even in the smallest limits. Here it should be especially emphasized that not only the possibility of interpersonal communication is achieved by the factor of teaching foreign language culture, but at the same time the spiritual world of the individual is being enriched on the basis of acquired knowledge of the language being studied culture, its history, literature, music, art, etc. So we can conclude that the process of intercultural competence formation should be organized in such a way that an information about the culture of the country and its originality should receive a systematic practical application in students' speech communication (or verbal behavior), i.e. in a social context. This contributes to the formation and development of intercultural competence among students and allows them to carry out full-fledged productive communication with native speakers of the language being studied.

Exploring the educational potential of the discipline "foreign language" in the development of intercultural competence of students in a multicultural education, it is necessary to implement a pedagogical approach, which is based on a variety of basic provisions that contribute to the holistic perception of young people of the cultural picture of the world, structured by languages and a unique system of values of each nation. Basic provisions we would call the following (see Fig. 1):

1. a methodological basis (philosophical and pedagogical doctrine of intercultural education, which implies the conscious development of intercultural competence and cultural viability of education);

2. didactic basis (general pedagogical principles of teaching as: visualization, accessibility, systematic and consistent, scientific, problematic, activity and consciousness of students in the development of the multicultural world, integrativeness and interdisciplinary knowledge, as well as a combination of theory and practice in learning a foreign language);
3. psychological basis (it refers to the combination of the main and minor types of communicative activity, as well as identifying the leading role of receptive types of communicative activity in the development of intercultural competence in the process of learning a foreign language);
4. linguistic basis (the adoption of language as a social phenomenon of a special system of signs, which reflects not only the cultural and historical heritage of the people, but also being a tool for communication between different groups of people).

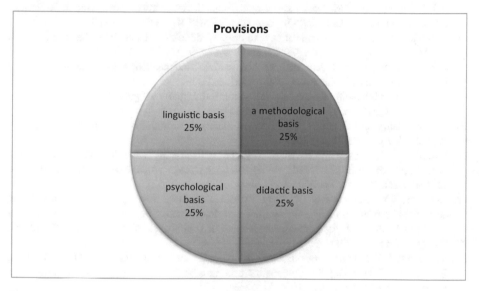

Fig. 1. The basic provisions that contribute to the holistic perception of young people of the cultural picture of the world.

It is thanks to the aforementioned fundamentals that the educational aspect of the educational process is implemented, where learning a foreign language is focused, on the use of a foreign language as a tool for communication in the dialogue of cultures and civilizations of the modern globalized world. This implies not only the multicultural, but also the sociocultural development of a young person by means of a foreign language for full-fledged preparation for intercultural communication in the process of his further professional activity, as well as the use of languages as a highly effective means of self-education in various areas of human knowledge.

References

1. Bilyalova, A., Salimova, D., Zelenina, T.: Digital transformation in education. In: Antipova, T. (ed.) Integrated Science in Digital Age, ICIS 2019. LNNS, vol. 78. Springer, Cham (2020)
2. Michael, B., Adam, N., David, S.: Developing intercultural competence in transition. In: Multilingual Matters. Languages for Intercultural Communication of Education, vol. 1, edn. 1. Computational Mechanics Publication (2001)
3. Brown, H.D.: Designing by Princess: An Interactive Approach to Language Pedagogy, p. 429. Person Education, New York (2001)
4. Common European Framework of Reference: Learning, Teaching, Assessment. http://www.coe.int/t/dg4/linguistic/Source/Framework_EN.pdf
5. Erickson, E.: Identity: youth and crisis / Tr. from English; general ed. and foreword A. V. Tolstykh. - M.: Progress (1996)
6. Feldstein, D.I.: Man in the Modern World: Trends and Potential Development, 15 p. Publishing House of Moscow Psychological and Social Institute: SPU Modek, Voronezh (2008). 1000 copies - ISBN 978-5-9770-0339-1, ISBN 978-5-89395-910-9
7. Zimnyaya, I.A.: Z-62 Pedagogical psychology. Textbook for universities, 2nd edn., p. 384. Logos Publishing Corporation (2000). ISBN 5-88439-097-1
8. Koryakovtseva, N.F.: Modern methods of organizing independent work of foreign language learners: a manual for teachers, p. 175. ARKTI (2002)
9. Language and intercultural communication: tutorial. http://www.gumer.info/bibliotek_Buks/Linguist/Ter/_Index.php. Accessed 16 Jan 2014
10. Lerning language/ English/ Fundamental method. http://www.college-training.ru/lang/english/74. Accessed 27 Jan 2014
11. Lyubova, T.V., Bilyalova, A.A., Evgarafova, O.G.: Grammatical and Communicative method - a new approach in the practice of teaching foreign languages. Asian Soc. Sci. **10**(21), 261–266 (2014). ISSN: 19112017 E-ISSN1911-2025
12. Lyubova, T.V., Gilfanova, G.T.: Scientific researches analysis in digital multitasking field of educational process in modern universities and determination of new conceptual boundaries of it. In: Antipova, T. (ed.) Integrated Science in Digital Age, ICIS 2019. LNNS, vol. 78. Springer, Cham (2020)
13. McGroarty, M.: Language attitudes, motivation and standards. In: MacKay, S.L., Hornberger, N.H. (eds.) Sociolinguistics and Language Teaching, pp. 3–47. Cambridge University Press, Cambridge (1996)
14. Mudrik, A.V.: Social pedagogy, 1st edn. M. (1999); ed. 2nd and 3rd. M., pp. 12–15 (2000)
15. Petrovsky, B.: Psychology and time. - SPb.: Peter, 448 p. (2007) (Masters of Psychology). ISBN 978-5-469-01675-5
16. Slobodchikov, V.I., Tsukerman, G.A.: Integral periodization of general mental development. Quest. Psychol. **5**, 38–50 (1996)
17. Vegh, J., Anh Nguyen Luu, L.: Intercultural competence developmental models – theory and practice through comparative analysis. People Int. J. Soc. Sci. **4**, 882–901 (2019). https://doi.org/10.20319/pijss.2019.43.882901

"Flipped Classroom" Technology in Teaching Foreign Languages

Emma Gilyazeva[1](✉)(iD), Olga Evgrafova[1](iD), Nailya Sharypova[2], and Raisa Akhunzianova[3]

[1] Kazan Federal University, Naberezhnye Chelny 423800, Russia
emma.giljazeva@mail.ru
[2] Kazan State Agricultural University, Kazan 420015, Russia
[3] Naberezhnye Chelny State Pedagogical University, Naberezhnye Chelny 423800, Russia

Abstract. The relevance of the research is determined by the dynamic growth of the share information technologies in modern society and the educational environment as an integral and most progressive part in the context of global integration processes. The article touches upon the problems of integration of "Flipped Classroom" technology into the educational process of teaching grammar of the second foreign language to students of the language faculty. The main features of the use of Google Classroom service within the "Flipped Classroom" technology are considered, and the advantages and disadvantages of Google Classroom based on practical activities are highlighted. In addition, the stages of working with the service and the opportunities provided by the resource are described. Based on the analysis of the results of the implementation of the experimental training, a number of guidelines for the use of the service Google Classroom within the technology "Flipped Classroom" in training were formulated. The analysis of the functionality of the service Google Classroom in relation to the "Flipped Classroom" technology allows us to conclude that the chosen service is an effective tool to improve the quality of students assignments, ensures the timeliness, easy availability and safety of tasks, reduces the time spent on their creation, editing and tracking, contributes to a better disclosure of the creative potential of students.

Keywords: Flipped Classroom · Google Classroom · Teaching technology · Teaching and learning foreign languages

1 Introduction

At the present stage of development of higher education in Russia, the problem of the optimal combination of classical and innovative teaching methods, including in the field of teaching foreign languages at the language department, is of particular importance. Reducing the number of classroom hours with a teacher and strengthening the role of independent work of students determines the need to find such teaching methods that would allow to achieve the goals set in educational programs.

Special problems arise in the study of grammatical phenomena of a foreign language due to the inability to properly assess the completeness and own level of theoretical

© Springer Nature Switzerland AG 2020
T. Antipova and Á. Rocha (Eds.): DSIC 2019, AISC 1114, pp. 229–240, 2020.
https://doi.org/10.1007/978-3-030-37737-3_21

knowledge, as well as the correctness of the practical application of self-studied grammatical material. It should be noted that the classroom and consulting hours in the training system are not always enough to obtain the necessary advice and provide the necessary control. As a result, the problems of structuring classroom time when teaching students are also quite acute. In this regard, the use of Internet capabilities and modern methods that would contribute to the intensification of the process of teaching foreign languages, especially in the classroom, is relevant in teaching a foreign language.

Widely practiced use of Internet resources and information technologies in the process of learning and teaching in educational institutions at different levels has long been a common norm and standard international practice in modern education [1]. Today, there are various services, websites and online platforms that allow to implement the technology of "flipped" teaching in practice. Rapidly gaining popularity web platforms e.g. Moodle, iSpring, edmodo, service Google Classroom allow to create online courses or upload tasks for independent work of students with the possibility of feedback. Various Internet designers for the development of interactive tasks (https://learningapps.org), interactive posters (https://edu.glogster.com), joint diagrams (https://cacoo.com) also allow to make part of the material to study or practice online.

The "Flipped Classroom" technology, which we have taken as a basis, is a modern, innovative, dynamically developing technology that allows students to study the topic independently at a convenient time and at a convenient pace for them, which allows the teacher to gain time and spend more time in the lesson to consolidate the material they have studied [2]. In this regard, we would also like to draw the attention of teachers to the innovative educational application developed by Google – Google Classroom. The article provides a detailed overview of the advantages of the educational process with the use of technology "Flipped Classroom" based on the use of the service Google Classroom in its integration into the learning process of an experimental academic group of students of the Department of Philology of Kazan Federal University (branch in Naberezhnye Chelny).

The theoretical basis of the study is connected with the works of J. Bergmann and A. Sams, revealing the theoretical foundations of innovative technology "Flipped Classroom" [2–4], as well as the experience of working on this methodology of domestic researchers [1, 5, 6], as well as the own results of work on this method in teaching German grammar to students of the language faculty.

The aim of our research was to study the effectiveness of using the "Flipped Classroom" technology in teaching 2nd year students the grammar of a second foreign language. In accordance with this goal, the following structure of the study was built: 1. an analyze of the scientific literature on the issue; 2. a review of resources for the organization of the "Flipped Classroom" technology; 3. systematization and generalization of the studied material; 4. development of tasks within the module "Grammatik aktiv: Präpositionen"; 5. the experimental testing the module "Grammatik aktiv: Präpositionen" developed on the Google Classroom platform; 6. the assessment of the effectiveness of the "Flipped Classroom technology" in teaching grammar; 7. the development of methodical recommendations on the use of the "Flipped Classroom" technology based on the Google Classroom platform.

2 Methodology

The study was conducted in 2 stages. The first stage included theoretical methods: (a) analysis of the literature on the methodology of teaching foreign languages using "Flipped classroom" technology; (b) systematization and generalization of the studied material.

The second stage included the analysis and synthesis of experimental work on the use of "Flipped classroom" technology based on the Google Classroom platform, assessment of "Flipped classroom" technology as a means of improving the efficiency of the process of teaching a foreign language. The main task of the pedagogical experiment was to test the "Flipped Classroom" technology in the experimental group and to conduct a comparative analysis of the assimilation of the material before the training module within the technology "Flipped Classroom" and after. The second stage included the following empirical methods: (a) pedagogical experiment; (b) testing, survey technique; (c) statistical processing of pedagogical experiment data; (d) descriptive-analytical method; (e) comparative method.

3 Theoretical Background of the Research

3.1 The Essence of the "Flipped Classroom" Technology, the History of Its Origin and Formation

"Flipped Classroom" is a modern teaching technology. In her book "Blended Learning in Grades 4–12" Catlin R. Tucker defines it as a mixed method in which a teacher combines traditional form of learning (classroom training) with distance learning. The use of this approach promotes personality-oriented learning [7, p. 10].

The "Flipped Classroom" technology was invented in 2008 by teachers Jonathan Bergmann and Aaron Sams [2]. It was used in high school, first to help students who miss classes, and then for all students in the class, who appreciated the opportunity to view the materials of lectures at home and better consolidate their knowledge. This gave teachers the opportunity to review their teaching methods and the system as a whole. Initially, J. Bergmann and A. Sams created Power Point presentations of their lessons with narration. Then the presentations were replaced by the author videos. So teachers quickly realized that the approach of preliminary online submission of theoretical material frees classroom hours, which are useful for more thorough study of educational material already with personal contact in the classroom [8].

After the transition to the system of "flipped" teaching, J. Bergmann became easier to interview students individually, to investigate various misconceptions about certain scientific concepts, as well as to clarify misconceptions about certain phenomena. In addition, such a system makes it possible to work on each lesson with each student individually. Thus, an individual approach to each student is implemented. J. Bergmann notes that now during classes he devotes more time to lagging students, who no longer refuse to do their homework, but rather work on difficult tasks in the classroom. Moreover, the need for successful work in the classroom creates an additional motive for the study of theoretical material at home. Meanwhile, top students have more freedom to learn on their own. J. Bergmann says that the new system of education improves the relationship of both teachers and students with each other, and increases the level of motivation of students [3].

"Flipped Classroom" is a relatively new term, and there is still no exact meaning, so teachers, researchers use it to refer to different forms of learning. Troy Cockrum, the author of "Flipping Your English Class to Reach All Learners", offers the following interpretation: this method is the use of modern technology to bring to the students asynchronously instructions, thereby freeing the classroom hours for personality-oriented learning [9]. He also notes that due to the many interpretations of the term, this model may look different in different classes. Using this technology in his work, the teacher frees time in the classroom, for example to work on a project, to help laggards or further work with students who quickly learn the material, as the theoretical material has already been studied by students at home [10]. Students can view the videos, presentations at any time of the day and work with the materials as long as they need. Of all Blended Learning technologies "Flipped Classroom" requires the least amount of time and resources in order to start working on it. However, here you can face a number of problems. All students at home must have a computer and a permanent Internet connection. As for the teacher, he will have to get used to changing his role. In the course of his work, he acts not only as a teacher, but also as a tutor [7, p. 3].

"Flipped" means "inverted", so the teacher, working on this technology "turns" the lesson. To avoid a number of problems associated with the coup, the transformation from the traditional to the inverted class is carried out gradually. It is important for the teacher to understand that his role is not to give a lesson, to convey and then to test knowledge. His role is to create an educational situation for independent cognitive-research activities of students [11, 12]. Such a situation, working in which they will be responsible for their training. That's when you can assume that the class is inverted. Having considered several definitions, we can distinguish similarities. The general principle of this technology is that students study the material online each on their own way, and classroom hours can be used for practice or to discuss the material studied. For example, students or students have attended a lecture outside the classroom, but they do their homework in the classroom, and the teacher, in turn, assists and answers questions.

3.2   Analysis of the Functionality of the Interactive Educational Platform Google Classroom

Google Classroom is a free service from Google, developed in 2014 for schools that sought to simplify the teaching process, namely: creating, distributing and evaluating assignments online. It was created as another Google service that can be used for education, as well as the already known Gmail, Docs and Drive. But Google Classroom is ready to provide users with a one-stop solution – by combining quick integration with Google Drive, a user-friendly interface, and new features that educators need. Since 2017, the service is available to all users with a Google account. This service has a large number of advantages (Table 1).

Table 1. The main advantages of the Google Classroom platform for different categories of users

User category	Opportunities
1. Teachers	Creating courses, assignments and their management, work with assessments. Rapid rating and commenting work in real time
2. Students	Track assignments and course materials. Information exchange and communication in the feed rate or by e-mail. Delivering of completed tasks. Receiving of teacher ratings and comments
3. Parents	Receiving letters with information about the student's progress, including overdue jobs and assignments that will need to be completed soon

The choice of this platform is justified by the following features:

1. Class setup. Each class has its own code that students can use to join the community. This process eliminates the need for pre-registries.
2. Integration with Google Drive. When a teacher uses Google Classroom, a folder "Class" is automatically created on their Google Drive with new attachments for each class they create.
3. Organization. When students use Google Classroom, a folder "Class" is created on their Google Drive page with subfolders for each class they join.
4. Automation. When you create an assignment as a Google Doc, the platform will create and distribute individual copies of the document for each student in the class.
5. Timing. When creating a task, the teacher can indicate the deadline for the work. When a student submits a task before the deadline, the status "View" appears on his document, which allows teachers to do the sorting.
6. Work/Correction. When students have begun their work, the teacher can provide feedback while the student is in the "Viewing" status. When the work is returned to the student, the student again switches to the "Edit" status and continues working on the document.
7. Convenient review. Both teachers and students can see all the tasks on the main Google Classroom screen. This allows controlling the work in several classes at once.
8. Communication. On the main page of the virtual classroom there are links by clicking on which the teacher will be able to create class announcements. During the performance of tasks, it is also possible for the student to contact the teacher (and vice versa) thanks to the integrated possibilities of commenting assignments [24].

In our case, communication with students was carried out through the "Comments" function in real time. To do this, the teacher must select only the part of the text that contains the error, and leave a comment.

Based on the functionality described above, the "Google Classroom" application can be used both for lectures (for example, video lectures, seminars for students with the task to comment or answer questions on the content) and for practical classes (dictation, translation, essay, tests, presentations, mini-forums on a given topic, implemented using the commenting option).

Topics are created inside the "classroom", and you can attach lesson documents, assignments, tests, YouTube Videos, and links to third-party sources to each topic. Assignment documents can be sent in three ways: an individual copy of the document to each student in the class, a document for general editing, and a document for viewing only. After completing the task, the student clicks the "submit" button and the document goes to the "view-only" status. The teacher checks the tasks, puts marks, using a convenient scale for them, can leave a comment. After checking the teacher can return the task for revision and then the document again goes into edit mode. Each new task can be limited in time or left for an indefinite period. After the deadline, the assignment for students becomes available only for viewing. Each action of the teacher is accompanied by an automatic notification to the e-mail of students. The service provides the ability to connect multiple teachers to one "classroom", as well as copying tasks from other classes [6]. "Classroom" allows teachers not only to give assignments, but also to send announcements or create thematic discussions. You can not worry about the fact that the student forgot his work at home or the "flash drive" is not readable - all documents are saved in a structured form in directories on Google Drive. Installing the Classroom mobile app is available on Android and iOC mobile operating systems for free.

Google services in training meet the principles of the modern education system. However, on closer examination, in terms of use in e-learning, these services have a number of disadvantages:

- all files are stored on a "foreign" server;
- a relatively small number of elements that can be used in the educational process.

Distance learning based on Google services will be easy to organize and operate, but the quality of such a system will be very different from Moodle and Efront. Google services is an MS Office Suite integrated into a website. As a control, it is proposed to use a calendar or tables (similar to MS Excel).

4   Testing the "Flipped Classroom" Technology on the Google Classroom Platform

4.1   Description of the Experiment on the Passage of the Training Module "Grammatik Active: Präpositionen"

The main task of the pedagogical experiment was to test the "Flipped Classroom" technology in the experimental group and to conduct a comparative analysis of the assimilation of the material before the training module within the technology "Flipped Classroom" and after.

Since all the necessary Google services are combined by one account, the experiment began with a preparatory stage. We conducted a conversation-survey of students of the experimental group for the presence of a mobile phone, tablet or computer with open access to the Internet. The next step was to create accounts and activate them for further work in the virtual classroom, which was shown to students on an interactive whiteboard to make sure they would know how to enter the course on their own. After all students have successfully joined the course in Google Classroom, there was a briefing on the main functions of the platform, such as sending assignments, commenting, communication with the teacher.

Thus, students of the experimental group had to systematically perform tasks on the course for one semester, studying the theory on their own, referring to the course materials as many times as they need. While viewing the material at home, students had the opportunity to check their level of assimilation of the material, with the help of exercises and tests placed in the tasks. Their work was checked either automatically or by a teacher online, allowing students to see their results almost immediately. In case they had any questions, they could ask for help during the pre-established working hours. Thus, instant feedback was organized. In addition, during the practical training, the theoretical material studied at home was fixed in the exercises.

The students of the control group were engaged in the traditional teaching methods: they studied theoretical material during classes, performing exercises at home.

As a result of passing the module "Grammatik aktiv: Präpositionen" control sections were carried out in the experimental and control groups to identify the effectiveness of the proposed technology.

4.2   The Data Obtained During the Pedagogical Experiment

Before starting the pedagogical experiment, we conducted a pre-test on the topics covered in grammar to make sure the equivalence of the selected control and experimental groups. The results of the pre-test, which included 50 questions, are shown in the diagram:

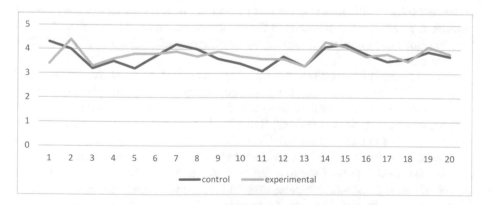

Fig. 1. Results of pre-test of the control and experimental groups

Figure 1 shows that there are no statistically significant differences between the experimental (mean score – 3.7) and control group (mean score – 3.8). This indicates the relative equivalence of the selected groups.

For one semester students of the experimental group were taught the second foreign language using the "Flipped Classroom" technology on the platform Google Classroom. As a result of passing the module "Grammatik aktiv: Präpositionen" control sections were carried out in the experimental and control groups to identify the effectiveness of the proposed technology. Figure 2 presents the results in comparison.

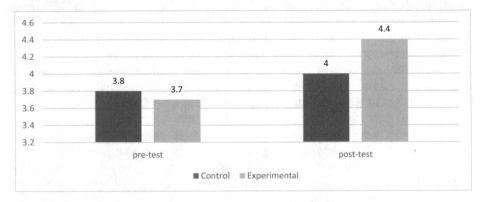

Fig. 2. Results of post-test of the control and experimental groups after passing the module «Grammatik aktiv: Präpositionen»

In order to assess the effectiveness of testing the "Flipped Classroom" technology, a questionnaire was conducted at the last lesson. The developed questionnaire included 7 questions, which were compiled using the Likert scale. The Likert scale was first proposed in 1932 by the American expert in the fields of organizational psychology and management Rensis Likert. This scale was developed by him during his postgraduate studies at Columbia University. Since then, this scale has been widely used to measure the attitude of respondents to the subject of the study. This method assumes that the respondent expresses his agreement or disagreement with each judgment (from the proposed set).

The "classical" Likert scale has 5 points. Ratings on the Likert scale can be expressed in categories of consent, frequency, importance, quality, etc., for example: (5) Strongly agree – (4) Agree – (3) Neutral attitude – (2) Disagree – (1) Strongly disagree; (5) Always – (4) Very often – (3) Sometimes – (2) Rarely – (1) Never; (4)Very important – (3) important – (2) To some extent important – (1) Unimportant; (5) Very good – (4) Good – (3) Acceptable – (2) Bad – (1) Very bad [13, 14]. In our study, students were required to evaluate the effectiveness of the inverted learning methods relative to the traditional method of learning, and also to express an opinion on the readiness to continue training in this new way. In the questionnaire, it was proposed to express their agreement or disagreement with the proposed statements on a 5-point scale of the following content: 1 – point meant "completely disagree"; 2 – "partially disagree"; 3 – "it is difficult to say, agree or disagree"; 4 – "partially agree"; 5 – "fully agree". All 20 students of the experimental group participated in the survey. Due to the small number of participants, it was decided not to carry out statistical processing of the results of these questions, compiled using the Likert scale, presenting them in Table 2.

Table 2. Analysis of the results of the survey

Statements	1 completely disagree	2 partially disagree	3 it is difficult to say, agree or disagree	4 partially agree	5 fully agree
The learning process for previous lessons on the "Flipped Classroom" technology was more effective than the traditional method (when new material is introduced in the classroom and fixed at home with the help of tasks/exercises)	1	–	5	12	2
Home preparation for classes in testing the "Flipped Classroom" teaching method took more time	1	3	4	3	9
It became easier to perceive the lesson material after completing tasks at home	–	3	7	6	4
"Flipped Classroom" allows you to learn the material better than the traditional method of learning	–	2	4	10	4
Interactive assignments and tests laid out on the Google Classroom platform contributed to better memorizing the grammar of the lesson on the topic under study	1	2	8	5	4
I liked the "Flipped Classroom" technology	1	1	2	11	5
I would like to continue learning the "Flipped Classroom" method in the future	1	3	4	8	4

In general, students are quite positive about the new method of teaching, noting the higher efficiency of classes on technology "Flipped Classroom". Diligent students noted that the preparation for classes by the method of "Flipped Classroom" requires more time. More than half of the participants in the experiment assessed their diligence by putting "good", using the proposed descriptors. Those who were conscientiously preparing for classes noted that it was easier to perceive material in the classroom than the usual way of introducing new material in the classroom. This, in turn, encouraged the teacher to focus more on linguistic and grammatical issues in the classroom and to develop communication skills through face-to-face discussion and work in pairs.

The method of inverted learning has shown its effectiveness in the short-term pedagogical impact on the experimental group.

4.3   Guidelines for the Use of the "Flipped Classroom" Technology Based on the Google Classroom Platform

Summing up the results of the work carried out on the use of the "Flipped Classroom" technology based on the Google Classroom platform, we have developed a number of proposals and recommendations:

1. At the very first introductory stage, it is recommended to conduct an introductory lesson in the classroom with computers. You also need to have an Internet connection. If it is not possible to hold an introductory event in the computer class, you can use tablets, students' smartphones, but there is a need to have a stable wireless Internet, so that students can enter the virtual classroom using their own devices. For accounts, we recommend that students create their own accounts according to the instructions provided by the teacher. When you create a google account, you need a phone number to confirm that it wasn't created by a robot.

2. Preparation of the course for students. We recommend that you create assignments in the virtual classroom before students join. If the terms allow, you can leave in a virtual classroom trial task that students will perform, in order to familiarize and simplify further work with the system.

3. After all the students have joined, the tasks are published by the teacher, you can immediately begin to implement them. During the assignments, the teacher should be periodically online and monitor how students perform the task. If errors occur, then use the "Comments" function to indicate to the students what they should pay attention to, refer to the presentation, video or other material so that the student looks again and corrects the error himself. As for the tasks themselves, it will be more effective if the teacher himself compiles knowledge tests based on the individual characteristics of students in each class, while theoretical material (the rules of grammar in this case) can be taken from other open resources.

4. A traditional lesson should be closely related to the online component, and this implies that students have mastered the basic grammatical knowledge on the topic, which they will use in their speech during the lesson. If you have an interactive whiteboard in your classroom, you can help your students display slides from a presentation they viewed at home or create a new one with diagrams. Use the system of evaluation of the work performed by the student in the virtual classroom, taking into account the assessment for the work on the traditional lesson on this topic.

5   Results

The analysis of the functionality of the service Google Classroom in its use in relation to the "Flipped Classroom" technology allows us to conclude, that the chosen service is an effective tool to improve the quality of assignments by students, ensures the timeliness of execution, easy availability and safety of tasks, reduces the time spent on their creation, editing and tracking, contributes to a better disclosure of the creative potential of students. In the experiment, the basic principle of this technology was implemented:

self-study of new material at home with the help of created electronic resources and in-depth consideration of more complex/interesting issues during the classroom. We recorded positive dynamics both in the traditional lesson, students actively worked and often gave the correct answers in the tasks and on the control test. This method has allowed not only to increase the level of formation of grammatical skills of students, but also to make the learning process more interesting and diverse. Students worked with interest on the tasks, each did at their own pace, to whom it was not clear, asked questions. Experimental work and its positive results allowed us to formulate methodological recommendations for the use of the "Flipped Classroom" technology based on the Google Classroom platform, which contribute to the development of grammatical skills in foreign language lessons.

6 Conclusion

The Google Classroom service made it easy and convenient to organize the process of blended learning within the "Flipped Classroom" technology due to the rapid integration with other Google services (Google Drive, Google Forms, Google Sheets, Google Docs, etc.), user-friendly interface and lack of complex registration. The interface of the service is minimalistic, and therefore it was not difficult for teachers and students to understand the navigation of this service. The conditions for the implementation of the technology were met, as the theoretical material remained completely on an independent remote study, which allowed students to devote the necessary amount of time to assimilate the information, and practical classes in the classroom helped to consolidate the material and deal with all the issues that cause difficulties. Thanks to this technology, students can study from a distance and take advantage of the teacher's advice. The results of the experiment show all these advantages.

It goes without saying that at the present stage of development of information technology training tasks described above in the context of the use of Google Classroom, can be more or less successfully implemented using a number of other applications and platforms, however, based on the experience gained in the practical application of information technology educational tools, their implementation within a single global platform, along with a high degree of integration, automation and systematization of the elements of the educational process, of course, makes Google Classroom, in our opinion, the most promising, effective and convenient to use.

References

1. Bilyalova, A.: ICT in Teaching a Foreign Language in High School. In: Procedia, Social and Behavioral Sciences, pp. 175–181 (2017)
2. Bergmann, J., Sams, A.: Flip Your Classroom: Reach Every Student in Every Class Every Day (2012). http://www.ascd.org/Publications/Books/Overview/Flip-Your-Classroom.aspx. Accessed 4 Oct 2018
3. Bergmann, J. How the Flipped Class was born (2011). https://flippedclass.com/the-history-of-the-flipped-class/. Accessed 6 Sept 2018

4. Spencer, D., Wolf, D., Sams, A.: Are you ready to flip? (2011). http://www.thedailyriff.com/articles/are-you-ready-to-flip-691.php. Accessed 10 Dec 2018
5. Evseeva, A.: Use of flipped classroom technology in language learning. In: Evseeva, A., Solozhenko, A. (eds.) XV International Conference "Linguistic and Cultural Studies: Traditions and Innovations, vol. 206, pp. 205–209. National Research Tomsk Poly-technic University, Tomsk (2015)
6. Tulina, E.: Vvedenie v Google Classroom [Introduction to Google Classroom]. NewToNew. https://newtonew.com/web/vvedenie-v-google-classroom. Accessed 4 Oct 2018
7. Tucker, C.R.: Blended Learning in Grades 4–12: Leveraging the Power of Technology to Create Student-Centered Classrooms (Corwin Teaching Essentials), First Edition, Corwin. 272 p. (2012)
8. Tucker, B.: The Flipped Classroom – Online instruction at home frees class time for learning. Education Next (2012). http://www.msuedtechsandbox.com/MAETELy2-2015/wp-content/uploads/2015/07/the_flipped_classroom_article_2.pdf. Accessed 6 Oct 2018
9. Cockrum, T.: Flipping your English Class to reach all learners. Routledge, London (2013). 145 pp
10. Westerberg, C.: The Flipped Class, What does a good one look like? (2011). http://www.thedailyriff.com/articles/the-flipped-class-what-does-a-good-one-look-like-692.php. Accessed 10 Dec 2018
11. Davies, R.S.: Flipping the classroom and instructional technology integration in a college-level information systems spreadsheet course. In: Davies, R.S., Dean, D.L., Ball, N. (eds.) Educational Technology Research and Development, vol. 61. No. 4, pp. 563–580 (2013)
12. Marshall, H.: Three reasons to flip your classroom. Marshall H. USA, TESOL (2013). http://newsmanager.commpartners.com/tesolbeis/issues/2013-08-28/6.html. Accessed 18 Jan 2019
13. Dubina, I.N.: Matematicheskie osnovy empiricheskih social'no-ekonomicheskih issledovanij: uchebnoe posobie [Mathematical foundations of empirical socio-economic research: study guide]. Publishing House of the Altai University, Barnaul (2006). 263 p.
14. Mackey, A.: Research methods in second language acquisition: a practical guide. In: Gass, S. M., Mackey, A. (eds.), p. 326 Blackwell Publishing, Oxford (2012)

Digital Learning as a Factor of Professional Competitive Growth

Lidiya I. Evseeva⬤, Olga D. Shipunova(✉)⬤, Elena G. Pozdeeva⬤,
Irina R. Trostinskaya⬤, and Vladimir V. Evseev⬤

Peter the Great St. Petersburg Polytechnic University,
St. Petersburg 195251, Russia
o_shipunova@mail.ru

Abstract. The article considers the problem of developing the university educational environment as the condition for the successful professional activity of a future specialist. The authors emphasize the competitiveness relation with digital competence that is a certain level of mastering new software products and with tacit requirement of high learning ability. In order to study the prospects for the competitive growth of future specialists, the authors have surveyed the attitude of Peter the Great St. Petersburg Polytechnic University students to the digital educational environment and its influence on the parameters of personal and professional competence. The monitoring results show that, as to the majority of students in the reference group, the digital educational environment is the main factor in increasing their competitiveness in the labor market, since it develops independence and responsibility for actions in the information sphere. At the same time, the assessment of the digital technologies impact on various indicators of activity and creativity reveals the ambiguous influence of the digital environment on students' intellectual self-development.

Keywords: E-learning environment · Competitiveness · Digital competence

1 Introduction

In today's world, professional transformation is caused by the needs of the digital economy and technology process improvement. The real economy sector makes new requirements for professional competency, knowledge, and skills. The implicit requirement for a high learning ability of professionals implies the ability for mobile adaptation to manufacturing improvements and a high level of individual intellectual potential [1]. Intelligence, creative efforts, a deep knowledge of virtual, communication, and digital culture are required for professionals to boost their competitive advantage. The system of HR education for the digital economy is a critical source of its development [2, 3]. In this context, the relevant challenge is to research the digital education environment with a view to enhancing the competitive margin of future professionals.

© Springer Nature Switzerland AG 2020
T. Antipova and Á. Rocha (Eds.): DSIC 2019, AISC 1114, pp. 241–251, 2020.
https://doi.org/10.1007/978-3-030-37737-3_22

1.1 Research Objectives

Today's world needs individuals who exceed the minimal criteria set by the market. Those educational institutions that focus on successful employment of their graduates must seek to meet the market requirements related to professional skills that are in line with the modern digital culture. Today's vocation training is challenged to develop the developmental educational environment focused on developing the vocational training of a new generation of professionals that meets the market needs in the growing digital culture [4].

The strategy of the education system, which is focused on the systemic identity development of future professionals, must bring together the digital technology of professional education and the humanitarian principle of personal development as a value and goal of education. Humanitarian technologies play a critical role in the development of professional competency [5, 6]. The developmental educational environment will facilitate the learning of skills included in the education programs and shape new skills related to self-development, self-management, and self-discipline. The problem of using the competency-based approach in professional training is widely discussed in the academic community [7, 8].

In the context of digitalization, the virtual education environment of higher education institution plays a critical role. It is formed on a flexible basis of using sophisticated computer technologies and software to develop content for students, teachers, and administration. The key objective is to develop such digital educational resources that will combine the goals of personal development with important social, economic, political, and cultural goals.

This article addresses expectations related to enhancing the competitive advantage of future professionals in digitalized higher educational institutions through the example of St. Petersburg Peter the Great Polytechnic University. To reach this goal, the authors monitored the students' assessment of digital technology effects on personal and professional skills.

1.2 Literature Review

Modern literature treats knowledge that is produced in the field of digital culture and using information technologies as one of the fundamental and evolving forces of evolution of the human civilization. Study of digital culture involves the consideration of practices resulted from the emergence of digital technologies and that are common among the young. In particular, these are: computer games, Internet, computer graphics, mobile phones, smart technology. In our work, we rely on UNESCO Recommendations, whereby the digital literacy of a contemporary student can be reviewed as part of daily culture of the information community participant and should also be governed in laws, codes of conduct and safety rules developed and approved by citizens of this community together [9].

The digital revolution of Century 21 is associated with dissemination of convergent technologies, artificial intelligence systems and neuron networks, as well as the Internet penetration in all areas of business [10]. A number of authors treat the digital culture study from the positions of mass media transition from analog knowledge presentation formats to digital ones, thus making the digital culture almost equivalent to the virtual media environment. This approach reviews the changes related to the shifts in the mass media system whereas the digital transformation processes go beyond this system [11]. The cultural transformations influenced by digital technologies are under-studied. In particular, it is emphasized that education is not regarded as one of the basic development tools of the information culture of the society. Nonetheless, it is information and communications technologies applied in education that facilitate changes in the cultural space of the information society. Digital communications technologies create a new habitat that is based on the new culture taking shape, values, standards, rules, and the role repertoire [12].

The studies of the tools for shaping the human information culture in the educational space are devoted to the analysis of education and culture co-influence in the context of informatizing of the contemporary community. A special interactive cyber space environment is generated and supported by virtual reality computer technologies, in which the user's orientations are mostly determined by the computer game. To describe a new communicative culture of Z generation, G. Jenkins suggested using the terms of "participation culture" and "digital literacy" [13]. The nature and typology of behavior when individuals are involved in network communications, when they co-participate in the Internet communities, rather than consume the information passively, is interpreted as the participation culture with 4 parameters: "Participation culture" measurement (1) affiliations: formal and informal membership in the online community; (2) expressions: creation of new creative forms (remakes, memes), (3) collaborations: joint team work; (4) broadcasting: shaping the flow of outgoing messages.

The issues of the digital culture influence on education are associated, first of all, with adaptation of students to new education technologies in the digital environment. In recent years, the term of "digitalization" became common in the scientific literature. In a broad sense, digitalization as the information packaging method describes a trend in efficient global development associated with lower costs and new opportunities. Provided that digital transformation of the information meets the following requirements [14]:

- it covers production, business, science, social environment and daily life of individuals;
- it is supported by efficient use of its findings;
- its findings are available to users of the transformed information;
- its results are not only used by professionals but by rank-and-file people;
- the digital information users are skilled in handling such information.

When a great number of digital environments is used, the personal repertoire of communications means and emotional registers is expanded for each person. For emergence of poly-media as an environment of communications opportunities, three conditions should be met: physical excess, availability (including financial) and media literacy [15].

Digital competence of a future professional as a factor of his/her competitiveness improvement envisages readiness of the personality to apply computer technologies confidently, efficiently, with critical attitude and safely in different life areas [16]. Today we can speak about the broad use of the virtual educational environment technologies that focuses on students [17]. In terms of humanistic approach, four components are singled out in the digital competence structure as a new landmark in professional education: knowledge; skills and abilities; motivation; responsibility (including, in particular, safety) in different lines of business in the Internet.

2 Methods

2.1 Theoretical Basis

Theoretical settings of the study are determined by the macro-systemic approach. From this point of view, the personality development process is included into the information and socio-cultural environment. The educational environment of a university having significant influence on the professional establishment and development of a personality was selected for a particular study.

The educational environment, in a broad sense, is a psychological and teaching reality containing specially created conditions for establishment and development of a personality. Its key feature is modality comprising opportunities/ lack of opportunities to develop a student's activity and independence in mastering professional competencies in the educational environment.

We regard the digital educational environment, into which a student plunges, as a developing education system.

2.2 Empirical Research Method

To study the prospects of improving future professionals' competitiveness, a sociological survey was conducted concerning the attitude of the employees studying at St. Petersburg Peter the Great Polytechnic University, to digital technologies and their impact on the personal and digital competence parameters.

We examined the following criteria of personal competencies: self-development, communication culture (flexibility), leadership, creativity, independence, and critical thinking.

Digital competency criteria included: information culture, responsibility, and self-control.

The main objectives of the sociological research included the following questions:

(1) Which competencies are best developed with the help of digital technology?
(2) What are the time perspectives of the digitalization of education?

The empirical research methods included questionnaire and online survey on the selected parameters of professional and personal competence. The survey was taken in March 2019 among the students of Peter the Great St. Petersburg Polytechnic University. Totally 115 respondents were interviewed (65% women, 35% men). In the research group there were some students with online learning experience and experience of communicating with teachers online. The sample was random based on the students' interest in discussing the advantages and disadvantages of remote education technologies and the prospects for developing the online environment in university educational process. Among respondents, 84% consisted of the first year students of various professional training programs, such as "Informatics and Computer science", "Mechanical Engineering, "Applied Mathematics and Informatics", "Jurisprudence", "Advertising and Public Relations", "Psychology". The dominance of freshmen among respondents is due to the fact that during the first academic year they learn a large number of disciplines of humanitarian, socio-economic, natural-scientific cycles in online format, which is a completely new experience for them. The desire of freshmen to participate in the survey is motivated by their encouragement to demonstrate an active position by expressing their attitude to online courses.

The questionnaire included the following questions:

1. What do you think of teaching some subjects online?
2. Indicate subjects that you learn online.
3. Which subject taught online did you like most and why?
4. Which subject taught online did you like least and why?
5. Which benefits of online learning do you value (you can select one or more options)?
6. Which drawbacks of remote learning did you encounter (you can select one or more options)?
7. Which kind of education do you think is the best?
8. Will online learning help to develop self-learning skills in students?
9. Which knowledge fields is it best to learn online?
10. Are you willing to pass your exams and tests online?
11. Please assess online learning level in your university (on a scale from 1 to 5, where 1 is the minimal value and 5 is the maximum).
12. Are you going to recommend online learning to your friends from other universities? Please indicate your willingness on a scale from 10 to 0, where 10 mean "I will definitely recommend" and 0 "I will never recommend".
13. Please assess the time prospects of full education digitalization (10–15 years, 30 years, and 50 years, undecided).

The survey is evaluated by the development of personal and digital competencies and, additionally, by the nature of estimate (yes, no, I don't know). The calculation results are multiplied by 100 and shown as a percentage. 100% means 10 points.

3 Results

The students' attitude to the digital environment in the Polytechnic University
Most students (69%) are positive about using digital technology in business communications and 22% remained neutral. In assessing the digital education performance of the Polytechnic University, about half of respondents (48%) put 7 or 8 points on a 10-point scale.

The survey showed that students are positive about online courses, electronic libraries, online conferences, and webinars (total 60% of answers "positive" and "on the positive side"). 16.5% of respondents are neutral to digitalization. More than 70% of respondents have no difficulty in dealing with online courses. Key challenges in working online include deadlines, technical failures (about 50%), and motivation in doing tasks (41%).

Most respondents said the digital environment makes communications easier (52%) and freer (38%). The benefits of digital communications include: convenience (89%), speed (83.5%), and multi-channeling (45%).

Most students (83%) are negative about the full digitalization of education and replacement of human teachers with digital ones. Most students (84%) are satisfied with their relations with teachers. Some students (61%) said they communicate with their teachers at seminars and project work (37%). Only 19% of respondents prefer to learn on their own. They do not contact or think they do not need to contact their teachers on any learning matters.

60% of respondents said the digital culture had positive effects on the nature and forms of relations in the education environment and 32% had the opposite view.

The monitoring findings of the students' attitude to development of personal and professional qualities in the digital educational environment
The results of study students' attitude to development of personal and professional qualities in the e-leaning are shown in Fig. 1.

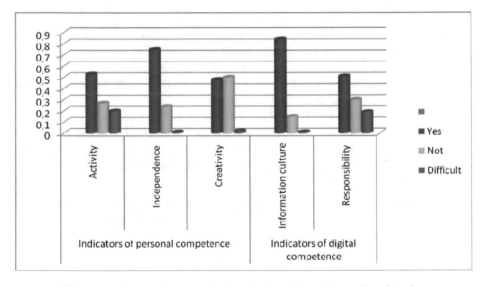

Fig. 1. Indicators of personal and professional development in e-learning

According to the monitoring, most respondents believe that digital technologies promote independence, information culture and responsibility but do not encourage activity and creativity. This result testifies to the unambiguous impact of the digital environment on the personal self-development and requires a further study.

Findings of students' opinion poll on temporary forecast of the university's educational environment digitalization

The respondents were invited to forecast comprehensive digitalization of education (item 13 in the common questionnaire). Monitoring findings of waiting for comprehensive digitalization of the university's educational environment in Fig. 2 are:

- 45% respondents believe that it will happen in 10–15 years
- 24% respondents think it will only occur in 30 years
- 11%, only in 50 years
- 20% respondents found it difficult to answer.

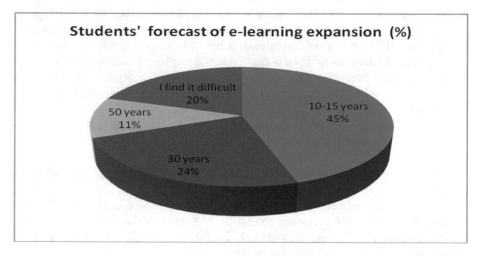

Fig. 2. Students' temporary forecast of e-learning expansion

4 Discussion

Features of digital educational developmental environment

SMART education notion became a standard in many countries [18, 19]. The SMART education development prospects consist in remote education and e-training technologies [20]. The educational space is transformed under the influence of digitalization, with educational problems and new management problems arising. Universities are actively involved in the study of the digital culture impact on the educational environment. The findings in the Chart (Fig. 1) testify to the positive average assessment of the education digitalization impact on the future professional's qualities as a specialist.

In the contemporary community, the distinctive features of the education system are its mobility and variability, which envisage the emergency of a multitude of diverse education methods and tools. The explosive development of Internet and mobile technologies will lay the groundwork for the new educational model [21], which enables to speak about the development of digital environments, in particular, in education.

The digital educational environment is an open set of information systems intended to support different education objectives.

The following fundamental principles of shaping the digital educational environment can be singled out:

- openness meant as the ability of the educational service consumer to use the information systems, to substitute them or to add new components;
- the availability principle means the unlimited digital OS functionality for a particular student via the Internet;
- the uniformity principle meant at coordinated use of digital technologies in the single educational and technological logic;
- the competiveness principle includes securing the freedom of full or partial replacement of the digital educational environment with competing technologies;
- the responsibility principle that envisages the right, duty and responsibility of an educational participant to address the objectives as part of own responsibility;
- the sufficiency principle is meant as securing the conformity of the information system composition to the goals, powers and abilities of the educational service consumer;
- the utility principle represents the shaping of new opportunities, reduction of labor costs due to introduction of digital educational environment [22].

The role of the digital learning environment in enhancing the specialist competitiveness

The new aspects of the real and virtual connections in the professional and educational environment put forward new requirements to the future professional's awareness of the multi-media information environment, independence and activity in decision-making, the intelligence and capacity to study.

In the modern context, expansion of the interactive educational environment is the priority in shaping of the individual digital competence as a future professional's competitiveness improvement factor [23].

The notion of "digital competence" and its components are widely discussed in the scientific community [24, 25]. Such key areas of digital competence should be distinguished: content, communication, techno sphere and consumption. Digital competence index can be measured in a standardized testing. In terms of a contemporary professional's competitiveness increase, we can speak about four types of digital competence:

(1) *Information and media competence* envisages knowledge, skills, motivation and responsibility related to search, understanding, arrangement and archiving the digital information, its critical re-evaluation, as well as the new content creation using the database resources.
(2) *Communicative competence* envisages knowledge, skills, motivation and responsibility as required for online communication in different formats and with different goals.
(3) *Technical competence* implies knowledge, skills, motivation and responsibility that allow for efficient and safe use of the computer and the respective software to address different objectives, including the computer network use.

(4) *Consumer competence* implies knowledge, skills, motivation and responsibility that allow for addressing various daily objectives using a computer and via the Internet, related with some particular life situations to meet different needs [26].

The extremely rapid changes in the social environment in the digital era, which shape the personality identification, lead to more questions than the answers the global scientific and teaching community can give. In the recent studies of the remote training and knowledge control format, new challenges associated with academic unscrupulousness emerge. The psychological problem of students' and teachers' adaptation to new technologies offered by the digital environment and, on the contrary, development of the systems taking into account the personal mental style and actions, should be in the focus.

The attributes of the developing educational environment that target greater competitiveness of a professional include:

- infrastructural indicators, in particular: perfection of academic curricula in connection with the operating scientific and production teams and market demand;
- educational indicators, including the intellectual potential and digital literacy;
- Student's readiness to self-actualization in each particular situation.

Taking these aspects into account, one should focus the educational space development vector on facilitating activity and independence in academic content selection and mastering.

5 Conclusion

A student's professional development depends on the conditions existing in the university, which allow for systemic shaping of a future professional's competencies towards his/her competitiveness in modern conditions of the digital economy.

The digital educational environment makes the knowledge acquisition process customized, increases students' activity and independence in mastering general and professional competencies.

The study suggests that students link their professional actualization and competitiveness opportunities with digital competence and the search for a better ratio between remote, electronic and conventional training format. Enhancing a professional's competitiveness envisages digital literacy as technical skills in programming and data processing plus social communications skills and high learning capacity.

Digital competence as competitiveness improvement factor should not only account for the operational level of mastering new technologies but also for the ability to assess their potential and risks. A future professional's readiness to constantly acquire new knowledge, new personal and professional competencies is an equally important competitiveness factor.

References

1. Shipunova, O.D., Berezovskaya, I.P., Mureiko, L.V., Evseev, V.V., Evseeva, L.I.: Personal intellectual potential in the e-culture conditions. ESPACIOS **39**(40), 15 (2018). http://www.revistaespacios.com/a18v39n40/18394015.html
2. Townsend, A.: Smart Cities: Big Data, Civic Hackers, and the Quest for New Utopia. W. W. Norton & Company, New York (2013)
3. Hammad, R., Ludlow, D.: Towards a smart learning environment for smart city governance. In: 9th International Conference on Utility and Cloud Computing (UCC), pp. 185–190. ACM, New York (2016). https://doi.org/10.1145/2996890.3007859
4. Baranova, T., Khalyapina, L., Kobicheva, A., Tokareva, E.: Evaluation of students' engagement in integrated learning model in a blended environment. Educ. Sci. **9**, 138 (2019). https://doi.org/10.3390/educsci9020138
5. Trostinskaia, I.R., Safonova, A.S., Pokrovskaia, N.N.: Professionalization of education within the digital economy and communicative competencies. In: 6th Forum Strategic Partnership of Universities and Enterprises of Hi-Tech Branches (Science. Education. Innovations), vol. 6, pp. 29–32. Institute of Electrical and Electronics Engineers Inc. (2017). https://doi.org/10.1109/ivforum.2017.8245961
6. Baranova, T.A., Gulk, E.B., Tabolina, A.V., Zakharov, K.P.: Significance of psychological and pedagogical training in developing professional competence of engineers. In: Auer, M., Tsiatsos, T. (eds.) The Challenges of the Digital Transformation in Education. ICL 2018. Advances in Intelligent Systems and Computing, vol. 917, pp. 44–53. Springer, Cham (2019). https://doi.org/10.1007/978-3-030-11935-5_5
7. Mulder, M.: Conceptions of professional competence. In: Billett, S., Harteis, C., Gruber, H. (eds.) International Handbook of Research in Professional and Practice-based Learning. Springer International Handbooks of Education, pp. 107–137. Springer, Dordrecht (2014). https://doi.org/10.1007/978-94-017-8902-8_5
8. May, D., Ossendorf, P.: ModellING Competences – Developing a Holistic Competence Model for Engineering Education. In: Frerich, S., et al. (eds.) Engineering Education 4.0. pp. 877–894. Springer, Cham (2016). https://doi.org/10.1007/978-3-319-46916-4_72
9. UNESCO Policy Guidelines for Mobile Learning. https://iite.unesco.org/pics/publications/ru/files/3214738.pdf. Accessed 16 July 2019
10. Schwab, K.M.: The Fourth Industrial Revolution. Translated from English. Exmo, Moscow (2017). (in Russia)
11. Deuze, M.: Participation, remediation, bricolage: considering principal components of a digital culture. Inf. Soc. **22**(2), 63–75 (2006). https://doi.org/10.1080/01972240600567170
12. Zakharova, I., Kobicheva, A., Rozova, N.: Results analysis of russian students' participation in the online international educational project x-culture. Educ. Sci. **9**, 168 (2019). https://doi.org/10.3390/educsci9030168
13. Jenkins, H., Clinton, K.., Purushotma, R., Robinson, A.J., Weigel, M.: Confronting the challenges of participatory culture: media education for the 21st century. In: An Occasional Paper on Digital Media and Learning. The MacArtur Foundation, Chicago, Ilinoes (2006). https://www.macfound.org/media/article_pdfs/JENKINS_WHITE_PAPER.PDF. Accessed 18 July 2019
14. Khalin, V.G., Chernova, G.V.: Digitalization and its impact on the Russian economy and society: advantages, challenges, threats and risks. Manag. Consult. **10**(118), 55–59 (2018). https://doi.org/10.22394/1726-1139-2018-10-46-63

15. Madianou, M., Miller, D.: Polymedia: a new approach to understanding digital media in interpersonal communication. Monitoring of Public Opinion: Economic and Social Changes **1**, 334–356 (2018). (in Russia) https://doi.org/10.14515/monitoring.2018.1.17

16. Biggins, D., Holley, D., Evangelinos, G., Zezulkova, M.: Digital competence and capability frameworks in the context of learning, self-development and HE pedagogy. In: Vincenti, G., Bucciero, A., Helfert, M., Glowatz, M. (eds.) E-Learning, E-Education, and Online Training. Lecture Notes of the Institute for Computer Sciences, Social Informatics and Telecommunications Engineering, vol. 180. Springer, Cham (2017). doi: https://doi.org/10.1007/978-3-319-49625-2_6

17. García-Álvarez, M.T., Pineiro-Villaverde, G., Varela-Candamio, L.: Proposal of a knowledge management model and virtual educational environment in the degree of law-business. In: Rocha, Á., Adeli, H., Reis, L., Costanzo, S. (eds.) Trends and Advances in Information Systems and Technologies. WorldCIST 2018. Advances in Intelligent Systems and Computing, vol. 746, pp. 1275–1286. Springer, Cham (2018). https://doi.org/10.1007/978-3-319-77712-2_122

18. Kim, T., Cho, J.Y., Lee, B.G.: Evolution to smart learning in public education: a case study of korean public education. In: Ley, T., Ruohonen, M., Laanpere, M., Tatnall, A. (eds.) Open and Social Technologies for Networked Learning. OST 2012. IFIP Advances in Information and Communication Technology, vol. 395, pp. 170–178 (2013). Springer, Berlin, Heidelberg. doi: https://doi.org/10.1007/978-3-642-37285-8_18

19. Budhrani, K., Ji, Y., Lim, J.H.: Unpacking conceptual elements of smart learning in the Korean scholarly discourse. Smart Learn. Environments **5**, 23 (2018). https://doi.org/10.1186/s40561-018-0069-7

20. Zhu, Z.-T., Yu, M.-H., Riezebos, Yu.P.: A research framework of smart education. Smart Learn. Environ. **3**(4) (2016). https://doi.org/10.1186/s40561-01-6-0026-2

21. Curran, J., Fenton, N., Freedman, D.: Misunderstanding the Internet. Routlenge, London (2012)

22. Methodological foundations of the formation of a modern digital educational environment: Monograph. NGO Professional Science, Nizhny Novgorod (2018). (in Russia). http://scipro.ru/conf/monographeeducation-1.pdf. Accessed 15 July 2019

23. Evangelinos, G., Holley, D.: A qualitative exploration of the EU digital competence (DIGCOMP) framework: a case study within healthcare education. In: Vincenti, G., Bucciero, A., Vaz de Carvalho, C. (eds.) E-Learning, E-Education, and Online Training. eLEOT 2014. Lecture Notes of the Institute for Computer Sciences, Social Informatics and Telecommunications Engineering, vol. 138, pp. 85–92. Springer, Cham (2014). https://doi.org/10.1007/978-3-319-13293-8_11

24. Ferrari, A., Punie, Y., Redecker, C.: Understanding digital competence in the 21st century: an analysis of current frameworks. In: Ravenscroft, A., Lindstaedt, S., Kloos, C.D., Hernández-Leo, D. (eds.) 21st Century Learning for 21st Century Skills. EC-TEL 2012. Lecture Notes in Computer Science, vol. 7563. Springer, Heidelberg (2012). https://doi.org/10.1007/978-3-642-33263-0_7

25. Ilomäki, L., Paavola, S., Lakkala, M., Kantosalo, A.: Digital competence – an emergent boundary concept for policy and educational research. Educ. Inf. Technol. **21**(3), 655–679 (2016). https://doi.org/10.1007/s10639-014-9346-4

26. Soldatova, G.U., Nestik, T.A., Rasskazova, E.I., Zotova, E.Yu.: Digital competence of adolescents and parents. The results of the All-Russian study. Internet Development Foundation; Faculty of Psychology, Moscow State University Mv Lomonosov (2013). (in Russia)

Digital Universities in Russia: Prospects and Problems

Darya Rozhkova[1]([✉]) [iD], Nadezhda Rozhkova[2] [iD],
and Uliana Blinova[1] [iD]

[1] Financial University Under the Government of the Russian Federation,
Moscow 125993, Russia
rodasha@mail.ru
[2] State University of Management, Moscow 109542, Russia

Abstract. The education sector is subject to significant changes due to the increasingly active spread of digital technologies. Usually, the trends in the implementation of digital technologies in educational and research activities are set by commercial organizations – private universities, business schools, corporate universities. But public universities and institutions start to think more and more about digital transformation.

By the end of the year, several Russian universities will receive grants for the creation and operation centers of development of digital university models. Anticipating the events, the article considers the model of a digital university, which consists of four blocks – university management information systems, online support for the educational process, key competencies of the digital economy and educational process management based on an individual educational path. The object of the study was the State University of Management and the Financial University under the Government of the Russian Federation. After examination of the current state of university digitalization, we have identified major problems which become an obstacle during digital transformation.

Keywords: Digital technologies · Digitalization · Digital university

1 Introduction

The modern world is constantly changing. Innovations are being introduced into various spheres of human activity, which, on the one hand, directs people to continuous development, enhancement their knowledge, skills, competencies, acquirement new types of activities in related sectors of economy. On the other hand, routine work is increasingly transferred to machines, and a person is required to be creative, actively collaborate with colleagues in a search for new solutions, and most importantly have an ability to evaluate information offered, both for its reliability and for logical embedding in current task.

International studies show the lack of highly qualified specialists in most countries. According to Korn Ferry, global human talent shortage will amount to 85.2 million people by 2030. Most (74%) of leaders of largest organizations in the world believe that technologies will make a greater contribution to creating business value than the

T. Antipova and Á. Rocha (Eds.): DSIC 2019, AISC 1114, pp. 252–262, 2020.
https://doi.org/10.1007/978-3-030-37737-3_23

human factor in the future. At the same time, highly qualified personnel are going to be in a high demand [1].

Certainly, innovative methods and technologies are increasingly being introduced into modern education system, allowing to strength the practical orientation [2]; as well as the emergence of a global knowledge system and information and communication technology represent global forces affecting the higher education [3]. At the same time, it can be noted that the informatization stage is at the final step, since all educational institutions are equipped with computer technology, and teachers and students use information technologies in the educational process (computer classes, internet, mobile applications) [4].

Nevertheless, a discussion about need to create an education system in Russia, which will most fully reflect the ongoing digitalization processes, as well as the creation of a platform economy has only recently begun [5, 6].

With these issues in mind, the present paper addresses the following set of research questions:

- What is the current stage of developing the concept "digital university in Russia", main problems?
- What are the key elements of digital university?
- Which are criteria of digitalization?

The paper proceeds as follows. Section 1 reviews model of digital university, first steps of its creation in the Russian Federation, states existing practical directions. Section 2 provides the experience of creating a digital university in leading universities in Russia, summarizes main route in the Financial University under the government of the Russian Federation and the State University of Management. In Sect. 3 we made an attempt to evaluate results, summarize main problems and prospects that need state regulation.

2 Model of "Digital University"

2.1 Digital Educational Environment in Russia: The State

Since 2016, the federal project "Modern Digital Educational Environment in the Russian Federation" has been launched, approved by the Government of the Russian Federation as part of the implementation of the state program "Development of Education" for 2013–2020. Within the framework of this project, it is planned to modernize the education system, bring educational programs in line with the needs of the digital economy, introduce digital tools of education and integrate them widely into the information environment, provide citizens with the opportunity to learn according to an individual curriculum throughout their lives [7].

In accordance with the "Strategy for the Development of the Information Society in the Russian Federation for 2017–2030", approved by presidential decree of May 9, 2017 No. 203, the primary goals and objectives of the state in the field of education and training of qualified personnel in the digital economy are:

- human development;
- the formation of the information space, taking into account needs of citizens and society in obtaining high-quality and reliable information;
- use and development of various educational technologies in the implementation of educational programs, including distance and online learning technologies;
- development and implementation of partnership programs of federal higher education institutions and Russian high-tech organizations, including on the issue of modernization and improvement of curricula;
- development of technologies for remote online interaction of citizens, organizations, state bodies, local authorities along with the preservation of the possibility of personal appeal (without the use of modern digital technologies);
- stimulation of digitalization in Russian firms and enterprises in order to create the necessary conditions for distance employment;
- creation of control and monitoring systems based on digital and communication technologies that would serve to optimize processes in all areas of public life.

In June 2019, the development of "digital university" models began in Russian universities. By the end of the year, several Russian universities will receive grants for creation and operation of centers in order to develop digital university models. The Ministry of Science and Higher Education will hold competitions for the provision of such grants in the second half of the year 2019. This year, money will be allocated for the work of five centers, in 2020 – for the work of another 15 centers. The grant will amount to approximately 100 million rubles.

The development of digital university models is provided for by the "Personnel for the Digital Economy" direction of the state program "Digital Economy". The performance indicators laid down in the program stipulate that universities will graduate 120 thousand people a year in areas related to information and telecommunication technologies by 2024. And 800 thousand graduates a year should have competencies in the field of information technology at the global average level.

The concept of a digital university model is expected to be officially presented in the second half of 2019. The framework for its formation will consist of four areas: a university management information system, online support for the educational process, key competencies of the digital economy and educational process management based on an individual educational path.

The university's information management system assumes that all the university's digital services will operate in a single window mode; students will be able to quickly receive the necessary information, certificates, etc. Along with online courses, it is planned to introduce into the educational process courses using virtual and augmented reality technologies (VR/AR technologies). It is planned to form an individual educational path using an artificial intelligence: it will process information about grades and other information received online (what subjects the student is studying, how he passed intermediate tests, attendance rate, etc.).

It is planned that elements of digital university models should be implemented in all Russian universities by 2024. Each student should have access to popular educational content, effective learning technologies, digital support services. A digital university should work for all stakeholders in the educational process: for students, and for researchers, for teachers, and administration. It is of fundamental importance that under the constantly changing demand of the Russian labor market environment, it would be possible to build personal development paths.

2.2 Digital University Conceptual Model

In our opinion, it would be a mistake to call any university digital only if it has introduced any digital technologies, or a university that trains personnel for the digital economy. Digital University involves the restructuring of internal business processes based on the introduction of modern digital technologies.

Since today in Russia there is no conceptual model of a digital university, we have formed a conceptual model of a digital university based on the work of G. Sidorov. Digital University Conceptual Model consists of five levels and a supporting platform [8].

The first level is the most important; it is represented by scientific and pedagogical staff, students, industry and academic partners of university, graduates and applicants. The first level is, in fact, internal and external stakeholders of a university.

The second level is represented by basic information services. Their task is to create a single information space for digital interaction within university using flexible tools. Examples of such services are video screens for lectures and seminars, wireless communications throughout the university (including dormitory), cloud storage for keeping and exchanging data, professional printing etc.

The third level includes services that greatly facilitate the lives of students and teachers in a modern university. For example, a digital library, international databases, etc.

The fourth level is the most resource-intensive in terms of implementation, but at the same time it allows a university to get the highest added value. It consists of services such as digital marketing, research project management, procurement management, interaction with applicants and students.

Digital marketing is a new field for Russian universities aimed at solving the following problems:

- organization of interaction with the teaching support staff, students, applicants, graduates using the entire modern spectrum of digital communication channels;
- monitoring changes in the perception of the university brand in the target markets based on the results of research and monitoring of social networks; conducting preventive and reactive measures to form a positive image of the university;
- stimulating the creation of new digital communities and innovations at all stages of the educational cycle, as well as communication of the content of educational programs and features of student activities for applicants;
- development of personalized marketing materials for target audiences based on analysis of data from various sources.

Interaction with applicants and students includes the following tasks:

- use of digital technologies to interact with applicants and inform them about the stage of processing applications for admission;

- usage of analytics to determine the most promising applicants and increase their enrollment rate;
- use of various communication channels both digital and traditional in order to provide applicants with the most complete information about the university. This task is most relevant for foreign applicants who cannot visit the university and want to form an idea of it using information from the Internet;
- using analytics to identify the most successful and least successful students;
- automation of work, for example, the creation of a "digital student office".

The fifth level consists of digital technologies, which are highly likely to be widely used in the university environment. Such technologies, for example, include drones (unmanned aerial vehicles). In this context, as a first step, universities will actively introduce drone technology into the internal educational and research space, purchasing equipment, setting up laboratories, encouraging students and researchers to test and work with new technology.

A transition to a digital university is impossible without supporting activities aimed at introducing changes at the university. Such events may include:

- development of an optional or compulsory module in the framework of training programs aimed at improving digital literacy among students;
- providing support to scientific and pedagogical workers who set trends in the development of digital skills and are engaged in the development of innovative teaching methods;
- encouraging the advanced use of learning platforms to ensure better student learning outcomes and improve the overall performance of the university;
- assisting teachers with less advanced digital skills.

In our opinion, a university should adequately work with all levels of the digital university model described before and constantly maintain feedback with key stake-holders - students, teachers, industry and academic partners, graduates, and applicants.

We emphasize that while creating a "digital university" it is necessary to create a new digital environment that provides learning mobility and new format for students to communicate with teachers and potential employers, university management. The main goal is to make this environment comfortable for everyone, to take into account the individual learning path.

The principles of digital university should be:

1. Educational trajectory. Online courses, blended learning, flexible educational paths.
2. Information content. Creation of services – an online schedule, a single authorization system, a digital portfolio, the integration of information systems into a single space; system of personal accounts, full automation of the university.
3. A new system for assessing achievements. Multivariate assessment based on data on training, knowledge of foreign languages, as well as soft and digital skills.

The components of a digital university should be:

- An electronic educational environment, including online courses, blended learning courses for the main educational programs of students, as well as a digital portfolio, electronic ordering of books, electronic individual curriculum, etc.

- Online courses for various categories: applicants, students, managers and teachers of educational organizations.
- Creation of the portal "Open educational space".
- Creation of an effective electronic university management system [9].

In general, it is necessary to create: a single educational ecosystem in which a system for assessing the quality of education will be created; personal educational portfolio based on blockchain technology, a system for assessing people's cognitive skills; service for building a personal development path taking into account professional orientation.

3 The Experience of Creating a Digital University in Russian Leading Universities

3.1 State University of Management (SUM)

In the framework of the digital university conceptual model presented in the Sect. 1, the State University of Management, as the leading university in management, implements all areas. As part of creation of the university's basic information services, a management system based on "1C: University" program was created. The system automates accounting, storage, processing and analysis of information about main processes: admission to the university, tuition, graduation and employment of graduates, calculation and distribution of a workload of the faculty, activities of teaching departments and deans, management of scientific work, publications and innovation, additional and postgraduate education, certification of scientific staff, personal accounts (students, teachers, entrants). An automated access control system has also been created, which allows keeping records of all students and teachers.

The university's information management system assumes that in the future all the university's digital services will operate in a single window mode - students will be able to receive quickly necessary information, certificates, etc. Along with online courses, it is planned to introduce courses using virtual and augmented reality technologies (VR/AR technologies). It is also planned to form an individual educational path with the help of artificial intelligence: it will process information about grades and other students' achievements.

A top management of many universities, including the State University of Management, considers it timely and necessary to introduce educational online technologies already at a stage of school education. A pre-university will be created by end of 2020. Its curricula will contain more than 30% of courses with online classes, adaptive interactive testing, modern IT equipment, machine teaching technology.

Within the framework of digital technologies, the State University of Management created a design and training laboratory "Digital Economy and High Technologies". The activities of the laboratory are related both to practical tasks and to scientific and theoretical directions. On the basis of the laboratory, initiative students and young specialists develop and implement their own projects aimed at solving problems in the field of a digital economy, create new digital products, which in the future will contribute not only to development of technological equipment of the University, but also to the global digitalization of society as a whole.

The laboratory is equipped with modern powerful equipment, which can significantly expand the range of capabilities of participants, primarily programmers. On the basis of the Laboratory projects of various orientations are being implemented, therefore, one of the key tasks is to attract young specialists from various domains in order to create diverse and multifunctional development teams.

A virtual reality is a very promising area in a digital era. This technology is not only a pleasant way of spending leisure time, but also a tool for training by creating virtual models of specialists in various fields and areas. Full immersion technology can be very useful in distance learning.

In 2018, a YurTech project was created. It is the platform solution with elements of machine learning, which will allow building algorithms for finding relationships and contradictions in Russian law. This project is aimed at simplifying and optimizing a work of lawyers. YurTech is developed using the most modern program codes. One of the technical solutions of the YurTech platform aims to utilize an artificial intelligence to create a virtual assistant for the most effective solutions of individual tasks.

Since 2018, a competition for students has been held in order to identify talented youth and solve real problems from a field of a digital economy. For a long time, a classic hackathon was held only for programmers, but modern tasks require the participation of specialists from different fields. Students studying humanities offer their own methods for solving the tasks that a team of programmers translates into program code, creating digital products that can perform socially significant projects.

A creation of a social digital environment for students is solved using "the Portal solution" with a system of virtual assistants, personal accounts with learning management systems (LMS) elements. The learning management system is a basis of the management of educational activities. It is used to develop, manage and disseminate online educational materials with shared access.

3.2   Financial University Under the Government of the Russian Federation (FU)

The Financial University under the Government of the Russian Federation has vast experience in developing a digital university. At the same time, the university uses both existing software solutions on the market and its own developments in technology.

In particular, the university has created a university network with a round-the-clock free wireless access for students and staff to the university's information resources and the Internet in all campuses. A university data center was formed as well as the organization of the provision of standard university information technology services for personal and personal computers and mobile devices of users with the preservation of all university security policies and data safety for official use. The development of the campus card system has led to a stimulation of the electronic payment system on the territory of the Financial University, a development of electronic commerce facilities.

The Financial University under the Government of the Russian Federation is actively developing its internal services for both university management and education services. An integration of corporate and educational portals of the university, as well as separate Internet resources of departments, led to the formation of a single university portal as an extensive structure covering all aspects of the university's activities, including informing the target audience about educational programs, the capabilities of scientific and analytical departments.

There was a transition to personalized "one-window" services in the personal accounts of entrants, students, teachers and employees. You can follow the schedule, generate electronic reports and sign them, make a note of the readiness to complete course and graduation work, receive automatic verification of posted works on anti-plagiarism, download presentations and files. The planned and actual workload of departments and, in particular, the computational workload of each teacher is also automated. You can work with the information and educational portal of the university through a mobile application as well.

An information field of education is also being created at the Financial University. Audio and video recordings of classrooms and scientific events are being carried out, followed by posting on the information in the educational portal; all students and teachers of the university has an access to entire methodological base for the activities of the university and the full methodological support of disciplines.

The distance education program is actively developing at the Institute of Digital Competencies; a number of distance learning courses were introduced for independent work of a student without visiting classes. "Networked Silver University" was formed as a part of a professional retraining program for the older generation.

The university's digital library provides students and teachers an access to scientific literature from any device, regardless of location and time. Many modern universities combine traditional and digital features. So, for example, in a traditional library you can find and read a book or magazine from a library computer, at the same time, any user can find the book in the electronic catalog of the library and get it when they come to campus. This convergence of traditional and new technologies at the Financial University provides a higher level of comfort for students and teachers and positively affects the image of the university.

4 Evaluation of the Results

The analysis outlined problems that, to one degree or another, concern all educational institutions trying to implement the digital university model.

In terms of the formation of an electronic university in Russia, on the basis of our research and model presented in part 1.2, we have formed the main eight criteria for the implementation of this concept presented in the Table 1. Using several criteria, we have done evaluation of current stage of digitalization of Financial University under the government of the Russian Federation (FU) and State University of management (SUM), assuming the maximum target of 100%.

Table 1. Current stage of digitalization

Criteria	FU	SUM	Target
1. The use of federal state educational standards or own standards in terms of requirements for the formation of digital economy competencies for higher education	90	90	100
2. The educational programs of all levels of education have been updated in order to use in educational activities common and professional digital tools	90	80	100
3. The legislative and regulatory legal framework has been updated in terms of organizational and methodological conditions, certification forms, subject programs, teaching materials, etc.	90	90	100
4. Creating a complementary education system to train competent professionals for the digital economy	80	70	100
5. Development or selection of an information system for supporting an individual student competency profile	50	30	100
6. Using e-learning technologies	80	70	100
7. Development and implementation of education programs, professional retraining, continuous professional development of teaching staff	70	60	100
8. Creating infrastructure for the management and activities of the university in the conditions of digital economy	90	80	100

The results of the analysis of achievement a target of 100% of two leading Russian universities are presented in the Fig. 1.

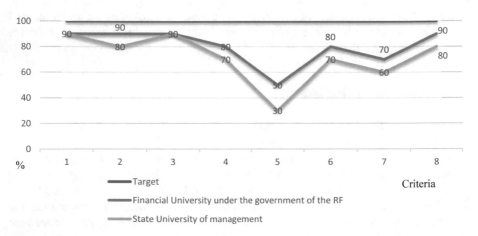

Fig. 1. Evaluation of digitalization of Financial University under the government of the Russian Federation and State University of management

The analysis shows that the process of forming an electronic university progressed quite successfully, especially in the field of the legislative and regulatory legal framework, creation of federal state educational standards which include formation of digital economy competencies for higher education (90% for both universities).

At the same time, one of the main problems is the creation of an individual learning path, personnel and technical problems. Level of development or selection of an information system for supporting an individual student competency profile in State University of Management is only 30%.

Of course, one of the areas of activity of digital universities will be networking and dissemination of best online courses, practices of using digital services and proven solutions for building digital architecture of universities.

The main problem is copyright protection and, in principle, the impracticability of the requirement "to open everything and share everything". The university system in Russia is built on the principles of market competition. But if you keep this in mind, it will immediately become clear that no university will ever present ready-made solutions (source of idiosyncratic competitive advantage) to another university.

One solution to the above problem may be a sale with some restrictions and limitation of usage. The emergence of consortia of universities applying solutions created within the consortium is highly probable. It is possible to use online courses of other universities. But all this will not be implemented on a free basis.

The second major problem is the lack of a single standard for digital solutions and formats - electronic courses, common platforms or requirements for the compatibility of individual services, the lack of harmonized requirements and quality standards for electronic content and online courses, as well as the unresolved issues of digital transformation of the education system.

The level of digitalization in Russian universities is very different. Universities seek to build these competencies and pool resources to increase competitiveness. However, there are still no unified standards for connecting data and services, and this seriously hinders the development of universities.

There is also the problem of a lack of personnel (both employees and teachers) who are ready and able to carry out an effective digital transformation.

5 Conclusion

In the context of globalization and the development of information technology, universities are becoming the center of the construction of new economic concepts based on knowledge and associated with the transition from the computerization of individual business processes to the digitalization of business models. During the study, we noted that the digitalization of education is at the early stages of implementation. The authors proposed the principles and components of digital university. The use of digital technologies in education is complicated by a number of unsolved problems, which are associated, first of all, with a lack of understanding of the mechanism for creating a system for implementing a digital environment in an educational institution. In particular, there are no developed universal criteria for assessing the quality of the digital format of teaching various disciplines; there is the problem of selecting highly qualified specialists with a sufficient set of competencies to provide the very same quality assessment of these disciplines; poorly developed implementation roadmap from the part the state.

References

1. https://www.kornferry.com/institute/talent-crunch-future-of-work. Accessed 21 July 2019
2. Ivanovic, Z., Milenkovski, A.: Importance of new approaches in education for higher education institutions. UTMS J. Econ. T. **10**(1), 67–76 (2019)
3. Altbach, P.G., Reisberg, L., Rumbley, L.E.: Trends in global higher education: Tracking an academic revolution. – BRILL (2019)
4. Bond, M., et al.: Digital transformation in German higher education: student and teacher perceptions and usage of digital media. Int. J. Educ. Technol. Higher Educ. T. **15**(1), 48 (2018)
5. Yudina, T.N.: Thinking about digital economy. Theoret. Econ. **3**(33) (2016). Юдина, Т. Н.: Осмысление цифровой экономики. Теоретическая экономика, **3**(33) (2016)
6. Rozhkova, D.: Digital platform economy: definition and operating principles. Manage. Econ. Syst. Electr. Sci. J. (10), 32–32 (2017). Рожкова, Д.Ю.: Цифровая платформенная экономика: определение и принципы функционирования. Управление экономическими системами: электронный научный журнал, (10), 32–32 (2017)
7. Modern digital education area in the Russian Federation. Passport or priority project. Approved by the Bureau of the Board of strategic development and priority projects under the government of the President of the Russian Federation (protocol dated 25.10.2016 № 9). Современная цифровая образовательная среда в Российской Федерации. Паспорт приоритетного проекта. Утвержден президиумом Совета при Президенте Российской Федерации по стратегическому развитию и приоритетным проектам (протокол от 25 октября 2016 г. № 9)
8. https://www.itweek.ru/idea/article/detail.php?ID=192831. Accessed 15 June 2019
9. Rozhkova, N., Blinova, U., Rozhkova, D.: The concept of management accounting based on the information technologies application. In: Information Technology Science. MosITS 2017, AISC 724, pp. 89–95 (2017). https://doi.org/10.1007/978-3-74980-8_8

Web-Based Environment in the Integrated Learning Model for CLIL-Learners: Examination of Students' and Teacher's Satisfaction

Baranova Tatiana(ID), Aleksandra Kobicheva$^{(\boxtimes)}$(ID), and Elena Tokareva(ID)

Peter the Great Saint-Petersburg Polytechnic University, St. Petersburg 195251, Russia
kobicheva92@gmail.com

Abstract. In this paper we examine the students' and teacher's satisfaction with the newly introduced web-based environment in integrated learning model. For the analysis both quantitative and qualitative methods are used. The results of survey show that students level of satisfaction is much higher than average, especially such advantages were noted as the availability, simplicity and clarity of use, as well as the presence of useful theoretical material. According to the interview with coordinator the process of preparing for classroom activities has become more time-consuming for students, but their final results on discipline became much higher. Both coordinator and students noticed that one of the problems was cross-cultural communications during the online project, in this case next semester it is extremely important to try alleviate it allocating participants from the same time zone in working teams. Overall, the developed web-based environment can be confirmed effective and successful.

Keywords: Web-based environment · E-learning · Blended learning · Digital competencies

1 Introduction

Due to rapid changes and challenges caused by new technologies and competitive pressures, universities are trying to innovate their services and increase their public image. Higher education is undergoing dramatic transformations now. Technologies play a powerful role in the life of modern students and higher schools can no longer satisfy their needs only with the help of classrooms. Universities are paying more and more attention to determining the right model for integrating technology into teaching and learning, in order to meet the needs of students and provide the education and skills necessary for a future society [1].

Blended learning is one of the ways that higher schools can prepare for the next era of education. It allows combining face-to-face and online teaching and learning. This includes various teaching or learning methods (lectures, discussions, guided practices, reading, games, case studies and modeling), various delivery methods (in real time in

T. Antipova and Á. Rocha (Eds.): DSIC 2019, AISC 1114, pp. 263–274, 2020.
https://doi.org/10.1007/978-3-030-37737-3_24

the classroom or through a computer), various scheduling (synchronous or asynchronous), and different levels of guidance (individual, instructor or expert under supervision, or group/social learning) [2].

To be competitive and not lag behind the technological trends we elaborated an integrated learning model [3, 4] that is built on a Content and Language Integrated Learning (CLIL) methodology as a framework and flipped classroom activities, project-based learning as pedagogic tools for creating a blended learning environment.

The aim of this paper is to evaluate the satisfaction with web-based environment in such integrated model from students' and teacher's perspectives.

2   Background

2.1   Description of Web-Based Environment in Integrated Learning Model

The design of integrated learning model we introduced into the 4th year bachelors studying on international profile is presented in the Fig. 1. The web-based educational environment includes two stages – online students' preparation and online project. We believe that such web-based environment could boost the efficiency of educational process and improve the overall results for CLIL students.

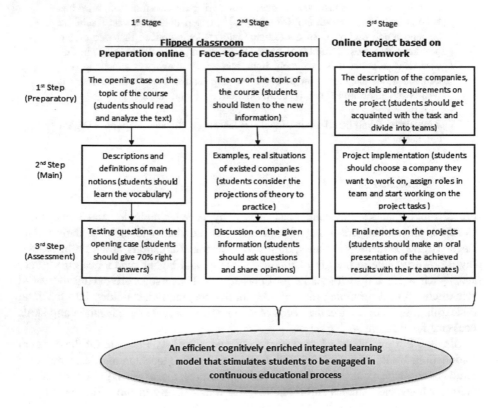

Fig. 1. Design of integrated learning model.

The online preparation is a first stage in the educational cycle of the integrated learning model and it consists of 3 steps:

- Preparatory step – the opening case on the topic of the course. At this step students should read and analyze the text offered by teacher via online platform.
- Main step – descriptions and definitions of main notions. This step assumes that students will learn the necessary vocabulary on the topic.
- Assessment step – testing questions on the opening case. To get positive result, students should give 70% right answers.

Wcb-course is elaborated on the base of Moodle educational platform, the access to preparatory materials is open for students the week before face-to-face classroom. All information on the assessment is transmitted to the CCs' accounts.

At the first classroom the CC described a new process of study and provided an access to the platform for all students. Students were informed that CC would control their activity on the web-course.

According to the curriculum students had two-hours classroom per week. At the day of first lesson students got an access to preparatory material for the first topic (see Fig. 2).

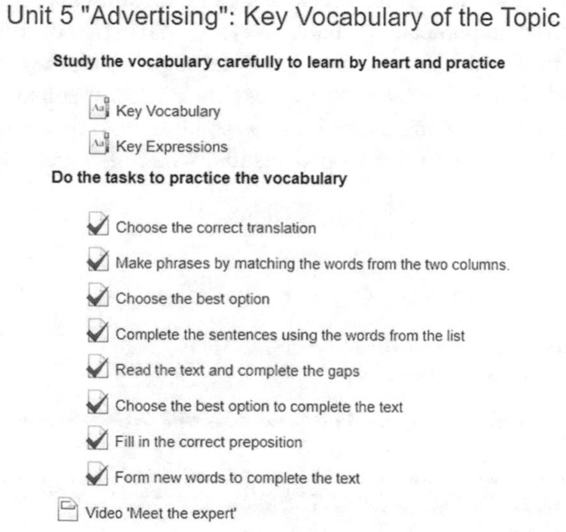

Unit 5 "Advertising": Key Vocabulary of the Topic

Study the vocabulary carefully to learn by heart and practice

- Key Vocabulary
- Key Expressions

Do the tasks to practice the vocabulary

- Choose the correct translation
- Make phrases by matching the words from the two columns.
- Choose the best option
- Complete the sentences using the words from the list
- Read the text and complete the gaps
- Choose the best option to complete the text
- Fill in the correct preposition
- Form new words to complete the text
- Video 'Meet the expert'

Fig. 2. Preparatory material.

Students should explore all preparatory material, read the text, translate all unknown terminology, and prepare questions to the teacher.

After studying all materials learners had to pass the test on the opening case (see Fig. 3). Those who got 70% right answers, were considered as prepared to the face-to-face classroom. There was unlimited quantity of attempts to pass the test. However questions were different in each try.

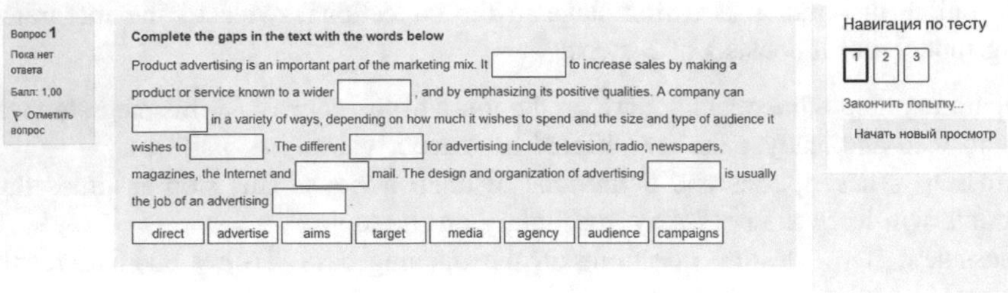

Fig. 3. Preparatory test.

The third stage of integrated learning model is online international project X-culture. It is elaborated for students studying International Business, International Marketing and Management and provided an opportunity to work on a real project in an international team online during 8 weeks. All information regarding the project students and coordinators receive via e-mail. A series of webinars based on Zoom platform are conducted about the grading/marking students' performance for coordinators and about the participation in a project or extra material on the companies for students.

When students are allocated into teams they can choose an appropriate way for their team members to communicate (usually Skype, WhatsUp or other messengers). Weekly students receive a link on e-mail going through which they have to undergo a small survey on their work process and assess their team members work as well as upload the materials they obtained (see Fig. 4) At the same time coordinators receive a consolidated excel table with all students' results, which they can use for evaluation.

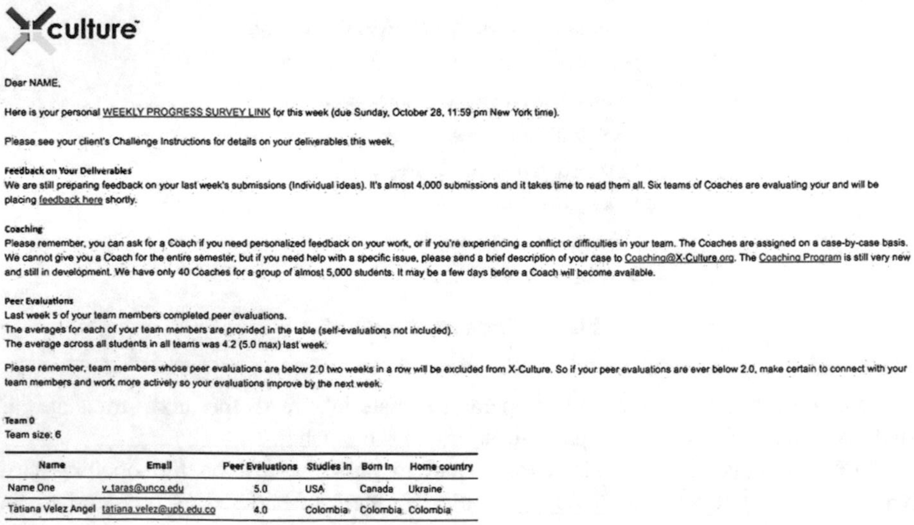

Dear NAME,

Here is your personal WEEKLY PROGRESS SURVEY LINK for this week (due Sunday, October 28, 11:59 pm New York time).

Please see your client's Challenge Instructions for details on your deliverables this week.

Feedback on Your Deliverables
We are still preparing feedback on your last week's submissions (Individual ideas). It's almost 4,000 submissions and it takes time to read them all. Six teams of Coaches are evaluating your and will be placing feedback here shortly.

Coaching
Please remember, you can ask for a Coach if you need personalized feedback on your work, or if you're experiencing a conflict or difficulties in your team. The Coaches are assigned on a case-by-case basis. We cannot give you a Coach for the entire semester, but if you need help with a specific issue, please send a brief description of your case to Coaching@X-Culture.org. The Coaching Program is still very new and still in development. We have only 40 Coaches for a group of almost 5,000 students. It may be a few days before a Coach will become available.

Peer Evaluations
Last week 5 of your team members completed peer evaluations.
The averages for each of your team members are provided in the table (self-evaluations not included).
The average across all students in all teams was 4.2 (5.0 max) last week.

Please remember, team members whose peer evaluations are below 2.0 two weeks in a row will be excluded from X-Culture. So if your peer evaluations are ever below 2.0, make certain to connect with your team members and work more actively so your evaluations improve by the next week.

Team 0
Team size: 6

Name	Email	Peer Evaluations	Studies in	Born in	Home country
Name One	v_taras@uncg.edu	5.0	USA	Canada	Ukraine
Tatiana Velez Angel	tatiana.velez@upb.edu.co	4.0	Colombia	Colombia	Colombia

Fig. 4. Weekly progress survey link.

At the end of the project students upload their final project reports into the indicated dropbox for the assessment. Then within a week each student receives a personal performance review, a personal recommendation letter (positive if did well, negative if did not do well), and a copy of print-ready high-resolution certificate (only to those who successfully completed the project). The certificates of participation in a project are sent to the "global educators" (coordinators) too.

This online project plays a significant role in the educational process evolving many competencies of students such as critical thinking, team work, digital skills and problem solving. Also it allows to build a holistic web-based environment supplementing the first and second stages of integrated learning model.

2.2 Literature Review

Satisfaction is defined as a person's attitude or feelings associated with various factors that influence a particular situation [5]. Student satisfaction is more accurately conceptualized as student perception, which develops from the perceived value of education and experience gained at an educational institution [6]. In the area of human-computer interaction, it is generally assumed that user satisfaction is a manifestation of affection achieved through communication [7, 8]. The concept of user satisfaction represents the level of correspondence between the information system used by users and their requirements [9].

E-learning involves the delivery of information using telecommunication technologies for education and training. In the modern education system, e-learning has become a new paradigm thanks to the tremendous progress in communications and information technology. Features of e-learning cover all the requirements of learning in the modern world, and e-learning has become more popular among enterprises and higher educational institutions due to this special quality. Offering all of its courses online, the Massachusetts Institute of Technology (Massachusetts Institute of Technology) has reported on the strategic importance of e-learning to all higher education institutions [10].

Several factors contribute to user satisfaction in an e-learning environment, which may include teacher, student, course, system design, technology, and environmental aspects [11–19].

Santha Saraswathy and Geeevargese Peter analyzed student satisfaction with the use of the Internet [20]. A study was undertaken at St. Peters College Kolenchery. For analysis, statistical tools were used, such as percentages and chi-square test. For most students, the level of satisfaction with the endless information received via the Internet, the latest news, and the provision of training files and applications was very high.

O. R. Carrasco conducted a survey to determine the level of student satisfaction with a master's program in higher education with a specialization in quality management [21]. The survey was designed as a type assessment tool with 37 questions that evaluated various points of perception of the program in question, the platform being taught, and the professionals who are involved in the learning process. Overall results showed that students were satisfied with the program. However, it is important to note that distance education is the biggest problem in the interaction between the staff of the institution that offers the program and the student. In addition, it was emphasized that the students participating in the program have a high motivation and desire to study. Students had access to the material, and they decided whether they want to study it and when they want to do it, which made the learning process a major part of their sole responsibility. This made students enrolled in such programs highly motivated, because otherwise they would fail in the process.

Muhammad Zaheer, Masroor Elahi Babar, Uzma Hanif Gondal, Uzma Hanif Gondal and Mubashar Majeed Qadri measured the satisfaction of students studying in e-learning in Pakistan [22]. A structured questionnaire was used to measure student satisfaction based on eight parameters, namely: assessment, course content and organization, teacher, learning environment and teaching methods, learning resources, quality of education, student input and teaching aids. The survey shows that most students are satisfied with their education in e-learning, which shows that e-learning has great potential in expanding higher education in a country like Pakistan, where university opportunities are limited.

Choy, McNickle, and Clayton identified the following 10 services that students most expect: (i) complete information on completion requirements for the course/ module; (ii) comprehensive course information; (iii) confidentiality of personal data in the institute's database; (iv) clear statements regarding students' expectations regarding instruction; (v) a helpful response from teachers; (vi) details for evaluation; (vii) interaction with teachers using various methods, such as face-to-face meeting, email, online chat; (viii) quick response from teachers; (ix) instructions regarding who to contact for help; and (x) enrollment guidelines. In addition, researchers found in subsequent interviews that "there are three key areas that students consider important… [including] regular contact with teachers, quick teacher response, and regular support for learning" [23]. The students in this study noticed that there is a need to improve not only the facilitation of the teacher, but also the technical systems.

Hara and Kling found that factors such as lack of quick feedback, technical problems, and controversial course instructions lead to student frustration and dissatisfaction [24]. McNickle and Clayton showed that the course should be designed in such a way that it stimulates not only student discipline, but also his/ her consistent approach to work [25].

Studies conducted on online courses have also revealed some problems, including quick and useful communication with the instructor; clear guidance on course expectations; support for enrollment; student assignments and requirements; and data security. These questions can increase student satisfaction if they are correctly addressed [26–28]. These areas can be divided into issues that relate mainly to the content of the program and its implementation.

In addition, to successfully complete the course, students on the Internet must be familiar with the technology used [26]. In distance learning, student satisfaction is mainly affected by access to technology [27]. Typically, those students are not satisfied with those who are frustrated with the use of technology in the course [24, 29].

3 Methodology

In our study 63 undergraduate students (4th-year) from Peter the Great St. Petersburg Polytechnic university took part. All of them were enrolled in a course based on integrated learning model. The average age of students was 20 years old. The group consisted of 37 girls and 26 boys. The research is based on quantitative and qualitative data. To get the data we conducted an online Student Satisfaction Survey, which included 27 items on a 5-point Likert scale and three open questions which defined students' satisfaction with web-based environment. The items were created on the base of literature review.

The CC views regarding the satisfaction of web-based environment were collated through two face-to-face interviews conducted by the two researchers. Both interviews with the CC were of 45 min duration and followed a semi-structured format. The interviews were recorded, fully transcribed and cleaned to ensure that the transcribed narrative fairly represented the CC's responses during the interview. Based on the research questions, we examined the transcripts, identified key issues and made a summary.

This paper is based on the following research questions:

1. Are students satisfied with the web-based environment in integrated learning model?
2. Does course coordinator (CC) feel satisfaction with the web-based environment in integrated learning model?
3. What can be improved from the students and teacher perspectives?

4 Results and Discussion

4.1 Students' Online Survey Results

About 77% of students completed an online satisfaction survey. The indicators of Standard Deviation on each of the items we researched were quite low, so we can say that the difference in answers was not significant. The results of the first part of the survey are presented in the Table 1.

Table 1. Survey results.

Group	Items	Mean	SD
Interaction	A web-based session keeps me always alert and focused	4	0,46
	Interaction is adequately maintained with the course coordinator	4,1	0,87
	I can ask coordinator a question at any time possible	3,97	0,67
	I am satisfied with the quality of interaction between all involved parties	4,2	0,7
	I am dissatisfied with the process of collaboration activities during the course	4,05	0,51
	I am satisfied with the way I interact with other students	4,1	0,56

(*continued*)

Table 1. (*continued*)

Group	Items	Mean	SD
Instruction	The use of web based environment in this course encourages me to learn independently.	3,9	0,9
	My understanding is improved compared to similar courses I studied before	3,7	0,82
	My performance in exams is improved compared to similar courses I studied before	3,4	0,74
	I am satisfied with the level of effort this course required	3,7	0,6
	I am dissatisfied with my performance in this course	3,9	0,7
	I believe I will be satisfied with my final grade in the course	3,7	0,78
	I am satisfied with how I am able to apply what I have learned in this course	3,9	0,47
	I am willing to take another course using the web-based mode	3,97	0,58
	I am satisfied enough with this course to recommend it to others	3,82	0,39
	Compared to face-to-face course settings, I am less satisfied with this learning experience.	3,3	0,91
	I enjoy working on assignments by myself	4,12	0,54
Coordinator	The coordinator makes me feel that I am a true member of the class	3,9	0,7
	I am dissatisfied with the accessibility and availability of the coordinator	3,5	0,64
	Feedback on evaluation of tests and other assignments was given in a timely manner	4,2	0,37
	The coordinator always takes attendance	3,91	0,8
	I attend webinars the same way I attend face-to-face classes	3,6	0,69
Technology	The instructor's voice is audible	3,5	0,73
	Course content shown or displayed on the smart board is clear	3,72	0,67
	The video images are always clear	3,7	0,8
	Technical problems are not frequent and they do not adversely affect my understanding of the course	4,11	0,79
	The technology used for web-based environment is reliable	4	0,47

Also we conducted t-test on the group of overall student satisfaction with web-based environment, to evaluate whether the mean was significantly different from 2.5 - an accepted mean for student satisfaction (mean*) [27]. The results are presented in the Table 2.

In all categories the mean was much higher than 2,5 and the level of satisfaction with the web-based environment was significantly higher than the average rate.

Table 2. Survey results.

Group	Mean*	Mean	SD	t-value
Interaction	2,5	4,07	0,63	15,3***
Instruction	2,5	3,76	0,67	9,8***
Coordinator	2,5	3,82	0,64	11,1***
Technology	2,5	3,8	0,69	10,8***

Note: * $p < 0,05$; ** $p < 0,01$; ***$p < 0,001$

At the end of the survey students had to answer on three open questions. There were following questions:

(1) Do you feel satisfaction with the course?
(2) What do you like mostly?
(3) What you would like to improve?

Most students noted satisfaction with the web-based environment. As the advantages, students highlighted the availability, simplicity and clarity of use, as well as the presence of useful theoretical material. Students also emphasized that course materials, assignments and feedback from instructors really facilitate learning. Student responses demonstrated that the online international project created a sense of community among students, where they were free to express their opinions. It can be considered a successful recreation of a classroom environment. Moreover, 96% of students agreed that the course encouraged them to discuss their ideas with other students.

However, it is also important to note that there are a noticeable number of students who think that they have not learned as much in the online class as in a face-to-face class.

Some students answered that it would be better to increase the number of face-to-face classes, as they did not have enough time to discuss al the material with the teacher. 15% of respondents noted that material was too complicated and they had to search additional information with more detailed explanation. Also, students complained about difficulties with cross-culture communication during their work in X-culture project (for instance, they faced problems in finding appropriate time to communicate with team members from USA, Canada, Mexico, etc., because of different time zones). But everybody confirmed that it was a great cultural experience at the same time.

4.2 CC's Interview Results

The CC described the web-based environment as "unlimited educational space with unlimited access." The CC also noted that such an environment allows students to independently manage their time, to study at a convenient time. Moreover, each student spends the amount of time necessary for him personally to work with the material.

CC was satisfied with the preparation of students for the classroom. Thanks to good online preparation, the classes became more informative, which allowed to discuss more detailed topics. In addition, the atmosphere in the audience became more comfortable, as all students had the same training and were not afraid to participate in discussions.

Conducting classes in the new mode motivated the teacher to work harder, to select more interesting and practical material in order to deepen the theoretical knowledge of students. However, the process of preparing for classroom activities has become more time-consuming.

It is worth noting that working with an online project has become a logical complement to the discipline course, as this allowed students to apply their knowledge in a real project. CC reported that "students noted difficulties with cross-cultural communication. Therefore, it is worth paying more attention to this aspect in the future."

Overall, the CC is satisfied with the course and expresses a desire and willingness to continue teaching using the developed model.

5 Conclusion

In our study we aimed to examine the students' and course coordinator's satisfaction with the introduced web-based environment in an integrated learning model that helps efficiently combine several learning methods which affect simultaneously cognitive functions, professional skills and motivation. The analysis of students' survey results showed that they were satisfied with all components of web-based environment such as interaction, instruction, coordinator and technology level. CC noted in an interview that such web based environment allowed to enhance the process of education and professional competence building.

In comparison with other studies students enrolled to integrated learning model expressed a higher level of satisfaction, they were less anxious regarding the online education mode and didn't feel a high necessity of more face-to-face classrooms. Among the negative points both students and teachers noted some problems concerning cultural aspects of communication in international project.

The theoretical work of other scholars in this field has been a useful resource for planning and design, and we expect that our study will provide something of value for future researchers too.

The next year of the International Business discipline delivery we are going to investigate which exactly factors influence students' and coordinator's satisfaction in the developed integrated learning model and therefore what are the key instruments for a successful educational course based on web-based learning system.

References

1. Shipunova, O., Evseeva, L., Pozdeeva, E., Evseev, V.V., Zhabenko, I.: Social and educational environment modeling in future vision: infosphere tools. In: E3S Web of Conferences, vol. 110, p. 02011 (2019). https://doi.org/10.1051/e3sconf/201911002011

2. Bauk, S., Scepanovic, S., Kopp, M.: Estimating students' satisfaction with web based learning system in blended learning environment. Educ. Res. Int. (2014). https://doi.org/10.1155/2014/731720
3. Baranova, T.A., Kobicheva, A.M., Tokareva, E.Y.: Does CLIL work for Russian higher school students?: The Comprehensive analysis of Experience in St-Petersburg Peter the Great Polytechnic University. In: ACM International Conference Proceeding Series. pp. 140–145 (2019). https://doi.org/10.1145/3323771.3323779
4. Baranova, T., Khalyapina, L., Kobicheva, A., Tokareva, E.: Evaluation of students' engagement in integrated learning model in a blended environment. Educ. Sci. **9**, 138 (2019)
5. Bailey, J.E., Pearson, S.W.: Development of a tool for measuring and analyzing computer user satisfaction. Manag. Sci. **29**(5), 530–545 (1983)
6. Astin, A.W.: What Matters in College? Four Critical Years Revisited. Jossey-Bass, San Francisco (1993)
7. Mahmood, M.A., Burn, J.M., Gemoets, L., Jacquez, C.: Variables affecting information technology end-user satisfaction: a meta-analysis of the empirical literature. Int. J. Hum.-Comput. Stud. **52**(5), 751–771 (2000)
8. Razinkina, E., Pankova, L., Trostinskaya, I., Pozdeeva, E., Evseeva, L., Tanova, A.: Influence of the educational environment on students' managerial competence. In: E3S Web of Conferences, vol. 110, p. 02097 (2019). https://doi.org/10.1051/e3sconf/201911002097
9. Cyert, R.M., March, J.G.: A Behavior Theory of the Firm. Prentice-Hall, Englewood Cliffs (1963)
10. Wu, J.P., Tsai, R.J., Chen, C.C., Wu, Y.C.: An integrative model to predict the continuance use of electronic learning systems: hints for teaching. Int. J. E-Learn. **5**(2), 287–302 (2006)
11. Arbaugh, J.B.: Managing the online classroom: a study of technological and behavioral characteristics of web-based MBA courses. J. High Technol. Manag. Res. **13**, 203–223 (2002)
12. Shipunova, O.D., Berezovskaya, I.P., Mureiko, L.V., Evseev, V.V., Evseeva, L.I.: Personal intellectual potential in the e-culture conditions. ESPACIOS **39**(40), 15 (2018). http://www.revistaespacios.com/a18v39n40/18394015.html
13. Aronen, R., Dieressen, G.: Improvement equipment reliability through e-Learning. Hydrocarbon Process. **80**, 47–57 (2001)
14. Chen, W.L.C., Bagakas, J.G.: Understanding the dimensions of self-exploration in web-based learning environments. J. Res. Technol. Educ. **34**(3), 364–373 (2003)
15. Hong, K.S.: Relationships between students' and instructional variables with satisfaction and learning from a web-based course. Internet High. Educ. **5**, 267–281 (2002)
16. Lewis, C.: Driving factors for e-learning: an organizational perspective. Perspectives **6**(2), 50–54 (2002)
17. Piccoli, G., Ahmad, R., Ives, B.: Web-based virtual learning environments: a research framework and a preliminary assessment of effectiveness in basic IT skill training. MIS Q. **25**(4), 401–426 (2001)
18. Stokes, S.P.: Satisfaction of college students with the digital learning environment. Do Learners' temperaments make a difference? Internet High Educ. **4**, 31–44 (2001)
19. Thurmond, V.A., Wambach, K., Connors, H.R.: Evaluation of student satisfaction: determining the impact of a web-based environment by controlling for student characteristics. Am. J. Dist. Educ. **16**(3), 169–189 (2002)
20. Peter, G., Saraswathy, S.: Student satisfaction with internet usage. Int. J. Manag., IT Eng. **2249–0558**, 7 (2017)
21. Rojas Carrasco, O.: Student satisfaction with distance study programs. RA J. Appl. Res. **2** (2018). https://doi.org/10.31142/rajar/v4i4.09

22. Zaheer, M., Elahi Babar, M., Gondal, U., Qadri, M.: E-learning and student satisfaction (2015). https://www.researchgate.net/publication/295400881_E-Learning_and_Student_Satisfaction. Accessed Jul 19 2019

23. Choy, S., McNickle, C., Clayton, C.: Learner Expectations and Experiences: An Examination of Student Views of Support in Online Learning. Australian National Training Authority, Leabrook (2002)

24. Hara, N., Kling, R.: Students' distress with a Web-based distance education course: An ethnographic study of participants' experiences. Center for Social Informatics, Bloomington, IN (2000). http.//www.slis.indiana.edu/CSI/wp00-01.html. 24 Aug 2001

25. Vonderwell, S., Turner, S.: Active learning and preservice teachers' experiences in an online course: a case study [Electronic version]. J. Technol. Teach. Educ. **13**(1), 65–85 (2005)

26. Belanger, F., Jordan, D.H.: Evaluation and Implementation of Distance Learning: Technologies, Tools and Techniques. Idea Publishing Group, Hershey (2000)

27. Bower, B.L., Kamata, A.: Factors influencing student satisfaction with online courses. Acad. Exch. Q. **4**(3), 52–56 (2000)

28. Chong, S.M.: Models of asynchronous computer conferencing for collaborative learning in large college classes. In: Bonk, C.J., King, K.S. (eds.) Electronic Collaborators: Learner-Centered Technologies for Literacy, Apprenticeship, and Discourse, pp. 157–182. Lawrence Erlbaum Associates, Mahwah (1998)

29. Giannousi, M., Vernadakis, N., Derri, V., Michalopoulos, M., Kioumourtzoglou, E.: Students' satisfaction from blended learning instruction. In: Proceedings of the TCC Worldwide Online Conference, vol. 1, pp. 61–68 (2009)

System Approach to the Formation of the Information Culture Levels in Cyberspace

I. Dzhalladova$^{(\boxtimes)}$ ⓘ, N. Batechkoⓘ, and Y. Gladkaⓘ

Kyiv National Economic University named after Vadym Hetman,
Kiev 03057, Ukraine
idzhalladova@gmail.com

Abstract. The article deals with the problem of formation of information culture of cyberspace by methods of system analysis. An analysis of modern approaches to the formation of the concept of "information culture" taking into account trends in the development of information and communication technologies is conducted. A methodology for forming information culture levels on the basis of the systematic approach is presented. The peculiarities of informational culture formation in conditions of uncertainty are specified.

Keywords: Information culture · Levels of information culture · System approach

1 Introduction

The modern information society development implies the transformation of information into one of the major strategic resources, the storage, development and rational use of which is a vital necessity for the national security of the state as a whole. Researchers are increasingly stating the growing role of the information domain, which is an aggregate of information, data infrastructure, entities that collect, compile, disseminate and use information, increasingly regulate public relations, in particular, information culture of a society in general and an individual recipient in particular.

Having analyzed some statistics, we can state that the global data storage volumes increased by 23% per year, telecommunications by 28% and computing capacity by 58% per year in recent decades. Such high growth rates of the information and communication technology (ICT) are capable to transform our entire lives in decades. The well-known British researcher, President of the Center for Economic Policy Research, founder of the Vox EU.org portal Richard Baldwin, called the large-scale changes caused by ICT transformational, revolutionary, and destructive [1].

In fact, "turbulent information flows render chaos to the economy, politics, technological and anthrosphere of our world, leaving no room for long-term forecasts" [2]. Raising of the information culture of an individual, which under the conditions of constantly improving information interaction technologies undergoes significant transformations, is of particular importance against this background.

© Springer Nature Switzerland AG 2020
T. Antipova and Á. Rocha (Eds.): DSIC 2019, AISC 1114, pp. 275–283, 2020.
https://doi.org/10.1007/978-3-030-37737-3_25

2   Problems of Formation of Information Culture Levels

2.1   Research Objective, Methods, Framework of the Research

A well-known scientific fact is the three-pronged nature of ICT, which is already enshrined in the term itself. Thus, in particular, I - means information capacities, which are determined by the cost of data processing and storage, C - communication capacities due to the progress in the data transmission, T is technology itself. The Moore's law, which stands for I in ICT, reads that the computing capacity of computers grows exponentially, for example, every 18 months the capacity of chips is doubled. T is described by two laws: Gilder' and Metcalfe' [1]. Thus, George Gilder noted that communication system bandwidth rises three times faster than the computing capacity of a computer, which is doubled every six months. This speeds up data transmission and reduces the computation associated with data processing and storing. Robert Metcalfe states the effect of a telecommunications network is proportional to the square of the number of connected users of the system. That is, the growth of new users is not a linear process. With each new growth of the network the scale of its further expansion increases, that is, an additional increase overlaps an additional increase.

If certain patterns have already been established in the information, communication and technological trends of the ICT development, threats they cause at different levels of social activity are still an open question. This is especially true for the study of patterns of enhancement of the information culture effect, both at the national and individual levels.

It is worth noting numerous research papers in the field of informatics, cybernetics, pedagogy, psychology, and philology. However, a systematic approach to the analysis of the formation of an "information culture" is now needed, which will lead to the study of both technological and psychological aspects of the phenomenon at various levels and interconnections.

The goal of this research paper is to apply a systematic approach to the study of the "information culture" phenomenon taking into account the modern aspects of the development of the information society at different stages of its lifecycle. To achieve this goal we used methods of analysis (systemic, problem-specific, structural), theoretical and functional-structural modeling, synergistic approach to analyze the system stability and predict its development.

2.2   Analysis of Modern Approaches to Formation of Information Culture

The term "information culture" in domestic publications first appeared in the 1970s. Initiators of the development and popularization of the corresponding concept were librarians.

Currently, information culture is increasingly being interpreted as a particular phenomenon of the information society (O. Dzuban, N. Semeniuk, O. Prudnikova). Depending on the object under consideration, researchers distinguish information culture from the society; individual categories of consumers of information; person [3–5].

By integrating different scientific interpretations, information culture can be regarded as an integral part of a general culture focused on information support for human activity. Information culture reflects the achieved levels of organization of information processes and the efficiency of creating, collecting, storing, processing, presenting and using information that enables a holistic vision of the world, its modeling, predicting the results of human decisions.

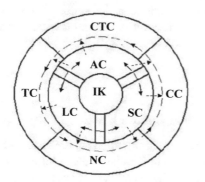

AC - audiovisual culture
LC - logical culture
SC – semiotic culture
CTC - conceptual and terminological culture
TC - technological culture
CC - communication culture
NC - network culture

Fig. 1. Components of information culture.

Information culture can be interpreted widely as throughout set of principles and mechanisms that ensure the interaction of ethnic and national cultures, their integration into the common experience of mankind; in a narrow sense it is an optimal way of handling information and presenting it to the consumer to solve theoretical and practical problems; mechanisms for improving the technical means of production, storage and transmission of information; development of training system, education of a person for effective use of information means and information.

The main trend in dynamics of the "information culture" concept formation relates to the fundamental and multidimensional nature of this phenomenon, not only determined by the conditions of scientific and technological progress, electronic means of processing, storage and transmission of social information, but above all - by an operating infrastructure that permeates all epochs and civilizations, all spheres of human activity and all stages of human development as a social unity.

Information culture is formed through information literacy and is an integral concept that includes such components as audiovisual, logical, semiotic, conceptual-terminological, technological, communication, and network culture (Fig. 1).

There are strong logical links between the information culture components, and it can be assumed that at some level and under certain circumstances, one or another component may prevail, which ultimately influences the development of the system and the whole.

A person perceives the major part of information through virtual communications. New working conditions create dependence of one person's awareness on information acquired by other people. According to Henry Kastler, information is a random choice made by a consumer from several possible and equal sources.

Therefore, it is not enough to be able to independently learn and accumulate information, and it is necessary to learn such technology of working with information when preparing and making decisions on the basis of collective knowledge. This suggests that a person must have a certain level of culture in dealing with information. The term "information culture" was introduced to reflect this fact. It is information culture that makes it possible to distinguish valuable or informed data in the information space.

The study of the phenomenon of information culture at its various levels is of particular importance in these circumstances, given that it undergoes constant significant transformations under the transient information technology.

3   Results

3.1   Methodology of Formation of Information Culture Levels

The complexity and versatility of the phenomenon of "information culture" necessitates the use of a systematic approach in the scientific search. If we consider culture from the point of view of a systematic approach, namely as the social subsystem together with the economy, politics, culture, we can state that, according to A.V. Prudnikova, formation of a new information culture makes it more closely interconnected with the external environment, it is linked to by many network connections.

The ability to manipulate these connections without losing touch with the internal culture of a man, the ability to use the necessary components for this without losing the main value settings of personality characterizes the degree of development of modern information culture [4]. So, in particular, network information technologies and neuroscience domain radically change the ability of a person to critically analyze, set the unquestionable belief in the searcher, the clipping consciousness, the trance of constant Internet surfing and gaming, etc. [2].

It is very easy to manipulate such people, he/she is an ideal citizen for the Machiavellian type of elites being constantly cultivated. An educated and thinking citizen is uncomfortable and even dangerous. Such a strategy of managing the society quite often leads to various social upheavals, which we have been increasingly witnessing in the recent period.

Therefore, in these circumstances, it is advisable to consider information culture as a systemic formation at the appropriate interconnected levels. In our case, it is quite possible to use a multilevel approach, which is "a model of interaction where a set of intellectual systems and their components interact and exchange knowledge" [5].

The concept of levels, as noted by A.V. Shcherbakova and G.S Fedorova, is one of the models used to divide complex systems into simpler components. With this approach, there is an upper level that describes the system as a whole; and a lower level that describes the categories, using the concepts of higher level, etc. Thus, each sublevel provides functionality for the next higher level, which in turn ensures existence for the upstream level [6].

However, the most difficult, according to scientists, is to define the content and limits of responsibility of each level. In our case, to overcome the complexity in the process of studying the "information culture" phenomenon, it is appropriate to consider this system formation in terms of traditional systems (integrity, complex of interacting elements, structural, hierarchy, interdependence of the system and the environment), and dynamic systems: with feedback and the presence of nonlinear effects.

This approach allows you to streamline the hierarchy of component systems - levels, to build a kind of hierarchy of relationships within the system, and to predict the behavior of systems in bifurcation states.

By establishing interconnections between levels in the system, we thereby determine its structure and mechanism of functioning. According to A.I. Ilyash, the structure is formed by the subjects and objects that are its elements, and the functioning mechanism is an appropriate set of rules (procedures, functions) that govern the actions of elements of the system in the course of its operation [7].

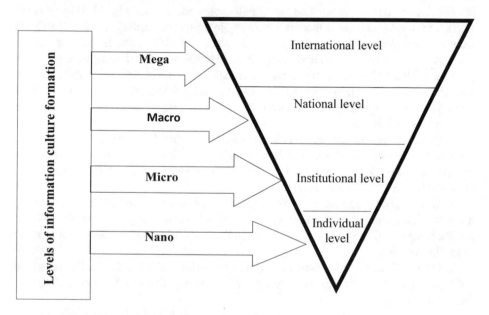

Fig. 2. Hierarchy of levels of information culture formation.

We would like to point out the presence of some particular special features of the information culture system that affect its general properties. First of all, it is resilience (the ability of the system to properly function and resist destabilizing factors); second, the presence of non-equilibrium states (gradual change of stability parameters of the system); third, the presence of dissipative structures (the cultivation of more balanced structures within the system), etc.

Note that the hierarchical structure of a system is often associated with the process of changing it, first and foremost, in order to increase its efficiency.

Thus, in view of the above, we suggest defining the information culture system as dynamic, multifunctional feedback structures, all components of which are distributed on levels, interact with each other, receiving a synergistic effect and aimed at achieving the main result of information culture in different levels.

3.2 Application of the Approach to the Formation of Information Culture Levels

Taking into account the multifaceted nature of the "information culture" phenomenon, the question arises of its reaching at the local level of personality (nano level), at the level of an individual institution (micro level), at the national - macro level, international - mega level (Fig. 2).

With regard to the mega level of information culture formation, the international community identifies the following priority areas, in particular: accessibility of information, preservation of information, information literacy, information ethics, and information for development. The most well-known and effective in this respect is the UNESCO Media and Information Literacy Program. According to the UNESCO website, these are "actions to give people the skills and abilities to critically perceive, evaluate and use information and media in their professional and personal lives." The purpose of the initiative is to create information-literate societies by creating and maintaining the educational literacy policy. Project participants work with teachers around the world, training them in information literacy and providing resources for use in their classrooms.

UNESCO publishes information literacy studies in many countries, looking at how information literacy is being taught, how it differs in various communities, and how to improve it. Educational programs for school boards and teachers with guidelines for their use have become quite common.

UNESCO introduces the new integrated concept of MIL - media and information literacy, which describes the competencies needed to survive and successfully address the challenges of life for individuals, communities, societies and humanity as a whole in the information age.

Later on, UNESCO has launched the Intergovernmental Program "Information for All", aimed at providing universal access to information that can be used to improve the quality of life.

An effective process of information culture formation in Ukraine requires the implementation of a number of projects at the national level, both governmental (introduction of new subjects) and non-governmental (for example, media resource osvita. com.ua). One of requirements for the effective formation of information culture at the macro level is the continuous improvement of the state mechanism and legal regulation in this field, in particular, the preparation of legal acts and norms for their application in the conditions of informatisation of the society.

In this regard, the acquisition of theoretical and practical skills in the basics of using modern methods of legal protection of state, official, commercial, professional, sensitive and personal data in computer systems is of particular relevance; as well as licensing and certification in the field of information security, development of practical skills and abilities to ensure the legal protection of information [8].

An important condition for organizing the work of the management apparatus is the determination of the legal status of each agency, its units, and employees. Today, with changes affecting all levels of management, departmental and job descriptions need to be substantially updated. A textbook "Educating the Internet User Culture. World Wide Web Security" was published to develop the skills of children to use the Internet resources competently and securely.

For many companies and businesses in Ukraine, the level of information culture has been already an integral part of the general corporate culture of the institution. Sometimes, employees do not have sufficient or complete understanding of the managerial information processing, they are unable to analyze the situation using available information, do not know the ways of its passage and lack the technical knowledge necessary to protect against unauthorized information capturing.

The level of information culture at the individual level significantly influences a person's life and extends freedom of action. Moreover, it becomes a major resource for raising social and professional status (along with education, economic and social status). Assumptions are made that very soon, the level of information culture would determine the future of an individual, as the ability to find, receive, process and adequately use the information is required not only in the professional field, but also in everyday life.

However, it is worth noting that researchers in this field usually focus on technological aspects of the interconnection of human and information society, while in our view, the psychological and pedagogical systems are not sufficiently developed.

In this context, N. Semeniuk expressed an interesting point of view, according to which the information culture of an individual as a condition of human adaptation to life in the information society evolves into the knowledge society. Therefore, the researcher believes that purposeful and comprehensive human training is needed, including a wide range of knowledge, skills and information competences related to the search and critical analysis of information [9].

As O.V. Prudnikova notes, the formation of information culture only through the study of computer science is insufficient, because it limits to the technical and software aspects of informatisation. At the same time, the level of information culture of a person will in the future determine the desirable knowledge and skills in the field of search and semantic processing of information. Therefore, they should be the subject of particular attention by educational institutions as a challenge of national importance as the foundation and universal tool of any national strategy and security.

3.3   Formation of Information Culture in Non-equilibrium States

It should be noted that the methodology proposed above is valid only for an equilibrium system of information culture formation with a certain margin of stability. Unlike equilibrium formations, so-called dissipative structures are implemented in non-equilibrium open systems. Such a process is a manifestation of fluctuations, i.e. deviations of the system parameters from the averages. The role of fluctuations increases and becomes decisive near the critical (bifurcation) points.

During the period of bifurcation perturbations, the system of information culture formation switches to the non-equilibrium state and acquires the characteristics of dissipative structure, which are formed due to flows of energy, momentum, and mass across the boundaries of the system. Then, the interaction of fluctuations grows and becomes so strong that the existing structure (in our case, the information structure system) is destroyed. These phenomena occur in times of social upheaval, cataclysms, etc. In such circumstances, the information culture system undergoes significant transformations and is eventually destroyed. At the same time, more stable structures occur that are called dissipative ones.

Note that in the process of information generation, as in other dynamic systems, the phenomenon of "mixed layer" occurs. After all, any development can be represented as alternation of dynamic and chaotic stages (more precisely stages of "a mixed layer"). In general terms, this phenomenon was mentioned two hundred years ago in the writings of Hegel, who proposed the famous triad:

thesis $\rightarrow$ antithesis $\rightarrow$ synthesis

Here, according to D.S. Chernavsky, "thesis" and "synthesis" can be considered as dynamic stages in which the feature "synthesis" contains more information. To the contrary, "antithesis" can be considered as the opposite of the dynamic (ordered) phase, that is, "chaos" [10]. This may well be attributed to the formation of "information culture", but it is a topic of separate scientific research for the above parameters.

4 Conclusion

The problem of information culture formation is considered in the research paper as a phenomenon, the study of which showed the existence of a vast empirical material. In order to systematize it, an attempt was made to form levels of information culture: mega (international, macro (national), micro (individual institution levels), nano (personal levels)).

International trends in the information culture formation cover accessibility of information, its preservation, information literacy, information ethics, and information for development. One of the major requirements for the effective formation of an information culture at the macro level is the improvement of the state mechanism and legal regulation in this field. A necessary condition for the formation of the information culture of an individual is purposeful and comprehensive training of the person, covering a wide range of not only information knowledge, skills, but also information competences related to the search and critical analysis of information.

Development of information technology bring new opportunities that are growing by the exponential law, but on the other hand, ICT innovations create new security threats, and it is the information culture at different levels of its formation that has the mission of preventing unauthorized threats.

References

1. Richard, B.: The Great Convergence: Information Technology and the New Globalization. The Belknap Press of keyword University Press, London, England (2018)
2. Budanov, V.: Methodology of Synergetics in Post-Neoclassical Science and Education. Synergetics in the Human Sciences, no. 7 (2017)

3. Dziuban, O.: Philosophy of Information Law: General - Theoretical Foundations: monograph. H., Maidan (2013)
4. Prudnikova, O.: Information culture and formation of the information person. Bull. Nat. Univ. "Yaroslav the Wise Law Academy of Ukraine", no. 2, pp. 154–165 (2016)
5. Kolesnikova, A.: Hybrid intelligent systems: theory and development technology. In: Yashin, A.M. (ed.) SPbSTU, SPb (2001)
6. Scherbakova, A., Fedorova, G.: A multi-level approach to the construction of a hybrid intellectual system. Data Process. Syst. **3**, 96–99 (2011)
7. Ilyash, A.: Transformation of the Social Security System of Ukraine: Regional Dimension: monograph. PAIS, Lviv (2012)
8. Dzhalladova, I., Batechko, N., Kolomiyets-Ludwig, E.: Modern Approaches to Information Security Legislation Modelling. Scientific Development and Achievements, pp. 125–142. SCIEMCEE, London (2018)
9. Semeniuk, N.: The need for information support of life-long education. Gilea: Scientific Bulletin: coll. research papers K VIR UAN, vol. 78, no. 11, pp. 294–297 (2013)
10. Chernavsky, D.: On the Methodological Aspects of Synergetics. Synergetic Paradigm. Nonlinear Thinking in Science and Art, no. 2, pp. 50–67 (2002)

Swift a New Programming Language
for Development and Education

Rostislav Fojtik(✉)

University of Ostrava, 701 03 Ostrava, Czech Republic
rostislav.fojtik@osu.cz

Abstract. The development of information and communication technologies also develops programming languages and development tools. Swift programming language is one of many emerging languages. The language was primarily designed for developing applications for iOS and macOS. Gradually, Swift becomes the primary language for developing iOS mobile apps. This paper describes the essential features, advantages and disadvantages of the Swift programming language and compares it with other languages. The article shows some measurement results and a comparison of Swift program speeds and some other programming languages. We find out from the tracking of students' work and practical projects that working with Swift is often faster, brighter and safer than the previously used Objective-C language. The article describes the results of the survey among students and demonstrates the advantages of language in the field of programming teaching. Another goal of the article is to show the possibilities, advantages and limitations of Swift's programming language in the development of mobile applications. The paper also shows the possibilities of Swift language in teaching.

Keywords: Comparison of languages · C++ · C# · Education · Java · Objective-C · Programming · Swift

1 Introduction

The development of mobile systems and applications requires appropriate programming languages and development tools. First, Objective-C was the primary language for iOS. In recent years, that Objective-C has been shown to cease to meet modern developments. That's why Apple has begun to develop the new Swift programming language. It is a compiled programming language for iOS, watchOS, tvOS, macOS, and Linux applications. It was created for more comfortable and safer code creation, but also easier learning programming. It is first introduced at Apple's 2014 Worldwide Developers Conference. Language has undergone dynamic development, and this has occasionally led to problems. Swift brings simpler syntax, shorter code, and safer programs. The language is not a purely object-oriented language but is hybrid. Apple says that Swift is 2.6x faster than Objective-C and 8.4x faster than Python [10]. Swift, together with Kotlin, is one of the fastest-growing programming languages [3].

© Springer Nature Switzerland AG 2020
T. Antipova and Á. Rocha (Eds.): DSIC 2019, AISC 1114, pp. 284–295, 2020.
https://doi.org/10.1007/978-3-030-37737-3_26

2 Swift Programming Language

The Swift programming language offers simpler syntax and some constructions than common programming languages. For example, Java provides a large number of collections, some of which have similar properties, and some are currently not recommended. In contrast, Swift offers only essential collections.

2.1 Swift Advantages and Disadvantages

During the development of applications and in the course Mobile Application Development, we have defined the following main advantages and disadvantages of Swift [2, 7].

Advantages of Swift:

- Uses a new, simpler and cleaner syntax.
- A more straightforward syntax leads to faster code creation and control.
- Reducing the number of SDK class reference data has led to more efficient memory usage.
- New features such as Tuples, Optionals, Closures, Generics, etc.
- Optional types help to check the existence of an object within the code, which makes it possible to avoid further accidents.
- An open-source language.
- The language is not for the only iOS.
- Swift has already been customised for Linux and used on Android as well.
- Other large companies are starting to support and develop Swift systems. For example, IBM, Lyft, LinkedIn, Airbnb etc. [8].

Example of use of Tuple:

```
let contact = ("Steve", 602345678)
let name = contact.0
let phone = contact.1
```

The significant advantage of Swift is more apparent and usually shorter code. We compared thirty simple projects written in Objective-C and Swift. The programmer who writes the code in Swift wrote an average of 25% of the code less than in Objective-C. Less code usually means fewer errors.

The following codes show a comparison of the simple distance transfer function in meters per mile. The first code is written in Objective-C, the second in Swift. The Swift programmer had to write only 76% of the characters in the Objective-C programmer [10, 12].

The sample code in Objective-C:

```
-  (IBAction)buttonConvert:(id)sender
{
    double miles = 0;
    double meters = 0
    meters = [inputText.text doubleValue];
    miles = meters / 1.609;
    NSString *textMile = [NSString stringWithFormat:@"Distance: %.3lf ml", miles];
    labelOutput.text = textMile;
}
```

The sample code in Swift:

```
@IBAction func buttonConvert(_ sender: UIButton) {
        var miles : Double
        var meters : Double
        meters = Double(inputText.text!)!
        miles = meters / 1609.0
        labelOutput.text = "Distance: \(miles) ml"
    }
```

Swift is designed to be safer than previous languages [6]. The following example of code in C++ lists the class in which we create a dynamic array. The programmer must ensure proper memory release in C++. He must manually release dynamic objects and have adequately prepared programmed destructors. If the dynamic array is not released in the destructor, the object will disappear after cancellation, but the array would remain in memory. C++ requires manual manipulation of memory from the program, which can lead to dangerous situations. Students are taught that they must also have a *delete* for each *new* one.

```
class MyClass{
private:
    int *array;
    int max;
    int count;
    //other attributes
public:
    MyClass(int max){
        this->max = max;
        count = 0;
        array = new int[this->max];
    }
//destructor!
    ~MyClass(){
        delete [] array;
    }
    //other methods
};
//create the object
MyClass *myObject = new MyClass(100);
//cancel the object
delete myObject;
```

Swift uses Automatic Reference Counting (ARC) to track and manage the app's memory usage. ARC automatically frees up the memory used by class instances when those instances are no longer needed. ARC is less demanding for processor performance than a garbage collector.

Disadvantages of Swift

- Continuous rapid development and frequent changes.
- Incompatibilities between versions.
- Less choice of libraries and frameworks than other languages.
- Some Cocoa Touch methods yet require Objective-C objects, which complicates the work with Swift.
- Smaller extensions.

A significant drawback to Swift is constant and rapid development. Individual versions of the language bring new features, significant changes, and are often not compatible in some areas. Therefore, you need to convert older versions. Between version 3.0 and 4.0 can still be auto-mailed, but some manual code modifications are required for some projects [1].

2.2   Development Tools Supporting Swift

The primary development environment for Swift is Xcode. It contains all the necessary tools for developing mobile applications. The disadvantage is available only on macOS. Swift development of Swift and changes in Xcode lead to worse stability than previous versions. Programmers occasionally complain about problems with Xcode. The development tool is free.

Another commonly used development environment is JetBrain's AppCode. IDE uses Xcode and SDK, is stable and programmers appreciated as very good. The development tool is free for study purposes only.

The programmer can also use CodeRunner, which allows you to run code in 23 different tongues. Other tools include Atom, Sublime Text, or IBM Swift Sandbox. For example, the IBM Swift Sandbox is an interactive website that let users write Swift code.

Interesting is the newly developed SCADE tool, which is lying on Eclipse. In this tool, Swift can be used not only to develop mobile applications for iOS but also for Android. The developer can develop multiplatform mobile apps. Similarly, multiplatform mobile applications can be produced in Xamarin or Visual Studio, but we need to use the C# language.

2.3   Speed Measurement and Comparison

Apple says Swift is a fast programming language. Tests have been performed to verify this claim [4, 5, 13]. The Computer Language Benchmarks Game performed a comparison of speeds and CPU and memory usage of the same programs written in different languages. Swift was faster than Java in six tests out of ten [9] (Table 1).

Table 1. Test binary-trees description [9].

Language	Seconds	Memory
Swift	5,50	192,172
Java	8,39	933,808
C++	2,62	130,212
C#	8,26	870,360

We've made several speed measurements of Swift, Java, C++, and C# programs. Measurements were performed on MacBook Pro (13-in. Retina Late 2013), Intel Core i5-4258U processor, 2.4 GHz, 8 GB 1600 MHz DDR3 memory, 250 GB SSD. Operating system macOS 10.13.5 (Build 17F58b). We measured the Fibonacci sequencing rate and worked with collections. All measurements were repeated. Programs have been created and tuned in development environments: Xcode 9.3, AppCode 2018.1.2, CLion 2017.2.3, IntelliJ IDEA 2017.2.6, NetBeans IDE 8.2 and Visual Studio for Mac 7.4.3.

First, we were investigating the rate of Fibonacci sequencing. A recursive algorithm solved the calculation. Swift did not miss the test very well. C++ and Java were faster than Swift (Fig. 1).

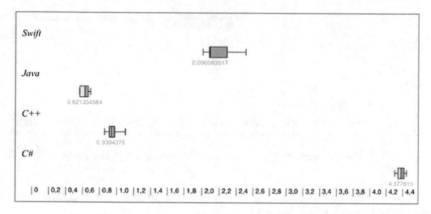

Fig. 1. Speed measurement of the Fibonacci sequence.

We carried out further measurements in collections programs. Swift offers three collections: Array, Dictionary, and Set. In this article, we will show the results of array rate measurement. We used the Array collection in Swift, in Java and C# we used the ArrayList collection and in the C++ we used the vector collection. We launched each program ten times. First, we measured the time to add 18050 elements to the collection. Elements were random numbers (Double). Figure 2 shows that Swift had similar results as C++ and C#.

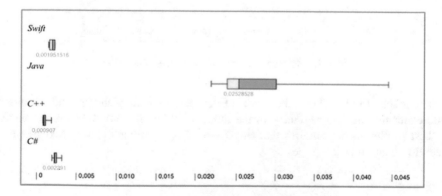

Fig. 2. Measuring the time of adding elements.

We also measured the speed of sorting elements in the collection using the sort() method. The fastest language was C++. Swift had similar results to Java. The results are shown in Fig. 3.

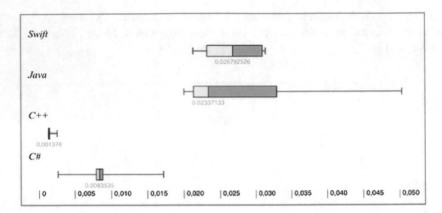

Fig. 3. Measuring the sorting of elements in the collection.

The following measurements measured the rate of the reverse() operation. The shortest time was done in C++ and C# (Fig. 4).

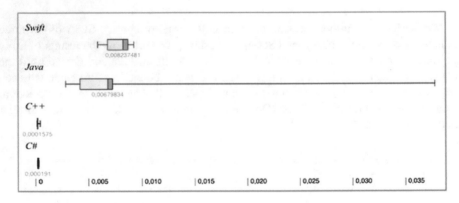

Fig. 4. Measure the time of the reverse operation.

The graph of Fig. 5 shows the results of the measurement of the gradual removal of 6000 elements from the beginning of the field. Swift had the worst results in this test.

Test results do not confirm that Swift would be significantly faster in programs execution than other languages.

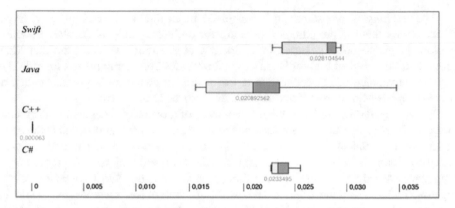

Fig. 5. Results of time measurement remove elements.

3 Teaching of Programming

Apple Swift continually develops, but also generates excellent support for teaching this language. Students can install the Playgrounds app on iPad. This app learns in the form of a game. Children perform tasks - they create programs that control the Byte character. The character walks through a changing flying island, collects gems, passes through portals, and the like.

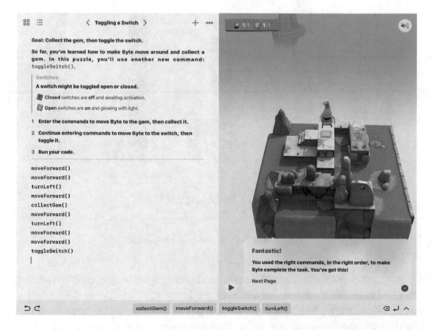

Fig. 6. Application playgrounds for the teaching of programming.

The teaching of programming is conceived in an interactive and fun way. Experience shows that the program is interesting for children. Students can independently explore language options. Based on children's observation, we can state that Playgrounds is an excellent tool for beginners and provides unique motivation for children. Another advantage of the application is the ability to create robot programs such as MeeBot, Lego Mindstorms, Sphero, Wonder Dash or Dron Parrot (Fig. 6).

We surveyed 48 students of the first semester of a bachelor's degree in Informatics at the University of Ostrava. Students generally have no more in-depth knowledge of programming. Students had acquired the essential features of Java. The students were asked to comment on five simple examples - the C++ and Objective-C programming languages. A case was pure console dump, work with the Sheet collection, cycle, creating and interrupting objects, and working with object methods. As could be expected, C++ was more readable and understandable because its syntax is similar to Java [11]. Figure 7 shows how students evaluated the code's clarity in each program.

We performed a similar experiment with a Swift code. Students were asked to comment on what the program is doing. The results were much better than Objective-C. None of the students knew to work with Swift. Still, 84% of students correctly commented on the code presented.

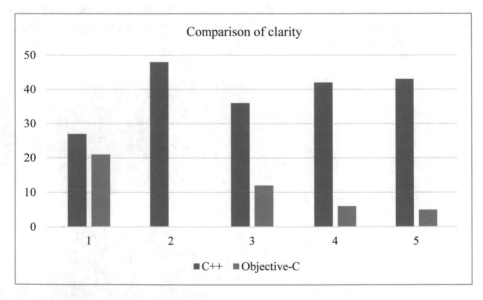

Fig. 7. Clarity of code in programs.

We used the method of the questionnaire and pedagogical experiment. We want using the results of research to determine whether students' test results are dependent on distance or full-time study. Students completed a paper version of the questionnaire, which contained eight questions and two tasks. The questionnaire contained content items, but also functional items, specifically items contact and control. Most of the items of the questionnaire were closed or semi-closed structured entries. The questionnaires were sent out, but the individual respondents were personally sought to fill, to ensure a full return. The evaluation of the data obtained were used statistical methods. For analysis of the results in each item was measured as having detected data variability.

Parts of the questionnaire were two practical tasks. The first task was to comment code in language Swift. This language is partly based on the C language, and the student is not yet acquainted with him. The second task was to rewrite the code from language Swift to language C. Students could get for each task 0 to 10 points.

We compared the results of students in courses. We used statistical methods. For comparison, the results of the two groups (distance and full-time students), we used the T-test: Two-Sample Assuming Equal Variances. Two sample t-test is used to compare the difference of two populations. This parametric test assumes that the variances are the same in both groups.

The questionnaire was completed, and 35 students attended the experiment. Respondents studied bachelor program Applied Informatics. Distant students were 8, of which 63% were men and women were 38%. Their average age was 39 years. Full-time students were 27; their average age was 21 years. 89% were men and women were 11%. Distance students had an average attendance of 94%. Full-time students had a 78% participation in present education.

Although students did not know the Swift language, they were able to interpret the code presented correctly. We conducted a similar experiment in the course of Mobile Application Development. In this course, students learned how to create mobile applications for iOS and Android. Students were able to develop the program even if they did not know the Swift programming language. They learned the language continuously during the course. After completing the course, we conducted interviews with students. 86% of respondents said Swift seemed simple. Students appreciated the simple language syntax, clear and straightforward libraries and collections. Although they knew Java, they found the Swift language better for mobile application development.

4 Results

Experience shows that the Swift programming language is suitable for both mobile application development and programming teaching. The main advantages of the language include its simplicity, modern features, simple and clear libraries. Another advantage is that Swift belongs to the Open Source category. Apple not only cares about the quality of the programming language but also offers many resources for teaching programming. The company is aware of the need to devote sufficient energy to support programming teaching. Experience to date has shown that Swift is popular with students and children alike.

5 Conclusion

Swift is a hybrid programming language that uses object-oriented, procedural and some functional features. Currently, it appears to be the most appropriate language for developing native mobile applications and systems for iOS, watchOS and tvOS. Despite rapid development and occasional backward compatibility issues. Experience and measurements show that Swift is a fast and safe language. Swift code is up to a third shorter than Objective-C, which leads to more rapid, clearer but also safer code creation. Experiments with students show that Swift is easier to read for students than Objective-C or C++.

Due to the short existence of Swift, there are, of course, problems with rapid development and language changes, less support and a wealth of libraries, frameworks and development environments.

Swift does not bring new programming approaches, but its creators have tried to incorporate it and combine the best features we can have in a specific range in other programming languages. For example, Closures (JavaScript), Generics (Java), Tuples (Python), Operator Overloading (C++), Optional types (Haskell), Protocols (Java) etc.

Experience shows that the Swift programming language is suitable for teaching programming. Students can work with Swift programming language very well. The simplicity and structure of the programming language contribute to a quick understanding of the language.

References

1. David, T.: The comparison between Swift and Java programming language. Bachelor thesis, University of Ostrava (2018)
2. García, C.G., Espada, J.P., Pelayo G-Bustelo, B.C., Cueva Lovelle, J.M.: Swift vs. Objective-C: a new programming language. Int. J. Interact. Multimed. Artif. Intell. (2015). https://pdfs.semanticscholar.org/a709/2e9eaf0490c3ff2982abf925dd0d7cf01220.pdf
3. Kinley, K.: Apple's Swift programming language is now top tier. https://www.wired.com/story/apples-swift-programming-language-is-now-top-tier/. Accessed 18 Mar 2018
4. Prechelt, L.: An empirical comparison of seven programming languages. Computer **33**(10) (2000). ISSN 0018-9162
5. Ray, B., Posnett, D., Filkov, V., Devanbu, P.: A large scale study of programming languages and code quality in Github. Commun. ACM **60**(10), 91–100 (2017). ISSN 0001-0782
6. Schmieder, A.: Swift, C++ performance (2014). http://www.primatelabs.com/blog/2014/12/swift-performance/
7. Team authors: 9 reasons to choose Swift for iOS App development. https://applikeysolutions.com/blog/9-reasons-to-choose-swift-for-ios-app-development. Accessed 26 July 2017
8. Timokhina, V.: Swift vs. Objective-C: 7 benefits of Swift development. https://dzone.com/articles/swift-vs-objective-c-7-benefits-of-swift-developme. Accessed 20 Feb 2017
9. The Computer Language Benchmarks Game. https://benchmarksgame-team.pages.debian.net/benchmarksgame/. Accessed 20 Mar 2018

10. The Good and the Bad of Swift Programming Language. https://www.altexsoft.com/blog/engineering/the-good-and-the-bad-of-swift-programming-language/. Accessed 28 Mar 2018
11. Vit, M.: Comparison of Objective-C and C++. Bachelor thesis, University of Ostrava (2012)
12. Wells, G.: The Future of iOS development: evaluating the Swift programming language. CMC Senior Theses, 1179 (2015). http://scholarship.claremont.edu/cmc_theses/1179
13. Parveen, Z., Nazish, F.: Performance comparison of most common high level programming languages. Int. J. Comput. Acad. Res. **5**(5), 246–258 (2016). ISSN 2305-9184

Digital Finance, Business and Banking

The Early Practice of Maritime Insurance Accounting in the First Proprietorship of Francesco di Marco Datini in Pisa in 1382–1406

Mikhail Kuter$^{(\boxtimes)}$ ⓘ, Marina Gurskaya ⓘ,
and Armina Papakhchian ⓘ

Economy Department, Kuban State University,
Stavropol'skaya st., 149, 350040 Krasnodar, Russia
prof.kuter@mail.ru

Abstract. This paper presents early examples of Maritime Insurance accounting. A detailed study of the books of the first of proprietorship of Francesco di Marco Datini in Pisa in 1382–1406 allowed to investigate the peculiarities of reflection transactions of insurance accounting. Describes the features of the reflection settlements of insurance premiums, description of insurance transactions in the accounting books, situations of cancellation of the insurance policies and accounting of financial result from insurance activities in the account in the "Profit on merchandise" account (the prototype of "Profits and Losses" account) in the final folios of book Merchandises (Mercanzie) prepared in 1406, in which the final financial result for the entire period from 1383 to 1406 was introduced. It should be noted considerable attention is paid to the issues of Maritime Insurance in economic historian literature, but, unfortunately, this issue until recently was not sufficiently described in the scientific accounting literature, it makes this study necessary.

Keywords: Maritime Insurance accounting · Medieval accounting · Francesco Datini

1 Introduction

The purpose of the present publication is the following: to show the results of the study carried out in the Datini archives by Russian scientists concerning the organization of settlements for insurance. Particular attention is paid to the study of methodological support of accounting operations under condition of application of single-entry bookkeeping as well as and double-entry bookkeeping. According to de Roover, "Premium insurance did not develop prior to 1300, but merchants in the thirteenth century were searching for a solution of the risk problem and were experimenting with different types of contracts that would offer protection". Taking into account the fact that from the survived documents to double-entry bookkeeping before Francesco Datini's books in Pisa (1382–1406) only three earlier companies can be referred to (J. Farolfi's company in Salon (1299–1300), Del Bene's company (1318) and the Libro

© Springer Nature Switzerland AG 2020
T. Antipova and Á. Rocha (Eds.): DSIC 2019, AISC 1114, pp. 299–313, 2020.
https://doi.org/10.1007/978-3-030-37737-3_27

of the massari of the city-commune of Genoa (1340)) and none of them in all appearance kept records of insurance transactions, it can be stated that we have before us one of the earliest examples of insurance premiums accounting, for the purpose of which a special account was opened, and the balance of account which, as a separate position item, was reflected in "Profit on merchandise" account.

Of course, we have presented only a single fragment (the results of the initial stage of the study). We have to study the books of accounts of other proprietorships and Datini's companies in Pisa and other companies during subsequent years. It is necessary to photocopy insurance policies and link them with accounting data. In addition, we should compare the accumulated materials with the ledger of Averardo di Francesco de' Medici (1395), where, according to R. de Roover, insurance activities were also carried out and "Insurance" account was opened. Thus, a full-scale study is expected, which, in our opinion, will be of certain scientific interest.

2 Review of Prior Literature

As it is known, insurance originated in ancient times. R. de Roover, Professor in Brooklyn College, paid much attention to the sources of its use in the Middle Ages in his publication "The Organization of Trade" [1]. In Sect. 1 "A General Picture" he introduces a general picture of the middle of the 14th century: "It is now an accepted view that the Black Death (1348) the end of a long period of demographic and economic expansion and beginning of a downward secular trend characterized by the of markets the recurrence of wars and epidemics and the contraction of the volume, of trade. Without challenging this view, it may be pointed out that no such setback is noticeable with respect to the improvement or business techniques. On the contrary, the fourteenth is one of continuous progress, innovation and experimentation. The draft form of the bill of exchange, for example, although known 1350, did not come into general use until that date. The same applies to marine insurance. Mercantile book-keeping, too, did not reach full maturity until 1400, as is clear if we pare, for the account books of the Peruzzi company (failed ill 1343) with Datini (1410). Another innovation introduced 1375 is the combination of partnerships similar to the modem holding best example of this is the Medici banking-house 1397. It is true that the foundations of all these new commercial institutions were laid ill the twelfth and thirteenth centuries. Nevertheless, they did not fully develop until later. Perhaps it could be argued that the decline which set in with the Black Death, by sharpening competition and profit-margins, spurred the merchants to improve methods, increase efficiency and reduce costs, with the result that only the fittest were to survive. Perhaps it is significant that no not the bank ever attained the size of the famous Peruzzi and Bardi companies, which both failed shortly before the Black Death" [1, p. 44].

Many scholars point out that modern insurance originated from Maritime insurance [2–5], and to be exact, from so-called "sea loan" [5, 6]. In Sect. 2 "The Travelling Trade before 1300" shows the role of the sea loan, the development of which contributed to the development of insurance: "Next to the commenda and the societas maris, the sea loan was frequently used to finance overseas ventures. It differed from a straight loan in that repayment was contingent upon safe arrival of a ship (sana eunte nave) or successful completion of a voyage. The risk of loss through the fortunes of the sea or the action of men-of-war was thus shifted from the borrower to the lender. Prior to the days of premium insurance, the sea loan performed, to a certain extent, the same function of protecting the merchant against loss through shipwreck or piracy. At any rate, in case of misfortune, he was relieved from any further liability which might otherwise have thrown him into bankruptcy" [1, p. 53].

By describing the economic situation in medieval Florence, Richard Goldthwait mentioned the important role of insurance for Florentine merchants. He had attention, that reduced the risks of transport, Florentines perfected the insurance contract. Among the earliest documentation for Florentines operating abroad is insurance they took out for shipment of merchandise—cotton from Genoa to Arezzo in 1214 and cloth from the Champagne fairs to Vercelli in 1215. The earliest insurance contracts, however, date from the first half of the fourteenth century and, as one might expect, are found mostly in port cities. They were notarized documents, usually following the model of a maritime loan or a purchase-sale contract to preclude any charge of usury [7].

"Premium insurance did not develop enough prior to 1300, but merchants in the thirteenth century were searching for a solution to the risk problem and were experimenting with different types of contracts that would offer protection. The role of the sea loan has already been mentioned in connection. Another type of contract was the so-called insurance loan by which a ship-owner made an advance to a shipper with the understanding that it was due, together with freight charges, only upon the arrival of the shipment at destination. Complete coverage was not achieved since such advances rarely exceeded 25 or 30% of the cargo's value" [1, p. 56].

"The earliest known examples of insurance loans date from 1287 and are found in deeds drafted by a notary in Palermo. Later, this form of contract is also encountered in Pisa (1317). Its rather late appearance may be explained by the fact that insurance loans were usually granted to merchants remaining ashore instead of travelling aboard the same ship with their merchandise, as had been the common practice hitherto" [1, pp. 56–57].

The examples of insurance loans date from 1287 as well as their further development are described in more detail in the work by Florence Edler de Roover [3]. She bases her reasoning on the appearance of insurance activities on the basis of statements by distinguished authors who examined in detail the issues of maritime trade in the Middle Ages and the Renaissance [9–11] and draws attention to the fact that "Most authorities on the subject agree that marine insurance was unknown to Greek and Roman antiquity. The first insurance contracts were made by Italian merchants of the Middle Ages to whom we apparently owe the invention of the bill of exchange, of double-entry bookkeeping, and of commercial banking, as well as other innovations in business procedure. The date of the earliest example of premium insurance is one of those moot questions on which the legal writers have failed to agree. Some of them consider certain ambiguous texts as sufficient evidence of the existence of premium insurance, but others are more cautious and question the value of anything that is not clear and explicit. At any rate, there is no doubt that genuine insurance was a product of the commercial revolution which occurred during the period from 1275 to 1325 or thereabouts" [8, p. 173].

"Since Palermo was a secondary centre, and since some of the underwriters were Genoese, it may safely be assumed that premium insurance was known prior to 1350 in Genoa, Pisa and perhaps Venice. In Genoa, however, insurance contracts continued to be disguised under the form of a mutuum, or gratuitous loan, and later of an emptio venditio, or sales contract. This practice may be due to the influence of the decretal Naviganti condemning the sea loan, although the moralists from the start were disposed to consider insurance as a contract made valid by the risk involved. Whereas, in Genoa, insurance contracts were entrusted to notaries, a different practice prevailed in Pisa and Florence where policies were drafted by brokers and circulated by them among prospective insurers until the risk had been completely underwritten. Despite high premiums charged by underwriters, the insurance business was not especially profitable. According to his records, Bernardo Camhi, a Florentine underwriter of the fifteenth century, paid out more in claims than he received in premiums. Presumably the business was highly competitive. Another trouble was that insurance lent itself easily to fraud. Ships were sometimes deliberately shipwrecked in order to claim insurance for goods that were not even on board. It also happened that shippers rushed to take out insurance after they had received secret intelligence of a disaster. Such frauds still gave rise to complaints in the sixteenth century. It is only much later that their perpetration was made more difficult by the organization of Lloyd's" [1, pp. 99–100].

Florentine merchants had as sophisticated a system of marine insurance as was to be found anywhere in Europe. The contract was simple and straightforward, not requiring a notary; a fairly coherent system of rates facilitated a rational a decision on the part of the insurer; many people in the population could be found who had the financial security and the willingness to underwrite policies; brokers existed to bring these potential insurers together with those who sought insurance; the merchant could insure his goods at about any value he wished by recourse to multiple subscriptions (although individuals limited the value they were willing to underwrite); and finally, the state allowed for the complete liberalization of the sector. Thus, Florentine merchants had all the protection they needed to operate their international network even though they, in general, did not directly control a transport system, except in the years when the state offered the limited services of its galleys [7].

In another later work, de Roover describes Francesco Datini's archives, the material of which we have to study for the present research: "The Datini archives today are unique for their completeness and fill an entire room in the old palace; they include about five hundred account-books and more than 100,000 business letters, not to mention several bundles filled with bills of exchange, insurance policies, early cheques, bills of lading, and other documents" [12, p. 140]. The insurance policies were made in Pisa and Florence (402 copies), from which comes the oldest certificate drawn up in Pisa, of 1379, while the first, in Florence, dates to 1385. However, both cities will soon issue policies that are increasingly rich in details, so as to be a model throughout the Mediterranean [13, p. 11].

Florence Edler de Roover also paid special attention to the documents of the Datini archives. Describing the medieval insurance practice and analyzing the differences between the Genoese, Sicilian, Pisan and Florentine rules of taking out insurance policies, she gives an example of a Florentine insurance policy, dated July 10, 1397, found in the Datini Archives, bears the names of eleven underwriters who, together, were answerable for 2,100 florins [8, p. 187].

Giovanni Ceccarelli studied the practice of marine insurance allowed late Medieval merchants to evaluate various of factors of risk involved in the sea trade, either structural or contingent, that has been studied through a detailed inquiry mainly based on the Datini archives in Prato. Although businessmen did not develop a notion of probability in a strict "statistical" sense, they made use of various levels of "probabilistic reasoning", depending on the degree of uncertainty that they had to face [14].

"A significant development in the matter of insurance was the building up of uniform customs and rules of law. This situation was undoubtedly favoured by the diffusion of Italian business methods and practices all through the Levant and Western Europe. Even the Bruges court often consulted leading Italian residents regarding the law merchant before deciding cases involving insurance, bills of exchange or other matters. Codification of the prevailing rules did not start until 1484 when the Barcelona customs on marine insurance were framed into a statute, printed in 1494 together with the Libro del Consolat del Mar, a collection of sea laws. This publication exerted great influence on similar legislation" [1, p. 100].

No less interesting for the further research will be a book of accounts that is not stored in Prato "Coeval with the Datini records is an extensive fragment of the ledger of Averardo di Francesco de' Medici (1395). Be it said at once that this ledger, contrary to the assertions of several writers, does not belong to the famous Medici bank, established by Giovanni di Bicci de' Medici in 1397, but to another firm founded earlier by Vieri di Cambio de' Medici, a distant cousin, and continued by Francesco di Bicci, Giovanni's elder brother, and his son, Averardo di Francesco. It was one of the leading banking houses in Florence and had branches in Pisa and in Spain. The extant fragment has 94 folios and contains several examples of Nostro and Vostro accounts, an insurance account, and sundry accounts for operating results, including an expense account (Spese di banco, bank charges), and a profit-and-loss account (Avanzi e disavanzi)" [12, p. 146].

Peragallo, by reference to Astuti [15, pp. 1968–69, 68 and 110–115] devoted two studies to the Ledger of Jachomo Badoer (Constantinople September 2, 1436 to February 26, 1440) [16, 17]: "Badoer, among other things, also ventured into marine insurance, which was a new industry in its initial phases of development. Badoer insured cargoes from Constantinople to all parts of the Mediterranean and the Black Sea. The first insurance he wrote was for a policy of 200 ducats for Aluvixe Arduine quoted below: Insurance contracted in Constantinople credit this 12th day or October for cash received from Aluvixe Arduini for an insurance or 200 ducats on his merchandise which was loaded on the ship or patron Mr. Piero di Belveder for the trip from Constantinople to Venice at a 6% premium, cash payment or 6 gold ducats and 21 perperi, the insurance, should anything occur, to be paid to Mr. Zuan di Priolli son of Mr. Nicholo....L.F.I6 perp. 40 car. 12" [17, p. 888].

In the future, Badoer himself became interested in insurance activities: "Badoer charged Arduini an insurance premium of 6%. This premium, however, varied from 3% to 19% depending upon the risk factor, one being the Turkish peril (pericholo de turchi) which he mentions on L.F.248. Over a 3-year period, Badoer collected 548 perperi 16 carati in premiums and suffered only one insurance loss of 330 perperi, netting him a profit of 218 perperi 16 carati. This speaks well of the protection which the Venetian navy afforded its merchantmen" [17, p. 889].

We should note that the events described by Peragallo took place almost a century after the events studied by us in the present paper.

How we can see, many economic historians gave attention to Marine insurance in the Middle ages, but it should be noted that this topic was not investigated in the way of accounting insurance transactions. We could not find examples of maritime insurance accounting in literature. It is why in our opinion this study will be very interesting and allow to give a contribution to accounting history research.

3   Research Method

The principal research method adopted in this study is archival. That is, it entails examining and critically analysing original sources which, in this case, are account books held in the State Archive of Prato. This research team has been working with this material for the past thirteen years and many of the records have been recorded and linked together using logical-analytical modelling. This is a flowchart-based approach that we developed for the purpose of enabling entries in the account books to be traced visually between accounts and books and from page to page.

By adopting this approach, we are able to see the entire accounting system, making entries and their sources clear in a way that is not possible if all that you have is the original set of account books. This enables us to consider each transaction in detail, trace its classification, and so explain the bookkeeping and accounting methods adopted without possibility of misinterpretation. This approach represents a new paradigm in how to analyse and interpret accounting practice for periods when there was no concept of either a standard method or a unified approach to either financial recording or financial reporting.

The main research method adopted in the present study is archival. That is, it involves the study and critical analysis of the primary sources, which in this case are books of accounts stored in the State Archives of Prato. This research team has been working with this material for the past thirteen years, and many entries have been recorded and linked together using logical-analytical modeling. This is an approach based on a flowchart that we have developed with the aim of provision of visual tracking of entries in books of accounts between accounts and books and from page to page.

Using this approach, we can see the entire accounting system, making the records and their sources clear in such a way even if all you have is an initial set of books of account. This allows to consider each transaction in detail, trace its classification and thus, explain the accepted methods of accounting and accounting without any opportunity for misinterpretation. This approach represents a new paradigm in the way how to analyze and interpret accounting practice for the periods when there was no concept of either a standard method or a single approach to financial statements, nor financial statements.

In this paper, we present (1) the accounting method adopted by the accountant in Pisa; (2) how the account of insurance premiums accounting was introduced and its purpose. As far as we know, the entries in the books of account included in this study have never been analyzed before.

4 The Development of Maritime Insurance Accounting

The study of Francesco Datini's archives made it possible to give the answer to the question that for decades did not even arise among scientists who did not see the point in modern studies of medieval books of account, motivating it with the thesis that all the main problems had been solved by early distinguished researchers in the first two thirds of the last century. As for the Datini archives, there was no doubt that the former director of the State Archives in Prato, Professor Federigo Melis, who left his mark on every page of the archival documents, could not ignore a single detail. Melis really did a tremendous job of safeguarding archival exhibits. After almost six centuries, he renumbered all the stored pages in each of the five hundred books, sorted them according to companies (territories and time) and generalized them in his fundamental work [18]. Somewhat later, individual descriptions of archival sources and their photographs will take place in another big book [19]. It is true, it will concern not only the Datini archives but will also include information from other archives. In the main, Melis did not undertake deep research into accounting development in Francesco Datini's companies: out of 133 of his publications, only 4 papers refer to this issue, while they concern single-entry bookkeeping.

Which exactly question was mentioned in the previous paragraph? According to Melis [19, pp. 176–177], the first enterprise was officially closed on August 19, 1386. We have provided convincing evidence that Personal accounts in the Ledger were opened in subsequent years. In this case, they reflected business transactions (only sales of goods) after 1386 (on "closing" the first enterprise), and after 1392 ("closing" of the second enterprise), as well as after 1400 (the announcement of complete liquidation of the company), up to 1406, when all the merchandise inventories in accounts opened during the period of 1383–1386 were completely sold.

Moreover, all the accounts in "Mercanzie" book were dated 1384–1386, and "Profit on merchandise" account (pro-type of "Profits and Losses" account) in the final folios of the book was prepared in 1406, in which the final financial result for the entire period from 1383 to 1406 was introduced. This information is very important since our subsequent statements will proceed precisely from "Profit on merchandise" account.

"Profit on merchandise" account consisted of three consecutive folios: 73v – 74r; 74v – 75r; 75v – 76r. The third indicator on the credit side of the first folio (74r) says: "And they must have f. 157 s. 10 d. 11 a oro for profits made on many insurances in this book at c. 38". Thus, it can be stated that in the first enterprise one of the income items was insurances.

Let us see what economic historians say in this regard, in particular, Marcello Berti, one of the authors of the monograph prepared under the editorship of Professor Giampiero Nigro [20]. As it is known, in the 80s of the 14th century Pisa was waging a war for independence from Florence. In addition, the plague epidemic raged in the city. All this was accompanied by the impoverishment of the population, which affected its purchasing power. However, even under those conditions, the town proved itself capable, too, of further significant innovations and an undaunted economic vitality, in the maritime insurance sector, in maritime joint venture accounting and in accounting methods. Historian described the origin of maritime и travel insurance. "Unfortunate events of another sort re-placed the plague. "Now there is bad weather", he wrote and left it at that.

Ginger, kermes, pepper and other such goods could not be sold and navigation was risky and brought to an almost complete halt: "I can see disorder everywhere so that we can never sail" and the insurance market suffered as well as a result of a shortage of insurers willing to insure alia fiorentlna - covering maritime risks in their entirety, that is – as Bruno di Francesco reflected bitterly lending himself to the completion of the cover, at 3%, of the ship of the Genoese Sisto Grillo. He managed to do this a few days later for a total sum of 800 florins and supplied the name of the insurer, the Genoese Battista Lomellino. Another example relates to the ship of Piero Gariga for whom the Pisa fondaco had asked for information from Genoa on the size of the insurance premiums payable on a journey from Aigues-Mortes to Porto Pisano. An insurer willing to underwrite it at 4% was, however, Genoese "because here, now, there is no foreigner willing" to do it. This is a further indication, it seems to me, of the difficulty of ensuring alia fiorentina in Genoa. It was not a very attractive proposition to the Genoese and foreigners, often Florentines, were not interested perhaps because of the particularly high risks involved in sea travel" [20].

"The ships left from Barcelona, Valencia, Peniscola, Tortosa, Majorca and from the "coast" – where the ships often stopped to complete their loads. Where possible, however, the ships preferred to concentrate on one or more of these ports in order to form convoys and reduce the risks. The Venetian ship which had come from Majorca "will leave from here within 8 days and then its going to Peftiscola and Tortosa to load up wool and then to Barcelona and from here, it will proceed, together with other ships, towards Pisa. In this case the fleet consisted of 4 ships" [20]. "The steady flow and large quantities of the Pisa re-export of fustian to Catalonia, in a sea characterized by frequent heated skirmishes between Genoese and Catalans, made insurance cover advisable" [20]. In more detail, the trade routes of Francesco Datini companies were investigated by Angela Orlandi [21].

As F. Melis wrote on this subject: "Already at that time Tuscan companies conducted active insurance operations (which were used to reduce the probability of risk) applying "Insurance" account, which implied both insurance premiums and their reversing as well as other operations. This meant economic accounts, in the literal sense of the word, which were included in the balance, and which were allotted from the main profit account" [19, p. 406]. In our example, only the maritime insurance is considered. Often, a mixed route was insured, both by sea and by land, for instance, from Pisa to Porto Pisano and to Avignon.

As for the accounting of insurance operations, we should draw attention to the fact that a dynamic line consisting of 4 accounts has been opened for their introduction: Prato, AS, D, 357, c. 333v, Prato, AS, D, 377, c. 9r, 13v and 38v. As you can see, in 1383 the line starts in Libro Grande Bianco A, and ends in the book of Merchandises (Mercanzie). Since profitable operations are applied in these accounts, then the debit is placed under the credit. On the credit side of the account approved income (insurance premiums) and the contra entry to the section of cash inflow in "Entrata e Uscita" book or with the debit of the account of the particular insured is calculated, and on the debit side of the account – the return of unearned insurance premiums the contra entry to the cash withdrawal section. Given that the data was sequentially accumulated in all four accounts, the last debit entry is the accumulated sum to be transferred; the same sum acts as the opening balance of the account in the credit entry of the continuation account.

Initially, it seemed to us that account Prato, AS, D, 357, c. 333v (the photocopy is shown in Fig. 1, the translation of entries in the account is introduced in Table 1) is the first account of the dynamic line. This was confirmed by the first sentence of the first entry: "On August 1st, here below we will write the profit we will do on insurances, may God give me salvation and profit", it was the typical form of the opening entries as an appeal to God [22]. However, the second sentence of the same entry ("There are f. 15 as in the Memoriale "A" at c. 95 for the old account") indicated that the opening balance of the account in account Prato, AS, D, 357, c. 333v was formed not by the current entry of one-time insurance but presents the accumulated total in the insurance activity account, which was conducted in 1382 before the official presentation of the first enterprise.

All other entries in account Libro Grande Bianco A, c. 333v are the entries of registration of current insurance transactions in 1383, starting from August 1st. The entries in the account have links to the page in "Entrata". The exception is the eleventh entry, where the contra entry is directed to the account of Prato, AS, D, 357, c. 174v (1), owned by Ambruogio di Bino, with whom stable business relations are maintained and for whom a personal account is opened in Libro Grande Bianco A (Prato, AS, D, 357) as a sequential dynamic line.

Figure 1 shows the installation of the photocopies of the first two accounts of the dynamic line of insurance indicators accounting. The accumulated sum of f. 15 from account Prato, AS, D, 366, c. 95r is transferred as the opening balance of account to account Prato, AS, D, 357, c. 333v.

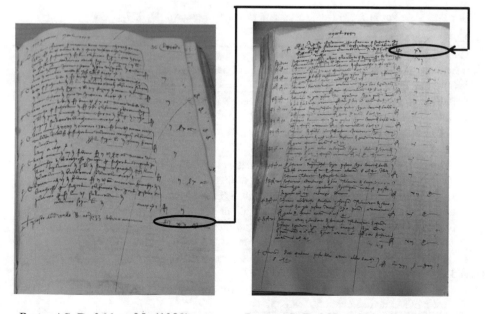

Prato, AS, D, 366, c. 95r (1382)   Prato, AS, D, 357, c. 333v (1383)

Fig. 1. Installation of the photocopy of the first accounts of the dynamic line of insurance indicators accounting

Let us see how the total was formed in account Prato, AS, D, 366, c. 95r (the translation is introduced in Table 2). The account was opened on January 30th 1382. The first entry begins with the words: "Insurances we made for maritime trade must have on January 30th 1382 f. 3 d'oro we had in cash …" Further it follows: from whom the sum was received, at what sum the insurance was performed, travel itinerary and description of the insured cargo, and also the ship on which the commercial trip was made. The insurance rates depended on many factors: time of year, state of war or peace, presence or absence of pirates along the sea routes, type of vessel, and other circumstances [8, pp. 189–190].

The entry is ended with the reference to the page in the cash Income book of "Entrata". All six entries contain the reference to c. 1. We could not find this book in the archives. And one more fact should be emphasized. If you sum up the sums of insurances in Prato, AS, D, 366, c. 95r, then we get f. 15 s. 8. The accountant entered the sum of f. 15 and that very sum he transferred to the continuation account of Prato, AS, D, 357, c. 333v.

Figure 2 presents the diagram of accounting of awarded premiums concerning insurance activities.

The characteristic feature of settlements with debtors and creditors in the 14th century consists in the fact that, as a rule, they were registered in accounting in the form of a complex system of mutual settlements. The exception is insurance operations. As is seen from Fig. 2, the great bulk of insurance operations were carried out by means of "Entrata e Uscita" (the book of Income and Outgoing of cash" (Prato, AS, D, 403).

Table 1. Modern translation of page Prato, AS, D, 357, c. 333v

1383	
On August 1st, here below we will write the profit we will do on insurances, may God give me salvation and profit. There are f. 15 as in the Memoriale "A" at c. 95 for the old account	f. 15
f. 100/we insured for Naples to Tieri and Lanberto di Domenico di Tieri, we had f. 6 as in the Memoriale "B" at c. 67	f. 6
f. 100/we insured for Aigues Mortes to Caccino di Francesco and Co., we had f. 3 s. 5 d. 8 *a oro* as in Entrata "B" at c. 22	f. 3 s. 5 d. 8
f. 100/we insured for *Bicholi* and Aigues Mortes in Porto Pisano wheat of Francesco Gittalebraccia as in Entrata "B" at c. 23	f. 3 s. 5
f. 50/we insured from Civitavecchia to Talamone linen of Giovanni Franceschi we had f. 3 as in Entrata labeled "B" at c. 28	f. 3
f. 100/we insured from Porto Pisano to Palermo Sienese cloths of Lello of Lodovico, we had f. 3 s. 10 *a oro* in Entrata "B" at c. 36	f. 3 s. 10
f. 100/we insured from Paniscola to Porto Pisano wool of Baldo Ridolfi and Co. we had f. 4 as in Entrata "B" at c. 37	f. 4
f. 50/we insured from Minorca to Porto Pisano o wool of Baldo Ridolfi and Co. we had f. 2 as in Entrata "B" at c. 37	f. 2

(continued)

Table 1. (*continued*)

1383	
f. 100/we insured from *Chridisi* at the *Schiuse* o in *Antone* wine, that is *malvagia* [malvasia] of messer Anfrione d'Aghuano of Genoa, we had f. 4 as in Entrata "B" at c. 37	f. 4
f. 100/we insured from Porto Pisano to Barcelona 1 bale of silk of Francesco di Lotti and Co., we had f. 4 *d'oro* as in Entrata "B" at c. 38	f. 4
f. 100/we insured from Paniscola to Porto Pisano wool of Baldo Ridolfi we had f. 3 *d'oro* as in Entrata "B" at c. 40, we insured them on the boat of Ghiuglieta Sala	f. 3
f. 200/we insured to Ambruogio di Bino on the boat of Bartolomeo Vitale from Porto Pisano to Palermo Milanese cloths as in this Book at c. 174 Ambruogio *must give*	f. 8
f. 100/we insured to Lodovico Stiancato and Co. on the boat of Simone Triec from Porto Pisano to Tunisi cloths, we had f. 4 *d'oro* as in Entrata "B" at c. 41	f. 4
f. 200/we insured to Tieri di Lamberto di Domenico on the boat of Chimento di Fazio of Genoa from Porto Pisano to Naples bales 5 of cloths and 2 bundles of tallow, we had f. 8 *d'oro* as in Entrata "B" at c. 41	f. 8
71.0.8	
They *gave* as we put they *must have* in the Libro Bianco "B" at c. 9	f. 71 s. – d. 8 *d'oro*

Table 2. Modern translation of page Prato, AS, D, 366, c. 95r

January 30 1382	
Insurances we made for maritime trade *must have* on January 30th 1382 f. 3 *d'oro* we had in cash on this day by Domenico d'Andrea and Co. for insurance of f. 50 on Lagino di Gorgetto Chostanzi from Livorno to Arles on 2 coffers and 2 bales of Florentine merchandise in the name and sign of Piero da Spigiano and Andreavino di Martino Marroffo as in Entrata at c. 1	f. 3
And they *must have* on this day f. 3 *d'oro* we had from the aforementioned Domenico for insurance of f. 50 from Livorno to Arles on the boat of [NOT READABLE] Micheli for 2 bales and 2 coffers of Florentine merchandise in the name and sign of Piero da Spigiano in Entrata at c. 1	f. 3
And they *must have* on this day f. 2 *d'oro* s. 10 *a oro* we had from the aforementioned Domenico for insurance of f. 25 we made him for Rome for 2 bales of Veronese cloths in the name and sign of Filippo di Ricco Caponi on the boat of Olbingane in Entrata at c. 1	f. 2 s. 10
And they *must have* on January 31 1382 f. 2 *d'oro* we had in cash from Baldo Ridolfi for insurance from Livorno to Naples on the boat of for f. 50 on bales of Florentine cloths in Entrata at c. 1	f. 2

(*continued*)

Table 2. (*continued*)

January 30 1382	
And they *must have* on February 4 f. 2 *d'oro* s. 10 *a oro* we had from Francesco di Bonacorso and Co. for insurance from Porto Pisano to Palermo for 1 bale of *sarge* and *pancali persi* on the boat of Bartolomeo Solavuglia in Entrata c. 1	f. 2 s. 10
And they *must have* on February 6 f. 2 *d'oro* we had from Francesco di Bonacorso for insurance we made them from Porto Pisano to Palermo for f. 50 *d'oro* on the boat, Catalan, on bales of in Entrata at c. 1	f. 2
Put in the Libro Giallo B at c. 333 they *must have*	f. 15

We should note that a cancellation entry can be referred to reverse entries. When calculating the insurance premium, the insured party's account was debited, which turned into a debtor, account "Income from insurance activities" was credited, and when canceled, it was vice versa: account "Income from insurance activities" was debited, and the account of the insured party was credited, in our example, that was Giovanni Francesco's account of.

The fourth account in the dynamic chain "Income from insurance activities" – Prato, AS, D, 377, c. 13v. The first entry in the account introduces the changing balance of account from account AS, D, 377, c. 9r in the amount of f. 160 s.10 d. 10. On the credit side of the account the sum of insurance premiums in the amount of f. 70 s. 18 d. 4 is accumulated. The total sum on the credit side of the account was f. 231 s. 9 d. 2. The third transaction in the amount of f. 3 s. 10 was canceled because the commercial trip did not take place. When closing the account the balance of account for transfer was f. 227 s. 19 d. 2.

Of particular interest is the final account of the dynamic line Prato, AS, D, 377, c. 38v (the photocopy is shown in Fig. 3). There are only 2 insurance premium entries in the credit entry of the account, which, together with the balance of account after the transfer (f. 227 s. 19 d. 2), form the total sum of insurance premiums accumulation amounted to f. 280 s. 15 d. 1.

On the debit side of the account (lower part) there are 3 sums:
f. 100 – payment of the insured sum for loss of the insured cargo (the ship sank);
f. 23 s. 4 d. 2 – refund of the sum of the transaction that was not made;
f. 157 s. 10 d. 11 – profits made on many insurances.

It was this very last sum that was transferred to "Profit on merchandise" account (the prototype of "Profits and Losses" account) in the final folios of book Merchandises (Mercanzie) prepared in 1406, in which the final financial result for the entire period from 1383 to 1406 was introduced. Here arises the question, which may require further research, both into the archival materials of the second enterprise and even into the materials of Francesco Datini's companies in Pisa. The fact is that the final account of the dynamic line Prato, AS, D, 377, c. 38v was opened in 1385, and the last entry was introduced in 1386, on the closing date of the first proprietorship, but "Profit on merchandise" account was introduced in 1406. The explanation of this must be offered.

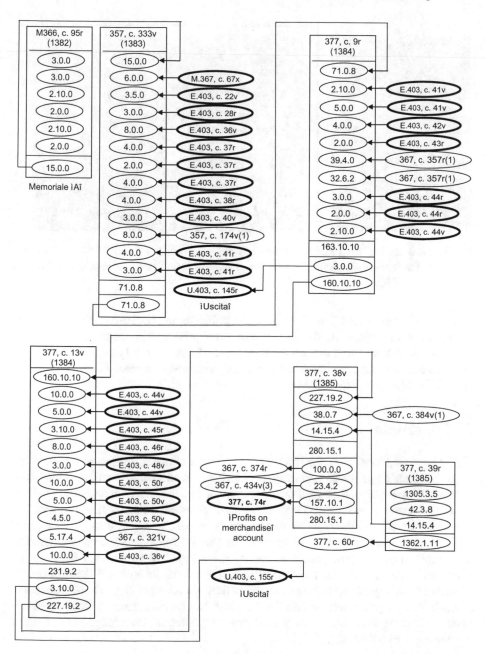

Fig. 2. Diagram of awarded premiums accounting concerning insurance activities

It should be mention, Datini in his activity could be as an insurer (assicuratore), as an insured (assicurato). The Archive has a contract [23] of marine insurance a bit special: a slave named Margarita was loaded onto a ship direct to Pisa to Barcelona. Is ensured for 50

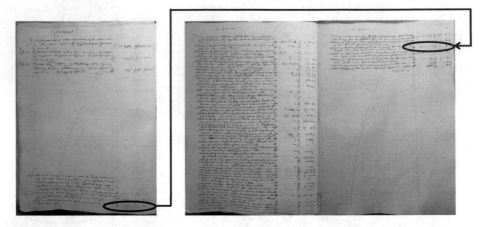

Fig. 3. Final account of "Profit on insurance" dynamic line (Prato, AS, D, 377, c. 38v) and transfer of the balance of account to the credit of the first folio "Profit on merchandise" (Prato, AS, D, 377, c. 74r)

gold florins, with award 7%. The insurance policy is in the Tuscan cities (Pisa and Florence) an inscription written in the vernacular private. This insurance policy was drawn up on 9 May 1402, the insurer is Michael Monduccio and the insured company Datini. The slave Margaret has to be transported from Porto Pisano in Barcelona with the ship led by Matthew Turo Catalan.

5   Conclusion

The review of literature in the direction of research allowed leading to a conclusion about the lack of publication concerning Maritime insurance accounting. Many economic history scholars described the development of Maritime insurance, but we have not the information on how the insurance transactions were reflected in accounting. The material considered allows us to summarize how perfect the system of differential accounting of profit and other income was in the 14th century. Insurance premiums accounting should be recognized as one of the components that form the financial result. Both the system of mutual settlements between the insured and insurers as well as the methodological support for the introduction of such operations in the accounts in Datini's first proprietorship in Pisa for that time can be considered sufficiently developed. This study is an important contribution to theoretical knowledge in the field of development of accounting.

References

1. De Roover, R.: The organization of trade. In: Postan, M.M., Rich, E.E., Miller, E. (eds.) The Cambridge Economic History of Europe. Volume III "Economic Organization and Policies in the Middle Ages", pp. 42–118. The University Press, Cambridge (1977)

2. Din, S.M.: Impact of cost of marine and general insurance on international trade and economic growth of Pakistan. World Appl. Sci. J. **28**(5), 659–671 (2013)
3. Piattoli, L.: Le leggi fiorentine sull' assicurazione nel medioevo. Archivio Storico Italiano, 90(Serie 7, vol. 18) (4 (344)), 205–257 (1932)
4. De Simone, E.: Breve storia delle assicurazioni, 2nd edn., Milano (2011)
5. Piccino, L.: Genoa, 1340–1620: early development of marine insurance. In: Leonard, A.B. (ed.) Marine Insurance: Origins and Institutions, 1300–1850. Palgrave Macmillan, Basingstoke (2016)
6. Hoover, C.: The sea loan in Genoa in the twelfth century. Q. J. Econ. **40**(3), 495–529 (1926)
7. Goldthwaite, R.A.: The Economy of Renaissance Florence. The Johns Hopkins University Press, Baltimor (2009)
8. De Roover, F.E.: Early examples of marine insurance. J. Econ. Hist. **5**(02), 172–200 (1945)
9. Byrne, E.H.: Genoese Shipping in the Twelfth and Thirteenth Centuries. The Mediaeval Academy of America, Cambridge (1930). pp. 12–19
10. Di Tucci, R.: Studi sull'economia genovese del secolo decimosecondo, La nave e i contratti marittimi, Turin, pp. 24–49 (1933)
11. Lane, F.C.: Venetian Ships and Shipbuilders of the Renaissance. The Johns Hopkins Press, Baltimore (1934). p. 115
12. De Roover, R.: The development of accounting prior to Luca Pacioli according to the account-books of Medieval merchants. In: Littleton, A.C., Yamey, B.S. (eds.) Studies in the History of Accounting. Sweet and Maxwell, London (1956)
13. Cecchi, A.E.: L'Archivio di Francesco di Marco Datini. Fondaco di Avignone. Inventario, Roma (2004)
14. Ceccarelli, G.: The price for risk-taking: marine insurance and probability calculus in the Late Middle Ages. Journal Electronique d'Histoire des Probabilités et de Statistique **3**(1), 437–455 (2007)
15. Austuti, G.: Le forme giuridiche della attivita mercantile nellibro dei conti di Giacomo Badoer (14361440). Annali di Storia del Dirito, XII–XIII (1968-69)
16. Peragallo, E.: The ledger of Jachomo Badoer: constantinople September 2, 1436 to February 26, 1440. Acc. Rev. **52**(4), 881–892 (1977)
17. Peragallo, E.: Jachomo Badoer, Renaissance Man of Commerce, and his Ledger. Special Accounting History Issue (1980)
18. Melis, F.: Aspetti della Vita Economica Medievale. Studi nell'Archivio Datini di Prato. Florence (1962)
19. Melis, F.: Documenti per la storia economica dei secoli XIII–XVI. F. Melis. Firenze (1972)
20. Francesco di Marco, D.: The Man the Merchant. Firenze University Press, Firenze (2010). Edited by Giampiero Nigro
21. Orlandi, A.: Between the Mediterranean and the North Sea: networks of Men and Ports (14th–15th Centuries). In: Nigro, G. (ed.) Reti marittime come fattori dell'integrazione europea: selezione di ricerche (Maritime Networks as a Factor in European Integration: Selection of Essays), pp. 49–70. Firenze University Press, Firenze (2019)
22. Adamo, S., Alexander, D., Fasiello, R.: Time and accounting in the Middle Ages: an Italian-based analysis. Acc. Hist. (2019). https://doi.org/10.1177/1032373219833140
23. http://datini.archiviodistato.prato.it/en/2014/04/24/assicurare-una-schiava-in-viaggio/. Accessed 22 Aug 2019

Multidimensional Model of Accounting

Aleksei Kovalev[⊠] [iD]

Novosibirsk State University of Economics and Management,
630099 Novosibirsk, Russia
lex2000@mail.ru

Abstract. This article describes the concepts of multidimensional accounting. It is based on a multidimensional (faceted) classification and a semantic multidimensional data model of the accounting. Instead of Accounts, the multidimensional accounting uses the categories of economic activity. The multidimensional data model is more flexible with respect to the hierarchical model of Accounts. The multidimensional data model allows you to extend the accounting capabilities, taking into account the diverse needs of users of accounting information. To create a multidimensional accounting system, identified categories of accounting, formulated the concept of double-entry and the balance in a multidimensional view of (pro-balance).

Keywords: Bookkeeping · Accounting · Semantic data model of accounting · Multidimensional accounting · Multidimensional data model · Coal classification · Categories of economic activity · Commercial activity · OLAP · Inside the sphere of accounting · Outside the sphere of accounting · Pro-balance

1 Introduction

In recent decades, corporate financial information, conveyed by complex quarterly and annual reports, has lost most of its usefulness to many users, also to investors, and is urgently in need of revitalization and restructuring. Given the crucial role of financial (accounting) information in fostering prosperity and growth of business enterprises and the economy at large, the serious deficiencies of this information, should be of great concern not only to investors, but also to managers and accountants [1].

Humankind uses accounting since 12–13 century. The basis of accounting was the work with information technology based on the use of paper. Accounting as an information system has not changed since that time. Modern information space formed by complex technologies, hardware, and software opens up new possibilities for the development of accounting and presents it to the new requirements.

We endure an epoch of formation of an information economy. Developed not only new equipment and new principles of information processing.

There is a growing value of information. The information itself is transformed into a new quality, more suitable for decision-making. This process is depicted as DIKW Pyramid. Standard DIKW Pyramid and consists of four levels: (D) data, (I) information, (K) knowledge, and (W) wisdom [2]. An extended version of DIKW Pyramid consists of 6 levels (see Fig. 1) measurement; facts; data, information; knowledge; and wisdom [3].

© Springer Nature Switzerland AG 2020
T. Antipova and Á. Rocha (Eds.): DSIC 2019, AISC 1114, pp. 314–324, 2020.
https://doi.org/10.1007/978-3-030-37737-3_28

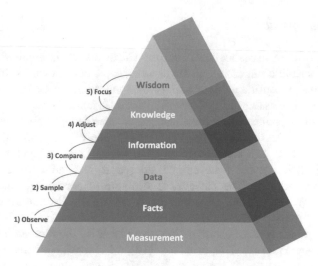

Fig. 1. Extended version of Data Information Knowledge Wisdom pyramid.

Accounting as information system cannot stand away from this process. But developing of accounting lags behind the growing demands of users of accounting information. The accounting system stay the level of "Data" and does not increase above. Accounting is the source of data for the level of "Information". At the level of "Information" are data warehouse. At the level of «Knowledge» of the class Business Intelligence (BI), for example, online analytical processing (OLAP) [4].

In accordance with the definition given by the American Association of Accountants' Accounting - the process of identifying information, calculation and evaluation of performance and reporting of information to users to create, study and decision-making "[5, c. 10].

The subject of accounting are the facts of economic activity, which in the accounting process are the identification, classification and registration. There are two basic types of classifications: hierarchical and multidimensional.

Hierarchical classification is very useful in terms of navigation and paper technology for information processing. But at the same time, a hierarchical classification has a rigid structure and changes in the structure entail incomparability of data and require time-consuming.

Multidimensional classification is most universal of two kinds of classifications. Multidimensional classification supposes addition of new signs of classification, change of interrelation of elements. But thus, work effective work with it is possible only with computer facilities application. It is possible to construct any hierarchical classifications of multidimensional classification within considered signs. In this sense multidimensional classification is base in relation to hierarchical classification and gives to the user the maximum freedom in work with the information.

In the multidimensional classification of features does not stand out major and minor. It provides possibility of development of system without reorganization of all tree of signs (in case of hierarchical classification), and also to build the reports containing any possible variants of hierarchical classifications within the limits of considered signs.

2   Previous Study

The important role of classification is mentioned in a large number of publications. References to scientific sources are given as an example of such works. In these works, the classification is considered in a variety of applications: classification of balance sheet items [6, 7], classification of articles of the statement of financial results [8], classification of accounts [9], Classification of assets and liabilities [7, 10, 11] accounting system classification [12, 13] Classification of the cash adjustments [11] Accounts referred to any element of the classification or assigns a classification function [14]. Schmalenbach developed analytical and mixed classifications of accounting accounts [15]. The categorical structure of accounting information was developed by W. McCarthy. He used three categories: Resources events agents (REA) [16].

The closest to the topic of classification as the basis for the formation of accounting information came Professor Shaposhnikov [17], who considered the use of classification in various tasks of accounting but his research has not received further development.

The study of classification as the basis for the creation of accounting information is considered for the first time.

3   The Main Body

3.1   Methods

As research methods were used multidimensional classification and OLAP.

Now the business accounting is based on use of accounts and sub-accounts. This system corresponds to the hierarchical model the first level of division in which are the business accounting accounts, the second a sub-account. Structure accounts and sub-accounts it is generated on the basis of a feature set reflecting interests of dominating group of users. Interests of groups of users in relation to the organization considerably differ. Occurrence of new requirements of users leads to necessity of labour-intensive process of reorganization of hierarchy or creation and simultaneous use of alternative books of accounts. This fact is one of the reasons of process of forming of various kinds of accounting observed now. Therefore, the most suitable to the business accounting organization is multidimensional classification which in computer science concerns multidimensional model of the data. Multidimensional model of the data used in technology OLAP (OnLine Analytical Processing - operative analytical handling). Basic purpose OLAP - handling of the information for analysis and decision making carrying out. OLAP refers to the level «Knowledge» in the system DIKW Pyramid.

In relation to other conceptual models of the data, OLAP has a number of advantages:

1. It is easily perceived by end users (not specialists in the field of information technology), occurrence of new aspects of use of the data and necessity of entering of new communications does not lead to re-structuring of all model of the data and a database as a whole [18].
2. In OLAP tools of the analysis of the data are brightly expressed. She allows to receive an intellectual estimation, i.e. to make generalization, grouping, removal of the excessive data and to raise reliability at the expense of an exception of errors. Advantage OLAP consists that inquiries here are formed not on the basis of rigidly set forms, and by means of the flexible independent approaches which are not demanding for modification attraction of the programmer [19]. Groupings can be built, freely choosing a detailed elaboration order in a cut of signs. OLAP provides revealing of associations, laws, trends, carrying out of multiple classifications, generalization or detailed elaborations, drawing up of forecasts, carrying out of the intellectual analysis («Data mining») i.e. are given by the tool for enterprise management in real time. Thus, execution of the widest scope of inquiries of users to information system of accounting is provided.
3. The Multidimensional model of the registration information supposes with the minimum costs expansion and increase in detailed elaboration of the data according to inquiries of users of the registration information. In multidimensional model it is possible to add new analytical attributes (to increase detail of the data) and numerical indicators without any changes in existing forms of input and reports [20; 129].

The classifier of a multidimensional classification consists of a of signs and values of these signs.

The multidimensional model allows to register signs in any combination and to receive any slices of the information within the limits of the considered signs, as reflects graphical representation of a 3-dimensional hypercube.

Now there are highly effective software of handling OLAP of cubes. One of the most accessible and intuitively clear means OLAP, is Pivot table in MS Excel.

3.2   Conceptual Bases of Multidimensional Accounting

We call accounting, which is built on a multidimensional data model, multidimensional accounting. Multidimensional structure and the corresponding hierarchical structure relate as a general to specific. Thus, and multidimensional accounting concerns the business accounting as a general to specific.

At a design stage of multidimensional model, it is necessary to comprehend domain knowledge of accounting. Features of the structure of the resulting model should not violate the rules of formal logic. Logically consistent structure of the model provides the benefits that arise when working with data. Disadvantages of the conceptual domain data model cause of errors and contradictions at work with accounting data. In this aspect, the multidimensional model is means of check of correctness of understanding of base concepts and the technician of the descriptions of the considered essence.

Therefore, development of a technique of accounting of the facts of economic activities in multidimensional model has demanded the critical analysis of bases of an accounting method.

Such bases consist of: a subject of accounting, the accounting account double record and balance, a ratio of analytical and synthetic accounting.

1. For a basis the point of view is taken that an accounting subject are the facts of economic activities affecting, directly or indirectly to the financial position the organization (capital). It is the point of view has been stated by Schär [21] and it is supported by modern writers, such as Sokolov [22] and etc.
2. Multidimensional accounting uses some the basic, most general concept base primitive things (or categories in philosophical understanding of the term) instead of business Accounts. Used categories are the most general concepts inherent in all economic operations. The semantic model of accounting consists of the three main categories: the object of economic operation, the entity of economic operation, the type of relation between object and economic entity. As usual, the relation is the relation of property between object and economic entity.

The object	The relation	The economic entity

For the basic accounting model, it is categories: the object of economic operation; the economic entity; the type of relation between object and economic entity; time of event and a measure. All main properties of economic operation are characteristics of the specified categories.

There are examples of the similar approach at the western accounting school. It is the system named REA (from the name of primitive things - resources, events, agents). For the first time concept REA has been entered by McCarthy in 1982 [16]. REA has a number of differences from the multidimensional accounting, most essential of which is absence of double-entry and balance sheet model in an explicit form. Even earlier, in 1975 role Michael A. Crew underlined a role of agents and resources in the description of the facts of economic activities. In his book "Theory of the firm," he wrote, that the facts of economic activity are combined into a few basic types of:

- Transfer a resource or money from one agent to another;
- Hand over the information from one agent to another;
- Transform a number of resources to other kind of resources.

Generally active business accounts concern the characteristic of objects of operations, and passive and is active-passive accounts concern the characteristic of entity of operations.

3. For multidimensional accounting the scheme of warehouse of the data "Star" most approaches. It provides storage of essential properties of economic operations concerning accounting categories (see Fig. 2):

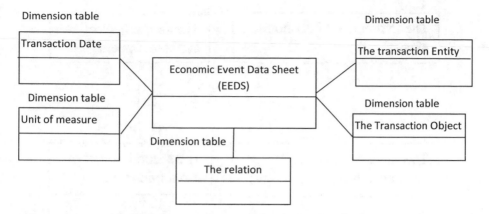

Fig. 2. The base scheme of a data structure in multidimensional accounting.

In solving practical problems of accounting schema "snowflake" replaces the schema "star", which admits the existence of hierarchical classifications in the dimension tables. Given the variability of analytical features for analytical dimensions must be applied bitemporal databases.

4. In accounting double entry uses a different combination of accounts. Application of categories in multidimensional accounting has provided a constancy of a design of double-entry. In multidimensional accounting double-entry contains constant quantity of elements for registration of all facts of economic activities. In a multidimensional accounting the double entry records changes in each category or sign. Changes between categories (or sign) are impossible. Equality of the sums on the number module in double-entry of multidimensional accounting provides relationship of cause and effect and possibility of reconstruction of events on the basis of the registration information. Double-entry defines the end of the previous state (quality) and the beginning of a new state (quality) of the object and the subject of accounting. Thus, the continuity is preserved signs and recording of change in their values (see Fig. 3). Double-entry is expressed in the completion of one situation and the appearance of new situation. This scheme is directly implemented in the double-entry of multidimensional accounting.

Instead of the debit and the credit of accounts in multidimensional accounting are used positive and negative numbers. Has offered and researched use possibilities in accounting of rational numbers of Tsygankov [23]. Positive and negative numbers with success are use in cameral accountings and single-entry accounting. At use of rational numbers, the basic scheme of registration of economic operations looks as follows (see Fig. 4):

Where: X - the sum of money and (or) in natural indicators. Values for the first and second line are equal modulo.

The first example concerns the events connected with change of the proprietor or moving of objects between divisions of one proprietor.

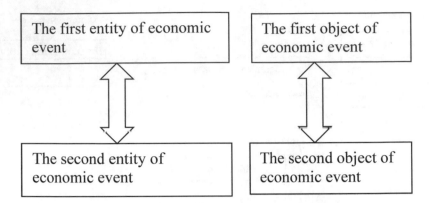

Fig. 3. Three. Double-entry multidimensional accounting.

1. Change of the economic entity.

№	Moment in time	Subject of the event	Relation	Object of the event	Value
1	YY.MM.DD	The economic entity **1**	Property	The Object 1	+ X1
1	YY.MM.DD	The economic entity **2**	Property	The Object 1	− X1

2. Change of the Object.

№	Moment in time	Subject of the event	Relation	Object of the event	Value
1	YY.MM.DD	The economic entity 1	Property	The Object **1**	+ X1
1	YY.MM.DD	The economic entity 1	Property	The Object **2**	− X1

3. Mixed type of event.

№	Moment in time	Subject of the event	Relation	Object of the event	Value
1	YY.MM.DD	The economic entity **1**	Property	The Object **1**	+ X1
1	YY.MM.DD	The economic entity **2**	Property	The Object **2**	− X1

Fig. 4. The main scheme of registration of economic events in multidimensional accounting.

The second example describes the scheme of events when one object of operation replaces another, such as: transfer of materials to production, output finished product etc.

The third example reflects the mixed type of events when there is a change of the economic entity and object as a result of the economic events.

The table containing double-entry of registration of the facts of economic events is called «Economic event data sheet».

5. The Following task consisted in development of a basis of balance sheet in multidimensional system. The first problem was practice of allocation accounts receivable to an asset balance.

In the traditional balance sheet, a positive balance of settlements with counterparties is reflected in the asset balance and negative balance in the passive. This principle at the beginning does not allow to determine what part of the balance sheet will reflect the result of the relationship with the counterparty at the balance sheet date. I.e. it is situational, not systematic and does not allow to allocate a steady sign on which the balance sheet parties are formed.

In multidimensional accounting the balance sheet model is organized in another way. In our opinion, the balance sheet model is reflection of isolation of property of the organization from set of the property rights existing in economy.

Principle of isolation follows of the direct reading of the definition of a legal entity in the Civil Code Russian Federation, paragraph 48 [24], in which isolated property described as the main criterion for the existence of a legal entity. Separate property is the economic basis of Organization on which behalf the accounting is. The other entities may become contractors of this Organization.

In multidimensional accounting the balance sheet is formed on the basis of values of a special sign: «accounting area». This sign is given for entities of operations. Generally, it has only two values: «Internal area» and «External area». Sign of "internal area" corresponds to the organization in terms of which is accounting. «The external area» is used for counterparts of this Organization.

We name the received design of the balance sheet "Pro-balance sheet".

The isolated property relates to the asset of the balance. We call this asset of the balance sheet of "pro-activ"

The other side of the balance sheet is detailed balance of any mutual settlements with all counterparts and owners. We call this other side of the balance of "Pro-passiv". Feature of the pro-passiv is that it reflects accounts receivable with the opposite sign.

Pro-activ and pro-passiv results are equal in absolute numbers.

The pro-balance sheet circuit diagram is shown in a Fig. 5:

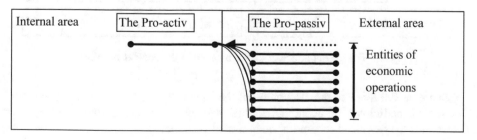

Fig. 5. Dichotomic division by values of accounting area.

The division of entities into two classes corresponds to the logical operations: the dichotomy. Using only a dichotomy, we have one article in the pro-activ, and many in the pro-passiv. Articles asset balance are formed by the decomposition of the analytical elements of the asset at an additional economic characteristic (see Fig. 6). In liabilities on the contrary, entities are arranged in groups on economic grounds.

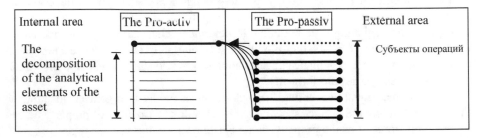

Fig. 6. Forming articles on pro-asset.

As a result of these operations, we have general balance sheet structure in the context of articles (see Fig. 7).

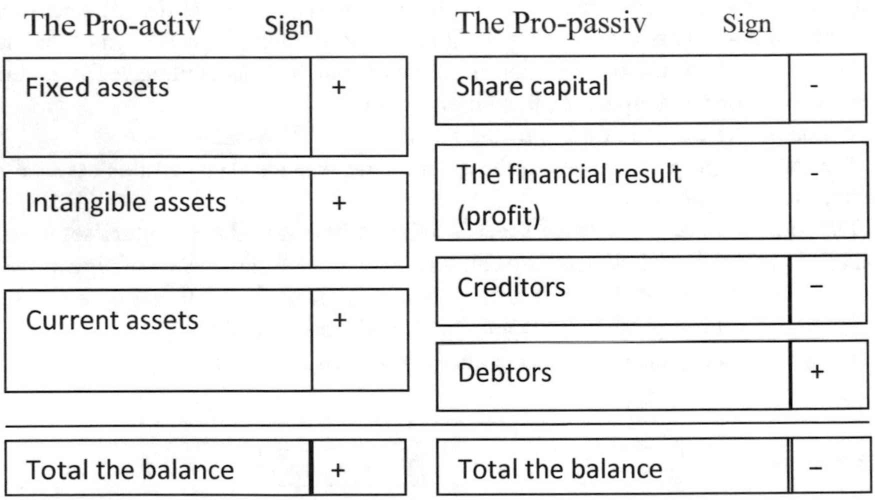

The Pro-activ	Sign	The Pro-passiv	Sign
Fixed assets	+	Share capital	-
Intangible assets	+	The financial result (profit)	-
Current assets	+	Creditors	–
		Debtors	+
Total the balance	+	Total the balance	–

Fig. 7. The overall structure of the balance-in the context of articles.

Share capital refers to the calculation of the founders. The financial result concerns the special participant of economic mutual relations - to the market. The market determines the margined income of commercial operations.

Feature of pro-balance is negative value of a result of a pro-passiv. The debt receivable is reflected in a pro-passiv with positive value. Total assets are equal to total liabilities in absolute numbers, respectively, when added up give a zero.

4   Conclusions

Development of a technique of multidimensional accounting has allowed to look at conceptual bases and technology of the business accounting in a new fashion. Multidimensional accounting eliminates rupture between development of the business accounting and modern information technology, allows to use all advantages OLAP to handling of the considered data.

Multidimensional accounting based on the most general terms - a category, it uses a multidimensional classification instead of hierarchical. This suggests that the multidimensional accounting refers to the traditional accounting as a general to specific. Accordingly, the concept of balance (the pro-balance) and double-entry accounting disclosed in the multidimensional accounting, and is applicable to conventional accounting.

Multidimensional accounting improves the accounting systems to the level of "Information" in the DIKW.

We have identified the presence of the categories of entity and object in the semantic content of accounts. But it also found that the accounts contain features related to the object and entity. It is these features define the structure of accounts. Accounts contain the reference to base categories of accounting and value of the selected signs of these categories. The name of the accounts contains value of the selected signs related to the basic categories. But they have not reference to the appropriate category, though implies their participation. This phenomenon is analogous to what in linguistics is called the "hided verbal message". Consequently, the accounts of accounting are complicated in their content and unite categories and their attributes. This shows the hierarchical system of accounts as well as the hierarchical structure of sub-accounts. In our opinion, the hierarchical system of accounts defines the rigidity of the accounting system, and it is the cause of the constant struggle of different groups of users to reflect their information requests in the structure of accounts.

Multidimensional accounting allows to form the consolidated bookkeeping. For this purpose, it is necessary to unite Economic event data sheet of the organizations of the consolidated group, to designate them as «internal area», and provide unity of analytical directories. Financial results from operations between these entities also concern "internal area" of accounting.

In the future, multi-dimensional account can be used to organize the planning and budgeting. With the entry of additional attributes in it will be possible to combine different types of accounting: financial accounting, managerial accounting, tax accounting, cameral accounting, etc.

For a small volume and simple structure of accounting information a multidimensional accounting can be realized in the environment of the spreadsheet Excel.

References

1. Lev, B., Gu, F.: The End of Accounting and the Path Forward for Investors and Managers. Wiley, Hoboken (2016)

2. Gu, J., Zhang, L.: Data, DIKW, Big data and data science. In: 2nd International Conference on Information Technology and Quantitative Management, ITQM (2014)
3. http://www.systems-thinking.org/dikw/dikw.htm. Accessed 21 Sept 2019
4. http://indico.cern.ch/event/276139/contribution/59/attachments/501004/691997/DWH_and_BI_Improve_strategic_decision_making.pdf. Accessed 21 Sept 2019
5. Drury, C.: Management and Cost Accounting, 8th edn. Cengage Learning (Emea) Ltd., Andover (2012)
6. Van Der Poll, H.M.: Towards a classification framework for accounting information. Submitted in fulfilment of the requirements for the degree doctor commercii (financial management sciences). University of Pretoria. Republic of South Africa (2007)
7. Van der Poll, H.M., Gouws, D.G.: Issues surrounding the classification of accounting information. S. Afr. J. Econ. Manage. Sci. (SAJEMS) 12(3), 353–369 (2009)
8. Fairfield, P.M., Sweeney, R.J., Yohn, T.: Accounting classification and the predictive content of earning. Acc. Rev. 71(3), 337–355 (1996)
9. Hedish, N.D.: Account classification and principle codification. Acc. Rev. 34(4), 660–662 (1959)
10. Bedford, N.M.: The foundations of accounting measurement. J. Acc. Res. 6(2), 270–282 (1968)
11. Kolitz, D.: Financial Accounting. A Concepts-Based Introduction. Routledge, New York (2017)
12. Maciuca, G., Socoliuc, M.: The role of accounting system classification in the optimization of international harmonization process. USV Ann. Econ. Public Adm. 2(18), 200–206 (2013)
13. https://www.researchgate.net/publication/327217155_Classification_and_Factors_Influencing_Accounting_Systems 2018. Accessed 21 Sept 2019
14. Foskett, D.J.: Construction of a faceted classification for a special subject. In: Proceedings of the International Conference on Scientific Information. National Science Foundation, pp. 867–888 (1959)
15. Schmalenbach, E.: Der Kontenrahmen. Leipzig, G.A. Gloeckner (1927)
16. McCarthy, W.E.: The REA accounting model: a generalized framework for accounting system in a shared data environment. Acc. Rev. (3), 554–578 (1982). https://www.msu.edu/~mccarth4/McCarthy.pdf. Accessed 21 Sept 2019
17. Shaposhnikov, A.A.: Classification models in accounting. Finance and Statistics (1982)
18. Shvetsov, V.I.: Database. Internet University of Information Technologies. Moscow (2009)
19. https://www.cfin.ru/vernikov/kias/vest.shtml. Accessed 21 Sept 2019
20. Thomsen, E.: OLAP Solutions. Building Multidimensional Information Systems, 2nd edn. Wiley Computer Publishing, Hoboken (2002)
21. Schär, J.F.: Buchhaltung und Bilanz: auf wirtschaftlicher, rechtlicher und mathematischer Grundlage, für Juristen, Ingenieure, Kaufleute und Studierende der Privatwirtschaftslehre mit anhängen über "Bilanzverschleierung" und "Teuerung geldentwertung und bilanz". Springer, Heidelberg (1922)
22. Sokolov, V.Ya.: Accounting as the sum of the facts of economic life. Master, INFRA M, Moscow (2010)
23. Tsygankov, K.Y.: Double entry, Ledger accounts and their alternatives. Siberian Financial School, no. 2, pp. 123–136 (2008)
24. Civil code of the Russian Federation, part 1., in the edition of the Federal law of 2010/05/08.. N 83-FZ

Factors Influencing Digital Bank Performance

Ekaterina V. Koroleva$^{(\boxtimes)}$ (iD) and Tatiana Kudryavtseva (iD)

Peter the Great St. Petersburg Polytechnic University,
Polytechnicheskaya, 29 St. Petersburg, 195251, Russia
plotnikova_ev@spbstu.ru

Abstract. Implementation of information technologies has led to the concept of the digital economy which is based on the use of the most advanced innovations. Digital transformation is realized in every economic sector including the banking sector. Nowadays, banking is one of the most important sectors which introduce and use the latest achievements in the field of information and communication technologies. Despite the increasing relevance of digitalization, surprisingly little is known about the determinants of digital banks performance. This paper attempts to fill this gap by evaluating the link between inner factors and bank performance in Russia. We use regression models on the dataset of 16 digital banks. Performance of digital banks is measured by return on assets in 2018. The results show that digital banks which have more customers and more transactions through digital communication, have higher performance. Our results remain rather inconclusive with respect to a share of loans and deposits through remote banking. These results imply that the identified inner factors of a digital bank play an important role while contributing to its success.

Keywords: Digital banking · Bank performance · Innovations

1 Introduction

To date, the priority direction of economic stabilization and the banking sector stability is introduction of advanced information technologies that create a competitive advantage and make banking services more attractive and accessible to the public. Digitalization has become a strategic priority for the banking industry worldwide. It is becoming more difficult for the banks operating according to the traditional business model to compete with digital banks, which are actively entering the banking services market. Therefore, in order to remain competitive, traditional banks have to introduce elements of digitalization into their systems of operation, which undoubtedly affects effectiveness of their activity.

In the Russian Federation, experts evaluated advantages of digitalization in the banking sector, and in 2017, the Digital Economy program was launched, one of the directions of which is to develop financial technologies aimed at promoting competition in the financial market, improving availability, quality, and a wide range of financial services, reducing risks and costs in the financial sector, as well as increasing competitiveness of Russian financial technologies in general.

© Springer Nature Switzerland AG 2020
T. Antipova and Á. Rocha (Eds.): DSIC 2019, AISC 1114, pp. 325–333, 2020.
https://doi.org/10.1007/978-3-030-37737-3_29

Digitalization of the banking sector brings benefits to both banking institutions and their customers. The main advantages of digital communication channels for banks are cost reduction, increased productivity and optimization of the bank's performance. Customers in such communication channels appreciate the possibility of real-time control, user friendliness, and time saving. Therefore, the share of customers who prefer digital channels of communication with the bank is steadily growing both in the Russian and global banking service market.

The demand of customers for digital banking services encourages banking institutions to actively introduce elements of digitalization into their systems. Every year, more and more credit institutions are transformed into digital banks by introducing new technologies into their systems, as digitalization is a necessary condition for being in the market. The emergence of fully digital banks, as well as modernization of the existing banking institutions, undoubtedly affects the banking sector and the development of the economy not only on a national scale, but also worldwide. As a result of different ways of digitization, the following business models are distinguished: digital bank brand, digital banking channel, digital subsidiary bank, true digital bank. Each of these models has a number of features.

One of the main tasks in the banking sector is evaluating performance of financial institutions. In this area, there are many directions and a large number of different methods have been created to evaluate the economic efficiency of the bank's activities. It should be noted that today, there are practically no studies that allow evaluating performance of a digital bank taking into account its business model. It is impossible to use the methods proposed for evaluation of traditional financial institutions, because every model of a digital bank has specific features that should be considered when evaluating its performance.

Thus, within the framework of the study, it is planned to build an econometric model that would allow evaluating performance of a digital bank taking into account its business model. Modeling will be carried out on the example of traditional banks with a digital banking channel operating in Russia. Recently, traditional banks have been facing increasing competition from the banks which work with clients using new technological solutions - mobile applications and chats. The reasons for the growing popularity of such banks are expansion of the Internet and smartphones in Russia (76% and 59% respectively), the quality of services provided by fully digital banks and a higher literacy rate of Russian users. Therefore, traditional banks are actively introducing elements of new technological solutions into their operating systems. Every year, there are more and more such traditional banks in Russia. According to the results of a global survey conducted by the CGN consulting group, by 2020, 68% of the total client base of banks will be using only remote banking services. The data of these surveys show that the public is gradually preparing for a full digitalization of the banking sector, which will lead to the reduction of physical bank branches. Therefore, special attention should be paid to evaluating performance of traditional banks with a digital banking channel operating in Russia. As a result of this objective, an econometric model should be obtained, allowing an analysis and evaluation of digital bank effectiveness depending on its digitalization model.

2 Theoretical Background of the Research

Currently, the topic of banking performance is not well developed in either national or global economic science. Today, there are views on certain aspects of the issue, and there is no generally accepted interpretation of banking performance and its measurement. As a rule, economic efficiency is interpreted as the following: the ratio between economic effect and costs [1–4]; achievement of the highest volume of goods or services with a given amount of resources [5, 6], or the ratio between the expenditure of rare resources and the volume of goods or services [7, 8]. According to the above definitions, economic efficiency is a relative parameter which can be measured by various types of indicators. Empirical studies use the following evaluations indicators of performance: revenue, profit, return on assets, return on investment, or survival duration [9, 10]. Other indicators, used less frequently, include value added, the growth of turnover, or the number of employees [11]. When evaluating performance of financial institutions, it is necessary to account for a number of features of financial institutions, including specifics of their operation, as well as complexity of the organization's system, which does not have clear distinction based on variables describing the results of the banking organization's activity [12–14]. Nevertheless, despite all the organizational complexity of a financial institution, its economic efficiency will be represented by the return on assets (ROA) indicator.

Having analyzed the theoretical and empirical studies of experts, it was found that traditional banks with a digital banking channel pay most attention to digitalization of the customer journey [15]. This fact allowed identifying the following factors that can affect the performance of banks that have set out on the digitalization path: the number of remote banking customers (Usersn); the share of deposits opened through remote banking (Deposits); the share of loans issued through remote banking (Credits); the share of transactions made via digital communication channels (Payments). In order to neutralize the impact of the bank's operation scale, it is necessary to normalize the indicators depending on the size of the bank. In this case, we normalize the indicator of the number of users of remote banking services.

In modern conditions, there have been formed three main approaches that allow conducting economic efficiency of a banking organization: financial, expert, and econometric models [16, 17]. The models based on financial reporting 5 have sufficient simplicity but give only a certain slice without a probabilistic assessment of the results. Expert methods are the most frequently applied in the field of performance evaluation, but along with their wide application they are not devoid of disadvantages, in particular, subjectivity in evaluation. Recently, econometric models have been used more and more often, and despite the difficulties in their construction, they allow us to obtain fairly objective data on the basis of which we can identify the dependencies of indicators and build long-term trends. There are two main groups of econometric models: regression models and models which determine the efficient frontier. Stochastic frontier models using a translogarithmic function are used in the construction of models that determine the efficient frontier [18, 19]. Creating the efficient frontier is a rather complex task and requires a lot of work. Construction of such econometric models allows evaluating performance of any financial institution, but creation of such models requires a large number of conditions that should be fulfilled during construction of the models, for example, in order to obtain the necessary results of econometric modeling, it is necessary to have a large amount of panel data, i.e. data distributed over a long period of time. Regression models are the simplest models in econometric studies, but they are quite effective in terms of analyzing trends and indicators affecting the evaluation of performance [20].

In order to evaluate performance of digital banks, it is more reasonable to use a regression model, which will allow estimating the impact of all parameters considered in the model on the ROA and does not require sufficiently stringent conditions and restrictions, as in constructing other models. Our research aims to evaluate the factors that influence evaluation of a digital bank's performance as a structure that has not been sufficiently studied and does not have standard approaches to research. Despite its relative simplicity, the regression model can be an effective tool for evaluating performance of financial institutions, which allows providing forecasting in the operation of a banking institution. On the basis of the created model it is possible to determine contribution of each of the parameters in the formation of ROA of a digital bank, and, consequently, to determine criteria for evaluating performance of a financial institution.

3 Data and Methodology

Within the framework of the study, 16 largest traditional Russian banks with digital banking services channel will be analyzed to build an econometric model: PJSC Sberbank, VTB Bank (PJSC), PJSC Alfa-Bank, PJSC Credit Bank of Moscow, PJSC ROSBANK, JSC Raiffeisenbank, PJSC Bank Saint-Petersburg, PJSC Post Bank, PJSC CB UBRD, PJSC Promsvyazbank, PJSC UniCredit Bank, PJSC Baltinvest Bank, PJSC Vozrozhdenie Bank, and Asian-Pacific Bank. Figure 1 shows the values of descriptive statistics for the indicators used in the model.

Variable	Obs	Mean	Std. Dev.	Min	Max
ROA	16	.0153362	.0105015	.000947	.040013
Usersn	16	6.08e-06	.0000106	1.60e-07	.0000376
Payments	16	.631875	.2069209	.1	.87
Deposits	16	.281875	.1215576	.1	.62
Credits	16	.32875	.1905387	.1	.75

Fig. 1. Descriptive statistics for the indicators used in the model

The average number of users of digital channels of communication with digital banks is 6,187,500 people. The average share of deposits opened via remote banking in the 16 banks under analysis is 28% of the total number of open deposits. The number of online loans is higher than the number of deposits; on average, it is 33% of the total number of loans issued. It can also be noted that the majority of transactions are made online, 63% on average. It is necessary to analyze the studied indicators for outlying cases, as their presence can have a negative impact on the model. According to the data obtained, it is clear that the value of the Usersn variable of such banks as Renaissance Credit, Russian Standard Bank, and Post Bank are sharply distinguished from the total amount of data. However, some researchers believe that availability of outlying cases among the design variable data is not critical and may not affect the result [21]. Therefore, it is decided not to get rid of outlying cases for this indicator.

When creating an econometric model with a variable structure for evaluating performance of digital banks, first of all, a study on the presence of multicollinearity in the model was conducted. The multicollinearity effect implies the presence of inter-relation between input variables, which in some cases may reduce the reliability of the obtained results based on econometric research. For this purpose, a correlation analysis of the input parameters of the variable structure model was carried out, which gave the following results presented in Fig. 2.

	ROA	Usersn	Deposits	Credits	Payments
ROA	1.0000				
Usersn	0.5654	1.0000			
Deposits	-0.0185	0.0366	1.0000		
Credits	-0.1222	-0.2488	0.7905	1.0000	
Payments	0.5577	0.1937	0.4661	0.3753	1.0000

Fig. 2. Correlation matrix

After analyzing the correlation matrix for the availability of multicollinearity, we can conclude that there is a close relationship between the indicators of Credits and Deposits, as evidenced by the partial coefficient of correlation equal to 0.79. There are a number of studies which hold the view that factors should be excluded from the model if the correlation coefficient is higher than 0.8 [22]. Therefore, it was decided to leave the explanatory factors unchanged.

The following conclusions were made after the analysis of factors for the availability of outlying cases and collinearity: there is no collinearity between the explanatory factors; the absence of outlying cases of data.

Thus, the following type of equation is assumed to be used to create an econometric model:

$$ROA = a * \text{Usersn} + b * \text{Deposits} + c * \text{Credits} + d * \text{Payments} + y \quad (1)$$

Seven models were created for the analysis of evaluation of banks' performance. The logic of model construction is as follows:

m1–m3 - creating a statistically significant model with step-by-step exclusion of the least significant factors;

m4–m7 - evaluating the impact of each factor on the resulting indicator individually.

4 Results

As a result of the study, an econometric model with a variable structure was obtained, which makes it possible to evaluate performance of traditional banks with a digital banking channel. Within the framework of the study, the presence of a linear dependence between banks' ROA and its parameters was confirmed, namely, the number of clients-users of remote banking services and the share of transactions made through digital communication channels. The results of the study are given in Table 1.

Table 1. The results of the study of the impact of factors on the evaluation of performance of digital banks' ROA

	m1	m2	m3	m4	m5	m6	m7
Usersn	468.995	454.043*	472.0094*	561.593*			
	(215.4941)	(182.7678)	(192.8358)	(218.946)			
Deposits	−0.0314	−0.0280				−0.0016	
	(0.0294)	(0.0176)				(0.0230)	
Credits	0.0028						−0.0067
	(0.0191)						(0.0146)
Payments	0.0313	0.0315*	0.0236*		0.0283*		
	(0.0111)	(0.0105)	(0.0098)		(0.0113)		
_cons	0.0001	0.0006	−0.0025	0.0119	−0.0025	0.0158	0.0175
	(0.0066)	(0.0064)	(0.0064)	(0.0026)	(0.0075)	(0.0071)	(0.0055)
N	16	16	16	16	16	16	16
R^2	0.6112	0.6104	0.5284	0.3197	0.3110	0.0003	0.0149
adj. R^2	0.4698	0.513	0.4558	0.2711	0.2618	−0.0711	−0.0554

Standard errors in parentheses
$^*p < 0.05$, $^{**}p < 0.01$, $^{***}p < 0.001$

As a result, the following regression model was created:

$$ROA = -0.003 + 472 Usersn + 0,02 \text{Payments}$$

Characteristics of the model, in particular, the coefficient of determination reaches values in the range from 0.61 to 0.52, which indicates an acceptable effect of factors on the resulting attribute. The significance of the determination coefficient is confirmed by the F-test, the value of which is significantly higher than the critical indicator, which confirms the qualitative parameters of the model.

As a result of the analysis of the obtained model of digital banks' performance evaluation, it was found that coefficient a is positive, which indicates that the higher the number of remote banking users is, the higher the ROA of banks. Coefficient d was significant, which indicates that for digital banks, the number of bank transactions has a direct impact on the ROA of banks. Based on the standardized coefficients of the econometric model of evaluation of performance of the digital bank, it was determined that the greatest impact on the bank's ROA is made by a number of clients of a financial institution.

Return on assets is not affected by such factors as "share of loans opened through remote banking" and "share of deposits opened through remote banking", as evidenced by p-values higher than 0.05.

In order to estimate significance of the revealed dependence, the obtained regression was diagnosed for checking the following conditions:

- linearity. Dependence of return on assets on the standardized indicator "number of users/assets" and "share of transactions made through digital communication channels" is linear. The amount of balances tends to zero;
- the balance distribution schedule is normal;
- for each value of the explanatory factor of the model there is homogeneity of balances dispersion.

Thus, having analyzed the significance coefficients of the model and balance distribution, we can judge about the adequacy of the model. Therefore, the chosen model well describes the relationship between return on assets, normalized "number of users/assets" and "share of transactions made through digital channels".

5 Conclusion

Digital banks are already firmly established in the banking market. Currently, most commercial banks are adapting digital technologies to a greater or lesser extent, allowing them to increase loyalty of their customers to the services they provide. To date, a fifth of credit institution customers prefer fully digital banks that do not have physical branches and provide services online or via mobile applications. According to experts, this indicator will grow every year. Performance evaluation of such financial institutions is under development.

During the research aimed at evaluating performance of digital banks, the authors studied several methods that ensure evaluation of banking institution performance. Having estimated the advantages and strengths of each method, the authors chose the regression analysis, which allows evaluating the impact of quantitative and qualitative factors on the performance of commercial banks. Return on assets was determined as an efficiency indicator, as this indicator reflects the bank's ability to efficiently dispose of assets under its control.

Having studied the impact of factors on the performance of the Russian banks, which are at the stage of transformation from traditional to digital, we can conclude that the performance of such banking institutions is positively influenced by the normalized indicator "number of users/assets". Its impact can be explained by the fact that the use of various gadgets with access to the Internet has become ubiquitous and clients of banks, having evaluated the convenience of controlling their accounts online, give preference to those banking institutions whose digital services are convenient. Attracting new clients, provided that the bank has a competent policy of work, it gets more profit and hereby increases the efficiency of its operation. Also, it is necessary to note the influence of another factor on the resulting indicator: "the share of transactions made through remote banking". When clients make online payments, banks can analyze users' preferences and offer them favorable conditions, bonuses, and promotions, hereby increasing customer loyalty and satisfaction with the use of banking products. And in the long run it will reduce the operating costs of banks. Also, the study revealed that performance of digital banks is not affected by the share of deposits and loans through remote banking. It can be assumed that banks actively attract clients to provide these services through digital communication channels in order to reduce the workload on physical offices.

This study obtained an econometric regression model with a variable structure, which allows estimating contribution to the evaluation of the digital bank's performance, expressed through ROA from the internal characteristics of the bank itself. It was also obtained that the largest contribution to the performance evaluation is made by the size of the client base of a digital financial institution, which shows a direct correlation between a number of clients of the bank and its ROA. Identification of the factors that influence the performance of digital banks will allow banking institutions to develop promising areas of their operation in the rapidly developing digital world of banking services.

Further research on evaluation of digital bank performance can be aimed at identifying the digital factors affecting the organization of the internal system of banks, data storage, as well as the impact of biometrics on the efficiency of banking institutions.

References

1. Borisov, A.B.: Great Dictionary of Economics. The Book World, Moscow (2009)
2. Gianiodis, P.T., Ettlie, J.E., Urbina, J.J.: Open service innovation in the global banking industry: inside-out versus outside-in strategies. Acad. Manag. Perspect. **28**(1), 76–91 (2014)
3. Rudskaia, I., Rodionov, D., Degtereva, V.: Assessment of the effectiveness of regional innovation systems in Russia. In: Proceedings of the 29th International Business Information Management Association Conference - Education Excellence and Innovation Management through Vision 2020: From Regional Development Sustainability to Global Economic Growth (2017)
4. Sun, P.H., Mohamad, S., Ariff, M.: Determinants driving bank performance: a comparison of two types of banks in the OIC. Pac. Basin Finan. J. **42**, 193–203 (2017)
5. Buevich, S.Y., Krivtsova, K.A.: Features of project evaluation, taking into account the efficiency of the risk factors and uncertainties. In: Science and Contemporaneity, pp. 57–61 (2014)
6. Klomp, J., Haan, J.: Bank regulation, the quality of institutions, and banking risk in emerging and developing countries: an empirical analysis. Emerg. Mark. Finan. Trade **50**(6), 19–40 (2014)
7. Reddi, S.K.: Disruptive innovation in banking sectors. Int. J. Sci. Res. Manage. **4**(2) (2016)
8. Zineldin, M., Vasicheva, V.: Banking and financial sector in the cloud: knowledge, quality and innovation management. In: Cloud Systems in Supply Chains, pp. 178–194. Palgrave Macmillan, London (2015)
9. Bianchi, A., Biffignandi, S.: A new index of entrepreneurship measure. J. Mark. Dev. Compet., 35–50 (2012)
10. Fried, H.O., Tauer, L.W.: An Entrepreneur Performance Index. Springer, New York (2015)
11. Ferrando, A., Pal, R., Durante, E.: Financing and obstacles for high growth enterprises: the European case. EIB Working Papers (2019)

12. Beck, T., Chen, T., Lin, C., Song, F.M.: Financial innovation: the bright and the dark sides. J. Bank. Finance **72**, 28–51 (2016)
13. Bircan, C., De Haas, R.: The limits of lending: banks and technology adoption across Russia (2015)
14. Busby, D.: Adopting the best approach for a digital banking solution: combine the benefits of the 'build', 'buy' or 'outsource'options. J. Digit. Bank. **2**(1), 43–50 (2017)
15. https://www.bcg.com/ru-ru/about/bcg-review/digitalization-client-way.aspx. Accessed 13 May 2019
16. Buevich, S.Y.: Algorithm and barriers to the implementation of the project management system in the banking sector. Compet. Glob. World Econ. Sci. Technol. (4–5), 53–61 (2017)
17. Karminsky, A.M.: Methodology for calculating the rating of dynamic financial stability of banks. Banking and financial technologies, International Center of Banking and Financial Technologies, pp. 134–142 (2003)
18. Chen, H., Cummins, J.D., Viswanathan, K.S., Weiss, M.A.: Systemic risk and the interconnectedness between banks and insurers: an econometric analysis. J. Risk Insur. **81** (3), 623–652 (2014)
19. Wheelock, D.C., Wilson, P.W.: The evolution of scale economies in US banking. J. Appl. Econ. (2017)
20. Mester, L.J.: Applying efficiency measurement techniques to Central Banks. Working paper, The Wharton Financial Institutions Center, 41 (2003)
21. http://sixsigmaonline.ru/baza-znanij/22-1-0-306. Accessed 17 May 2019
22. https://docviewer.yandex.ru/view/0/?page=1&*=Yzndh1=ru. Accessed 13 May 2019

Development of the Conceptual Framework for Financial Reporting in the Context of Digitalization

Elena Dombrovskaya(✉) [ID]

Financial University under the Government of the Russian Federation,
Leningradsky pr., 49, Moscow 125993, Russia
den242@mail.ru

Abstract. The Conceptual Framework for Financial Reporting, or IFRS Framework, has undergone many changes since its initial adoption. Amendments to the Conceptual Framework were due to various reasons that are related not only to the development of reporting data processing and presentation approaches, but also to changes in the global economic system, the emergence and development of modern data processing technology. The information technology development and the digital economy have a significant impact on the business units' information system. The impact has resulted in a significant increase in the amount of business data used. With increased information availability, development of new processing methods and algorithms, the approaches to financial reporting need to be adjusted and modified. All changes to the basic principles of financial reporting have been included in the new version of the Conceptual Framework. It will come into force on January 1, 2020 for all business units developing their reporting policies under IFRS. The paper overviews the amendments and innovations introduced by the new version of this document. The Conceptual Framework dialectics has been reviewed based on all document versions since 1989. The paper is of practical importance and helps to assess various areas of system development of the IFRS methodology in the context of digital economy.

Keywords: Conceptual framework · Financial reporting · Digitalization

1 Introduction

The system of international financial reporting standards is based on uniform approaches and principles determining the logics and goals of building such system, its key elements, their features and interaction. Conceptual Framework for Financial Reporting, or IFRS Framework, establishes the logics for building up the reporting information system. It determines the most important basic concepts of IFRS system. Not being a standard, the Conceptual Framework is the basis for all standards and deals with the key aspects of financial reports.

The approaches to financial reporting have changed with the changes in the economic environment; they reflected the processes in the global and national economies [1]. Principles of financial reporting in the public sector have been developed [2].

© Springer Nature Switzerland AG 2020
T. Antipova and Á. Rocha (Eds.): DSIC 2019, AISC 1114, pp. 334–344, 2020.
https://doi.org/10.1007/978-3-030-37737-3_30

The realities of recent years included the new information technology, namely the digital one, the intensive development of which led to creating a brand-new concept of "digital economy". It is represented by various levels and has a significant impact on the business units' information system. A digital environment is being formed, which encourages development of platforms and technology, as well as effective interaction between various market players and economy sectors (areas of activity).

Digital technology has provided an unprecedented amount of business information used. As the result, for the last few years, the amount of information created has exceed more than twice that created for the entire human history [3].

The result was the growth of analytical and forecast value of reporting information about the business units' activity, the intensive use of which brought essential and substantive changes in the basic system of financial reporting [4]. Principles of valuation of accounting objects have undergone major changes [5]. In this regard, considering the dialectics of the Conceptual Framework for Financial Reporting is of great interest.

2 Developing Financial Reporting Approaches According to the International Accounting Standards Board (IASB)

2.1 Historical Background

Adopted by the International Financial Reporting Committee in 1989, the first version of the document under consideration was entitled "Statement of Principles for Financial Reporting". In 2001, after the reorganization of the Committee into the International Accounting Standards Board, the document was adopted as a backbone for IFRS. It was entitled "Conceptual Framework for the Preparation and Presentation of Financial Statements" [6].

The problems of development of the principles for financial reporting are touched upon in the works of many authors. Hendriksen and van Breda noted that the relevance is the most important qualitative characteristic of financial reports [7]. Needles and Powers Ethics defined that ethics is an important problem in preparing financial reports [8]. "Users of these reports must depend on the good faith of the people involved in their preparation. Users have no other assurance that the reports are accurate and fully disclose all relevant facts". It is possible to carry to number of researches of the early reporting [9–11].

The process of amending the Conceptual Framework has been launched in 2002, when the Memorandum of Understanding was signed and the convergence of international financial reporting standards and US financial accounting standards was started. The amendment process was associated with a series of bankruptcies and scandals in the business community, when it turned out that the current IAS system failed to ensure the proper quality level of the reporting information created. As a result, the principles and basic concepts of financial reporting needed to be revised.

The work on amending the Conceptual Framework to update and bring it closer to the approaches used in the US GAAP was finished in 2010 with the document entitled "Conceptual Framework of Financial Reporting" adopted. Structurally, the new Conceptual Framework included the text of the previous document version with some amendments and two new chapters: "Objective of General Purpose Financial Reporting" and "Qualitative Characteristics of Useful Financial Information". That is, the new document was not adopted in the final version, and work on it has continued.

In addition to a shorter title, the most important amendments to the Conceptual Framework as of 2010 included clarification of the financial statements' purpose and the changes in its qualitative characteristics. In particular, the objective of the financial statements, which is to provide useful financial information, has been preserved in the new version; however, the composition of users and the content of the reporting information have been adjusted. Existing and potential investors, lenders, and other creditors were listed as its users. In the 1989 version, the financial statement data included information about the financial standing, financial results, and changes in the financial standing. In the 2010 document, their list is not detailed, and the purpose of financial reporting was described as providing users with useful financial information to make decisions about earmarking resources for the relevant enterprise. At the same time, the focus was on the fact that users need information about the future cash flows with the enterprise [12].

Amendments were also made to the composition of the financial statement's qualitative characteristics. In the 1989 version, they included clarity, relevance, reliability, and comparability. The Conceptual Framework of 2010 identified two groups of qualitative characteristics: the basic ones and those improving the usefulness of the reporting information. The basic ones were relevance and truthfulness, the improving ones were comparability, verifiability, timeliness, and understandability. As for the changes in qualitative characteristics, it should be noted that the reliability has been replaced by veracity: prudence (caution) was removed from the list of financial statement characteristics with a new characteristic of "verifiability" introduced. Only one limitation was left, that is the cost of providing useful financial information. Therefore, with the growing need for forecast information in the financial statements, there were higher veracity and verifiability standards.

2.2 New Version of the Conceptual Framework

In March 2018, the International Accounting Standards Board released a new, revised version of the Conceptual Framework. It entered into force immediately for IFRS developers, and for those engaged in developing IFRS-based accounting policies, it will come into force on January 1, 2020. The main reasons for adopting the new concept document were the inconsistency with modern thinking, the need to restore some previously excluded important provisions, and the lack of clarity. All these reasons were mentioned by those criticizing the 2010 version of the Conceptual Framework. The expansion of public information scope in the context of the digital economy development resulted in another transformation of the requirements for the same, with new version of the Conceptual Framework for Financial Reporting adopted.

In Russia, the new version of the Conceptual Framework was published on the official website of the Ministry of Finance on November 8, 2018. The Ministry of Finance of the Russian Federation confirmed that this publication was official for applying the Conceptual Foundations in the Russian Federation. A review of amendments and innovations introduced by the new document version is of practical importance and helps to assess the direction of the IFRS methodology's systemic development in the digital economy.

In the new version, the IFRS Conceptual Framework once again changed its title. Now, it is entitled "Conceptual Framework for the Presentation of Financial Reports". Given that one of the previous versions entitled "Conceptual Framework for the Preparation and Presentation of Financial Statements" was not fundamentally different in terms of its content, we can conclude that the changes in the title make it possible to distinguish two document versions. See Table 1 for a comparison of the structure of the two Conceptual Framework versions (2010 and 2018).

Table 1. Substantial differences in the structure of the two versions of the conceptual framework

Conceptual framework for financial reporting (2010)	Conceptual framework for financial reporting (2018)
Chapter 1. The objective of general purpose financial reporting Chapter 2. Absent Chapter 3. Qualitative characteristics of useful financial information Chapter 4. The Framework (1989): the remaining text	Chapter 1. The objective of general purpose financial reporting Chapter 2. Qualitative characteristics of useful financial information Chapter 3. Financial statements and the reporting entity Chapter 4. The elements of financial statements Chapter 5. Recognition and derecognition Chapter 6. Measurement Chapter 7. Presentation and disclosure Chapter 8. Concepts of capital and capital maintenance

The Table above shows the changed in the Conceptual Framework for Financial Reporting of 2018 from the 2010 version. The new Conceptual Framework is more clearly structured; there are more chapters. Although the new version includes all the issues listed in the old document version, the latest format approved in 2018 is easier to search information. We can assume the 2018 Conceptual Framework has become complete as compared with the previous version, which was an intermediate stage of the work on the document.

The objectives of the Conceptual Framework in the old and new versions have also undergone some changes in terms of structuring and detailing (Table 2).

Table 2 compares the objectives set by the Conceptual Framework in the old and new versions. It should be noted that the composition and number of objectives as defined for the Conceptual Framework have been significantly reduced in the new version. The IASB excluded detailing in defining the goals and objectives of the Conceptual Framework, which was present in the 2010 version, mainly due to the targeted use of IFRS by various users. The integration of objectives shall be recognized as an indicator of the maturity of the IASB's position.

Table 2. Objectives for two versions of the conceptual framework for financial reporting

Conceptual framework for financial reporting (2010)	Conceptual framework for the presentation of financial reports (2018)
(1) Assisting the Board in developing future IFRS and revising the existing IFRS (2) Assisting the Board in promoting the harmonization of regulations, accounting standards, and procedures for the presentation of financial statements, by creating a basis for reducing the cases of alternative accounting procedures allowed by IFRS (3) Assisting national standard development bodies in the development of national standards (4) Assisting financial statement compilers in the application of IFRS and consideration of issues that have not yet become the subject of an IFRS (5) Assisting auditors in creating an opinion on whether financial statements are consistent with IFRS (6) Assisting financial statements users in interpreting the data contained in financial statements that are prepared in accordance with IFRS (7) Providing information on the Board's approach to developing IFRS to those interested in the IASB's activity	(1) Assisting the International Accounting Standards Board (the Board) in the development of IFRS (the standards) based on consistent principles (2) Assisting financial statement compilers in developing accounting policies in cases where none of the standards regulates an operation or other event, or when the standard allows for choosing an accounting policy (3) Assisting all stakeholders in understanding and interpreting the standards

In both versions, the IASB indicated that the Conceptual Framework does not prevail over standards. In the 2010 version, the Board recognized that, in a limited number of cases, there may be a contradiction between the Conceptual Framework and any of the IFRS. In such situations, IFRS provisions prevailed over those of the Conceptual Framework. As IFRS developed dynamically, the Board planned to reduce cases of conflict between the Conceptual Framework and IFRS over time.

The 2018 version generally preserved the subordination of the Conceptual Framework provisions to the relevant IFRS. However, the Board reserved the right to introduce, in some cases, specific requirements that deviate from the principles of the Conceptual Framework. Such deviation is justified by the objectives of presenting general purpose financial reports and shall be explained in the conclusions of the relevant standard.

Therefore, on the one hand, the influence of subjective factors in the IASB's activities is growing. On the other hand, it is obvious that the diversity of business units' activities does not always fit into the rigid principles established by the Conceptual Framework. This is especially true for the situation with the growth of information about business units in the context of digitalization. In particular, Big Data allows for processing virtually unlimited amounts of information.

The IASB tried to highlight its public mission and responsibility, which is to develop standards to ensure transparency, accountability, and efficiency of financial markets worldwide. The 2018 version of the Conceptual Framework emphasized that the Board acts for the benefit of the public, promoting trust, growth, and long-term financial stability within the global economy.

In the new document version, the objective of general purpose financial reports and the composition of their users have not changed. The concept of management activity for the benefit of owners, excluded from the 2010 document, was restated in the Conceptual Framework as of 2018. The information on the use of economic resources is provided in a separate paragraph. The Board emphasized the importance of such information for assessing the management's performance in terms of responsible data management, as well as for predictive assessment of the entity's prospects. This information is provided separately; however, it is as important as the information required by users to estimate the future cash flows.

The Conceptual Framework's section dealing with the qualitative characteristics of financial reports has been radically revised in the 2010 version (Macve R. 2015). In 2018, no major amendments were made thereto. As before, qualitative characteristics have been divided into fundamental ones, previously called basic, and those that increased the usefulness of information. Disclosing the contents of these characteristics has been preserved in the 2010 format. The most important change in this chapter was that prudence has been re-added to the list of financial statement characteristics (Efimova, O., Rozhnova, O. 2019). Prudence is related to reporting neutrality and means that caution shall be used when using judgments under conditions of uncertainty. The use of prudence means that assets and income shall not be overstated, and liabilities and costs shall not be underestimated. This is done to prevent underestimating assets or income, or overestimating liabilities or costs. At the same time, the need for an asymmetric approach to different items is denied, while recognizing asymmetric requirements in some standards to choose the most relevant information.

Therefore, the restoration of prudence, traditionally used in the recognition and evaluation of accounting items, demonstrates consistency and continuity of the principles of financial reporting under IFRS. At the same time, an increase in the amount of information processed inevitably results in higher requirements for its qualitative characteristics, toughening individual principles to prevent information distortion and provide an objective basis for its auditing.

As a result of the increased amount of information processed, its limits have expanded in terms of using both reporting and forecasting data. In this regard, the IASB acknowledged in the 2018 version that the information usefulness might be reduced because of a high uncertainty of the estimates. In such cases, preference should be given to information that was less significant, but more clearly defined in the estimates. A landmark point was restoring the concept of substance priority over form in the Conceptual Framework, which was excluded in 2010. Adhering to these conceptual approaches has one goal: to truthfully present information in financial statements.

A completely new chapter has been introduced in the 2018 Conceptual Framework: "Financial statements and the reporting entity". It describes the purpose and composition of financial statements, introduces the concept of a reporting entity, and considers features of consolidated and unconsolidated financial statements. Now, statements can be presented not only by an independent legal entity, but also by its part, which makes the information more valuable in the context of high risk and uncertainty.

The terms and definitions used in the new version of the Conceptual Framework have also been amended. In particular, the concepts of assets and liabilities have been updated. References to compulsory confidence in the inflow (outflow) of economic benefits were excluded from their definitions while recognizing reporting items. This allows extending the list of items included in the statements, which is very important for generating reliable information about the resources of the business unit in the context of big data processing.

At the same time, this will improve the accuracy of information about the assets and liabilities of the company, because the coverage of recognized objects will increase. The predictive nature of financial reporting, which would reflect a broader list of facilities, would also improve. The list of objects reflected in the financial statements will also be expanded by reflecting all the rights that the reporting entity has to benefit from the use of the asset and the obligation to transfer the benefits.

An important provision of the Framework is the evaluation of elements of financial reporting. The conceptual framework is established by two groups of estimates - "historical value" and "present value." The fair value may be fair value, reflecting the market value; the cost of use, representing the discounted value of future cash flows generated by the object; the fair replacement value, which is the price of acquisition of the equivalent asset at the valuation date. Fair value assessment requires the reporting entity to undertake serious work related to the collection, processing and analysis of various information. Large amounts of data must be used. The need to use predictive values complicates the valuation even when it comes to determining historical value. For example, the acquisition cost of fixed assets (IAS 16) or rights to use a leased asset (IFRS 16) includes the cost of liquidating the asset at the end of its use. In order to determine the capitalized value, it is necessary to establish: cost amounts, which will allow for its liquidation, which is expected to take place in many years, the establishment of a discount rate, and the subsequent continuous monitoring of possible changes in these figures, as well as the date of liquidation, will be necessary. It is digital technologies that provide relevant and reliable information and reduce the risks of non-recognition of objects in financial reporting.

The relevance and reliability of the reporting information depends to a large extent on the quality of the source data, their adequacy, compliance with the task to be solved, reliability, as well as the speed and depth of their analysis, including their classification into different groups. In this case, it is no longer limited to financial information, information of a non-financial nature is also necessary, it helps to obtain a complete picture of events and to verify financial data. In this way

At the same time, real achievement of relevance and reliability of reporting information, timely and truthful reflection only of real assets and liabilities, income and expenses, becomes possible only in conditions of application of digital technologies. It is the development of these technologies, such as advanced analytics tools using BIG DATA, that allows the fundamental provisions of the Financial Reporting Framework to be implemented in practice.

3 Problems of Financial Reporting in the Conditions of Digitalization

In order to identify problems of financial reporting in the digital economy, it is necessary to consider the impact of digital transformations on market participants and on the business environment.

In the context of digital technologies, there is a tendency to converge, deepen relations with users of reporting information. Working with them is individualized, the company is involved in the interests of investors and creditors, sensitive to their preferences. In these conditions, the flexibility of the reporting information is important, and it is possible to fill it with the data that is currently needed by a particular user. The format of financial reporting is increasingly expanded by non-financial information. Digital technologies save transaction costs and provide new potential for the development of a reporting information system. In this regard, it makes sense to assess the feasibility of the general user concept of financial reporting and its replacement with individual reporting. Financial reporting should not be driven only by the requirements of laws, regulations and standards. This inevitably leads to the depreciation of reporting information and formalizes the process itself. Admittedly, digital technology has provided unprecedented openness to today's business, and financial reporting has ceased to be the sole source of business information. The content of reporting should be based on information requests and expectations of specific users, provide them with real tools to solve their problems.

Digitalization of the economy has significantly changed the nature of competition. From costs, competition has shifted to creativity. Creativity, innovation, non-standard thinking can give competitive advantages to companies. Creativity is evident not only at the technological and organizational level. Accounting and financial reporting are becoming creative. The development of a creative approach in accounting and financial reporting is related to the complexity of business processes, the absence of appropriate regulations for new accounting objects and new technologies. Creativity in financial reporting is manifested in the use of an accountant's professional judgment to report innovations. The purpose of creative financial reporting is to overcome the limitations of existing accounting standards, to include forecast estimates and really necessary and useful information in the reporting. The conceptual basis of financial reporting should introduce into the normative plane both creative accounting, professional judgment of the accountant, and observance of professional ethics.

Expansion of information boundaries of financial reporting in conditions of digitalization, inclusion of forecast information and information resulting from professional judgment of the accountant, bring to the fore the problem of subjectivity and risk of fraud in reporting. The value of financial information under such conditions can be significantly reduced [13]. In order to protect the interests of users, it is necessary to establish reasonable boundaries for the creativity of financial reporting. The concept of accurate and honest reflection of information continues to be the philosophical basis of accounting. The return to the Financial Reporting Framework of the discretion requirement underscores this. However, objective strengthening of creativity in reporting requires guaranteed observance of the right of users to quality information.

A fair reflection of creative reporting information can be considered a reflection that does not mislead users or lead to financial losses.

The largest consequence of digitalization for financial reporting is the use of blockchain technology. Many companies are already introducing blockchain into their resource planning systems to manage purchases. Blockchain technology allows safe registration of transactions, ensures transparency and reliability of information. However, its implementation in companies is hampered by the complexity and lack of a regulatory framework. Accountants have yet to understand the new trend and assess its impact on financial reporting.

4   Results

When it comes to financial reporting in a digitalized economy, the most important thing is to ensure its qualitative characteristics, not to lose user confidence in the reporting information. Therefore, recognition of reporting items is linked to the fundamental characteristics of the reporting information: relevance and reliability. The IASB did not take the path of increasing the financial information uncertainty and riskiness, which could manifest themselves in the context of lower limits for recognizing individual items. Quality content of financial statements remains the most important task for informed economic decision-making.

Digitalization of the economy has had a significant impact on the reporting information system. Increased communication with users and the development of non-financial reporting require greater flexibility and user-oriented financial reporting. This can be facilitated by the concept of the individual usefulness of financial reporting, aimed at meeting the interests of specific groups of users. This approach will stop the impairment of financial reporting information that is currently taking place.

Digital technology has made creativity a competitive advantage. Financial reporting has also become creative, based on the professional judgement of the accountant and the observance of professional ethics. These issues should be included in the Financial Reporting Framework in order to enable the legal field to make a decision on the generation of reporting based on professional experience and expertise. To limit the risk of fraud in financial reporting, it is necessary to establish the boundaries of creativity. The main criterion may be the absence of financial losses for users of information.

The use of blockchain technologies in financial reporting is subject to serious reflection in the future. Technical difficulties and lack of regulatory frameworks have so far limited the ability to assess the impact of this innovation on financial reporting. But it is likely that these issues will be the subject of professional debate in the years ahead.

5   Conclusion

The purpose of this paper was to review the dialectics of the Conceptual Framework for Financial Reporting and to assess the impact of the economy digitalization on the development of financial reporting principles. The analysis of the various versions of the Conceptual Framework and, primarily, that of 2018, showed that amendments made to the latest version have made the document more convenient. Structuring by chapters better reflects the document content and makes it easier to find the necessary information. Work is underway to formulate and summarize the Conceptual Framework objectives, the number of which is reduced to three. With significant expansion of the information processed in the context of economy digitalization, as well as the influence of the subjective factor needed more rigorous reporting approaches. Therefore, prudence has been re-added to the list of financial statement characteristics. This was done to prevent distortion of the reporting information and increase its relevance. The use of forecast information in the statements shall not reduce its veracity. Preference shall be given to information based on more specific estimates.

Further development of the Financial Reporting Framework should take into account the processes taking place in the economic and technological environment. They are related to changes in the reporting information system, promotion of non-financial reporting, orientation to specific information requests of users, expansion of the scope of application of professional judgment, ensuring reliability of financial reporting in conditions of digital transformations.

The study could not cover all existing problems. But consistent work to address them will make significant progress in the area under consideration and determine the role and place of financial reporting in a changed world.

References

1. Henderson, S., et al.: Issues in Financial Accounting. Pearson Higher Education AU, London (2015)
2. Kachkova, O.E., Vakhrushina, M.A., Demina, I.D., Krishtaleva, T.I., Sidorova, M.I., Dombrovskaya, E.N., Klepikova, L.V.: Developing the accounting concept in the public sector. Euro. Res. Stud. J. **21**(1), 636–649 (2018)
3. Flower, J., Ebbers, G.: Global Financial Reporting. Macmillan International Higher Education, New York (2018)
4. Efimova, O., Rozhnova, O.: The corporate reporting development in the digital economy. In: Antipova, T., Rocha, A. (eds.) DSIC18 2018. AISC, vol. 850, pp. 71–80. Springer, Cham (2019)
5. Alexander, D., Bonaci, C., Mustata, R.: Fair value measurement in financial reporting. Proc. Econ. Financ. **3**, 84–89 (2012)
6. Macve, R.A.: Conceptual Framework for Financial Accounting and Reporting: Vision, Tool, Or Threat? Routledge, Abingdon (2015)
7. Hendriksen, E., van Breda, M.: Accounting Theory. Richard D. Irwin, Homewood (1992)
8. Needles, B., Powers, M.: Principles of Financial Accounting. Cengage Learning, South-Western (2011)
9. Gurskaya, M., Kuter, M., Aleinikov, D.: The early practice of analytical balances formation in f. datini's companies in avignon. In: Antipova, T. (ed.) ICIS 2019. LNNS, vol. 78, pp. 91–102. Springer, Cham (2020). https://doi.org/10.1007/978-3-030-22493-6_10

10. Kuter, M., Gurskaya, M., Andreenkova, A., Bagdasaryan, R.: The early practices of financial statements formation in medieval Italy. Acc. Historians J. **44**(2), 17–25 (2017). https://doi.org/10.2308/aahj-10543

11. Gurskaya, M., Kuter, M., Bagdasaryan, R.: The structure of the trial balance. In: Antipova, T. (ed.) ICIS 2019. LNNS, vol. 78, pp. 103–116. Springer, Cham (2020). https://doi.org/10.1007/978-3-030-22493-6_11

12. Khoruzhy, L.I., Gupalova, T.N., Katkov, Y.N.: Putting in place a system of integrated reporting in organizations. Int. J. Innov. Technol. Explor. Eng. (IJITEE) **8**(7), 748–755 (2019)

13. Tassadaq, F., Malik, Q.A.: Creative accounting & financial reporting: model development & empirical testing. Int. J. Econ. Financ. Iss. **2**, 544–551 (2015)

Data Analysis as a Toolkit for Construction and Evaluation of Consistency of Banks Ratings

Alexan Khalafyan⑩, Igor Shevchenko⑩,
and Svetlana Tretyakova$^{(\boxtimes)}$⑩

Kuban State University, 149 Stavropolskaya Street, Krasnodar 350040, Russia
dean@econ.kubsu.ru

Abstract. At present, the role of the banking sector has increased significantly, not only in the development of national economies, but also of the world economy as a whole. In this regard, the special importance acquires an objective comparative assessment of large financial institutions by groups indicators characterizing their financial activities. This actualizes the development of methodological approaches to analyzing the activities of banks based on the formation of ratings. In the article reviewed the features of the formation of rankings, distance and individual ratings of banks and their area of application. It is shown how with the help of computer analysis of data using the author's methodology, it is possible to build mathematically substantiated ratings of banks that characterize various aspects of their activities, followed by an assessment of their consistency using positional analysis. The methodology is based on the metric approach, which involves assessing the similarities and differences between banks through the distance between them as points of multidimensional space. Calculations were implemented in the environment STATISTICA 10 (Tibco, USA) package. In this paper, as an example, ratings of financial capabilities and efficiency of the activities of the 30 largest banks in the world according to Forbes magazine, based on data for August 2018, were formed.

Keywords: Rating of banks · Computer analysis · Metric approach

1 Introduction

The urgency of building ratings of banks is currently due to the active use of the sovereign credit ratings in financial markets. In accordance with the recommendations of the Basel Committee on Banking Supervision, requirements for assessing creditworthiness based on credit ratings are increasing.

At the same time, the modern capabilities of computer processing of large amounts of data with using probabilistic statistical methods implemented in modern software providing, allow to significantly increase the information content, reliability and objectivity of ratings.

The purpose of this research paper is to demonstrate the capability of computer analysis in the process of formation and examination of the mathematically validated ranking characterizing a variety of aspects in bank industry. The methodology for

T. Antipova and Á. Rocha (Eds.): DSIC 2019, AISC 1114, pp. 345–359, 2020.
https://doi.org/10.1007/978-3-030-37737-3_31

archiving this goal is formulated and presented in this research. It includes, firstly, the generation of bank rating applying metric approach, which implies the measurement of the disparity between the best-ranked bank and other banks, where the bank functioning indicators are used as the points in multidimensional space ranking. The indicators, in their turn, are ranked according to the extent of proximity to the indicators of the highest graded bank. Further steps comprise the evaluation of coherence (non-contradiction) of ranking based on position analysis, and demonstration of its advantages over the traditional usage of Kendall's coefficient of concordance.

The questions of building ratings using computer analysis and of their consistency are disclosed in the works of Russian and foreign scientists: Frolova I.V., Nikitina E.B., Atabekyan R.A., Bambaeva N.Ya., Baidak V.Yu., Bannikov V.A., Khalafyan A.A., Shevchenko I.V., Berg T., Koziol P. Construction of the banking ratings based on the CAMEL model and its development are reflected in the writings of such scientists as Ferrouhi E.M., Lopez J.A., Rostami M., Misra S., Aspal P., Shaddady A., Moore T., Parrado-Martinez P., De Moor L., Ozturk H. Problems in the qualitative assessment of the results of activities of commercial banks reflected in the publications of such domestic and foreign scientists as Gitinomagomedova A.M., Belousov A.S., Zhumabekov K., Kuprina I.V., Filimonov S.V., Jungherr J., Forte G., Feng G.

One of the examples is the case, where the "STATISTICA" toolkit was used: for the investigation about cluster structure of the 30 largest banks in the world; for the ranking of banks' financial capability and effectiveness; and for the estimation of coherence between the created ranking and Forbes ranking. It was found that the method of k-means clustering was utilized for banks classification based on homogeneity; and the method of complete-linkage was used for visualization of banks hierarchical clustering. The estimation of ranking coherence was conducted using two methods: Kendall's coefficient of concordance and Cronbach's alpha, which is applied for the position analysis.

2   Construction and Assessment of Consistency of Banks Ratings

2.1   Comparative Characteristics of Ratings and Rankings of Banks

The stability and sustainability of the banking sector largely determine the development dynamics of the entire the country's economy. Rightly and inverse the statement. High growth rates of the national economy of the state lead to an increase in the scale of activity of commercial banks of such a country. A striking example of this is the leading position of Chinese banks in the list of the world's largest financial intermediaries.[1]

---

[1] Rating of the largest organizations of Forbes magazine. URL: https://www.forbes.com/global2000/list/#tab:overall.

In this connection, economists devote much attention to methods for evaluating the activities of both individual credit institutions and the banking system as a whole. The relevance of the qualitative assessment of the activities of commercial banks and their transparency increased after the global financial crisis of 2008 [21].

Currently, such tools as ratings and rankings are often used to assess the activities of a commercial bank. Between them there are a number of significant differences. The comparative analysis of ratings and rankings was quite fully conducted by Frolova [11, p. 38].

The rankings are usually based on the analysis of quantitative information that is publicly available. This increases their objectivity, since expert evaluation is practically not used. In addition, the data on which the ranking is based may be useful for the user to conduct their own analysis and draw conclusions. The result of applying the ranking is the ranging of objects in descending or increasing one quantitative indicator.

The building of the rating is based on a combination of both quantitative and qualitative indicators. Depending on the goals of forming the rating, various methodics can be used, with a significant role being assigned to expert evaluation. This allows you to comprehensively evaluate the activities of the bank, it is possible to predict the dynamics of its development and even use the results of the rating for the purposes of regulation by the supervisory authorities. Currently, discussion on the questions of the relationship and inter influence of credit ratings and financial markets has activated [21].

In turn, there mark the two types of banking ratings - remote and individual [9, p. 88]. Distance ratings, as well as rankings, are based on public information. As a rule, quantitative indicators are used and the impact of expert opinion is minimal. The advantage compared to the ranking, in our opinion, is the possibility of obtaining a quick qualitative assessment of the bank's activities based on the analysis of quantitative indicators. In this connection, these ratings are built using modern computer analysis capabilities, as evidenced by research by a number of Russian scientists [1–4] and studies by the authors of the article [12–14]. Most often, as a result of building a remote rating, we see a list of objects with assigned ranks.

Individual ratings are calculated on the specific bank, including on the basis of non-public information. The role of expert assessments and specific assessment methodics is great here. Formation of an individual rating usually occurs on the order of the credit institution itself, since this is a labor-intensive process. A classic example of individual ratings are banks' creditworthiness and reliability ratings assigned by specialized agencies: Standart & Poor's, Fitch[2], Moody's[3]. The result of such a rating is the assignment of a rating, usually of the letterin, without building a general list in ascending or descending order.

---

[2] Methodology for assigning ratings banks of the Fitch agency. URL: http://www.fitchratings.ru/media/methodology/banks/Bank%20Rating%20Methodology%20201911008%20RUS.pdf.

[3] Ratings of financial stability of banks: the methodology for assigning ratings on Moody's global scale. URL: http://v3.moodys.com/researchdocumentcontentpage.aspx?Docid=PBC_103929.

The principles of the combinations of quantitative and qualitative indicators and the formation of the rating based on expert judgment also underlie the methodics based on the CAMELS model [18, 23–25, 30].

At present, the traditional approach is being modernized, for example, on the basis of CAMELS-DEA, which allows you to take into account the impact of financial regulation on the stability of banks [29].

The introduction into the international and domestic practice of the recommendations of the Basel Committee on Banking Supervision has led to the emergence of new integral indicators of the activities of banks and the assessment of the likelihood of their default based on a risk assessment. To these indicators include the SYMBOL (Systemic Model of Bank Originals Loss) [27].

Individual ratings are in demand not only by the counterparties of banks in the financial market, but and also by regulators and central banks to assess the reliability and stability of commercial banks.

Also, rating approaches are used to assess the creditworthiness and probability of defaults of individual countries - these are sovereign ratings. Wherein, the main issue under discussion is the level of subjectivity and methods for its reduction, including through the use of mathematical modeling [16, 26].

It should be noted that all three methodical approaches we have noted have their own advantages and are used for different assessment purposes.

In the present study, it will be formed us the ratings of the largest commercial banks in the world in indicators of financial capacity and efficiency based on the author's methodic. In contrast to the methodology, which is utilized in Forbes, the proposed methodology includes metric approach. Furthermore, comparing two estimation methods such as absolute metrics of the banks' financial capability and relative metrics of banks effectiveness, the last one is considered as a more objective model of comparison. Moreover, this method could be applied not only for the investigation of the world largest financial organizations but also for other institutions. To assess the financial capacity of banks selected indicators: revenue, profit, assets, market value. To assess the effectiveness of the activity, two profitability indicators selected us: return on assets and return on sales. This choice is due to some factors. First, the data on the basis of which these indicators are calculated are public, which increases their objectivity. Secondly, the profitability of activities in the banking sector is extremely important. This is noted by both domestic [5–8, 10] and foreign [22, 28] economists researching the issues of analyzing the activities of commercial banks. At present, the impact of profitability indicators of the activities on investment strategies of banks, a comparative analysis of the profitability of banks of the individual countries and regions is being actively researched abroad [17, 19].

By increasing profits, a growth in the capital adequacy indicator is ensured, and the possibilities for attracting new shareholders and investors are growing. The high level of profitability indicates the effective management of bank risks. Consequently, profitability indicators accumulate in themselves all the facets of the effective operation of a credit institution.

2.2   Methods of Forming a Rating of Banks

The methodological rationale of metric approach is provided below. The ranging of objects assumes their comparison by the principle - similar objects in the rating are locating side by side, dissimilar - at a distance in accordance with the degree of their distinction. But similarity and difference in multidimensional analysis are evaluated through the concept of distance, measured on the certain metric. Therefore, the objects to be ranked should be represented by points of the multidimensional space in the coordinate system of quantitative criteria characterizing their quality, and measure the distance between them and the object - the leader with the best indicators, which will be in first place in the rating. Subsequent places are distributed according to the degree of proximity to the leader. If there is no such leader, then a virtual leader is created, consolidating the best values of the quality criteria - the greatest, or the smallest, depending on what the increase in their values leads to - to increase or decrease the quality of objects. As an example, taking into account the page format, let us build the ratings of the first 30 banks from the list of the 2000 largest financial companies of Forbes publication in August 2018, presented on the journal's website[4], although modern data analysis technologies make it possible to easily build a rating all 2000 companies. The data required to build a rating of banks is given in Table 1.

It is natural to compare similar with similar, therefore, the ranging of objects is directly related to the research of their cluster structure, which allowing to identify groups of homogeneity (similarity) of objects - clusters, if they exist in the analyzed population. With the help of the k-means clustering procedure of the Cluster analysis module of the STATISTICA package, 2 clusters were allocated, the numbers listed in the Clusters column of the table. The reasons for choosing the k-means clustering approach is that it is more advantageous in the process of objects clustering which relies on the combination of standard indicators and measures within the total natural randomness.

Note that the first 7 banks of the list formed a cluster of 1 - a group of leading banks in the world regarding indicators characterizing their financial capabilities. From the graph of average values of standardized indicators (see Fig. 1), it can be seen that, according to all four criteria, banks of cluster 1 are "on average" significantly higher than banks of cluster 2. The superiority "on average" means, that according to the separate indicators, the banks of the cluster 1 may be inferior to the banks of the cluster 2.

Since the first 7 banks of the Forbes list formed a cluster of the world's leading banks and, given the small number of banks in the clusters, we will make a total rating of 30 banks in terms of their financial strength. According to Table 1, it is easy to establish that by the aggregate of 4 criteria - revenue, profit, assets, market value there is no bank leader. Therefore, in accordance with the metric approach for ranking, a virtual bank should be created, let's call it a bank consolidated leader (BCL), uniting the highest values of indicators for all 30 banks: revenue = 165.3; profit = 43.7; assets = 4210.9; market value = 387.7.

---

[4] Official Forbes magazine website. URL: http://forbes.net.ua/business/1416864-forbes-global-2000-krupnejshie-kompanii-mira.

>Table 1. Ratings of 30 banks from the Forbes list by indicators of their financial strength.

Bank names	A country	Forbes rating	Revenue billion dollars	Profit billion dollars	Assets billion dollars	Market value billion dollars	Claster	Euclidean distance	Euclidean distance rating
The largest banks in the world									
ICBC	China	1	165,3	43,7	4210,9	311	1	0,74	1
China Construction Bank	China	2	143,2	37,2	3631,6	261,2	1	1,55	2
JPMorgan Chase	USA	3	118,2	26,5	2609,8	387,7	1	2,40	3
Agricultural Bank of China	China	4	129,3	29,6	1301,1	184,1	1	3,67	7
Bank of America	USA	5	103	20,3	2328,5	313,5	1	3,17	5
Wells Fargo	USA	6	102,1	21,7	1915,4	265,3	1	3,46	6
Bank of China	China	7	118,2	26,4	3204,2	158,6	1	3,06	4
HSBC Holdings	Great Britain	8	63,2	10,8	2652,1	200,3	2	4,45	8
BNP Paribas	France	9	117,8	8,5	2353,9	93,6	2	4,68	9
Santander	Spain	10	56,1	8	1769,1	106,3	2	5,42	11
China Merchants Bank	China	11	49,9	11	993,7	112,4	2	5,66	13
Mitsubishi UFJ Financial	Japan	12	51,8	8,9	2774,2	86,2	2	5,21	10
Bank of Communications	China	13	59,1	10,7	1472,9	66,6	2	5,58	12
Royal Bank of Canada	Canada	14	40,5	8,8	1040,3	113,4	2	5,85	16
Itaъ Unibanco Holding	Brazil	15	62,3	7,5	437,6	87	2	6,08	21
Sberbank	Russia	16	46,3	13,4	470,9	86,3	2	5,99	18
TD Bank Group	Canada	17	35,7	7,9	1028,1	107,8	2	5,98	17

(continued)

Table 1. (*continued*)

Bank names	The largest banks in the world								
	A country	Forbes rating	Revenue billion dollars	Profit billion dollars	Assets billion dollars	Market value billion dollars	Claster	Euclidean distance	Euclidean distance rating
Postal Savings Bank Of China (PSBC)	China	18	56,1	7,6	1466,6	55,3	2	5,82	15
ING Group	Netherlands	19	56,6	5,5	1016,1	62,2	2	6,07	20
Sumitomo Mitsui Financial	Japan	20	49,1	7,2	1847,7	58,3	2	5,76	14
Intesa Sanpaolo	Italy	21	42,5	8,3	956,9	63,1	2	6,12	23
Industrial Bank	China	22	48	8,8	1023,1	53,5	2	6,05	19
Banco Bradesco	Brazil	23	76,5	4,7	370,5	61,3	2	6,24	26
Shanghai Pudong Development	China	24	48,3	8,2	974,8	50,7	2	6,11	22
Commonwealth Bank	Australia	25	33	7,6	752,4	93,5	2	6,22	25
China Citic Bank	China	26	43,6	6,5	894	46,5	2	6,30	27
Westpac Banking Group	Australia	27	29,4	6,4	668,8	76,2	2	6,44	29
Lloyds Banking Group	Great Britain	28	33,6	4	1098,6	65,4	2	6,35	28
ANZ	Australia	29	28,7	5,3	717,3	61,2	2	6,54	30
Mizuho Financial	Japan	30	29,7	5,1	1850,4	46,2	2	6,16	24

Sequentially, the columns of the table indicate the name of the bank, country, and rating of banks in the Forbes list. Indicators of banks (billions of dollars) characterizing their financial capabilities are displayed in the first four columns of the table: revenue, profit, assets, market value.

In order to exclude the influence of differences in the order of indicators on distances, for example, assets are 2 orders of magnitude greater than profit, one should also calculate the distances from their standardized values. The Joining tree clustering procedure of the Cluster Analysis module allows using different metrics - measures of proximity of objects of multidimensional space (in our case, 4-dimensional space by the number of analyzed indicators).

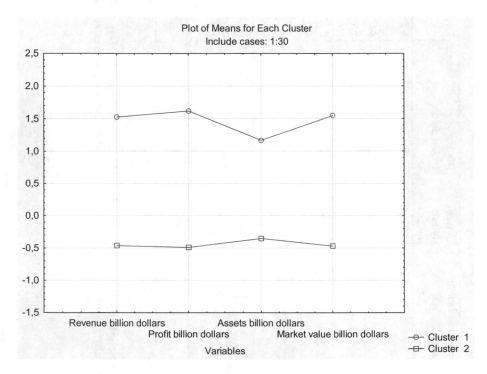

Fig. 1. Graphs of average standardized values of indicators

These are the euclidean distances, square euclidean distances, city-block Manhattan distances, Chebychev distances metric, and others. Let's use the most frequently used and intuitively more understandable Euclidean metric, which is also called the geometric distance. Joining Tree Clustering involves calculating a matrix of paired distances between all objects - in our case, banks. The distances between the BCL bank and the other banks are displayed in the Euclidean distance column of Table 1.

With the help to rank procedure, banks are ranked according to the principle that the smaller the distance, the higher the ranking. The ranks are displayed in the Euclidean distance rating column, Table 1. As can be seen from the table, the first 7 places were also taken by the banks of cluster 1, predominantly save their places, only the Agricultural Bank of China moved from 4 places to 7.

For cluster 2 banks there are also differences in ratings, which is quite reasonable, since the ratings were based on different methods, but on the whole they are non-contradictory. The consistency of ratings is also indicated by the value of the Spearman's rank correlation coefficient, equal to 0.95, indicating a strong and statistically significant relationship between the two ratings (significance level $p < 0.05$).

The cluster structure of banks in the rating is more understandable and interpretable if we use tree clustering using the Complete linkage method of the Joining tree clustering procedure. The graph, called the dendrogram, is displayed in Fig. 2 From the dendrogram, it follows that the greatest similarity between the two Chinese banks is Shanghai Pudong Development and Industrial Bank, since the distance between them is the smallest. When increasing the distance of the union, marked on the axis OY, the cluster of these two banks are consistently joined by the Italian bank Intesa Sanpaolo and China Citic Bank. Next, forming a new cluster, they are joined by the British bank Lloyds Banking Group and two Australian banks - ANZ and Westpac Banking Group.

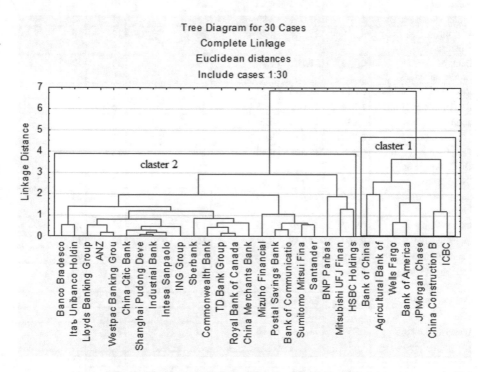

Fig. 2. Dendrogram of tree-like classification of banks.

Thus, when of a tree-like classification, an increasing the distance of the union entails to the parallel formation of new cluster structures, which ultimately merge into one cluster - an aggregate, consist of all banks. The figure shows that the leaders-banks also form their tree-like cluster, with this two leaders - Chinese banks ICBC and China Construction Bank, who took first and second place in both ratings, are at a considerable

Table 2. The ratings of the banks in terms of profitability

Bank names	The largest banks in the world					
	Return on assets	Return on sales	Euclidean profit margin	Rating in terms of profitability	Forbes rating	Euclidean distance rating
ICBC	1,038	26,437	2,81	3	1	1
China Construction Bank	1,024	25,978	2,85	4	2	2
JPMorgan Chase	1,015	22,420	3,04	8	3	3
Agricultural Bank of China	2,275	22,892	1,37	2	4	7
Bank of America	0,872	19,709	3,44	14	5	5
Wells Fargo	1,133	21,254	2,96	6	6	6
Bank of China	0,824	22,335	3,32	10	7	4
HSBC Holdings	0,407	17,089	4,28	21	8	8
BNP Paribas	0,361	7,216	5,38	30	9	9
Santander	0,452	14,260	4,49	25	10	11
China Merchants Bank	1,107	22,044	2,93	5	11	13
Mitsubishi UFJ Financial	0,321	17,181	4,39	22	12	10
Bank of Communications	0,726	18,105	3,77	19	13	12
Royal Bank of Canada	0,846	21,728	3,32	11	14	16
Itaь Unibanco Holding	1,714	12,039	3,43	13	15	21
Sberbank	**2,846**	**28,942**	0,00	1	16	18
TD Bank Group	0,768	22,129	3,41	12	17	17
Postal Savings Bank Of China (PSBC)	0,518	13,547	4,48	24	18	15
ING Group	0,541	9,717	4,88	29	19	20
Sumitomo Mitsui Financial	0,390	14,664	4,53	26	20	14
Intesa Sanpaolo	0,867	19,529	3,46	15	21	23
Industrial Bank	0,860	18,333	3,57	16	22	19
Banco Bradesco	1,269	6,144	4,66	27	23	26
Shanghai Pudong Development	0,841	16,977	3,72	18	24	22
Commonwealth Bank	1,010	23,030	3,01	7	25	25
China Citic Bank	0,727	14,908	4,08	20	26	27
Westpac Banking Group	0,957	21,769	3,16	9	27	29
Lloyds Banking Group	0,364	11,905	4,84	28	28	28
ANZ	0,739	18,467	3,72	17	29	30
Mizuho Financial	0,276	17,172	4,45	23	30	24

In the columns of Tables 1 to 6, the following are displayed: return on assets (ratio of profits to assets); return on sales (ratio of profit to revenue); Euclidean profit margin; rating in terms of profitability. The last two columns show the Forbes rating and Euclidean distance rating.

distance from other banks of the cluster - merger distance $\approx$3.75 that is easy to explain by a significant excess of the values of their indicators. The combination of both clusters into a common an aggregate of all 30 banks also occurs when a significant merger distance of $\approx$6.8, which is an additional confirmation of the high degree of difference of banks amongst the selected clusters on their financial capabilities.

Thus, in absolute indicators, a rating of banks was compiled characterizing their financial capabilities or financial potential. Obviously, to compare banks in terms of the effectiveness of their activities, one should use the given or relative indicators describing the profitability of banks. Profitability indicators: return on assets (profit/assets); return on sales (profit/revenue), calculated by indicators from Table 1 are shown in Table 2. Of the 30 banks analyzed, we managed to single out a bank that surpasses all other banks in terms of profitability criteria - this is Russian Sberbank with profitability of assets - 2,846 and sales - 28,942. Euclidean profit margin and Rating in terms of profitability are presented in the same columns of the table. It can be seen that the smallest distance - 0 has Sberbank, who took the first place in the rating, and the greatest - 5.38 has the French bank BNP Paribas, who took, the last, 30 place in the rating.

The degree of remoteness of banks in the Euclidean metric space can be illustrated using the Multidimensional scaling module. The scatterplot constructed by the module on the matrix of pair distances between all banks in the conditional two-dimensional coordinate system - measurement 1, measurement 2. In this coordinate system saved the order of the distances between banks - the closer (remote) banks according to the distance matrix calculated by the tree-like classification remain in the conditional coordinate system as well.

2.3   Assessment of the Consistency of Ratings

Since there are different approaches to the compilation of ratings, it is advisable to evaluate the degree of their consistency, or noncontradiction [13, 15]. Let us show by the example of three ratings from Table 2, how this problem can be solved with the help of the STATISTICA package. We first apply the Spearman's rank correlation coefficient (r), the values of which are presented in Table 3, taking advantage of the obvious fact that the stronger the correlation relation between the ratings, the higher their consistency.

Statistical significant correlations are in bold italics. In accordance with [12], we assume that if $\mid r \mid \leq 0.25$, then the correlation is weak, if $0.25 < \mid r \mid \leq 0.75$ - moderate, if $\mid r \mid > 0.75$ - strong. From the table it follows that the high consistency between the Forbes rating and the rating of the financial strength of banks Euclidean distance rating - r = 0.958. The consistency of both ratings with Rating in terms of profitability is moderate, close to weak - r = 0.427 and r = 0.337.

But coefficients of the Spearman's rank correlation characterize pair consistency, to assess group consistency, one should use the Kendall concordance coefficient, the calculation of which is implemented in the Fridman ANOVA & Kendalls concordance procedure. To use this procedure, you must previously create a table in which the columns denote banks, and the rows - ratings. The calculated value of the Kendall coefficient was 0,802, which indicates the presence of high group consistency between the ratings.

Table 3. Spearman's rank correlation table

Variable	Spearman rank order correlations (banks return) MD pairwise deleted Marked correlations are significant at p < ,05000		
	Rating in terms of profitability	Forbes rating	Euclidean distance rating
Rating in terms of profitability	1,000	**0,427**	0,337
Forbes rating	**0,427**	1,000	**0,958**
Euclidean distance rating	0,337	**0,958**	1,000

The table of correlation coefficients is symmetrical with respect to the main diagonal - cells with correlation coefficients equal to 1. Therefore, it suffices to interpret the values of the coefficients above the diagonal.

In work [14], it was proposed to use the Reliability and Item Analysis module, which was developed for building reliable questionnaires (scales) in psychology[5], as an alternative method for assessing the consistency of expert assessments. Each respondent of the questionnaire should express his attitude to the statements in the point scale of a fixed dimension, for example, a ten-point one. The key criterion in the positional analysis is the Cronbach alpha (α). statistics, which is calculated by the formula:

$$\alpha = (n/(n-1))\left[1 - \sum_{i=1}^{n} \left(s_i^2\right)/s_{sum}^2\right] \qquad (1)$$

where s2i - is the sample variance of the i-th statement, s2sum - is the sample variance of the total scale (the sum of points set by the respondents), n - is the number of statements (positions) in the questionnaire. When a Cronbach alpha (α), equal to 0, there is no consistency of respondents' opinions. When a Cronbach alpha (α), equal to 1 - complete consistency of the respondents' opinions. Applying to assessing the consistency of banks ratings - respondents, these are banks, and the statements, or the positions of the questionnaire, are the ratings of banks, the total scale - is sum of ranks of banks ratings. The results of calculations on assessment of the consistency, at the same time of all the three ratings, conducted by the Reliability and Item Analysis module are presented in Table 4. From the information part of the table shows that Cronbach alpha has adopted a value, equal to 0.613. The fact that average inter-item correlation was estimated as 0,689 then started to be measured in diapason between 0,25 and 0,75, it could be said, that regarding both metrics Cronbach alpha and average inter-item correlation, there is moderate coherence among ranks.

As we see, the estimates of consistency by both methods—the Kendall coefficient of concordance and positional analysis are not inconsistent. But the Reliability and Item Analysis module has a significant advantage over other statistical criteria for evaluating consistency - you can evaluate the contribution of each rating to the overall

---

[5] StatSoft: an electronic textbook on statistics. URL: http://statsoft.ru/home/textbook/modules/streliab. html#general.

Table 4. Results of positional analysis

Variable	Summary for scale: Mean = 76,967 Std. Dv. = 44,907 Valid N:30 (Banks return) Cronbach alpha: 0,613; Standardized alpha:, 796148 Average inter-item corr.: 0,689				
	Mean if deleted	Var. if deleted	StDv. if deleted	Itm-Totl Correl.	Alpha if deleted
Rating in terms of profitability	61,467	1593,849	39,92304	0,406	0,647
Forbes rating	31,000	200,333	14,15392	0,831	0,504
Euclidean distance rating	61,466	1308,582	36,17433	0,904	0,352

In the columns of the table successively show the results of the position analysis after removing the rating from the analysis, whose name is indicated in the row name: mean of the total scale (Mean), variance (Var.), Standard deviation (StDv.), General positional correlation (Itm-Totl Correl), Alpha Cronbach.

consistency of the Cronbach alpha values after its removal from the consistency assessment procedure, which are displayed in the last Alpha if deleted column. The rule is simple - if the value of Cronbach alpha after deletion exceeds the total Cronbach alpha = 0.613, then the rating reduces the consistency, if less, it raises. The table shows that the rating by profitability significantly reduces the consistency, both of the others, especially the Euclidean distance rating, significantly increase. Assessment of the consistency of ratings allows banks to be characterized not by rating alone, but by an ensemble of agreed ratings.

3 Results

The authors for the first time proposed mathematically justified statistical methods implemented in statistical packages for analyzing and constructing ratings of banking activities, assessing their consistency. Ratings building includes the application of a metric approach, the implementation of which is possible through the Joining tree clustering procedure, presented in the Cluster analysis module of the STATISTICA package. To assess the consistency of ratings, the Cronbach alpha criterion is proposed, the calculation of which is available in the Reliability and Item Analysis module of the STATISTICA package. To study the homogeneity of banks as objects of multidimensional space, followed by the allocation of clusters used the k-means clustering method of the Cluster analysis module.

Considering the potential of modern data analysis tools, it is possible to effectively, with minimal labor and time resources, build and analyze ratings of an unlimited quantity of banks when an arbitrary number of quantitative indicators characterizing them. Wherein, given the availability of alternative methods for building ratings, it is advisable to carry out the assessment their consistency, which ultimately will serve as a justification for choosing an ensemble of non contradictory ratings for a comprehensive and reliable characterization of the banks' activities.

4 Conclusion

To summarize, the proposed purpose of the research is achieved and the proposed methodology is implemented successfully. Based on 30 examples of leading banks, according to the Forbes rank August 2018 and limited indicators for their ranking, there is illustrated the possibility of formation and evaluation of the ranks coherence utilizing the mathematical rationale of metric approach and position analysis. Moreover, the appropriateness of the study the banks clustering within the comparative analysis is demonstrated. Additionally, the bank ranking was built applying the metric approach, which characterizes the financial capacity and effectiveness of banks. The moderate ranks coherence was indicated among the formed ranks and Forbes ranks.

Within conducting a comparative analysis of banks' functioning, the metric approach and the position analysis could be realized in a digital application, which would automate the entire operation of all necessary modules of statistic software. This would facilitate essentially the ease of utilization the data analysis for users which could be low-skilled in this area.

References

1. Atabekyan, R.A.: Application of multivariate statistical methods for classification of commercial banks of the Russian Federation by activity scale. Civ. Aviat. High Technol. **173**, 185–189 (2011)
2. Bambaevam, N.Y., Atabekyan, R.A.: To a question of construction of a rating of commercial bank. Civ. Aviat. High Technol. **167**, 181–185 (2011)
3. Baydak, V.Y.: Methodology for assigning the rating to banks by the boundary method. Russ. Econ. Online-J. **1**, 4–13 (2012)
4. Bannikov, V.A.: Computer technology for bank classification in the Russian banking sector. Econ. Manag. **7**(117), 70–79 (2015)
5. Belousov, A.S., Vinakurova, S.A.: Methodological approaches to the financial analysis of the activity of a commercial bank. Proc. Petrozavodsk State Univ. **5**(118), 108–111 (2011)
6. Gitinomagomedova, A.M., Omarova, O.F.: Methods of analysis of financial condition of commercial banks in Russian practice. Curr. Issues Mod. Econ. **2**, 77–85 (2015)
7. Zhumabekov, K.: Basic methods and receptions applied in analysis of financial accountability of a commercial bank. Bulletin of the Naryn State University named after S. Naamatov, vol. 2, pp. 113–116 (2015)
8. Kuprina, I.V., Stepanova, M.N.: On methods of analysis of financial results of the bank's activities. Mod. Bus. Space Curr. Probl. Prospects **1**(4), 167–169 (2015)
9. Nikitina, E.B.: Features of banking and rating assessment. Bull. Perm State Univ. **3**, 87–92 (2011). Series: Economics
10. Filimonov, S.V.: Problems of information support analysis of the financial condition of a commercial bank. Econ. Manag. Probl. Solutions **5**, 150–153 (2016)
11. Frolova, I.V.: Theoretical foundations and practical techniques for rating the investment potential of the regions of the Southern Federal District. Financ. Credit **24**(360), 38–50 (2009)
12. Khalafyan, A.A.: STATISTICA 6 Statistical Data Analysis. Beanom, Moscow (2010)
13. Khalafyan, A.A., Shevchenko, I.V.: Compiling and evaluating the consistency of bank ratings through computer analysis. Financ. Credit **28**(748), 1655–1677 (2017)

14. Khalafyan, A.A., Temmerdashev, Z.A., Yakuba, YuF, Kiseleva, N.V., Guguchkina, T.I., Antonenko, M.V.: Positional analysis as a method to assess the consistency of the expert estimates. Factory laboratory. Diagn. Mater. **12**, 69–78 (2015)
15. Berg, T., Koziol, P.: An analysis of the consistency of banks' internal ratings. J. Bank. Finance **78**, 27–41 (2017)
16. De Moor, L., Luitel, P., Sercu, P., Vanpee, R.: Subjectivity in sovereign credit ratings. J. Bank. Finance **88**, 366–392 (2018)
17. Feng, G., Wang, C.: Why European banks are less profitable than U.S. banks: a decomposition approach. J. Bank. Finance **90**, 1–16 (2018)
18. Ferrouhi, E.M.: Moroccan banks analysis using CAMEL model. Int. J. Econ. Financ. Issues **3**, 622–627 (2014)
19. Forte, G., Gianfrate, G., Rossi, E.: Does relative valuation work for banks? Global Finance J. (2018). https://www.sciencedirect.com/science/article/pii/S1044028317304787
20. Jorge, J.: Feedback effects between credit ratings and financial markets. Econ. Model. (2018). https://www.sciencedirect.com/science/article/pii/S0264999318306011
21. Jungherr, J.: Bank opacity and financial crises. J. Bank. Finance **97**, 157–176 (2018)
22. Spong, K., Sullivan, R., Young, R.: What makes a bank efficient? Financ. Ind. Perspect. 1–18 (1995)
23. Lopez, J.A.: Using CAMELS ratings to monitor bank conditions. FRBSF Economic Letter, Federal Reserve Bank of San Francisco, issue June (1999)
24. Rostami, M.: CAMELS' analysis in banking industry. Global J. Eng. Sci. Res. Manag. **2** (11), 10–26 (2015)
25. Rostami, M.: Determination of Camels model on bank's performance. Int. J. Multidisciplinary Res. Dev. **10**, 652–664 (2015)
26. Ozturk, H., Namli, E., Ibrahim, H.: Modelling sovereign credit ratings: the accuracy of models in a heterogeneous sample. Econ. Model. **54**, 469–478 (2016)
27. Parrado-Martinez, P., Gomez-Fernandez-Aguado, P., Partal-Urena, A.: Factors influencing the European bank's probability of default: an application of SYMBOL methodology. J. Int. Financ. Mark. Inst. Money (2019). https://www.sciencedirect.com/science/article/pii/S1042443118304700?via%3Dihub
28. Selvavinayagam, V.: Financial analysis of banking institutions. Eur. J. Acc. Auditing Financ. Res. 5–44 (1995)
29. Shaddady, A., Moore, T.: Investigation of the effects of financial regulation and supervision on bank stability: the application of CAMELS-DEA to quantile regressions. J. Int. Financ. Mark. Inst. Money **58**, 96–116 (2019)
30. Misra, S., Aspal, P.: A camel model analysis of state bank group. World J. Soc. Sci. **4**, 36–55 (2013)

Bankruptcy: Nature of Emergence, Diagnostic Problems

Elena Mamiy$^{(\boxtimes)}$ ⓘ, Aleksandr Pyshnogray ⓘ, Aleksandr Chulkov ⓘ,
and Aleksandr Korzun ⓘ

Kuban State University, 149, Stavropolskaya st., Krasnodar 350040, Russia
eamamiy@yandex.ru

Abstract. The study focuses on an assessment of the bankruptcy's probability. In the economic analysis the universal nature of emergence of company's bankruptcy and decline in its solvency had not yet been studied carefully. The existing analytical mechanism has variety of estimation models, but does not have the same degree of diversity among methodological approaches. The models considered have a single etiology, differing by private elements of the structure and coefficients. In recent years in Russia, the problem of bankruptcy has increased and acquired high growth rate. An insolvency of companies is systemic and becomes the new part of country's economic reality. Therefore, this article contains the background of setting up the universal mechanism to estimate company's risk of bankruptcy. As a result, there is formed the background to create a universal system for assessing the probability of bankruptcy. Moreover, there is increased the role of operational data in the analysis that is specific to each organization. It shows the case for forming clear algorithm for assessing the probability of bankruptcy on the basis of constant monitoring of the financial condition, combined with the calculation system assessment in relation to payment breaks and the complex evaluation of the coefficient system.

Keywords: Solvency · Bankruptcy · Payment discipline · Financial analysis · MDA model · Logit model

1 Introduction

This article is a research of identifying universal background to determine the natural occurrence of bankruptcy and analysis of existing methods of assessing the probability of bankruptcy.

The universal nature of company's bankruptcy and the deterioration of its solvency in economic analysis are studied rather modestly, the existing analytical mechanism has a variety of evaluation models, but does not have different methodological approaches. The phenomenon of bankruptcy in Russia has increased in recent years and gained the high growth rate. Insolvency of enterprises is systemic, becoming a new part of the economic reality of the country. In this regard, this article discusses the background to create the universal mechanism for assessing the risk of company's bankruptcy.

The research process identified different vulnerabilities of the existing analysis and assessment methodology of bankruptcy. It lies in the founded sophisms that negatively

© Springer Nature Switzerland AG 2020
T. Antipova and Á. Rocha (Eds.): DSIC 2019, AISC 1114, pp. 360–373, 2020.
https://doi.org/10.1007/978-3-030-37737-3_32

affect the success of the assessment and analysis of this phenomenon. Moreover, this reasoning is manifested in the presence of logic errors in the construction of a bankruptcy model, leading to the formation of a number of wrong ideas about the nature of bankruptcy. In addition, there were noted systematic contradictions that appear when the company studies the development of the probability of bankruptcy. The increase in the number and duration of gaps in solvency increases the degree of indignation of creditors, which subsequently increases the risk of bankruptcy.

2 Analysis of Basic Approaches to Forming Models of Bankruptcy's Forecast

2.1 Integral Forecast Methods of Bankruptcy

Economic system in general way is the chain of interdependent elements, which serves as a root system of society and its inviolability defines the inviolability and integrity of all society, in which is realized this economic mechanism. The gap of this chain shows that it (the chain) has experienced some signs of instability and regularly had gone wrong and then these signs disabled this chain by creating significant difficulties both for the state and for the closest partners in this system. Effective, qualitative analysis and in good time predicted the risk of bankruptcy can save the whole system or creditors in particular from partner's bankruptcy. Predicted risk can reduce the costs of uncertainty – a pay for potential systemic risk, for example, an interest rate on credit. Therefore, the relevance of this research is to identify the universal background that defines pre-bankruptcy and analyses existing methods of assessment of the bankruptcy's probability.

To achieve these goals, it will be necessary to:

- to investigate the level of current development in the area of methodology of assessment of the bankruptcy's probability;
- to give an assessment of existing methodology of assessment of the bankruptcy's probability;
- to describe the background to creating universal mechanism of assessment of the bankruptcy's probability.

Such researchers as Popova N.S., Ryabceva L.V., Stepanov I.G, Zykova K.E. discussed definitional breakthrough that related to the problem of bankruptcy. Nowadays there is some confusion in terms of bankruptcy, solvency, insolvency, payment discipline, cash gap and solvency gap for this reason, some terms need extra interpretation. Therefore, bankruptcy is neglected, legal process of economic result of company's solvency. Negative solvency is expressed as deterioration of payment discipline – the problem when creditor or partner is forced to have a reason to sue for restitution [1].

Nowadays the current development of economic analysis has moderate framework that is consists of several groups of methods of assessment of the bankruptcy's probability. Notably, the most famous international authors who deal with this problem are foreign authors: E. Altman, W. Beaver, G. Springate, A. Toffler, G. Tisshaw,

J. Ohlson, G. Sobato, Dzhu-Ha Tehong, J. Lego, D. Fulmer, J. Konan, M. Golder, etc. These authors used only two types of methodology of assessment of the bankruptcy's probability: MDA and Logit models that are built by means of the mathematical statistics on the base of ratio analysis that is on the base of training sample of companies [2–5]. Financial ratio analysis is the commodity for construction of models. This analysis is also the heart of company's economic analysis. Each of the authors combines ratios in their own way accordingly they get this or that model with different weight ratios. The great quantity of analytical ratios can help to create the great quantity of models [2–5]. Their scientific experience completely become transparent and successive for Russian researchers such as O.P. Zajceva, G.V. Davydova, A.YU. Belikova, E.A. Fedorova, G.A. Hajdarshina, R.S. Sajfullina, G.G. Kadykova, A.V. Kolyshkin, G.V. Savickaya, etc. So, obviously, majority of Russian and foreign authors follow the same way of ratio analysis and creating on its base some models. This process is almost endless and besides it increases the quantity of models [6]. For example, from a historical perspective of the economic analysis the main contribution of E. Altman is that he is the first who used multiple regression analysis in the assessment of the bankruptcy's probability. Statistical MDA analysis does not belong to any of authors of these models. The essence of the author's newness of each models is that the new training sample of companies was chosen [7, 15].

Some authors are specialized especially on recapitulative nature, for example, Stroganova E.A. carries out the analysis of data of Russian Statistics Committee in terms of the number of bankrupts in recent years ranking them by industries and identifying annual rates of additions.

It is submitted that construction sector and distributive trade are more exposed to bankruptcy than other sectors.

Moreover, in the material is noted that more companies become a bankrupt in Russia than in Europe in average [9]. The problem of bankruptcy is so relevant that by October 2017 Russia set a record over the past 10 years by the numbers of bankrupt-companies – more than 3200 bankrupts over the third quarter of 2017.[1]

Nowadays the large number of existing models is built according to common methodology and the growth of quantity of these models is making many uncertainties about company's analysis. However, it concedes that there is no way of knowing how effective these models work abroad, since there is only general information within the abstract numbers. For example, the efficiency of Fulmer's model is about 84% and Altman's model is only 62% [6]. At the same time, for example, the version given by Romanova O.V. and Zykova K.E. is that the efficiency of Altman's model is 95% [9]. Thereby, authors of scientific publications make a significant number of inconsistencies and judgment by making contradictions to this information that is crucial.

There are particular problems in Russia with database of companies in category "bankruptcy". L.M. Rabinovich notes in her research that it is connected to the fact that representative statistical database of bankrupties is formed in countries for decades, however, in Russia this database is only at the stage of development [3]. Also there is

---

[1] The official website of news agency "RBC" – URL: https://www.rbc.ru/economics/12/11/2017/5a05ca979a79475a8a66f090 (date of the request 28.11.2017).

no research center of such phenomena as bankruptcy and company's solvency, and all studies are being carried out randomly. Furthermore, existing models have only one approach to study bankruptcy – to compare investigated companies with other bank-rupt- and no bankrupt-companies by using ratio analysis. There are no other significantly distinct approaches, that's why there is quasi-choice (fictitious choice) of models that use the same tool in various ways. It is therefore necessary to improve and extend knowledge in the nature of emergence of company's bankruptcy and increase methodological mechanism of analysis and assessment.

First of all bankruptcy is debtor's inability, that is recognized by the arbitral tribunal, to meet fully creditor's demands of a monetary obligation. The nature of emergence of company's bankruptcy and company's solvency are two related and inseparable concepts.

Solvency is generally known as debtor's ability to pay the debts within a prescribed amount and term.

In this case debtor's insolvency is a situation, when the borrower has stopped paying (he has no means of paying) the debts in the prescribed amount and terms. The conclusion is that the bankruptcy procedure is exclusively a legal process of recognition of the debtor's insolvency, and the final point of development of insolvency is expressed in the formal (legal) bankruptcy procedure [11]. These are two sides of the same process – the inability to fulfill economic responsibilities that previously have been already assumed by the subject of economic relations. Meanwhile, the concept of insolvency refers more to the economy and mathematics, and bankruptcy - to the law.

There are situations when the company deliberately does not extinguish its own debts, while it has the potential solvency and resources to pay off debts [12]. Often it is associated with the facts of deception of creditors and has a different nature of bankruptcy, which is not considered in this material. The aim of the study is to identify the factors that determine the natural occurrence of bankruptcy in comparison with other methods of analysis and assessment. The observable mechanisms for assessing the probability of natural bankruptcy do not have natural universals, but have only artificial universals of bankruptcy, which are created within the framework of the training sample that formed the basis of the analytical model. This refers to the universal features that would be inherent in each bankruptcy individually and all at once or none of those who are inherent in bankruptcy a priori. This is necessary to strictly formulate the universal mechanism of bankruptcy for all enterprises in General [13]. Before proceeding to these concepts and the results of the conclusions, it is necessary to briefly consider the existing methods of assessing the probability of bankruptcy.

In the economic literature, there are several classes of models for the diagnosis of the risk of bankruptcy:

- MDA models;
- Logit/Probit models;
- Rating models;
- Scoring models.

Table 1. Summary table of integral methods for assessing the probability of bankruptcy

Authors	Formula	Interpretation of meaning
1 E. Altman (MDA model)	$Z = 1{,}2 \times X1 + 1{,}4 \times X2 + 3{,}3 \times X3 + 0{,}6 \times X4 + X5$	$Z < 1{,}81$ – probability of bankruptcy 80–100%; $Z = 1{,}81{-}2{,}77$ – probability of bankruptcy 35–50%; $Z = 2{,}77{-}2{,}99$ – probability of bankruptcy 15–20%; $Z > 2{,}99$ – probability of bankruptcy 15%.
2 R. Lis (MDA model)	$Z = 0{,}063 * X1 + 0{,}092 * X2 + 0{,}057 * X3 + 0{,}001 * X4$	$Z < 0{,}037$ – probability of bankruptcy is high; $Z > 0{,}037$ – probability of bankruptcy is small
3 J. Ohlson (Logit-analysis)	$R = 1/(1+eT)$; $T = -1{,}32 - 0{,}47SIZEt + 6{,}03(TLt/TAt) - 1{,}43(WCt/TAt) + 0{,}0757(CLt/CAt) - 2{,}37(NIt/TAt) - 1{,}83(FFOt/TLt) + 0{,}285X - 1{,}72Y - 0{,}521[(NIt - NIt{-}1)/(|NIt| + |NIt{-}1|)]$	R accepts values between 0 and 1, where 1 – is the maximum probability of bankruptcy, and 0 – is the minimum
4 Selezneva-Ionova (rating model)	$N = 25 * N1 + 25 * N2 + 20 * N3 + 20 * N4 + 10 * N5$	If integral N is 100 and more, the financial situation can be considered good, if it is less than 100 – она the financial situation causes to think about bankruptcy

In the Table 1 are collected for one the most known representative of an integral method that include MDA models, Logit models, rating methods. It should be noted that in each group there are integral methods [15]. For example, in scoring integral models are the most unsuccessful and inefficient, so they are not considered in this table.

2.2   Non-integral Forecast Methods of Bankruptcy

In addition, there are non-integral methods of assessing the probability of bankruptcy. For example, the A. V. Kolyshkin rating model is represented by three independent MDA equations and interpretations to them, which refer the company to one or another category. This tool should be used differently than in the case of one equation. Such value in each of the equations is compared with the standard values, and then it is discarded in a particular class according to the final value.

1. $0{,}47K1 + 0{,}14K2 + 0{,}39K3$;
2. $0{,}61K4 + 0{,}39K5$;
3. $0{,}49K4 + 0{,}12K2 + 0{,}19K6 + 0{,}19K3$

The Table 2 presents one of the most well-known models in the scoring of solvency assessment. The popular Russian model belongs to Savitskaya G. V. [17]. Methods of its application is reduced to a fragmented calculation of the ratios and their assignment to a particular class, and then follows the calculation of the final score.

Table 2. Savitskaya G. V. scoring model

Indicator	1 class	2 class	3 class	4 class	5 class
Return on assets, %	30 and more (50 points)	29,9–20 (49,9–35 points)	19,9–10 (34,9–20 points)	9,9–1 (19,9–5 points)	Less than 1 (0 points)
Current ratio	2 and more (30 points)	1,99–1,7 (29,9–20 points)	1,69–1,4 (19,9–10 points)	1,39–1,1 (9,9–1)	1 and less (0 points)
Equity to total assets	0,7 and more (20 points)	0,69–0,45 (19,9–10 points)	0,44–0,3 (9,9–5 points)	0,29–0,2 (4,9–1 points)	Less than 0,2 (0 points)
Limits of the classes	100 points	99–65	64–35	34–6	0 points

After the calculation, the data is interpreted by reference to a particular class:
1 class > 100 points; 2 class 65–99 points; 3 class 35–64 points; 4 class 6–34 points; 5 class 0 points. The more points, the less risk of bankruptcy.

The first group – MDA models, the founder of this class of models in the field of their application to the company's bankruptcy – E. Altman, who in the 60s of the last century in the United States began to produce models for different markets and for different market participants.

In particular, the model under study relates a company to one of 3 classes by an integral value. The mechanism of this class is based on a training sample of bankrupt and non-bankrupt companies for the same time interval. Thus, in 1968 E. Altman created the first known MDA model to diagnose the risk of bankruptcy at the company, but the training sample he took from 1945 and older.

Following E. Altman in the next twenty years, many MDA models were created both by Altman himself, and by followers. For example, another MDA model created by R. Lis in 1972 was built for the British economy and was based on a sample of European companies. The equation classifies companies into only two categories (bankrupt/non-bankrupt) and, unlike the previous one, uses four main ratios, rather than five, as in Altman model. Sometimes the Internet is often confused models and their authors, rearranging the places. Most of this is due to the inattention of the speakers, as MDA models are easy to confuse due to their external similarity [4].

2.3   The Main Problems of Bankruptcy Forecasting Models Application

The general algorithm for constructing MDA models is as follows, it makes sense to make out for a detailed analysis of their actions. First of all, a sample of bankrupt and non-bankrupt companies is equally created. The next step would involve selecting a range of possible-acceptable ratios that could potentially describe the pre-bankruptcy in a universal way. Then the coefficient analysis of each enterprise is performed. With the help of the statistical tool "multiple discriminant analysis" there will be a need to identify the ratios that the most well predict the onset of bankruptcy within the given sample. The result of this mathematical processing is a model of 5–8 ratios with weights for each of them. The process of selecting the most suitable ratios and weights for them reflects the essence of identifying universal signs of bankruptcy at the interval of the training sample, but not beyond its bounds. Nevertheless, if we take two or more absolutely any MDA models, compare their ratios and weights with them, it becomes obvious that they have nothing in common, since they are literally different equations. As a result, it is concluded that if the truth embodied in each of the equations would be universal, it would not be necessary to create such quantities of locally objective models describing the bankruptcy of the company. Therefore – the very existence of variety of MDA models today devalues the objectivity of each of them. Within this thesis, it is convenient to consider a thought experiment. If each of the MDA models was based on some training sample, it would be effective only within this sample and in this period of time, and its use in other territories and time intervals would give a randomly guessed probability of bankruptcy – it would be just a random value, which, of course, cannot be relied upon. Therefore, the smaller the training sample is, the more consistent models can be created in the finished market, but the more models there are, the less objective and valuable each of them is, since it is always possible to increase the sample and extend the training of the model. But, since the economy is varied and always contradictory (some of its regions and sectors can show opposite results based on varied resources of the sector), it follows that the fewer models there are, the more each of them is universal. So it does not take into account any exceptional features, which are so eager to small, but qualitative samples (each company has its own characteristics from region to region). Thus, it will not be able to accurately classify the company and the regularity of the qualitative analysis will again acquire a random feature. This idea implies that each company is so complex and individual that it is completely ineffective to try to investigate it using this generalized method.

No common and unified model within this method can be created, because if there was only one model, it would reflect the whole truth and consistency of the economic nature and bankruptcies, in particular, that a priori impossible (the existing number of models proves the systemic inconsistency of signs of bankruptcy). However, since researchers are always forced to build another model to better describe the reality of solvency, these models need to be built indefinitely, and each of them loses its value and objectivity.

But the nature of bankruptcy cannot be different for different companies, because this phenomenon is either an absolutely subjective process and can not be systematized, or all the universals that exist within each of the models are not such in reality, otherwise, there is a logical contradiction.

In addition to this reason, there is also a psychological reason in so many existing models. With the help of them, it is relatively easy to express yourself as a practical scientist who not only writes scientific abstractions, but also deals with very specific and tangible matters, thereby creating the ground for writing many works.

Now consider another class of linear models – Logit/Probit models. Their idea is very similar to the MDA model, but the authors use a different distribution law and a different interpretation of the data: logistic and normal distribution, respectively, and the integral result lies within 0 and 1 – the so-called dichotomous choice. In probit models is used the normal distribution law – it is the only difference from logit analysis.

The MDA model can classify incoming data into several groups, and logit event classification can be in a known interval and the division occurs only in two conditional classes, where the dividing line takes a value of 0,5 – above this value – the company is probably bankrupt, and less – rather – non-bankrupt. Researchers-practitioners noted that this model also involves a training sample of bankrupt and non-bankrupt companies, on average, judging by their reports, about 500 or more companies are required.

For example, the logit model by James Ohlson, which was developed later in 1980 than the Lis or Altman models. According to some reports, it is indicated that J. Ohlson used about 2000 companies.

The training sample is inside the "T" parameter. As in the MDA models, here using the statistical apparatus the most effectively predicting bankruptcy indicators are selected and involving in the "T". The calculation of the parameter "T" is similar to the MDA model by the method of creation, respectively adopting all the features and disadvantages of MDA models that have already been specified [8, 16].

The following type of methods for assessing the probability of bankruptcy is rating models. It is difficult to distinguish from the general array of models exactly rating, because its appearance is not installed, but has a number of distinctive features. Thus, rating models, with rare exceptions, do not have an integral value. These models have a multiple ranking of companies, but sometimes they do not even have this, and only the established values of the ratios, which are not integrally attributed to the company either in one class or in another in so doing, different ratios can be attributed to the company in opposite classes.

As you can see, according to these algorithms it is quite difficult to classify the company unambiguously. In addition, several MDA models are used at the same time; their values often in the calculations can be opposite.

Sometimes there is an integral rating in Table 1 at number 4, but with weights that were selected not in a standard way – through the statistical data processing apparatus, but rather logically. For example, the rating model of Selezneva-Ionova, which is an integral analysis tool, but contains a rating instead of probability – with the growth of the number N, the probability of bankruptcy grows too [17]. For calculating this model, it is necessary to use the updating of the ratios (Table 3) to their normative value, in which the calculated value is divided by the standard. The interpretation is as follows: if the integral value of N is equal to 100 or more, the financial situation can be considered good, if less than 100 – it causes to think about bankruptcy.

One of the most popular Russian models is G. V. Savitskaya model [17].

The peculiarity of scoring is that in this case there is a distribution of points in connection with the assessment of various facts of the company using ratios. The general idea is that: the higher the ratio is, the more points the company receives and the less the probability of bankruptcy is. The method is extremely simple and it is based on the sum of positive and negative factors that conditionally pull the scales with their weight, and the winner is the one whose sum is greater.

Table 3. Modifying factors for the Selezneva-Ionova model

Ratio	Standard
Inventory turnover	>3
Current ratio	2
Debt ratio	1
Return on total assets	>0,3
Return on sales (efficiency of managing the company)	0,2

Now the group of sophisms will be considered, which prevent to conduct a qualitative analysis of the subject of bankruptcy with the help of existing methods, some flaws of which have been considered.

The idea of modernizing the assessment of the probability of bankruptcy is based on historical, mathematical and logical background regarding the current situation around the economic analysis of the financial condition and bankruptcy of companies, which consist of a common base of information that after a significant period becomes false in the process of its using.

3   Results

The idea of describing the nature of the company in a few ordinary fractions is too simplified and accessible to everyone; from it, economists and managers believe in the success of such study, because it is feasible for every economist and creates some illusion about the quality of the results. Usually it happens like that: the researcher after ascertaining an isolated fact about the company, for example, after learning that the current liquidity ratio is less than one, he begins to draw many conclusions about the fact that the company has difficulties with paying for debts. Then the researcher gradually comes under the control of creditors due to the lack of guaranteed financial protection, suffers from a shortage of working assets, etc. In other words, the researcher often deals with sophistic statements, which are expressed in pseudo-truthful results of calculations that some ratios of primitive database generated within the company can effectively predict the behavior of the company in the future.

The logical foundations are hidden more deeply and become clear only as you delve into the subject of each individual company. They concluded that the principle of construction of models brings into the perview of the analysis of the bankruptcy, by searching for signs of a state status (bankrupt/nebancar) using the results of the activities of other companies. This means that by studying each individual company, researchers are looking for their assumptions about a particular financial condition, through companies whose financial affairs are inevitably different from their own affairs. This should always be more difficult to occur than the coincidence of the results of calculations of ordinary fractions. As already mentioned, the old American and European models were built on even older companies. It is logically clear that the probability of bankruptcy of your own company on the basis of other bankrupt and non-bankrupt companies has complicated effectiveness's assessment and consists of the following ideas:

1. Companies are evaluated on the basis of public accounting, which is always characterized by an exceptional generality of data due to the free approach to its formation. The basic law says that it is advantageous to make it more attractive to involve some interested parties (for example, investors, potential demand, etc.) [14] or to discredit the company, scaring off other interested parties, for example, for the purpose of debt cancellation. In each case, public accounting is a variable [10].
2. The individual nature of the payment discipline should correspond to some universals, which could be understandable and universal for each company individually and for all companies in general. This is necessary in order to avoid binding to the evaluation of other companies for better understanding of their own. The nature of the payment discipline should be investigated in order to assess the quantity and duration of payment gaps that lead to bankruptcy (the ultimate point of insolvency).

The main conditions for the emergence of universals and their general rules should fix the signs of bankruptcy in all companies and in particular in each company, in order for the search for signs of bankruptcy was directed inside each company separately and did not go beyond it. This framework will be considered the company itself and all its partners with whom it is engaged in economic activities, since the financial exchange of the company and the construction of its payment discipline takes place within this framework.

Here are the basic pre-requisites for universals:

1. Bankruptcy is a systematic and natural decrease in the solvency of the enterprise, not a sudden incident;
2. The activity of any company seems to be conditionally permanent, without a specific final point, by the onset of which it would be necessary to fully settle all property and financial claims with interested parties;
3. Debtors can logically and legally file a claim for initiation of bankruptcy proceedings if they do not receive timely and full payment of accounts payable;
4. Personalization of debt does not play any role, that is, it does not matter before whom there is a debt and during of this process the decrease of payment discipline is fixed.

So, these are general prerequisites that make possible the creation of a universal mechanism for assessing the risk of bankruptcy, which is based on the following natural laws:

- it seems that every time, when there is a solvency gap one after another in a row, there is an increase in the probability of problems with creditors (the risk of bankruptcy);
- the longer the gap is lasting, the higher the creditor's indignation is increasing and therefore the higher the probability of initiating bankruptcy proceedings is.
- different creditors have different exponent of the perturbation, that is, until some point the creditor's indignation practically does not develop, but from that moment the curve of the perturbation is steeply rushing up.

– the increase in gaps with the overall distribution of debt payments indicates an increased probability of initiating bankruptcy proceedings or, in any case, the problems with creditors.

In other words, there is a need to investigate the signs preceding the initiation of bankruptcy proceedings through the quantity and quality of gaps in the solvency of the company. Whereas, the more often and longer the debt is not paid, the more pressure creditors is exerted (growing indignation of creditors), that in the final stage is expressed in the initiation of the arbitration process to recover the debt, if other instruments of pressure do not help.

Analysis of solvency gaps (chronology of specific payments and specific requirements at any one time) just leads us to the area where the potential assessment of the probability of initiation of bankruptcy proceedings. Moreover, the facts of solvency gaps are universal and have consequences. The fact is that this already has indirectly incorporated the rest of the necessary information about the company: the ability to repay debts; generation of cash flows; dependence on creditors, etc. Since the main conditions of the market existence and development of the company are self-sufficiency and the ability to repay its obligations, its payment discipline and analysis of compliance with this discipline can provide an answer to the question of bankruptcy.

Such fixing of cash gaps and, as a consequence, earlier detection of bankruptcy probability, aimed at preventing risks and generating negative consequences of risks is possible only in the conditions of carrying out an integrated assessment, on the basis of a system of key indicators in a number of directions. Only this approach that will not only reveal the cash gap, but will also enable it to be identified at the early stage of formation and, possibly, prevent it completely or reduce its scale. In general, the algorithm for conducting such an evaluation should consist of a series of consecutive stages and the following key indicators:

1. Performance assessment:

   – financial results and profitability indicators: Net profit growth rate (loss reduction rate), income and cost growth rate correlation ratio, sales profitability, economic profitability, economic profitability growth rate
   – business activity and resource utilization indicators: asset turnover rate, the rate of growth of the turnover factor, the ratio of turnover of receivables and payables, the rate of growth of the payback (all spheres except trade), the rate of growth of the coefficient of use of production capacity (for production).

2. Cash flow analysis. It is one of the most important stages of tentative forecast of financial difficulties of the company, which can lead to bankruptcy. It is advisable to carry out such analysis in a continuous order with monthly detailing to achieve the maximum effect in the framework of operational monitoring. Analysis of net cash flows from the main activity is aimed at formation of sound management decisions within the framework of financial management of the company, and also it is able to determine the possibility of cash breaks.

3. Probability's estimation of increase of lag of cash breaks. Tracking of the dynamics of the occurrence and duration of cash breaks will allow to determine the first signs of the upcoming bankruptcy of the company, which can be identified in the process

of comparing incoming payments by date and by volume in accordance with the terms of repayment of liabilities in dynamics. The assessment is carried out taking into account the characteristics of the payment discipline, which is always individual.

4. Calculation system. A consistent analysis of cash gaps necessarily intersects with the analysis of the company's settlement system in the context of the matching of receivables and payables. The assessment is carried out taking into account the ratio of terms and payment volumes. It is advisable that such an analysis is also possible only on the basis of operational accounting data and in the breakdown of decades, which will allow timely action and incentive measures for timely payment of receivables and repayment of accounts payable. As a result of the estimation of company's calculation system, the definition of its external manifestations-financial stability and solvency.

5. Financial stability indicators: financial activity ratio, correlation ratio of the debt and equity capital growth rate, equity maneuverability ratio.

6. Solvency indicators: the ratio of security of working assets by its own working capital, the coefficient of current liquidity, the rate of change of the current liquidity ratio.

This approach will allow continuous monitoring of the financial and economic condition, to carry out earlier identification of problem zones and to develop measures to overcome and prevent the financial crisis in clearly identified areas of problem formation.

4 Conclusion

Historical assumptions are that any MDA or Logit model has a time, territorial and industry expiration dates (restrictions), based on the specified method of their construction. At the same time, inevitably and inseparably in the formed MDA or Logit model remains the imprint of the economic situation, the economic cycle, as well as the sectorial, temporal and historical features of the region where the sample was made at that period. Therefore, for example, it is impossible to apply the five-factor model of E. Altman for modern Russia, created in the 1970s in the USA based on a sample of companies existing much earlier than the 70s. It sets in motion the territorial, temporal and sectorial conflict with the reality in which this model is applied.

Mathematical conclusions, hidden in the existing methods of assessment of the probability of bankruptcy, consist in the fact that a very primitive ratio analysis gives only a limited set of fragmented facts about the activities of the company while there is no significant difference if they are selected in a single formula or are used dispersed. The process of bankruptcy is complex and has deep financial and management roots. It cannot just be described by five ordinary fractions, because of it, crushes and incorrect interpretations of facts begin to occur.

All this leads to the need in the process of analyzing the probability of bankruptcy to emphasize operational events. That is, along with the use of a comprehensive methodology for assessing the probability of bankruptcy, for its earlier detection, to analyze cash gaps in a clear relationship with the analysis of the calculation system, which is presented in the brief review of the concept of the proposed evaluation methodology.

While gaps in solvency should be fixed, firstly, in cases where payment has not been made or has not been fully implemented at any particular time of the creditor's claim for payment of the debt – this is called the identified solvency gap.

Gaps in solvency should be fixed, secondly, when it is necessary to get rid of assets to pay part of the debt (to lose the volume of liquidity of the company) – this is a hidden gap in solvency, because it would mislead our system of analysis regarding the ability of the company to repay debts.

This approach, of course, necessitates using complex methodology of bankruptcy risk assessment. Due to the dialectical and open nature of various methods of studying the bankruptcy of companies, this approach should be taken into account as the original and its originality correlates with the economy of companies and the existing reality of interaction between the creditor and the debtor. Moreover, it correlates with the principles of probability distribution of random events, in particular, for example, with the principle of Poisson distribution and with indicative distribution, when it comes to the disturbances of creditors. Relatively, it is possible to assess the probability of occurrence of a certain number of gaps in solvency and their duration, and thus assessing the payment discipline in the company, the practical implementation of which often leads to difficulties with creditors or lack thereof.

Last of all, it should be said that the existing MDA models are effective in a certain sense, as the ratio analysis alone is effective with the factor methods attached to it, when it comes to a company that is obviously experiencing financial problems. For example, the debts of the company exceed the assets several times, and the equity has dried up and formed an uncovered loss and there are almost no current assets. In these cases, or close to them, a large proportion of integrated models sends the company into the category of potential bankrupt. The reverse is the case when the company has virtually no debts - it has a large amount of liquidity and cumulative retained earnings. Both these cases are well and synchronously reflected in most models, but in situations where the company has some of the features from each case, the ratio analysis and integral models begin one after another to refer the company to different classes, confusing the researcher.

References

1. Popova, N.S., Ryabceva, L.V., Stepanov, I.G.: K voprosu ob opredelenii i sushhnosti ponyatij nesostoyatelnosti bankrotstva. Fundamentalnye issledovaniya **7**, 171–174 (2016)
2. Fadejkina, N.V.: O yuridicheskoj i ekonomicheskoj nesostoyatelnosti organizacij i metodah diagnostiki veroyatnosti ih bankrotstva. Sibirskaya finansovaya shkola **107**, 145–151 (2015)
3. Rabinovich, L.M., Fadeeva, E.P.: K voprosu ob ocenke veroyatnosti bankrotstva. Aktualnye problemy ekonomiki i prava **2**, 107–115 (2011)
4. Bobryshev, A.N., Debelyj, R.V.: Metody prognozirovaniya veroyatnosti bankrotstva. Finansovyj vestnik: finansy, nalogi, strahovanie, buhgalterskij uchet **1**, 48–53 (2010)

5. Fedorova, E.A., Fedorov, F.Y., Hrustova, L.E.: Prognozirovanie bankrotstva predpriyatiya na primere otraslej stroitel'stva, promyshlennosti, transporta, sel'skogo hozyajstva i torgovli. Finansy i Kredit **715**, 14–27 (2016)
6. Fedorova, E.A., Gilenko, E.V., Fedorov, F.Y.: Modeli prognozirovaniya bankrotstva rossijskih predpriyatij: osobennosti rossijskih predpriyatij. Problemy prognozirovaniya **2**, 85–92 (2013)
7. Shirinkina, E.V., Valliulina, L.A.: Formalizaciya modeli prognozirovaniya riska nesostoyatel'nosti predpriyatiya. Aktual'nye problemy ekonomiki i prava **4**, 169–180 (2015)
8. Rudnev, M.Y., Rudenko, M.N.: Analiz sovremennyh metodik vyyavleniya priznakov prednamerennogo bankrotstva. Rossijskoe predprinimatelstvo **17**, 2831–2844 (2015)
9. Tkhagapso, R., Kuter, M.: Retrospective analysis of Institute of Bankruptcy in Russia. In: Antipova, T., Rocha, A. (eds.) Digital Science 2018, DSIC18 2018. Advances in Intelligent Systems and Computing, vol. 850, pp. 296–305. Springer, Cham (2019)
10. Kuter, M., Gurskaya, M.: The early practices of financial statements formation in Medieval Italy. Acc. Hist. J. **44**(2), 17–25 (2017)
11. Skeel Jr., D.A., Triantis, G.: Bankruptcy's Uneasy Shift to a Contract Paradigm. Legal Scholarship Repository, Law University of Pennsylvania (2018)
12. Baird, D.G., Casey, A.J., Picker, R.C.: The Bankruptcy Partition. https://www.law.uchicago.edu/news/baird-casey-and-picker-their-working-paper-bankruptcy-partition. Accessed 25 May 2019
13. Chapman and Cutler LLP: Bankruptcy Desk Reference for Equipment Lenders and Lessors. Chapman and Cutler LLP, Chicago (2018)
14. Mears, P.E.: The winds of change intensify over Europe: recent European Union actions firmly embrace the "Rescue and Recovery" culture for business recovery. Pratt's J. Bankr. L. **10**, 349 (2014)
15. Trabelsi, S., He, R., He, L., Kusy, M.: A comparison of Bayesian, hazard, and mixed logit model of bankruptcy prediction. CMS **12**(1), 81–97 (2014)
16. Jones, S., Hensher, D.A.: Predicting firm financial distress: a mixed logit model. Acc. Rev. **79**(4), 1011–1038 (2004)
17. Oskina Yuliya Nikolaevna, Baeva Elena Aleksandrovna Obzor metodik analiza finansovyh rezultatov. Socialno-ekonomicheskie yavleniya i processy, no. 4(050) (2013)

What the Study of the Early Accounting Books in F. Datini's Companies in Avignon Has Given

Dmitry Aleinikov$^{(\boxtimes)}$ (iD)

Economy Department, Kuban State University,
Stavropol'skaya st., 149, 350040 Krasnodar, Russia
west888west@mail.ru

Abstract. The paper is a brief overview of the results of research conducted in recent years in the Archives of Italy, where a large number of accounting books, documents, personal and official letters are collected, allowing to trace the history of the development of the accounting system and its basic methodological techniques. The use of modern digital and information technologies allows scientists to study medieval accounting practices without being for a long time directly in the archives. Thanks to their use by Russian scientists, the method of logical and analytical modeling has been developed and used in research, which contributes to the most complete perception of the features of medieval accounting. This method is aimed at reconstruction and restoration of the lost accounting information, and also allows to model accounting systems, for the purpose of their deeper and detailed study. Special attention in the paper is paid to the peculiarities of the accounting system in the company of Francesco di Marco Datini in Avignon. Features of opening and closing of the enterprises, reflection on accounts of settlements with debtors and creditors, features of accounting of money, an order of identification of financial result and that is the most important, preparation and formation of the first and only known at that time, synthetic balance.

Keywords: History of accounting · Single-entry bookkeeping · Double-entry bookkeeping · Francesco Datini's companies · Avignon

1 Introduction

The application of modern digital science and information technology expands significantly the possibilities to carry out the archival research to study the history of the emergence and development of accounting, both single-entry and based on the principles of double entry. Of course, one cannot deny the role of the first researchers of the late XIX – early XX centuries that created the basics of accounting history as an independent area of economic science. This authority list begins with Fabio Besta and the representatives of his school [1–11]. Undoubtedly, there were scientists, who studied the archives of medieval accounting books before Fabio Besta. However, such studies have acquired scientific approaches including not only the statement of scientific facts, but also generalizations and relevant conclusions for the first time in the Venetian school of science. It is impossible not to mention the role of linguists who

© Springer Nature Switzerland AG 2020
T. Antipova and Á. Rocha (Eds.): DSIC 2019, AISC 1114, pp. 374–386, 2020.
https://doi.org/10.1007/978-3-030-37737-3_33

sought to preserve the medieval heritage through translations of texts. The merits of scientists of the mid-20th century are especially great, they significantly developed the scientific views of their predecessors. It is necessary to take into account American scientists' studies whose views and conclusions are very important.

What is the peculiarity of the modern stage of the research? Digitization of large volumes of primary sources stored in the archives of Italy, the creation of significant in volume digitized databases on modern compact media, the development of modern methods for the presentation of archival documents allow to carry out more in-depth studies of previously studied documents and, most importantly, to subject significant amounts of archival data which remained previously without due attention to detailed investigation and synthesis. Among such objects of research are the archives of the medieval merchant Francesco Datini which are stored in the state archive of the city of Prato near Florence. Datini's accounting books are a prime example of origin and development of both single- and double-entry bookkeeping.

The purpose of this article is to show the results of the research carried out in the archives of Datini by the scientists of Kuban State University with the direct involvement of the author.

2   Review of Prior Literature

As it has been found, not a single Russian (pre-revolutionary, 1917) scholar had worked with the medieval accounting books from the archives of Italy and other countries [9, 10]. The attempts to study and describe accounting books took place exactly a century ago, and this is connected with the name of an authoritative Russian scientist – Professor Sivers [11]. Initially Sivers E.E. demonstrated a negative attitude to archival research. He believed that very few trading books survived, their findings are random; each finding gives an idea of specific moments in the history of the origin and development of accounting but would never let you get a picture of this development in all the proper fullness.

Such an opinion of the authoritative scientist is understandable as the number of known publications on this issue was extremely small. The time of studying the medieval archives of accounting books not only had failed to reach Russia, but had not yet come on a full scale in Europe [12].

By that time, the archival monuments of accounting history had been little known. In 1865, Desimoni who was the Director of the Archives of Genoa, discovered an accounting book dating back to 1340. This is a book of massari (financiers) in the municipality of the commune city. In 1906–1907, Professor H. Sieveking studied this book and published articles about it for the first time. Later on, for many years, the Genoese medieval book was considered to be the earliest example of the application in accounting not only the double entry method, but double entry bookkeeping.

In the late eighties of the XIX century, the founder of historical accounting science Fabio Besta began the era of scientific archival research; his first publications on the study of the archives of Venice appeared.

Why did one of the most respected scientists of his time choose the archive of the city of Venice? First of all, Besta headed the department at the University of Venice.

Secondly, although Besta did not like Pacioli and accused him of all earthly sins including plagiarism, at the same time, he believed that if the «father of accounting» described in the treatise the "Venetian model", therefore, bookkeeping originated in Venice. That is why he was looking for early examples of the origin of double-entry bookkeeping in the archival books of Venetian merchants. A bit later after studying a book by the massari of the municipality of the commune city of Genoa (1340), F. Besta changed his mind in favor of Genoa.

Besta's main and fundamental work [1, 2] was published in 1909. It consisted of three volumes. The third volume is of most interest for our study. It covers the period of the accounting formation starting from the XIII century. In the monographic study, several chapters were devoted to the origin of the terms «debit» and «credit», to the emergence and development of single- and double-entry, early Opening and Trial balances, posted to the General Ledger (balances in the book of Barbarigo since 1430, had been perceived as the earliest), reverse records and accumulations. Opening balances were formed by the reverse side rule, that is, debit account balances were placed in the balance on the right side and credit balances were on the left side. Opening of accounts in the General Ledger was carried out not by the direct method, but in compliance with the requirements of double entry: debit balance in the balance sheet was credit balance on the opened account, credit in the balance sheet was debit balance on the opened account.

The teacher's views were shared by his disciples Vittorio Alfieri, Carlo Ghidiglia, Pietro Rigobon. All of them were unanimous about the idea that double-entry bookkeeping in 1340 originated in Genoa and then spread to Florence, Venice and Milan, which were well-known by their developed trading activities.

And only one of his disciples, the youngest, a native of Florence, Ceccherelli [3] held a different opinion. It was he who dealt a crushing blow to the generally accepted hypothesis questioning the Genoese origin of double-entry bookkeeping. He put forward the Tuscan hypothesis making Florence a rival to Genoa.

The interest in the archival research spread to other European countries.

Perhaps this was the reason for the Russian Professor Sivers as well to revise his attitude to the study of archival accounting books; in 1915, he wrote about «not in vain past years in finding trading books as invaluable sources of bookkeeping». [11, pp. 28–29].

In the article mentioned earlier, E.E. Sivers noted that many accounting books were found in the archives of the cities of Bruges, Venice, Genoa, Naples and others allowing to make judgement on bookkeeping (practical accounting), which had existed both before the treatise by L. Pacioli, and some time after the publication of this work.

According to Sivers, the largest number of such books had been preserved in the archives of Italian cities. In some countries (for example, in Germany), on the contrary, there are no such archives, and if some documents have been preserved, they are purely family in nature.

It is appropriate to mention one important remark of the Russian Professor. He pays attention to the fact that some of the discovered books have already been published and this allows you to get acquainted with them without a personal visit to the archives in which these sources are stored.

Thus, there are three possible options for conducting research on the medieval accounting books stored in the archives:

- directly from the archival documents;
- studying the descriptions of the sources made by other researchers;
- studying the researches published by other authors.

Sivers carried out a profound study of the historical sources stored in the archives of Ragusa. In it, Sivers expresses a wish so that Russian scientists "familiar with the Latin language", would familiarize the Russians with "the found rich materials of values" in more detail [11, p. 31]. The scientist died in 1917. Unfortunately, this only became possible in a century.

The above material allows you to recognize Sivers E.E. to be the first Russian researcher of the medieval accounting books.

Studying the Russian school of accounting history is unthinkable without mentioning three names – Professor O.O. Bauer, Professor S.F. Ivanov and Professor A.M. Galagan.

The first, ethnic Baltic (Latvian) German, is known as one of the three scientists (O.O. Bauer, E.G. Waldenberg, A.M. Wolf) who in 1893 responded to the idea of preparing the Russian translation of the text of «the father of accounting» to the 400th anniversary of the publication of the Pacioli's treatise (1894).

The name Bauer is also associated with the publication of the book in 1911 which collected translations of early accounting treatises.

By the way, if Sivers [11] did not immediately discern the need and expediency of the search and study of the medieval trading books (since at that time just single copies of the archival documents were known and it did not seem possible to recreate a holistic picture of the origin and development of accounting according to them), then, according to Bauer, in 1911 in Europe, especially in Italy, full-scale studies of «monuments of sacred antiquity» were carried out, and the accounting books belonging to individuals and institutions in the 13th–15th centuries survived to the present day, can serve as a further example.

O.O. Bauer concludes his arguments with the words which are fully relevant to the present day when some theorists and historians of accounting have formed an opinion that the solution to all the problems is already over: I consider it a duty to mention another fact that I am far from thinking of giving my humble work exhaustive significance and I will be fully awarded if it encourages at least one colleague to discover new sources in the history of accounting which we have so far poorly developed.

Galagan should be recognized to be one of the most respected researchers in the history of accounting in Russia. Galagan focuses on the research of Professor A. Ceccherelli [3, 12].

Galagan comes to the same conclusion reached previously by Sivers E.E. and O.O. Bauer. He claims: "A particularly rich collection of the accounting books from the early Middle Ages was found in Italy, as expected; the studies of prominent historians of accounting (Prof. Besta, Prof. Alferi, Prof. Rigobon, Prof. Ceccherelli) have found that in Italy, long before the appearance of the first printed bookkeeping work, many trading enterprises kept their accounting books according to all the rules of the counting art" [12, p. 56].

The journal «Accounting History» published the article written by the scientists of Kuban State University: «Alexander Galagan – Russian Titan of the Enlightenment in the History of Accounting» [10]. Its authors not only have given a profound analysis of the works of the great Russian scientist, a native of the Ukraine, but they have also revealed some previously unknown facts from the biography of one of the brightest educators of accounting history and theory to readers. In particular, for the first time, a reliable description of the last days of his life is given. For us, Galagan A.M. remains the last Russian and Soviet researcher of the history of accounting.

Further research cannot be carried out without explaining the influence of the US scientist R. de Roover's publication on the Russian accounting science [13]. In 1956, the collective monograph «Studies in the History of Accounting» was published in Great Britain and the USA edited by A.Ch. Littleton and B.S. Yamey. This book views the stages of the development of accounting from the Ancient World to the end of the XIX century prepared by leading scientists of the planet at that time.

A special place in the book is devoted to the Middle Ages, that de Roover's publication «The development of accounting prior to Luca Pacioli according to the account-books of Medieval merchants» [11, p. 114–174] was about. It was this chapter that A. Mukhin translated into the Russian language. In 1958, this chapter was published by «Gosfinizdat» as a separate brochure. Its volume was 67 pages and the circulation was 1000 copies.

In 1996, the capital work of Sokolov was published [7]. The monograph has become the reference book for many Russian scientists and practitioners. At the same time, Sokolov, although he highly valued the archival research of the medieval accounting books, had no direct relation to them.

Many authors wrote about the archives of Francesco Datini both in the XIX and XX centuries. The first references to the enterprises and companies of Francesco di Marco are found in the works of Besta [2, p. 317–320] and his disciples G. Corsani and A. Ceccherelli.

We should also note the publications of the authors who did not focus on accounting issues, but described the history of Datini's business or the history of the archive itself. These include Bellini, Carradori, Guasti, Livi, Nicastro and Bensa. Of course, the most in-depth studies were carried out in the second half of the 20th century (Melis [14], de Roover [13], Martinelli [15]).

Many authors continue to write about Datini's archives today as well, though, mostly about the history of the creation of the archive, its content, the personality of Datini and his life path.

However, the comparison of the results of our research with literary sources shows incompleteness and unfinished nature of earlier studies as regards double-entry bookkeeping, so single-entry.

Thus, it can be stated that neither in the XIX nor XX centuries did Russian scientists work with the archival accounting books in general, and Francesco Datini's archives have not been examined in detail neither by Western scientists, primarily Italian, nor Russian scientists even more so.

3   Statement of Basic Materials

In February 2007, the Russian scientists began a real study of the archives of the medieval accounting books in Italy for the first time. A great help in organizing these works was rendered by an authoritative Italian scientist Professor from the University of Parma Giuseppe Galassi.

Since that moment, two electronic databases have been formed at the department:

- a library of books by foreign researchers devoted to the history of the origin and development of accounting;
- a digital archive of the photocopies of accounts from early accounting books.

Great help in collecting the books for the library has been provided by the Italian scientists Galassi G, Sargiacomo M, Servalli S, Ciambotti M.; by French scientists J.-G. Degos, Y. Lemarchand, J. Richard; by English scientists St. Walker, R. Macve, K. Hoskin, A. Sangster, G. Stoner, B.S. Yamey; by American scientists G. Prevec, R. Becker and others.

We have managed to collect and translate more than 200 publications, the number of works by R. de Roover exceeded three dozen.

While electronic archiving, more than 60 thousand bookkeeping accounts have been digitized from the saved early accounting books which are stored in the archives of Venice, Genoa, Florence, Prato. Some photocopies were digitized at the Central Library of Florence. Many records on the accounts of the XIV century imprinted on the photocopies have been translated into English and Russian. Using photocopies of real accounts, schemes of accounting procedures were constructed.

At the same time, two major areas of research have emerged:

- the development of single-entry;
- the origin and development of double-entry bookkeeping.

The first area is based on the study of the accounting system of companies in Avignon (1363–1410) and is the subject of this article.

The information base for the second area was the archival materials relating to the accounting systems in Pisa (1382–1406) and Barcelona (1385–1410). Here we have managed to find the accounting tools and accounting procedures which no researcher previously wrote about. These include: early fully retained «Profits on merchandise» accounts, a prototype of the «Profits and Losses» account [16, 17]; early internal Trial balance performing both the control function and the function of closing the old General Ledger and opening the new one [17, 18]; early financial interim and final reporting [17]; periodic impairment accrual of long-term assets (as opposed to the Western scientists' statement about depreciation) [19–21]; early examples of the formation and use of reserves from profit [22]; early corrective entries (on «Profits on merchandise» [23]; «Accounts of household expenses» were examined in detail [24]; and so on.

In the first company of Francesco Datini in Barcelona (1393–1395), the Russian scientists discovered the simultaneous preparation of an internal Trial Balance and external reporting [17, 18]. In the second company (1395–1397), the accountant, for some reason, refused to carry out the internal balance having been content with the external balance. In the third company, an accountant named Simone Belandi first accrued the depreciation on office equipment by means of distributing the value of long-term property over useful periods: 10 percent per annum [19, 20].

Let us dwell in more detail on the studies of the development of single-entry according to the archival material of the accounting system in Avignon.

As we all know, medieval accounting began with the establishment of real amounts of receivables and payables. However, back to 1300, in the company of Giovanni Farolfi one calculated the financial result in the accounts of double-entry bookkeeping, and during 1304–1306 this was done in Alberti's company as part of single-entry. In the companies of Francesco Datini, the information about the settlements took place only at the First Company in Avignon (1363–1365). But in the next company, in the base of accounting was a financial result.

It was found that in the first companies in Avignon, companies closed irregularly. This happened for two reasons: change in ownership or the need to withdraw profits.

Functioning of the enterprises and companies of Francesco Datini in Avignon can be divided into periods for several reasons: before returning to Italy and after that returning (1363–1382; 1382–1410); by organizational forms of business entities (companies – 1363–1373, Datini's proprietorships – 1373–1382, companies – 1382–1410); accounting methods (on the mingled accounts – 1363–1398, on the Venetian accounts – 1398–1410). The Table 1 below sets the dates and time of the companies functioning at the first stage. As you can see, the first eight companies had not only different owners, but they also did have not the same time of activities significantly different in terms. From 03/21/1373 to 11/30/1382 (3571 days), Datini's individual enterprise had been successfully operating.

Before leaving for Tuscany, Datini closed his business and started a company that lasted from December, 1, 1382 to December, 31, 1385 (1136 days).

All subsequent companies even with permanent owners closed every year, as a rule, on December, 31 and reopened on January, 1.

Table 1. The dates and time of F. Datini's companies functioning in Avignon at the first stage

№.	The date		Number of days
	of Opening	of Closing	
1.	13.07.1363	31.12.1364	565
2.	17.01.1365	31.12.1366	540
3.	01.01.1367	24.10.1367	311
4.	25.10.1367	17.09.1368	357
5.	18.09.1368	31.12.1369	468
6.	01.01.1370	28.02.1371	424
7.	01.03.1371	31.12.1372	306
8.	01.01.1373	20.03.1373	79
9.	21.03.1373	30.11.1382	3571

On January, 1, 1386, the principle of a constant accounting cycle started working (equal to the calendar year - 365 days, from January, 1 to December, 31). This principle had been valid for 25 years and extended to 25 companies.

However, the performance of the companies should be given a separate assessment.

The matter is that the principle of the constant accounting cycle does not correspond, as it is accepted today, with the principle of an existing enterprise. In fact, the principle of enterprise liquidation is present. True, the liquidation was purely legal, formal. The company continued to operate at the closing date (December, 31), and the liquidation was pure conditional, the purpose of which was the withdrawal of profits under the revived Roman law requirements – profits could be consumed for remuneration only after the closure of an enterprise and payment of debts.

The bookkeeping in the accounts of owners did not differ from the bookkeeping in the accounts of ordinary debtors or creditors of the company. Moreover, a debit balance could have been formed on them.

The Tuscan form accounts – mingled account, or 'Paragraph' were used in the company in Avignon. The upper part of the account was assigned to the debit on current accounts with debtors and credit records were kept at the bottom. Current accounts with creditors were kept in the opposite order.

In the article which is coming out at present [25], we argue that current accounts were opened twice for each counterparty (separately for each type of the debt). This contradicts the literary sources where it was argued that the opening of accounts was carried out for each counterparty (debtor or creditor). Wherein accounts receivable were placed in the initial part of General Ledger and payables were placed in the second one. The study refutes the statement of reputable scientists, in particular, de Roover's statement [12] that the smallest balance sheet transfer to the account with an excess surplus for calculating the final result was carried out when General Ledger closed. Proven by examples such procedures took place periodically (but after filling out the Ledger page in full), as a rule, at the accountant's discretion, this allowed to see the real state of calculations. It is important to note that the identified closing balance was transferred to a new account in the same part of General Ledger and served as the basis for continuing calculations and later on the opposite side of General Ledger, a new account was also opened to record the calculations of the opposite polarity. The surplus calculated on the account being closed, was not in all cases recorded on the account to continue the first record but could be in any part of it.

Most regrettably, until now, not a single book «Entrata e Uscita» of Francesco Datini's company in Avignon from 1363 to 1411 was not the subject of a comprehensive and detailed study by not a single researcher.

The first Ledger to record monetary funds appeared at the Datini's company in Avignon in 1367. In 2014, it became the subject of study by the Russian researchers [26]. The technique used by the first accountants of the company turned out to be a high-level technique, similar to those that would be used in Pisa and other Datini's companies two decades later.

As a rule, these were two separate sections «Entrata» and «Uscita» which could be placed in the «Entrata e Uscita» book or separately, but together they provided a single cash flow accounting mechanism. The technique of intermediate cash balance withdrawing was demonstrated in the accounting books. The results of cash income and outgoing were displayed periodically for each section of the accounting book. The result in «Uscita» was transferred to «Entrata» and was recorded after a similar total for the same period. Cash balance was calculated which served as the first record of a new period in «Entrata», and accounting in «Uscita» starts from scratch. Noteworthy is the provision according to which cash outflow records registered at the Memoriale, were individually transferred to «Uscita», whereas the accumulated data for days or for several days combining several pages of the Memoriale, were transferred to «Entrata».

As for the Datini's company in Pisa in 1410, it seems that 11 years after the introduction of the first Ledger «Entrata e Uscita», and in connection with the change of persons responsible for accounting, the organization of cash accounting in the Ledger somewhat changed and not for the better. The cash balance amount withdrawal (f. 1316 s. 9 d. 7) in the «Entrata e Uscita» book, as they would say today, does not look transparent, and what is not clear at all: the amount of cash balance in the analytical notebook (f. 212 s. 10) was far from the amount shown in the cash book. Most likely, the information about Francesco's death and the transition of the company to a new jurisdiction (according to Datini's will) made themselves felt.

Of course, the most interesting is the system of the formation of the financial result [27]. At Datini's companies since 1367, stocktaking had been used for these purposes for goods for sale, inventory, cash, food stocks, and receivables, on the one hand, and on the other hand, invested capital, accumulated (undistributed) profits and debt obligations. The first amount was called «They must give», the second amount was called «They must have». The difference between the first and second amounts represented the financial result. The Prato archive stores document folders (analytical notebooks – Quaderni di Ragionamento) as separate non-stitched sheets on which the inventory results are shown line by line for each group of homogeneous objects. The results are calculated for each page (for control purposes) and for each section. Debit entries come first, then an empty page follows and credit records come next. After calculating the results according to "They must give" and "They must have", they are transferred to the free page which, as if, personifies the «Account of financial result». Since 1398, to account for invested and reinvested capital, accounts receivable and payable (including managers' wage arrears) they had begun to use accounts of the Venetian form and double-entry which were conducted in the Ledger and Secret Ledger.

In 1410, after the death of Francesco Datini, the person responsible for accounting Tieri di Benci prepared a synthetic or condensed balance on a separate sheet of paper. It consisted of two sections «Debtors» ("They must give") and "Creditors" ("They must have"). The first section is not in doubt, it has the same five indicators and it is fully consistent with the Quaderno di Ragionamento [27]. Melis described the synthetic balance as follows: "The condensed balance hides the capital between the creditors of the company, as it follows from this very unique copy, which is probably the earliest example of the discovered ones so far (if talking about the degree of syntheses). The 'Analytic notebook' of the same company has also survived, i.e. the analytical budget where even 11 creditors are listed, which make up the only entry (with profit) in liabilities side, does not single out that part of Credit which is owed to "*corpo*" (initial investment) and "*sovraccorpo*" (reinvestment capital, formed by the shareholders of the company at the expense of the profit which was meant for the liquidity replenishment)" [13, p. 162].

Indeed, in the second section of the Synthetic or Condensed balance, there are only 2 indicators:

- Profits of a year ended December 31, 1410;
- 11 creditors, in Ledger and Secret book counting Goods and creditors of this enterprise and profits; all in all.

Today, after six centuries, it is difficult to find out what motives prompted Tieri di Benci to take such actions: either the thirst for search in order to improve and simplify financial statements or hiding attempts. Territorial distance from the Main Office and excessive trust, especially the death of the main owner, could provoke any action. For us today, important is the fact that for the first time in history, analytical (elementary) indicators were grouped by homogeneous signs into synthetic indicators.

4 Conclusion

The main purpose of this article is to show the results of the research carried out in the archives of Datini by the scientists of Kuban State University with the direct involvement of the author. Here we have managed to find the accounting tools and accounting procedures which no researcher previously wrote about. These include: early fully retained «Profits on merchandise» accounts, a prototype of the «Profits and Losses» account; early internal Trial balance performing both the control function and the function of closing the old General Ledger and opening the new one; early financial interim and final reporting; periodic impairment accrual of long-term assets (as opposed to the Western scientists' statement about depreciation); early examples of the formation and use of reserves from profit; early corrective entries (on «Profits on merchandise»; «Accounts of household expenses» were examined in detail; and so on.

In the first company of Francesco Datini in Barcelona (1393–1395), the Russian scientists discovered the simultaneous preparation of an internal Trial Balance and external reporting. In the second company (1395–1397), the accountant, for some reason, refused to carry out the internal balance having been content with the external balance. In the third company, an accountant named Simone Belandi first accrued the depreciation on office equipment by means of distributing the value of long-term property over useful periods: 10% per annum.

The functioning of the enterprises and companies of Francesco Datini in Avignon can be characterized as follows.

On January, 1, 1386, the principle of a constant accounting cycle started working (equal to the calendar year - 365 days, from January, 1 to December, 31). This principle had been valid for 25 years and extended to 25 companies. True, the liquidation was purely legal, formal. The company continued to operate at the closing date (December, 31), and the liquidation was pure conditional, The bookkeeping in the accounts of owners did not differ from the bookkeeping in the accounts of ordinary debtors or creditors of the company. Moreover, a debit balance could have been formed on them.

We argue that current accounts were opened twice for each counterparty (separately for each type of the debt). This contradicts the literary sources where it was argued that the opening of accounts was carried out for each counterparty (debtor or creditor). The study refutes the statement of reputable scientists that the smallest balance sheet transfer to the account with an excess surplus for calculating the final result was carried out when General Ledger closed. Proven by examples such procedures took place periodically (but after filling out the Ledger page in full), as a rule, at the accountant's discretion, this allowed to see the real state of calculations.

The first Ledger to record monetary funds appeared at the Datini's company in Avignon in 1367. The technique used by the first accountants of the company turned out to be a high-level technique, similar to those that would be used in Pisa and other Datini's companies two decades later. The technique of intermediate cash balance withdrawing was demonstrated in the accounting books.

Of course, the most interesting is the system of the formation of the financial result. At Datini's companies since 1367, stocktaking had been used for these purposes for goods for sale, inventory, cash, food stocks, and receivables, on the one hand, and on the other hand, invested capital, accumulated (undistributed) profits and debt obligations. The first amount was called «They must give», the second amount was called «They must have». The difference between the first and second amounts represented the financial result. Since 1398, to account for invested and reinvested capital, accounts receivable and payable (including managers' wage arrears) they had begun to use accounts of the Venetian form and double-entry which were conducted in the Ledger and Secret Ledger.

In 1410, after the death of Francesco Datini, the person responsible for accounting Tieri di Benci prepared a synthetic or condensed balance on a separate sheet of paper. It consisted of two sections «Debtors» ("They must give") and "Creditors" ("They must have"). The Balance Sheet included totally 7 indicates: the first section included 5 indicators and the second one – the other two. Today, after six centuries, it is difficult to find out what motives prompted Tieri di Benci to take such actions: either the thirst for search in order to improve and simplify financial statements or hiding attempts. For us today, important is the fact that for the first time in history, analytical (elementary) indicators were grouped by homogeneous signs into synthetic indicators.

Our future action will be intended to the clarification of controversial provisions of this study, firstly, the that related to preparing of synthetic balance sheet.

The second goal pursued by the Russian scientists is to revive the interest in the archival research of medieval accounting books. As the results show, Francesco Datini's archives represent a storehouse of secrets and answers of the history of the development of single- and double-entry bookkeeping. Other archives of Italy are of no less interest.

References

1. Sargiacomo, M., Servalli, S., Andrei, P.: Fabio Besta: accounting thinker and accounting history pioneer. Account. Hist. Rev. **22**(3), 249–267 (2012)
2. Besta, F.: La Ragioneria. 3, 2nd edn. Facsimile Reprint, Rirea (2007)

3. Antonelli, V., Sargiacomo, M.: Alberto Ceccherelli (1885–1958): pioneer in the history of accounting practice and leader in international dissemination. Account. Hist. Rev. **25**(2), 121–144 (2015)
4. Basu, S., Waymire, G.: The Economic Value of DEBITS = CREDITS (2018). https://ssrn.com/abstract=3093303
5. Sangster, A.: The genesis of double entry bookkeeping. Account. Rev. **91**(1), 299–315 (2016)
6. Stefano, A., Alexander, D., Fasiello, R.: Time and accounting in the middle ages: an Italian-based analysis. Account. Hist. (2019). https://doi.org/10.1177/1032373219833140
7. Ryabova, M.: The account books of the Soranzo Fraterna (Venice 1406–1434) and their place in the history of bookkeeping. Account. Hist. J. **45**(1), 1–27 (2018)
8. Datini Francesco di Marco: The Man the Merchantedited by Giampiero Nigro, Fondazione Istituto internazionale di storia economica F. Datini Prato. Firenze University Press, Firenze (2010)
9. Sokolov, Ya.V.: Bukhgalterskii uchet: ot istokov do nashikh dnei (Accounting: From the Beginning to the Present Day), Audit, YuNITI, Moscow (1996)
10. Kuter, M., Gurskaya, M., Kuznetsov, A.: Alexander Galagan – Russian Titan of the enlightenment in the history of accounting. Account. Hist. **24**(2), 293–316 (2019)
11. Sivers, E.E.: Gorodskoy arkhiv v Raguze i ego znachenie dlya kommercheskogo obrazovaniya (City archive in Ragusa and its significance for commercial education). Commer. Educ. **7**, 28–31 (1915)
12. Galagan, A.M.: Schetovodstvo v ego istoricheskom razvitii (Accountkeeping in Its Historical Development). Gosizdat, Moscow (1927)
13. De Roover, R.: The development of accounting prior to Luca Pacioli according to the account-books of medieval merchants. In: Littleton, A.C., Yamey, B.S. (eds.) Studies in the History of Accounting, London, pp. 114–174 (1956)
14. Melis, F.: Aspetti della vita economica medievale (studi nell'archivio Datini di Prato). Monte dei Paschi di Siena, Siena (1962)
15. Martinelli, A.: The origination and evolution of double entry bookkeeping to 1440. ProQuest Dissertations & Theses Global pg. n/a 1974. Melis, F. Storia della Ragioneria. Cesare Zuffi, Bologna (1950)
16. Sangster, A.: The determination of profit in medieval times. In: Antipova, T., Rocha, Á. (eds.) Information Technology Science, MOSITS 2017, Advances in Intelligent Systems and Computing, vol. 724, pp. 215–224. Springer, Cham (2017)
17. Kuter, M., Gurskaya, M., Andreenkova, A., Bagdasaryan, R.: The early practices of financial statements formation in medieval Italy. Account. Hist. J. **44**(2), 17–25 (2017)
18. Gurskaya, M., Kuter, M., Bagdasaryan, R.: The structure of the trial balance. In: Antipova, T. (eds.) Integrated Science in Digital Age, ICIS 2019. Lecture Notes in Networks and Systems, vol. 78, pp. 103–116. Springer, Cham (2020)
19. Gurskaya, M., Kuter, M., Musaelyan, A., Andreenkova, A.: Specific features of depreciation accounting at the end of the 12th–Early 13th centuries. In: 5th International Conference on Accounting, Auditing, and Taxation (ICAAT 2016) (2016). https://doi.org/10.2991/icaat-16.2016.10
20. Kuter, M., Gurskaya, M., Bagdasaryan, R., Andreenkova, A.: Depreciation accounting in Francesco Datini's companies. In: 5th International Conference on Accounting, Auditing, and Taxation (ICAAT 2016) (2016). https://doi.org/10.2991/icaat-16.2016.26
21. Kuter, M., Gurskaya, M., Andreenkova, A., Bagdasaryan, R.: Asset impairment and depreciation before the 15th century. Account. Hist. J. **45**(1), 29–44 (2018)
22. Kuter, M., Sangster, A., Gurskaya, M.: The formation and use of a profit reserve at the end of the fourteenth century. Account. Hist. (2019). https://doi.org/10.1177/1032373219870316

23. Kuter, M., Gurskaya, M., Bagdasaryan, R.: The correction of double entry bookkeeping errors in the late 14th century. Contabilità e cultura aziendale – Account. Cult. **XIX**(1), 7–30 (2019)
24. Kuter, M., Gurskaya, M.: Accounts of household expenses in the medieval companies. In: Antipova, T., Rocha, A. (eds.) DSIC 2018, AISC, vol. 850, pp. 286–295 (2019)
25. Aleinikov, D., Kuter, M., Musaelyan, A.: The owners' accounts in the early Datini companies. In: St. Petersburg University Economic Papers: to the Anniversary of Economic Science at the University: Proceedings of the Third International Economic Symposium, pp. 354–370. St. Petersburg University, St. Petersburg (2018)
26. Aleinikov, D., Kuter, M., Musaelyan, A.: The early cash account books. In: Antipova, T., Rocha, Á. (eds.) Information Technology Science, MOSITS 2017. Advances in Intelligent Systems and Computing, vol. 724, pp. 195–207. Springer, Cham (2017)
27. Gurskaya, M., Aleinikov, D., Kuter, M.: The early practice of analytical balances formation in F. Datini's companies in Avignon. In: Antipova, T. (ed.) ICIS 2019, LNNS, vol. 78, pp. 91–102. Springer (2020)

Generating Risk-Based Financial Reporting

Irina Demina(✉) ⓘ and Elena Dombrovskaya(✉) ⓘ

Financial University under the Government of the Russian Federation,
Leningradsky Avenue 49, 125993 Moscow, Russia
demina_id@mail.ru, den242@mail.ru

Abstract. An unstable economic environment requires more transparent data on financial reports, especially with regard to risk disclosures. Nevertheless, a lot of companies build their risk management systems for records and, as a consequence, they come to be ineffective. Currently, neither national, nor international laws or standards contain clear requirements for a risk-based system to maintain accounting records and generate financial reports. When preparing financial reports, a company must take into consideration any and all existing and potential risks that may affect both its financial statements and decision-making by users of these financial statements and disclose such risks and their impact on its reporting or statements to disclose complete and reliable information about its financial position and bottom lines. The risk-based approach implies that financial reports should be generated in compliance with all requirements of international financial reporting standards, while the greatest focus is made on individual items with the maximum risk exposure. Therefore, the standardized usage of a risk-based financial reporting system (FRS) can be viewed as a cornerstone of transparent and most informative reporting data. The paper discusses both theoretical and methodological provisions about how companies can employ the risk-based approach to generate their financial reports.

Keywords: Financial reporting · Risks · Accounting policies · Internal control

1 Introduction

In the market environment, the significance of financial statements depends on their forecasting component. A situation of political instability and economic volatility implies a greater risk, namely, the risk of default and unforeseen losses, with more deceived investors and shareholders who failed to timely assess the real financial standing, threats or risks of a business, which became bankrupt as a result of intentional data misstatement for the benefit of its owners, inefficient resource management or external negative factors. A worse business performance often results from ignoring economic, social and legal consequences of risk.

Account analysis can be used as a tool for monitoring certain types of risks, such as credit, currency, or liquidity risk, although it should not be the only tool as every business is exposed to many risks and their impact is not attributed to corporate processes. Risk management tools applied under a corporate risk management policy

© Springer Nature Switzerland AG 2020
T. Antipova and Á. Rocha (Eds.): DSIC 2019, AISC 1114, pp. 387–399, 2020.
https://doi.org/10.1007/978-3-030-37737-3_34

should be disclosed on financial reporting along with detailed information about any risks identified.

Despite multiple scientific interpretations, widely studied nature, meaning and characteristics of risk, the economic literature provides no unambiguous understanding of what risk is so far, because it is quite complex and versatile and is used to denote a variety of economic concepts.

Its scientific definitions can be grouped based on similar understandings of the concept of risk. The first group includes definitions proposed by those researchers whose understanding of risk comes down to its identification with various kinds of uncertainty, actions in circumstances of uncertainty and its quantitative measurements, including formulations of Cantillon [1] and Knight [2]. The disadvantage of this approach is that the definition of risk fails to cover a potential outcome of the impact the risk may have on a business, whether it is an additional profit or additional losses incurred. Within this approach, Knight's opinion that risk is a quantitative measurement of uncertainty is of importance, since uncertainty with regard to patterns of relationships or future events can only be stated as a phenomenon, while risk helps quantify the magnitude of the expected threat by assessing its probability and potential damages it may involve.

The second group includes those who argue that risk is an opportunity to profit (Keynes, Pigou). When defining the concept of risk, they emphasize that the risk-taking should be viewed as a source of additional income, because profits would include exposure fees [3, 4]. The disadvantage of this approach is that the definition of risk only touches a positive aspect of the risk-taking and fails to reflect the economic nature of risk and possible negative consequences it may bring.

There are researchers who believe that risk is a cause of incurred damage, additional losses and all kinds of failures. They include Knight [2] and others. Representatives of this tradition, which can be also referred to as the classic tradition, treated economic risk as merely a potential damage that may occur as a result of economic action and a quantitative expression of estimated losses, but they did not take into consideration that risk could also cause additional income.

The above approaches are argued to be limited, because those researchers choose to only focus on a certain risk dimension, and their definitions fail to embrace the integral content of the concept of risk. Representatives of the next group propose the most comprehensive definitions, which contain the above characteristics of risk. Their interpretations cover uncertainty in decision-making, threatening losses and wish to make extra money [5].

It is possible to carry to number of researches of the early reporting [6–8].

Therefore, the concept of risk is always associated with a few factors, firstly, including danger and uncertainty due to unforeseen circumstances whatever, secondly, the expectation of some losses due to the impact this risk may have and possibilities to assess the event probability, and, finally, the risk is always associated with the subject and decisions (choices) the subject makes in the hope of a better outcome and additional income.

There are no uniform standards for risk management. Each company will develop its risk management system on an individual basis depending on the risks it has identified, their significance and degree of impact they may have on its business performance. There are several approaches to risk management, including risk mitigation through reducing the likelihood of an undesirable event or reducing damages it may cause; the risk-taking with financial losses minimized through risk insurance or by virtue of additional external sources; risk avoidance, or having less transactions associated with significant risks; and risk assignment to another company that can take advantage of those risks to generate income.

An effective risk management will essentially involve a comprehensive approach, including compliance with all requirements of accounting and financial reporting standards, risk management policies, an array of tools to manage each type of risk, adequate reserve accounting, and risk disclosures on financial reporting.

2 Basic Approaches to Risk-Based Financial Reporting

2.1 Role of Accounting as a Risk Monitoring Tool

Risk monitoring is a comprehensive procedure carried out by businesses at all stages of their growth throughout the entire period of their operations; it is applied at all stages of risk management and is the backbone of any risk-based financial reporting system a company may employ. A corporate risk monitoring system should be developed based on both qualitative and quantitative risk assessment techniques and should provide data about the dynamics and nature of any identified risk to classify risks depending on their nature and potential implications in order to eliminate the cause of such implications and determine possible countermeasures.

There is a variety of risks and factors that may maximize risk of misstatement (e.g., the changing business environment or unstable industry in which a company operates, amendments in the related regulatory and legal frameworks, senior management changes, personal interest of managers, excessive dependence on the parent or customer company, etc.), thats why companies should elaborate a comprehensive list of tools to be employed in their risk monitoring process.

Each stage of the monitoring process should involve a certain list of tools, while they are to be modified and combined depending on business operations, legal form, life cycle and financial standing, previous risk management experience, etc. Some essential tools used at the basic stages of the monitoring process are shown in Table 1 below.

Table 1. Risk monitoring tools

Types of risk monitoring tools	Characteristics of risk monitoring tools
1. Risk identification	1. External environment screenings, review of key contractors and regulatory frameworks and identification of risk signs and symptoms
2. Impact evaluation	2. Application of generally accepted assessment models depending on the type of risk under given conditions
3. Response measures	3. Development of a risk management strategy, including risk avoidance, assignment, reduction, and acceptance

Risk monitoring tools vary greatly and business entities should use multiple tools that are most likely to identify risk in order to do profound and comprehensive risk monitoring. In present-day circumstances, it is not enough to be guided by accounting data, although its significance in financial risk monitoring should not be underestimated. Table 2 describes kinds of financial risks, which should be assessed on the basis of accounting data.

Table 2. Features of accounting data for financial risk monitoring

Kinds of financial risks	Accounting data to be used in financial risk monitoring
1. Credit risk	1a. Total receivables 1b. Dynamics of receivables repayment 1c. Overdue debt time 1d. Provisions for bad debts
2. Currency risk	2a. Currency transactions, if any
3. Liquidity risk	3a. Amount of free cash 3b. Other current assets 3c. Current liabilities

Table 2 shows that the credit risk should be identified and assessed with the use of accounting data about the number of debtors, the total amount and dynamics of accounts receivable (in particular, in comparison with the amount and dynamics of accounts payable) to identify overdue receivables and delay periods and to further assess debtors' solvency, including provision for bad debt, to manage this type of risk. In addition, accounting data should be used to assess liquidity risk, namely, the amount of free cash and short-term assets on the balance sheet at each reporting date. Accounting data analysis in terms of unusual transactions and untimely statements is also a good tool option for monitoring risk of intentional misstatement.

At the same time, a business is exposed to many risks, which are not attributed to internal processes. Effects of such risks can only be explained by external factors, such as changes in the market or political environment. These risks cannot be evaluated through the use of accounting data, and international rankings or indices and analysis of external information will be the most suitable tool options [9].

Financial risk monitoring should include an array of tools to identify credit, operational, market, and liquidity risks. These risks will be monitored through both qualitative and quantitative analysis, including assessments of risk impact and absolute losses in certain circumstances. Financial risk monitoring is one of the most complex procedures within the risk-based financial reporting approach, because comprehensive and successful risk identification, assessment and management must cover a lot of factors, for example, a market risk assessment should take into consideration changes in any market parameters, such as interest rates, exchange rates, or stock prices. A company may be less exposed to market risk if it has no financial instruments among its assets, although it cannot level out the effects of exchange rate volatility as a manifestation of market risk. In the real sector of the economy, it is precisely market risks that have a greater impact on income/expenses from core and other business operations, profits, cash flows, and many items on a statement of financial position, therefore, evaluation of such impact is of importance. A market risk may be evaluated through a lot of well-known techniques (e.g., VaR, Shortfall, or analytical approaches) or with the use of a combined model [9].

The timely identification and effective management are essential in credit risk monitoring, because credit risk is closely associated with risk of a borrower's defaulting and large financial losses for the debtor. In order to identify an impact this risk may potentially have; companies must continuously analyze the financial standing of its debtors for possible debt collection. Credit risk monitoring is especially relevant for companies operating in industries with a high percentage of receivables among assets. In order to receive a debt timely and minimize bad debts, companies may deploy a receivables management system and implement related monitoring procedures as a tool for the credit risk leveling. Financial statements should include details of major debtors and provide their reliability assessment and delay-based receivables classification [10].

Monitoring of operational risk as part of financial risk monitoring involves assessments of risk of business disruption due to technical or informational defaults or failures, human errors, or external circumstances. Absolute losses associated with this kind of risk may exceed those from any other risks.

Liquidity risk is the risk that a company may be unable to operate in future. Liquidity risk monitoring consists in setting and continuously maintenance of adequate liquid assets that will be sufficient to pay off obligations in due time, both under normal and stressful conditions of operation. The scenario approach is thought to be necessarily used in order to maintain plenty of liquid assets, while widely used and well-known models (Altman, Taffler, or Liss models) can be applied to assess own creditworthiness.

Financial reporting is also a good financial risk monitoring tool. The importance of reliable and complete disclosure of risk information on financial statements is explained by the purpose users of such statements pursue, which is to draw a complete picture of a company's financial position and financial results. This requires disclosure of indicators and explanatory notes of potentially significant business risks the company is exposed to. These disclosures are essentials of any internal system for monitoring economic events.

In addition to general information about risk exposure, management policies, monitoring goals and tools, and responsible business departments, every business entity should show on their financial statements the following details for each type of risk for the main stages of monitoring:

- tools used for risk identification;
  - tools used for impact estimation;
- impact estimation results and forecast values of the financial indicators, which may be affected by a particular risk;
- a selected risk management strategy;
  - risk management tools used within the selected strategy;
- risk management results and evaluation of the selected strategy.

Additional indicators and risk explanations can be made out as a separate explanatory section to financial statements or otherwise by including them in explanatory notes for corresponding indicators of corporate reporting on individual assets, liabilities, income, expenses, or cash flows. A company should also prepare a separate risk disclosure report and incorporate it in its financial statements by reference, provided that it is available for review by all reporting users.

Risk monitoring is required for making timely management decisions to ensure undisturbed operations and uninterrupted growth for every business. Users are able to assess whether a company is responsible and reliable to the extent that the risk monitoring process and its outcomes are stated on the company's financial reporting.

2.2   IFRS Requirements to Risk-Based Financial Reporting Systems

When generating financial reporting a company should take into consideration all existing and potential risks that may affect both financial statements and the decision made by reporting users in order to disclose complete and reliable information about its financial position at the reporting date and its financial results for the reporting period. Financial statements must disclose any existing risks and describe effects they may produce. In other words, a company should employ a risk-based financial reporting system.

The movement towards standardized and unified principles of accounting and financial reporting is becoming increasingly significant, therefore, this suggests that the regulation of risk disclosures by every business entity will also be widely discussed in the nearest future, and an IFRS risk-based financial reporting system is thought to be the best method to disclose risk information so that it meets concerns of multinational companies, the global economy, and financial markets.

The risk-based approach to financial reporting involves the preparation of reports and statements in compliance with all IFRS requirements with the greatest focus on individual components of financial statements, which are exposed to maximum risk. In this case, we are talking about the risk of misstatement and data distortion due to the influence of certain circumstances. A risk-based system also implies the most comprehensive disclosures of all existing and potential exposures, which, if known, may affect the decision-making by users of such reporting.

International financial reporting standards do not contain any specific requirements for risk-based financial reporting systems, nor do they provide definitions and descriptions of this concept, although they regulate this matter in an indirect manner.

IFRSs set forth no step-by-step procedure for risk identification at the stage of financial reporting. Every company should, for its own advantage, do qualitative and quantitative assessments of those risks that may significantly affect some important reporting indicators depending on its business operations and industry sector.

The Conceptual Framework for Financial Reporting stipulates that when preparing financial reporting any company must be guided by the fundamental assumption of uninterrupted business operations. This can only be accepted if all existing and potential risks are identified and reliably assessed. Therefore, the Conceptual Framework for Financial Reporting requires use of a risk-based financial reporting system to comply with the fundamental principle.

Individual IFRSs require disclosure of information about risks, which may affect both reporting indicators and the decision-making by reporting users, together with related risk management and mitigation policies. The standards also often govern how a company's assets or liabilities must be reported so that their evaluation takes into account multiple risk impacts as widely as possible. See Table 3 below for the list of regulatory IFRSs.

Table 3. IFRSs regulating individual components of risk-based financial reporting

IFRS	Regulations
1. IFRS 1 First-Time Adoption of International Financial Reporting Standards	1. Recognition of the valuation allowance for expected losses
2. IFRS 7 Financial Instruments: Disclosures	2. Disclosure of risk information and risk classification, including credit, market and liquidity risks
3. IFRS 13 Fair Value Measurement	3. Statements of assets and liabilities on financial reporting with their valuation, which would be a market valuation
4. IFRS 16 Leases	4. Tenant's exposure to lease risks
5. IFRS 17 Insurance Contracts	5. Measurements of a group of insurance contracts at initial recognition to cover impacts of both financial and non-financial risks
6. IAS 1 Presentation of Financial Statements	6. Disclosure of financial risk management goals and policies
7. IAS 34 Interim Financial Reporting	7. Other information, including a risk report, on financial statements
8. IAS 36 Impairment of Assets	8. An unbiased asset reporting, taking into account the impact of a potential impairment
9. IAS 37 Provisions, Contingent Liabilities and Contingent Assets	9. Creation of resources

IFRS 1 First-Time Adoption of International Financial Reporting Standards stipulates that a company shall recognize a provision for expected credit losses for any financial asset, lease receivable, loan contract or loan commitments and financial guarantee contract, to which the impairment requirements apply. In addition, at the date of transition to IFRS, a company must determine the difference between credit risk at the date of initial recognition of its financial instruments (or at the date on which the company became a party to a contractual obligation, without the right to cancel it in the future, in the case of loan commitments and financial guarantee contracts) and credit risk at the date of its transition to IFRS.

IFRS 7 Financial Instruments: Disclosures contains a requirement to disclose risks together with risk classification, including but not limited to, credit, market and liquidity risks, as well as to disclose information about what procedures are used by a company for risk management and assessments.

Application of IFRS 13 Fair Value Measurement implies that assets and liabilities are stated on a company's financial reporting with their valuation, which would be a market valuation. Therefore, when measuring the fair value of its liabilities, a company shall consider risk of default and a variety of factors that could affect its amount.

IFRS 16 Leases requires a lessor to disclose on its financial reporting additional information regarding the nature of the lease so that reporting users could understand the exposure to various risks arising from the lease.

IFRS 17 Insurance Contracts regulates that an entity must measure a group of insurance contracts upon initial recognition to include financial risks associated with future cash flows and risk adjustment for non financial risk.

IAS 1 Presentation of Financial Statements regulates disclosure of non-financial information, e.g., a company's financial risk management goals and policies, in the notes to its financial statements.

Paragraph 16A of IAS 34 Interim Financial Reporting says that, in order to disclose additional information, an entity may incorporate into its interim financial reporting a cross-reference to another report or statement, including a risk report, which is available to users of financial reporting under the same terms and at the same times as its interim financial reporting.

IAS 36 Impairment of Assets also regulates the application of the risk-based approach to financial reporting in an indirect manner since there are many external and internal factors that can significantly reduce the estimated value of assets. In this case, a company should take into consideration the impairment risk and state the value of its assets in light of the effect it may produce, without bias.

IAS 37 Provisions, Contingent Liabilities and Contingent Assets regulates the procedure for the creation and reporting of a reserve, as well as the procedure for its value measurements, taking into account the number of possible outcomes and the likelihood of their occurrence. In assessing the value of such a reserve, a company will take into account both existing risks and uncertainties, and future events that may affect the amount of the created reserve.

Accounting policies of major companies often include a section that discloses approaches to assessing existing and potential risks, evaluation of their impact on components of financial reporting, and describing related management strategies to improve the quality of financial information provided. An accounting policy should also regulate goals and objectives of risk management policies, and the results accompany intends to achieve.

A company must carry out continuous risk identification and assessment with risk classification depending on the nature of risks and possible impacts. The primary objective of the risk-based approach in doing business and reporting is to take into account those risks that have been identified and assessed in business planning, elaboration of a strategy, and management decision-making. It also seems advisable that financial reporting should disclose impacts of potential risks through sensitivity analysis, inter alia, in order to explain future decisions that would mitigate potential negative effects [10], while, in order to reflect effects of external forces beyond the control of a company, its reporting may be generated with the application of the scenario-based approach reflecting impacts of external risks that can be taken into account and evaluated.

The question of whether to apply the risk-based approach to financial reporting is the most urgent at the moment in light of tough economic time, which manifests itself in the aggravation of contradictions in the economic interests of countries, mutual economic and political sanctions, and a decline in profitability of many businesses in general.

It is worth noting that even if a company discloses information about all risks it faces in compliance with all requirements of international standards, this does not guarantee a complete picture of its business operations at the reporting date and in the foreseeable future.

Therefore, international standards regulate the risk-based financial reporting system de jure and inadequately for the most unbiased and reliable disclosure of information so that the goal of financial reporting could be achieved. We can state that risk-based financial reporting systems are currently applied to a greater or lesser extent at the discretion of the management of reporting companies and following auditor's recommendations.

2.3   Gist, Principles and Objectives of Risk-Based Financial Reporting

Along with the assessment of the current situation and business performance, the value of financial statements lies in its forecasting component, the main area of which is providing information on potential risks that may have implications for a company's business and financial reporting [11].

The main requirement for financial reporting is that it must be reliable, while its fundamental quality characteristics are comprehensibility, relevance, reliability, and comparability. The risk-based approach to financial reporting will make it possible to bring financial statements as close as possible to the requirements, since statement of risk impacts on accounting and financial reporting increases the reliability of presented information by showing the real value of a company's assets, liabilities and future prospects, as well as helping users recreate the full picture of the internal and external environment that affects the company's business. Moreover, this approach, if applied by all companies, promotes the comparability of all statements, since financial reporting generated as part of a risk-based system may not represent a company in the best way compared to other companies that fail to employ a risk-based approach [12].

We propose that risk-based financial reporting should be financial reporting generated in light of risk assessment and showing consequences of risk impacts and related countermeasures. Internal regulatory documents of each company do contain information on risk management, but not every company file statements and reports that are risk-based, since the risk-based approach to financial reporting requires compliance with certain principles.

The fundamental principle is a sharper focus on risk monitoring throughout the operational cycle. Assessment of the risk affecting business operations is based on the information available to the company, while it is obtained, analyzed and processed on an ongoing basis. At each stage of risk assessment, a company should use newly emerging information, make adjustments, update estimated indicators and a benign level, because this will determine not only the final decision on whether to recognize a risk-related business event, but also the reliability of indicators on accounting reporting. A company's objective in terms of risk management is not to reject all business processes that are risk-exposed, but to properly manage them and make economically sound decisions, while taking into account the credible and reliable statement of such exposures on its financial reporting and in reliance on the risk-based approach [13]. In other words, it should not avoid risk in accounting, but it should be able to assess it, set forth some financial security limits and apply risk mitigation strategies. A common mistake of companies is that management processes are applied upon the assessment of damage caused by a risk, which was identified at the final stage of a certain business process, although with an effective management system within a risk-based system possible impact of this risk would have been evaluated prior to any actually occurred damage, and management processes would have been applied in due time in such a way as to minimize or prevent financial damage.

The next principle of the risk-based approach is fair financial statements, namely, excluding asset overstatement or capital or liabilities understatement, to meet the reliability requirements. In order to comply with this principle, companies need to measure their assets at fair value, check them for any impairment and state them with the value that is current at the reporting date, and justify their statements accordingly.

One of the most common and effective risk mitigation strategies in accounting is the creation of reserves. The creation of reserves for potential risks is an economic event designed to evenly distribute losses in relation to financial results, regardless of whether risk has been realized or not. Reserves will reduce profits without causing an outflow of funds, while preserving them in an indirect manner. A company should assess whether it needs to create a reserve, not only at the time when it is preparing its financial statements, but also as a result of an ongoing assessment of risks of additional costs. For each reserve created, it should develop a reserve creation procedure, including a related calculation methodology depending on the impact of a particular risk. The creation of reserves directly affects the amount of items on financial reporting. Accordingly, the risks for which reserves are provided have a direct impact on financial reporting. The creation of reserves helps increase the relevance and veracity of statements, which are the fundamental qualitative characteristics of any financial reporting according to the Conceptual Framework for Financial Reporting.

One of the principles of the risk-based approach is a focus on the risk of intentional misstatement. Companies often develop an effective risk management system assessing impacts of both external and internal operational risks, but fail to take into account the fact that financial reporting is at risk of deliberate misstatement due to achievement of certain managerial goals, including profit targets set by the parent company or industry average profitability indicators to attract investment and maintain a good reputation, or high solvency indicators to receive loans, or a certain level of other financial reporting indicators to obtain licenses or to comply with legal requirements [14]. The risk of excessive management intervention in the financial reporting process and the risk of dependence of a company's chief accountant on CEO, who gives reporting instructions, for example, to understate profits in order to reduce the tax burden or otherwise maximize profits in order to attract investors, is difficult to identify and evaluate. Nevertheless, it is the above risks that can have a decisive influence on financial reporting and the adherence to the principle of business continuity.

Another important principle of the risk-based approach is that financial statements should disclose information about impacts of all kinds of risks, namely, in the notes thereto. Risks may be disclosed in other documents or reports referred to in financial statements, but should also be an integral part of the statements themselves, since it is financial statements that are the fundamental source of adequate information about the financial and property status of any business entity, and this status cannot be analyzed separately from analysis of existing and forecasted risks [15]. In the notes to financial statements, a company must not only indicate the fact of risk identification, but also describe a possible impact, evaluation methods and processes used to manage the risk identified. The concept of risk is closely related to the concept of damage. Damage resulting in financial loss arises from unaccounted risk and for financial reporting users important information is not the risk itself, but rather potential financial consequences of its impact.

In accordance with the principles of risk-based financial reporting, its primary tasks are as follows:

- a reliable statement of financial position and results on financial reporting as result of the consistent application of the risk-based approach throughout the reporting period; and
- the most complete information provided to users about risk impacts and related financial consequences, including the risk of intentional misstatement.

3 Results

The risk management process applied in companies on a voluntary basis is a rare occurrence, since owners only associate the use of a risk-based system with additional costs, but fail to estimate the promising benefits it may provide. In this case, users of financial reporting cannot be sure that its generation involved identification, assessment and consideration of any and all existing and potential risks. Each company aims to fully and reliably state its financial position at the reporting date and financial results for the reporting period for users to make decisions, including to increase their level of trust, attract new customers or investments. In order to achieve this goal, a company should set itself the task of using a risk-based financial reporting system as widely and efficiently as practicable, and, if unavailable, it should be developed and deployed, even if neither national, nor international standards require so.

The procedure for its employment should be formalized in the relevant section in the order regarding accounting policies to govern risk recognition criteria, risk identification and assessment of risks and their impacts on financial statements, as well as risk management policies. Since any accounting policy defines principles of accounting and financial reporting, the risk-based approach should be the frameworks for the elaboration of accounting policies. Only in this case, financial reporting could fully meet IFRS requirements regarding the completeness and reliability of provided information.

A risk-based financial reporting system is related to a company's internal control system, which should also have a risk-oriented focus. The purpose of any internal audit system for risk-based financial reporting is to monitor whether all provisions of accounting policies and other regulatory documents are met by responsible departments in relation to risks.

4 Conclusion

The heart of financial reporting generated in light of the risk-based approach should be requirements of international financial reporting standards with the greatest focus on its individual components exposed to maximum risk. Therefore, a standardized application of a risk-based financial reporting system is thought to be a cornerstone of financial statements that are transparent, high-quality and most informative for users.

A company needs to implement a single risk based accounting, financial reporting and management system in order to move to a new qualitative level of financial reporting.

A risk-based financial reporting system is a necessary prerequisite for successful operations of a company in future. Its implementation, together with an improved risk management system, will make it possible to correctly rank risks and, as a result, effectively focus management efforts on the most crucial challenges and opportunities. Detailed and consistent disclosure of information about identified risks and the related risk management system on financial reporting will help its users draw informed conclusions about the current financial standing and development prospects of the company, including for the purpose of investment.

Therefore, a risk-based financial reporting system is a necessity of the present day. Market conditions, the processes of globalization and integration of the world economy and the ever-increasing importance of international financial reporting standards have determined the need for the risk-based approach to financial reporting as the only possible way to achieve the reliability, completeness and relevance of the information presented in financial statements.

References

1. Cantillon, R.: Essay on the Nature of Commerce in General. Routledge (2017)
2. Knight, F.H.: Risk, Uncertainty and Profit. Courier Corporation (2012)
3. Keynes, J.M.: The General Theory of Employment, Interest, and Money. Springer, Heidelberg (2018)
4. Pigou, A.: The Economics of Welfare. Routledge (2017)
5. Schumpeter, J.A.: Theory of Economic Development. Routledge (2017)
6. Kuter, M., Gurskaya, M., Andreenkova, A., Bagdasaryan, R.: The early practices of financial statements formation in medieval Italy. Account. Hist. J. **44**(2), 17–25 (2017). https://doi.org/10.2308/aahj-10543
7. Kuter, M., Gurskaya, M., Bagdasaryan, R.: The structure of the trial balance. In: Antipova, T. (ed.) ICIS 2019, LNNS, vol. 78, pp. 103–116. Springer, Switzerland (2020). https://doi.org/10.1007/978-3-030-22493-6_11
8. Gurskaya, M., Aleinikov, D., Kuter, M.: The early practice of analytical balances formation in F. Datini's companies. In: Antipova, T. (ed.) ICIS 2019, LNNS, vol. 78, pp. 91–102 (2020). https://doi.org/10.1007/978-3-030-22493-6_10
9. Adrian, T., Covitz, D., Liang, N.: Financial stability monitoring. Annu. Rev. Financ. Econ. **7**, 357–395 (2015)
10. Minnis, M., Sutherland, A.: Financial statements as monitoring mechanisms: evidence from small commercial loans. J. Account. Res. **1**, 197–233 (2017)

11. Efimova, O., Rozhnova, O.: The corporate reporting development in the digital economy. In: Antipova, T., Rocha, A. (eds.) Digital Science, DSIC 2018. Advances in Intelligent Systems and Computing, vol. 850, pp. 71–80. Springer, Cham (2019)
12. Davidson, R., Dey, A., Smith, A.: Executives "off-the-job" behavior, corporate culture, and financial reporting risk. J. Financ. Econ. 1(117), 5–28 (2015)
13. Sunder, S.: Social Norms, Risk and Financial Reporting (2016)
14. Knechel, W.R., Salterio, S.E.: Auditing: Assurance and Risk. Routledge (2016)
15. Martínez-Ferrero, J., Garcia-Sanchez, I.M., Cuadrado-Ballesteros, B.: Effect of financial reporting quality on sustainability information disclosure. Corp. Soc. Responsib. Environ. Manag. 1(22), 45–64 (2015)

Digital Health Care, Hospitals and Rehabilitation

Microscopic Image Recognition by Medical Multi-agent Systems

Tatiana Istomina, Viktor Istomin, Elena Petrunina$^{(\boxtimes)}$,
Anatoliy Nikolskiy, and Aleksandr Beloglazov

Federal State Budgetary Institution of Higher Inclusive Education
«Moscow State University of Humanities and Economics»,
Losinistrovskaya Str. 49, Moscow 107150, Russia
petruninaelenav@gmail.com

Abstract. The issues of modeling multi-agent systems and their use in the field of biomedical applications are considered.

The classical approaches to the functional classification of autonomous agents are described and the current state of the problem of the development of multi-agent systems (MAS) is analyzed.

The systematization of MAS for a specific subject area, biomedicine, as the most relevant area of their application, is proposed.

The modeling process of MAS is considered using the example of searching for diseased cells in medical microscopic images. The modeling technique was tested in the Matlab environment using test data obtained using a digital microscope with a USB output.

Using additional morphological operations, the procedure for transforming the original image based on the selection of the contour by the Canny detector has been improved, a combined method for selecting contours has been developed, and a flowchart for the recognition of medical microscopic images has been created.

To implement a simulation and analytical description of the behavior of self-organizing AA groups using computational methods, medical MAS was programmatically implemented in the Matlab environment.

A UML component diagram has been developed that describes a software implementation of the behavior modeling of medical AAs.

Based on the analytical description of the behavior of self-organizing groups of medical AA, a simulation was conducted in the Matlab environment.

Studies conducted on testing the results of modeling the behavior of medical AA based on recognized cell boundaries on samples of microscopic images of cell structures obtained and transmitted to a computer using a digital microscope will increase the reliability of predicting the behavior of medical AA in a biological object, as well as the effectiveness of recognizing various pathologies.

Keywords: Multi-agent systems · Biomedical systems · Behavior of autonomous agents · Swarm artificial intelligence · Medical microscopic images

© Springer Nature Switzerland AG 2020
T. Antipova and Á. Rocha (Eds.): DSIC 2019, AISC 1114, pp. 403–416, 2020.
https://doi.org/10.1007/978-3-030-37737-3_35

1 Introduction

To date, several concepts of mathematical description and modeling of multi-agent systems have been developed, known from the works of Wooldridge and Jennings [1, 2], Rao and Georgeff, Subrahmanian, Shoham and Leyton-Brown. In these works, including, industrial tools for implementing multi-agent systems are presented. However, despite this, the issues of their use in the field of biomedical applications have not yet been given due attention.

In works [3–5], technical applications of the theory of multi-agent systems based on the behavior algorithms of multi-agent living systems were considered; publications [6–8] describe approaches to the creation of microsystems for highly efficient applications in healthcare, biotechnology, and ecology.

However, in most known methods and algorithms for describing the behavior of autonomous agents (AA), the characteristics and functioning conditions of elements of robotic systems inside bio-objects are not taken into account. Besides, formalized specialized algorithms for the behavior of groups of micro-robots when they operate in biomedical systems (BMS) have not yet been found, while the use of universal algorithms in such complex systems is usually impossible.

All this necessitates research on the search for means of an adequate description of information models of multi-agent systems and the development of algorithms for computer modeling the behavior of groups of autonomous agents, which will improve the reliability of predicting their behavior. Thus, the research topic is relevant, and the solution of the problem under consideration requires the systematization and modeling of multi-agent systems, as well as a comprehensive study of their use in the field of biomedical applications.

2 Materials and Methods

The concept of swarm artificial intelligence appeared in 1989 and was proposed by the authors G. Beni and J. Wang in Swarm Intelligence in Cellular Robotic Systems [9]. It is based on the idea that to solve some problems it is impossible to build a centralized system, and all information is distributed between its separate elements. In each of these elements, only part of the information is stored, and the full effect is achieved only by using all the data stored in separate information elements, called autonomous agents.

An information model based on groups of autonomous agents (AA) successfully describes the behavior of both natural and man-made systems. To solve such problems, a transition from centralized to distributed management is required, taking into account the number of group members at the micro and nanoscale levels, which can reach hundreds of thousands. Thus, the urgent task is to build information-structural models based on AA groups to describe biological and technogenic systems for various purposes. In particular, BMSs are of particular importance, potentially allowing the treatment of cancer of a wide profile.

In recent years, biomedicine has become one of the most promising areas for the potential use of distributed self-organizing robotic systems. It is assumed that the robotic system, consisting of AA groups in the form of miniature biomedical micro-robots, will be able to solve the complex tasks of remote diagnosis, therapy, and surgery that modern high-tech medicine faces.

The AA in this paper refers to a wide range of concepts, from modified microorganisms to synthesized molecular objects. Since such biomedical systems consist of a large number of micro-robots, it is possible to carry out only simulation modeling that describes only AA functions. Since the procedures for conducting experimental studies of systems consisting of autonomous agents in the process of their design and testing are very laborious, it is justifiable from both a technical and an economic point of view to conduct studies of multi-agent systems using simulation modeling. Since computer methods for simulating and modeling the behavior of biomedical micro-robots are currently still imperfect and insufficiently developed, the results of their application are often inaccurate. Therefore, the development of the theory and technical principles of the construction of MAS, including AA groups, the improvement of their models and methods for modeling their behavior, is relevant.

The concept of creating autonomous agents implies the use of concepts from the field of psychology and sociology that are new for specialists in computer science and artificial intelligence (AI), and, first of all, concepts from the theory of activity and the theory of communication. At the same time, activity and AI are understood as processes that are recursively dependent on each other, which ensures their generation and implementation. AI within a separate AA acts as a subsystem for managing its activities, allowing it to organize and regulate its actions or influence the actions of another AA. At the same time, AI has a communicative nature and is formed in the processes of interaction of this AA with other agents, while the need for communication is associated with the implementation of purposeful activities.

Russell and Norvig [10] understand by an agent any entity that is in a certain environment, perceives it through sensors, receiving data that reflect events occurring in the medium, interprets these data and acts on the medium through effectors. Thus, the authors distinguish four initial agent-forming factors – environment, perception, interpretation, action.

The key characteristics of any AA as "artificial figures" are autonomy and focus. We are talking about the autonomous performance of certain actions based on focused, problem-oriented reasoning. The main signs of AA are decision-making and environmental impact [11].

The corresponding functional classification of autonomous agents (see Fig. 1) can be constructed according to the following two criteria: the degree of development of the internal representation of the external world and the way of behavior.

According to the first sign, cognitive and reactive agents are distinguished. Cognitive agents have a richer representation of the environment than reactive ones. This is achieved due to their knowledge base and solution mechanism. Cognitive agents are characterized by developed appropriate behavior in the agent community, quite independent of other agents. On the other hand, reactive agents, as the name implies, work mainly at the level of stimulus-reactive bonds, having a strong dependence on the external environment (community of agents).

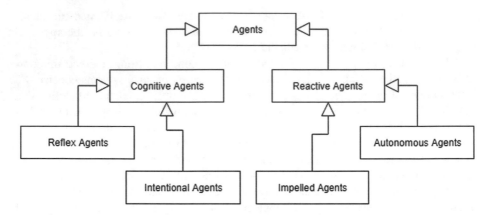

Fig. 1. Functional classification of autonomous agents

In the literature was repeatedly discussed the approach of developing algorithms for the behavior of elements of self-organizing systems based on modeling [12]. Prediction based on a model with results that can be obtained quickly enough is intended to support development at an early stage, before implementation on micro-robots. The search for optimal values can even be carried out using complete sets of possible parameter values. Besides, the development and application of such models can provide a better understanding of the useful effects of processes related to a general understanding of the theory of self-organizing systems and their specific applications.

Prediction of the behavior of self-organizing groups is based on their control algorithm using the mathematical apparatus of the theory of swarm artificial intelligence (SAI). In this paper, we used the results related to the modeling of self-organizing MAS.

SAI describes the collective behavior of a decentralized self-organizing system. It is considered in AI theory as an optimization method.

SAI systems, as a rule, consist of many agents that make up the MAS. Simple agents show signs of self-organization because of joint local interaction between each other and the environment. SAI systems, consisting of self-organizing groups of micro-robots, are nonlinear and can exhibit complex forms of behavior.

The AA system should exhibit the following abilities [13]:

– to learn and develop in the process of interaction with the environment;
– adapt in real-time;
– quickly learn based on a large amount of data;
– step by step adapt new ways to solve problems;
– have a database of examples with the possibility of replenishment;
– have parameters for modeling fast and long memory, age, and so on;
– analyze yourself in terms of behavior, error, and success.

A simple agent program can be mathematically described as an agent function that projects any suitable perceptual result onto an action that the agent can perform, or into a coefficient, feedback element, function or constant that can affect further actions.

The software agent, in contrast, projects the result of perception only onto the action. The term "perception" describes the signals received at the inputs of the agent at any time.

All AA can be divided into five groups, based on the degree of their autonomy and ability to perceive information [14]:

- agents with simple behavior;
- agents with model-based behavior;
- targeted agents;
- practical (rational) agents;
- learning agents.

To solve problems that are difficult or impossible to solve with the help of a single agent or a monolithic system, it is advisable to use MAS – systems formed by several interacting autonomous AAs, independent, trained and adapt to changing circumstances. MASs are related to self-organizing systems striving for an optimal solution to a problem without external interference.

In MAS, agents have several important characteristics [15]:

- autonomy: agents, at least partially, are independent;
- limited representation: none of the agents has an idea about the whole system or the system is too complex for knowledge of it to have practical application for the agent;
- decentralization: no agents are managing the entire system [16].

Based on the analysis of the existing AA and MAS concepts and the development of well-known scientists in the direction of their classification, the authors systematize them for a specific subject area – biomedicine, as the most relevant field of application of MAS. The proposed systematization, taking into account the main features of AA classes for BMS, is presented in Fig. 2. In the first place is the property on which all other classes depend, namely, the reality or virtuality of AA. The second or most important sign is the presence or absence of AA interaction with the external environment. The following AA classification properties that are important for classification are their adaptation and their motivation (goal setting). Note that targeted activity (proactivity) is the ability of agents not only to respond to stimuli coming from the environment but also to carry out targeted behavior, taking the initiative. Besides, it is proposed to distinguish AAs with centralized or distributed types of control, to subdivide them according to their behavior in time into passive (fixed) or active (mobile) ones, and to determine the presence or absence of self-learning properties in agents. This property of AA is directly related to the possibility of their autonomous functioning, which does not require external control.

This systematization makes it possible to attribute the considered MAS to the recognition of microscopic images to specific classes by the type of AA used for BMS and to assign a seven-digit code to it, which is generated based on binary symbols assigned to AA classes.

Many MAS have computer implementations based on step-by-step simulation. MAS components typically interact through a weighted request matrix and a response matrix. Agents can share their knowledge using some special language and obeying the

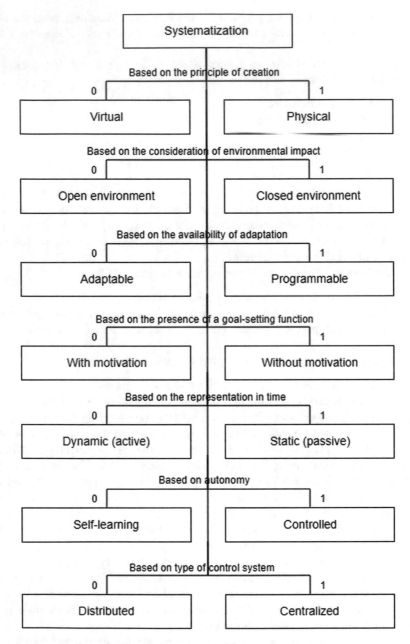

Fig. 2. Systematization of MAS for BMS

established rules of "communication" (protocols) in the system. Examples of such languages are the Knowledge Query Manipulation Language (KQML) and the FIPA's Agent Communication Language (ACL).

Consider the process of modeling MAS using the example of the search for affected (for example, cancer) cells in medical microscopic images. The modeling technique was tested in the Matlab environment using test data obtained using a digital microscope with a USB output. Positive modeling results allowed us to use real cells as objects, which required the refinement of the cell boundary recognition algorithm based on the Canny method using real microphotographs.

Contouring is one of the main methods of image segmentation. Most of the contouring methods and algorithms based on these methods use linear operators with subsequent threshold processing. Some of these operators calculate the first derivative, such as, for example, the Roberts, Sobel operators, or the second derivative (Laplace operator). However, these methods do not satisfy the requirements of continuity and the minimum thickness of contour lines. Gradient methods with simple threshold processing [17] are inferior in the quality of contouring to the method proposed by Canny.

The Canny Boundary Detector [18] includes the following steps. After filtering the noise on the image with a Gaussian filter with a given smoothing parameter σ, a gradient is calculated at each image point, which is characterized by the value of the modulus:

$$g(x, y) = \left[G_x^2 + G_y^2 \right]^{\frac{1}{2}},\qquad(1)$$

and direction:

$$a(x, y) = arctg\left(\frac{G_x}{G_y}\right).\qquad(2)$$

In the resulting array of gradients, analysis and tracking of areas with maximum gradients that form the ridges is performed. In the process of this analysis, points not lying on the ridges are assigned a zero value. A feature of this procedure is that the assignment of a value of zero is performed only if the gradient value at this point does not exceed the magnitude of the intensity jump at two neighboring points in the same direction of the gradient. As a result of this procedure, which is called non-maximum suppression, a thin line is obtained that lies on the crest of the image intensity differences. The resulting image is subjected to threshold processing using two thresholds T1 and T2, with T1 < T2. The pixels of the ridge, the intensity value of which is greater than T2, are called strong, and the pixels whose values fall in the interval [T1, T2] are called weak. The contour formation algorithm ends with a morphological operation, during which weak pixels are added to strong pixels, which are 8-linked to strong pixels.

As shown in [19], smoothing the intensity values increases the noise immunity of the Canny detector, but reduces the accuracy of selecting the contours of image objects. Therefore, the authors of [20] proposed to use wavelet transform, which preserves noise immunity of the methods of contour segmentation of images subjected to smoothing, and to a lesser extent reduces the accuracy of selecting contours of objects to emphasize differences in image intensity. The Pratt criterion was used as an indicator of the noise immunity of the proposed method, and the accuracy of edge detection was evaluated by the proximity indicator between the boundaries of the test ideally segmented image and the image segmented by the proposed processing method [21]. The authors of [22] showed that the positive effect of the proposed wavelet transform is achieved only for special cases when the signal-to-noise ratio in power exceeds a value of 5, and also under the condition that the boundary difference width exceeds 3 pixels. In other cases, the Canny method gives better results than the contour segmentation method, however, it is not accurate enough for medical applications [23].

To develop an algorithm that describes the recognition of medical microscopic images, it is necessary to analyze the expected results of the implementation of this algorithm. In this case, it is supposed to find out whether AA will act as expected from them, whether the boundaries of the cells in which their actions are modeled by the developed algorithm are reliably described. Besides, it is necessary to obtain information on how much the degree of efficiency differs when modeling using an advanced algorithm, compared with the original version of the method. This takes into account some restrictions introduced to take into account the specificity of using microscopic images of medical pathologies.

Based on the consistent use of the Canny detector, as well as on a number of morphological image processing operations, a combined contouring method was developed for recognizing cell boundaries in medical microscopic images for the purpose of subsequent collective search and neutralization of affected cells, intended for use by medical AA groups inside the human body. The block diagram of the recognition algorithm shown in Fig. 3 illustrates the essence of the developed combined method.

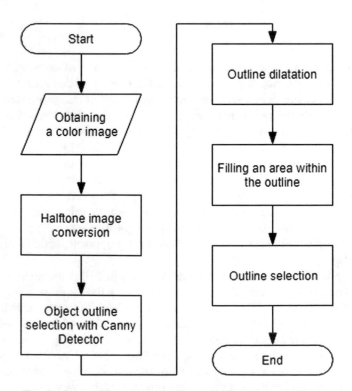

Fig. 3. Recognition algorithm for medical microscopic images

As follows from the diagram, the developed algorithm includes the procedure for converting the original image, in general, the color image, into a grayscale image, the procedure for selecting the contour with a Canny detector and additional morphological operations and procedures, including:

– operation of dilatation of the binary image of the contour obtained by the Canny method;
– the procedure for filling the area inside the circuit;
– contour selection procedure.

It is known that the operation of dilatation (escalation) of a binary image, in this case, the boundary D, by the structural element (primitive) B1 is determined by the ratio:

$$A = D \oplus B_1 = \left\{ z | \left(\hat{B}_1 \right) \cap D \neq 0 \right\}. \tag{3}$$

In the proposed algorithm, it is proposed to use the structure-forming element B1 in the form of a round-shaped primitive with a diameter of 5 pixels.

The procedure for filling an area is based on the sequential performance of operations of dilatation, addition, and the intersection of sets. At the beginning of the procedure, one of the elements inside the contour is assigned a value equal to 1, after which the entire area is filled with units in accordance with the following recurrence formula:

$$X_k = (X_{k-1} \oplus B_2) \cap A^c \quad k = 1, 2, 3 \ldots, \tag{4}$$

where Xk are the internal elements within the contour described by the set A; B2 – symmetric primitive in the form of a cross with a dimension of 3×3 pixels; Ac is the set inverse to A.

The last operation of the procedure for filling the area – the formation of the set C, consists in combining the sets Xk and A, namely, C = Xk $\cup$ A.

The procedure $\beta(C)$ of isolating the contour of the set C describing the binary image consists in performing the erosion operation C according to the primitive B3, and then obtaining the difference set between C and the result of its erosion, i.e.:

$$\beta(C) = C \backslash (C \ominus B_3),$$

where B3 is a square-shaped primitive with a dimension of 3×3 pixels.

Based on the proposed combined method for isolating the contours of microscopic images, the structure of an improved system for modeling medical AA is developed using the algorithm for recognizing medical microscopic images, as well as the relationships, characteristics, and functions of the objects that make up it. The structural implementation is based on the UML class diagram. A UML collaboration diagram has also been developed, which reflects the relationship between different subsystems of the medical AA behavior modeling system and the place of the developed algorithm for recognizing medical microscopic images in it. In this diagram, the researcher uses the described system to simulate the behavior of medical AA during the collective search and neutralization of affected cells. In the form of a diagram of UML activity, the authors present the process of using the algorithm for recognizing medical microscopic images when modeling the processes of collective search and neutralization of affected cells by a group of medical micro-robots. In addition, a sequence diagram has been created that describes the progress of recognition of medical images over time using the UML language. Thus, a package of UML diagrams has been developed that allows testing the possibility of using medical microscopic images of pathologies in a system for modeling the behavior of medical nano robots.

3 Results

To implement a simulation and analytical description of the behavior of self-organizing AA groups using computational methods, the corresponding MAS was programmatically implemented in the Matlab environment. The UML diagram of the components describing the software implementation of the simulation of the behavior of medical AAs is shown in Fig. 4.

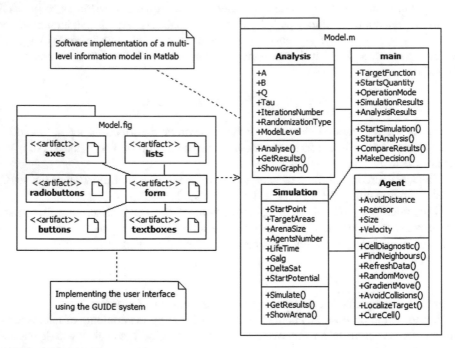

Fig. 4. A UML component diagram describing a software implementation of medical AA behavior modeling

Based on the previously developed model of the behavior of self-organizing groups of medical AA [24], a simulation was conducted in the Matlab environment, the results of which are presented in Fig. 5.

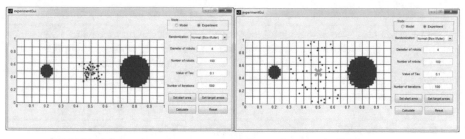

a) start of simulation   b) start of target area localization

Fig. 5. The results of simulation modeling the behavior of AA

The principle of operation is that AAs move along a limited rectangular field on which predefined arbitrary zones have located that act as clusters of affected cells.

In this figure, the target areas are visible, highlighted in red, and the cubic shape AA, whose brightness reflects the value of the potential field. Each AA cannot measure field boundaries and determine its coordinates. Nevertheless, the self-organizing AA group, using the trophallaxis strategy, can collectively localize even complex, scattered and loosely connected areas.

In the experimental part of the work, the efficiency of the pattern recognition process for all test image samples was also evaluated. The probability of image recognition was estimated by the frequency of making the right decisions about whether test images belong to a certain class. The recognition probability score, expressed as a percentage, as determined by the formula:

$$P = \frac{N_r}{N_0} \cdot 100\%,$$

where Nr is the number of positive outcomes when presenting test images in the test, and N0 is the number of presentations of test images in the test.

Tables 1 and 2 presents the recognition probability values when presenting test images in which the contours were distinguished by the Canny method, as well as by the combined method. In particular, Table 1 shows the results of experiments in which images obtained from reference images by changing the scale and rotation were used as test images.

Table 2 shows the results of experiments with test images obtained by adding additive noise with different standard deviation values.

From the results presented in Table 1, it follows that the recognition algorithm, in which the combined method is used to select borders, is more resistant to zooming. Rotating the image equally reduces the likelihood of recognition, both when highlighting the borders using the Canny method, and for the combined method. As the analysis showed, the decrease in the recognition probability during image rotation is due to distortions introduced by the image rotation algorithm in the contour description.

Table 1. Effect of image scale and rotation on recognition probability

Type of test sample	Recognition probability when selecting contours using the Canny method, %	Recognition probability when selecting contours using the combined method, %
Reference images	100%	100%
Image scale factor 0.75	82,5%	95%
Image scale factor 0.5	50%	70%
Rotate images at an angle of 20°	90%	90%
Rotate images at an angle of 30°	90%	90%
Rotate images at an angle of 60°	90%	90%
Rotate images at an angle of 80°	90%	90%

Table 2. The influence of the value of the standard deviation of additive noise on the recognition probability

Noise standard deviation	Recognition probability when selecting contours using the Canny method, %	Recognition probability when selecting contours using the combined method, %
0,036	85,0	90,0
0,062	52,5	87.5
0,073	35,0	85,0
0,086	32,5	80,0
0,101	25,0	67,5

From Table 2 it follows that an increase in the standard deviation of noise in the image reduces the probability of recognition for both algorithms. However, the recognition algorithm, in which the combined method is used to extract boundaries, is more resistant to additive noise.

Figure 6 shows an example of the cell structure used to model the behavior of medical micro-robots based on recognized cell boundaries obtained from a microscopic image and transferred to a computer using a digital microscope.

Studies on the search for means of an adequate description of biological objects, in particular, cells with pathologies and the development of recognition algorithms for medical microscopic images, in the future will improve the accuracy of predicting the behavior of medical AA in a biological object, as well as the recognition efficiency of pathological microstructures.

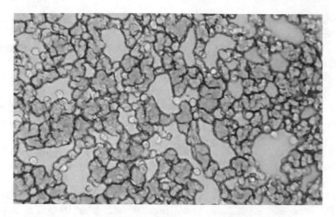

Fig. 6. An example of the cell structure used to model the behavior of medical AA

4   Conclusion

Based on the research, the following conclusions can be drawn:

– the proposed combined method for isolating contours in comparison with the Canny method provides a higher efficiency of recognition of cell boundaries in algorithms based on Fourier descriptors;
– the proposed recognition algorithm can be used to predict the behavior of medical AA systems in a biological object;
– studies conducted to test the results of modeling the behavior of medical AA based on recognized cell boundaries on samples of microscopic images of cell structures obtained and transmitted to a computer using a digital microscope will improve the reliability of predicting the behavior of medical AA in a biological object, as well as recognition efficiency various pathologies.

The scope of this study has broad prospects. Thus, in the framework of the considered task of neutralizing affected cells, further statistical studies of collective localization of cells by agents are planned using the relevant database of microscopic images of real pathologies, as well as with adaptive adjustment of the parameters of simulated micro-robots. Besides, there is areas of bio-objects in which the use of an improved system based on developed algorithms could bring a significant effect.

References

1. Jennings, N.R., Wooldridge, M.: Applications of Agent Technology. Agent Technology: Foundations, Applications Markets. Springer, Berlin (1998)
2. Wooldridge, M., Jennings, N.R.: Intelligent agents: theory practice. Knowl. Eng. Rev. **10**(2), 115–152 (1995)
3. Snooks, R.: Behavioral Formation: Multi-agent Algorithmic Design Strategies, p. 16. RMIT University, Australia (2014)

4. Engelbrecht, A.P.: Particle swarm optimization with crossover: a review and empirical analysis. Artif. Intell. Rev. **45**(2), 131–165 (2016)
5. Werfel, J., Petersen, K., Nagpal, R.: Designing collective behavior in a termite-inspired robot construction team. Science **343**(6172), 754–758 (2014)
6. Sitti, M.: Robotic collectives inspired by biological cells. Nature **567**(7748), 314–315 (2019)
7. Sitti, M.: Mobile Microrobotics. MIT Press, Cambridge (2017)
8. Wang, W., Giltinan, J., Zakharchenko, S., Sitti, M.: Dynamic and programmable self-assembly of micro-rafts at the air-water interface. Sci. Adv. **3**(5), e1602522 (2017)
9. Beni, G., Wang, J.: Swarm Intelligence in Cellular Robotic Systems, Proceed. NATO Advanced Workshop on Robots and Biological Systems, Tuscany, Italy (1989)
10. Russell, S.J., Norvig, P.: Artificial Intelligence: A Modern Approach, 3rd edn. Prentice Hall, Upper Saddle River (2010)
11. Franklin, S., Graesser, A.: Is it an agent, or just a program? A taxonomy for autonomous agents. In: Proceedings of the Third International Workshop on Agent Theories, Architectures, and Languages. Springer (1996)
12. Hamann, H.: Space-Time Continuous Models of Swarm Robotic Systems, Cognitive Systems Monographs, vol. 9. Springer, Heidelberg (2010)
13. Kasabov, N.: Introduction: hybrid intelligent adaptive systems. Int. J. Intell. Syst. **6**, 453–454 (1998)
14. Wooldridge, M.: An Introduction to Multi-agent Systems. Wiley, Hoboken (2002)
15. Jones, M.T.: Artificial Intelligence: A Systems Approach. Infinity Science Press LLC, Hingham (2008)
16. Panait, L., Luke, S.: Cooperative multi-agent learning: the state of the art. Auton. Agent. Multi-Agent Syst. **11**(3), 387–434 (2005)
17. Criffith, A.K.: Edge detection in simple scenes using a priori information. IEEE Trans. Comput. **22**(5), 551–561 (1971)
18. Canny, J.: A computational approach to edge detection. IEEE Trans. Pattern Anal. Mach. Intell. **8**(6), 679–698 (1986)
19. Deriche, R.: Using Canny's criteria to derive a recursively implemented optimal edge detector. Int. J. Comput. Vis. **1**, 167–187 (1987)
20. Gebäck, T., Koumoutsakos, P.: Edge detection in microscopy images using curvelets. BMC Bioinform. **10**, 75 (2009)
21. Lindeberg, T.: Edge detection and ridge detection with automatic scale selection. Int. J. Comput. Vis. **30**(2), 117–154 (1998)
22. Kimmel, R., Bruckstein, A.M.: On regularized Laplacian zero crossings and other optimal edge integrators. Int. J. Comput. Vis. **53**(3), 225–243 (2003)
23. Zhou, P., Ye, W., Wang, Q.: An improved canny algorithm for edge detection. J. Comput. Inf. Syst. **7**(5), 1516–1523 (2011)
24. Istomin, V.V.: Information model of behavior of self-organizing groups of autonomous agents for biomedical systems. Ph.D. thesis. Penza State Technological University, Penza (2013)

System to Support Multidimensional Assessment of Human Functioning

Verónica Pacheco Rocha[1] , Joaquim Sousa Pinto[2] ,
and Nelson Pacheco Rocha[2(✉)]

[1] Siemens Portugal, 4455-491 Porto, Perafita, Portugal
[2] University of Aveiro, 3810-191 Aveiro, Portugal
npr@ua.pt

Abstract. Considering the importance of the assessment of human function, the present paper reports a research study that aimed to develop an information system to support a comprehensive set of clinical instruments. Following the requirements analysis, the system was implementing according the Model-View-Controller architecture. The system allows the selection of the most appropriate assessment instruments for a given patient the registration of the assessments carried out, as well as the access to consolidated information for correct decision-making, and was evaluated using quantitative and qualitative methods. The quantitative evaluation consisted in a usability assessment using the Post-Study System Usability Questionnaire, while the qualitative evaluation consisted in a focus group to determine the perceived utility of the developed system to support the care of older adults. The results show the potential of the system to support the care of older adults.

Keywords: Human functioning · Functioning assessment · Assessment instruments · Older adults

1 Introduction

As a result of the epidemiological transition associated with the increasing of life expectancy, health care, once focused on acute conditions, has evolved to respond to noncommunicable diseases that are the leading causes of mortality and have a strong economic impact [1, 2].

Many of these noncommunicable diseases last for long periods of time, are chronic in nature and progress slowly [3]. Consequently, integrated approaches to care are needed, not only from the clinical point of view, but also considering a range of interventions that are essential for maintaining the quality of life of the patients, often older adults living alone within the limits of autonomy and independence [4]. Thus, in addition to the provision of health care from clinical teams, social care is needed, such as home care services [4, 5].

In this context, quality of life, autonomy and independence of the patients are relevant and, therefore, it is necessary to assess their functioning levels. Human functioning is related to the individuals' ability to perform their daily activities with autonomy and independence, as well as to actively participate in the society, and is

T. Antipova and Á. Rocha (Eds.): DSIC 2019, AISC 1114, pp. 417–428, 2020.
https://doi.org/10.1007/978-3-030-37737-3_36

conditioned by multiple factors [6]. Therefore, for the assessment of the different dimensions of functioning there are diverse clinical instruments.

The objective of the research study reported in this paper was to develop an information system to support multidimensional functioning assessment by recording and analyzing the results of the application of clinical instruments selected by caregivers. The developed system consists of a web platform that supports a comprehensive set of instruments for the assessment of both global and specific functioning dimensions. In order to allow the analysis of consolidated information for correct decision-making, a special emphasis was given to the implementation of a Dashboard that allows the visualization of dynamic graphs to present results from multiple assessments.

The following sections present the framework of this study, the methods, the results in terms of system requirements, system development and system evaluation, and, finally, some conclusions.

2 Background

As a result of the population ageing of the last decades there was a significant increase of chronic conditions, which require continued care in the community.

According to the World Health Organization (WHO), the main types of noncommunicable diseases are cardiovascular diseases (e.g., hypertension, heart attacks or stroke), oncological diseases, chronic respiratory diseases (e.g., chronic obstructive pulmonary disease or asthma), diabetes and mental illness (especially dementia and depression) [3]. As a result, a significant percentage of people over 65 need help with day-to-day activities [3].

The transition from institutional care to community care increases the complexity of services as patients are accompanied by multiple providers from different organizations [4, 7–10].

Thus, informal caregivers and formal caregivers, both health and social caregivers, must work as a team with common goals to provide a coordinated response to patients' needs [11, 12]. However, the organizational and geographical dispersion of the institutions promotes the fragmentation of the care services [4]. In addition, due to the nature of institutions, various factors (e.g., organizational or cultural factors) contribute to what the scientific literature referred as the health and social divide [11] and make it difficult the care integration [13].

The characteristics of the health and social sectors can also explain the technological differences between the institutions of both sectors. Indeed, despite the impressive development in information services to support health care, the information services of social care institutions are usually designed for bureaucratic purposes rather than to support professional practice [14]. Consequently, access and sharing of patients' information becomes difficult [13, 15]. This can result in errors, adverse events, complications and increased burden among care recipients and their supporting networks [8, 9, 16, 17].

From the patients' perspective, a relevant aspect to consider is their quality of life [18]. Regardless the perspective being considered to assess the quality of life of the individuals, a key aspect is their functioning, understood as their ability to perform daily living activities (e.g., eating or personal hygiene) with autonomy and independence, as well as the ability to have an active social participation.

The promotion of quality of life in a phase characterized by the change of body functions involves multidimensional factors [19]. These factors include: (i) physical or mental deficiency; (ii) depression; (iii) chronic pain [20]; (iv) musculoskeletal problems; (v) cognitive problems (e.g., impairment of memory or executive function) [21]; (vi) low frequency of social contacts; (vii) low physical activity; (viii) high or low body mass index; (ix) comorbidity of various diseases; (x) incorrect self-rated health conditions; (xi) incorrect self-assessment of visual function; or (xii) excessive alcohol consumption.

There are several theoretical models that explain human functioning (e.g., the Nagi Disability model [22] or the International Classification of Functioning, Disability and Health [6]). Consequently, for the assessment of the different dimensions of functioning (e.g., physical, mental, emotional or social dimensions), there are diverse clinical instruments such as the World Health Organization Disability Assessment Schedule (WHODAS 2.0) [23], the Barthel Index [24], the Lawton Instrumental Activities of Daily Living (IADL) Scale [25] or the Katz Index of Independence in Activities of Daily Living [26], among others.

The importance of good communication to enable person-centred care must be emphasized. When care provision is integrated, individuals might receive different services from different sectors (e.g., health care and social care) and the collected information must be available and useful to those who have to decide at the personal, clinical, managerial, and public policy levels [9, 27]. In this respect, well-designed information systems to collect and communicate the required information might have a considerable impact.

The research related to health and wellbeing applications using information technologies to support older adults is extremely active. There is a significant number of research studies related to ehealth, ambient assisted living, mobile health, pervasive health, technologies for ageing in place or Internet of Things applications as it has been pointed out by several systematic reviews (e.g., [2, 28–32]). However, scientific reports of applications to support the assessment of human functioning are scarce and focus specific pathologies (e.g., [33–38]). To the best of our knowledge, the first initiative to develop a software platform to allow a cross mural measure of functional independence across home, residential and hospital settings was BelRAI [9]. For that, BelRAI supports multidimensional assessments using the inteRAI [10].

However, interRAI instrument is not disseminated by the Portuguese health care and social care institutions. Moreover, currently, in the Portuguese social care institutions the assessment of the different dimensions of human functioning is mainly paper base or, in some cases, using special purpose spread sheets. Therefore, the aim of the present study was to develop an information system to support multi-dimensional assessment of human functioning considering the requirements of the caregivers of the Portuguese social care institutions.

3 Methods

In terms of development and validation of the information system, the authors followed a user-centered methodology. After a review of the literature, a set of requirements were identified.

To achieve the identified requirements, the system was implemented according the Model-View-Controller (MVC) architecture. The components of this architecture (i.e., Model, View and Controller) are interconnected with each other as well as with a database for information persistence. The Controllers display and control the state of the Views and may additionally update the Models so that these updates are reflected in the Views.

Within the available frameworks that support the MVC architectures, the ASP.NET MVC 6 is one that allows a good management and connection between the visualization and the server and, therefore, was selected for the implementation. In terms of information persistence, a Structured Query Language (SQL) database was used.

The system evaluation included two phases. The first phase consisted in a quantitative evaluation that was performed using a validated usability evaluation instrument, the Portuguese version [39] of Post-Study System Usability Questionnaire (PSSUQ) [40]. In turn, the second phase consisted in a qualitative evaluation performed by a focus group.

The PSSUQ was developed by the International Business Machines (IBM) [40]. It consists of 19 items rated on a seven-point likert scale (from "strongly agree" −1, to "strongly disagree" −7) [40]. The PSSUQ addresses five usability characteristics of a system: (i) rapid completion of the task; (ii) ease of learning; (iii) high quality documentation; (iv) online information; and (v) functional adequacy [40]. According to the authors of the original version, the score of the PSSUQ can be specified by three sub-scores, system utility (SysUse), items 1 to 8, information quality (InfoQual), items 9 to 15, and interface quality (IntQual), items 16 to 18. Finally, the PSSUQ sub-scores are the average of the scores of the respective items (excluding the items the participant fails to respond or that are classified as not applicable), and higher scores indicate lower usability and vice-versa.

For the qualitative evaluation, a focus group was conducted. Focus group is a data collection technique involving a small number of participants in an informal discussion focused on a specific subject [41], including information systems' design [42]. The focus group was composed with formal careers and experts in social care services and the discussion was supported by relevant personas and scenarios that were developed to highlight the problems to be solved [43, 44]. The aim of the focus group was to determine the perceived utility of the information system to support the care of older adults. Participants were asked to give their opinion, among other topics, on whether the information system could promote patient-centered care and efficiency and efficacy of caregivers' activities.

4   Results

4.1   System Requirements

A broad range of non-functional requirement were considered, including ethical and regulatory requirements, quality related requirements, interoperability requirements, compliance requirements, or use and deployment requirements.

Following the applicable regulations (e.g., EU Directive 95/46/EC on the protection of individuals) regard to the processing of personal data and on the movement of such data, several principles were considered to guarantee the privacy of the individuals, namely openness, transparency and inviolability of the information being gathered.

Concerning quality requirements, the following requirements were considered: (i) user experience; (ii) availability, accessibility and integrity of the information; (iii) performance and reliability of the system; and (iv) regulatory compliance. Although quality of user experience is perceived as subjective, it determines the usefulness of the information system [45, 46], which implies that its user interfaces must be customized and personalized, with user-friendly graphics and fonts and must ensure a consistent look-and-feel interaction and navigation across all the functions.

In what concerns interoperability requirements, the information system must: (i) support multiple caregivers collaborating in teams; (ii) respect a multitude of legal, regulatory and organizational policies; (iii) ensure continuity independently of the cross-professional, cross-organizational and cross-jurisdictional nature of the care services; and (iv) accommodate constant changes related to various factors, including technological, legislative or social factors. In this context, standardization plays an important role not only in terms of the technology being used, but also in terms of a common understanding of the collected information, processes, activities and policies. Moreover, international standards must be considered when implementing the interoperability components and these components must be tested against agreed interoperability conformance requirements.

Finally, in terms of use and deployment requirements, a primary goal is to make the system technically feasible and economically viable to conceive, design and deploy. For that, it is essential to guarantee extensibility, reusability, configurability, flexibility, and maintainability.

In what concerns the functional requirements, in a simple way, the following actors and respective uses cases were considered: (i) clinician responsible for the care provision (management of the caregivers and patients, clinical instruments definition, information access, notifications and secondary use of information for planning); and (ii) caregivers (clinical instruments selection, provision of patients' information, information access, and notifications). Moreover, it was considered a priority that the information system must: (i) support a comprehensive set of clinical instruments able to assess both the overall human functioning and the respective physical, mental, emotional and social dimensions; (ii) enable the caregiver to select the most appropriate instruments for multidimensional assessment of the functioning of a particular patient; (iii) record all assessments made to each patient; provide mechanisms to access all the records of each patient; (iv) provide the caregiver with a Dashboard able of integrating the global functioning assessment with the assessment of the respective dimensions;

and (v) allow the caregiver access to all information necessary for decision-making regarding the interventions to be carried out.

4.2   System Development

The architecture of the information system is presented in Fig. 1.

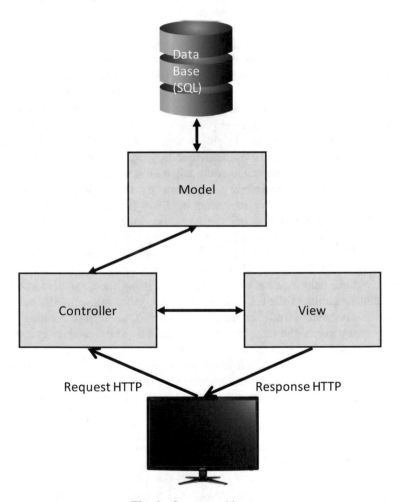

Fig. 1. System architecture.

The Controller of an ASP.NET MVC 6 application uses the C Sharp language to communicate with the Model, select the View to display, process the interaction with the users and query the database. In addition, ASP.NET MVC 6 uses a NuGet package manager that provides a variety of tools and services needed to develop an application [47]. These packages include Bootstrap [48] and jQuery [49]. Bootstrap is a framework that is responsible for developing web application interface and front-end components using HyperText Markup Language (HTML), Cascading Style Sheets (CSS) and JavaScript to ensure that pages are interactive, aesthetically pleasing, easy to use and responsive. In turn, jQuery is a JavaScript library that can be used for transaction management in HTML documents, event handling, and Asynchronous JavaScript and XML (AJAX) animations and interactions.

Regarding database access, this is done through queries according to Language Integrated Query (LINQ), which is a component of ASP.NET MVC 6. Queries range from simple access to a Model that consists of only its primary key to complex data operations that are part of the Views, and their syntax includes extension methods, according either the Enumerable class or the Queryable class, and the lambda expressions.

In order to map the structures of a relational database using an object-oriented programming language, Object-Relational Mapping (ORM) [50] was used. ORM has an advantage over other traditional connection techniques between object-oriented language and relational database, which is the reduction in the amount of code to be written [51]. On the other hand, using ORM implies a very high level of abstraction that can mask what is happening in the implementation code, so special care is needed so that their use does not result in incorrectly designed databases. There are two distinct methods: (i) the Code First method, in which models are thought abstractly and then converted into a relational database; or (ii) the Data First method, where the relational database is first created and data models result from the conversion of relational database tables. To avoid database design problems, the implementation followed the Data First method.

The ORM used for.NET was the Entity Framework 6 (EF6), which allows access to database structures as if they were.NET objects. On the other hand, using ASP.NET Identity User Profiles allows the integration of accounts with individual user authentication in a pre-existing database, as is the case.

Application development can also include extra packages such as FontAwesome, which provides icons that can be used in Views, and Microsoft AspNet Identity for user management and role assignment.

For each specific clinical instrument, a Model was created. Each Model contains the fields required to identify the clinical instrument and to register the patient, the caregiver and the assessments being performed. Optionally, since some clinical instruments have a total score, a score field might also be fulfilled. Additionally, there are integrating Models containing assessment lists associated with various clinical instruments.

All the fields defined for the Models contain the 'set' and 'get' methods in their declaration, which allow elements to be called by the Controller ('get') to be presented by a View and then, after being processed, to be persisted in the database through a 'set'. In turn, there is one Controller for each data Model whose purpose is to manage a set of Views for creating, editing, removing or viewing the respective data.

Finally, Views are responsible for part of the interface and interaction of the application and consist of built-in C Sharp language HTML pages to access data sent by the Controller. In addition, they also contain JavaScript for the management of individual pages.

One of the biggest challenges in developing this platform was the creation of an interactive Dashboard with graphs that show the relationship between the results of the various assessments belonging to a given patient. For this purpose, it was necessary to create a Classified Model containing the instrument and the list of assessments per patient. Using JavaScript, HTML, and built-in C Sharp language, it is possible to aggregate the assessments of the patients considering the diverse clinical instruments. The main objective is to draw graphs that relate the date of the performance of the different assessments with the respective total results, thus allowing to verify if there were significant changes over time.

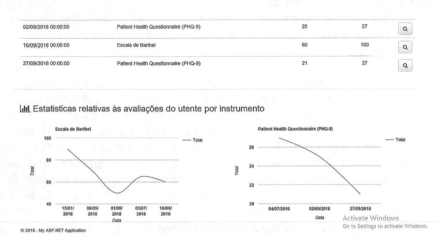

Fig. 2. Patient details.

Figure 2 shows the Dashboard page that uses Google Charts [52] to present a graph by instrument. This View differs from the common standard of other Views in that it serves as an interactive Dashboard that allows a comprehensive view of the assessments made by each patient. Charts update is dynamic as it changes whenever a new assessment is entered. If an instrument with total scores is used in the assessment of the patient, a new chart will be automatically added on the Dashboard. This allows better control and visualization of the results of each patient.

For each of the assessments, check whether the instrument contains a total score. If this condition is met, the different entries are sorted by date and the date of completion and valuation total fields become the x and y axes of the graph, respectively. For instruments that do not have total scores, their results are not presented as graphs, however, the assessments are organized in the list of assessments made by date and can be easily and interactively consulted.

4.3   System Evaluation

For the quantitative assessment using the PSSUQ, the sample consisted of ten participants, being all of them providers of social care institutions located in Aveiro, in the north of Portugal. Most participants were female (nine participants) and their aged vary from 35 to 64 years.

The experimental study took place in May 2019 and the results of the PSSUQ subscales were:

- System Utility (SysUse): average of the responses from items 1 to 8 = 2,41 (SD = 1,33).

- Quality of information (InfoQual): average of the responses from items 9 to 15 = 2,91 (SD = 1,39).
- Interface Quality (IntQual): average of the responses from items 16 to 18 = 2,63 (SD = 1,41).

Looking at the subscales scores of the PSSUQ, the best-performing score was the system utility (SysUse). The second best-performance score was the quality of interface (IntQual) and the worst-performance score of the information quality (InfoQual).

These results seem to suggest that participants consider that the information system might benefit their professional practice. Moreover, the participants were satisfied with the mode the information is presented and the general usability of the system.

In a more detailed analysis, the question that obtained the best result was "I felt comfortable using this system", an item related to the System Utility (SysUse) subscale. Moreover, the question that obtained the worse result was "whenever I made a mistake using the system, I could recover easily and quickly", an item related to the quality of information (InfoQual) subscale.

In turn, the focus group was composed by four elements and took place in May 2019. In terms of analysis, the relevant results of the focus group where extracted and were classified according to their positive or negative aspects.

Considering the positive aspects, the participants were unanimous about the potential of the information system to promote patient-centered care. Moreover, they considered that the Dashboard might be very useful, and the interfaces are well designed. Finally, they considered secondary analysis of the collected information particularly relevant.

In terms of negatives aspects, the participants considered that there are major barriers to the dissemination of the information system, namely the mystification of technological complexity by the caregivesr, concerns about security and privacy issues, and the reluctance that professionals might have in sharing their assessments with other professionals. Moreover, the participants suggest the inclusion of additional instruments and possible secondary information analysis.

5 Conclusion

We shared details of the requirements, technical implementation and evaluation of an infrastructure that was developed to support the assessment of human functioning. The information system presents the social caregivers with several clinical instruments that allow multidimensional assessments of functioning together with an interactive exploration and generation of arbitrary visualizations.

Although the information system that was developed in the context of the study reported by this paper has all the essential components to support the proposed objectives, there is room for improvement and additional functions tailored to the needs of careers. Moreover, a major future work concern is the implementation of a longitudinal pilot both to evaluate the impact of the system in the quality of care provision and to determine good practices of how it can be disseminated.

References

1. Abegunde, D.O., Mathers, C.D., Adam, T., Ortegon, M., Strong, K.: The burden and costs of chronic diseases in low-income and middle-income countries. Lancet **370**(9603), 1929–1938 (2007)
2. Chiarini, G., Ray, P., Akter, S., Masella, C., Ganz, A.: mHealth technologies for chronic diseases and elders: a systematic review. IEEE J. Sel. Areas Commun. **31**(9), 6–18 (2013)
3. World Health Organization: World report on ageing and health. World Health Organization (2015)
4. Santana, S., Dias, A., Souza, E., Rocha, N.: The domiciliary support service in Portugal and the change of paradigm in care provision. Int. J. Integr. Care **7**(1), e01 (2007)
5. Hägglund, M., Scandurra, I., Koch, S.: Studying intersection points–an analysis of information needs for shared homecare of elderly patients. J. Inf. Technol. Healthc. **7**(1), 1–20 (2009)
6. World Health Organization: International classification of functioning, disability and health (ICF). World Health Organization, Geneva (2001)
7. Coleman, E.A., Boult, C.: Improving the quality of transitional care for persons with complex care needs: position statement of the American Geriatrics Society Health Care Systems Committee. J. Am. Geriatr. Soc. **51**(4), 556–557 (2003)
8. LaMantia, M.A., Scheunemann, L.P., Viera, A.J., Busby-Whitehead, J., Hanson, L.C.: Interventions to improve transitional care between nursing homes and hospitals: a systematic review. J. Am. Geriatr. Soc. **58**(4), 777–782 (2010)
9. Vanneste, D., Vermeulen, B., Declercq, A.: Healthcare professionals' acceptance of BelRAI, a web-based system enabling person-centred recording and data sharing across care settings with interRAI instruments: a UTAUT analysis. BMC Med. Inform. Decis. Making **13**(1), 129 (2013)
10. Carpenter, I., Hirdes, J.P.: Using interRAI assessment systems to measure and maintain quality of long-term care. In: A Good Life in Old Age? Monitoring and Improving Quality in Long-Term Care, pp. 93–139. OECD Publishing (2013)
11. Rigby, M.: Integrating health and social care informatics to enable holistic health care. In: Blobel, B. et al. (eds.) pHealth 2012, pp. 41–51. IGI Global (2012)
12. Reamer, F.G.: Social work in a digital age: ethical and risk management challenges. Soc. Work **58**(2), 163–172 (2013)
13. Cameron, A., Lart, R., Bostock, L., Coomber, C.: Factors that promote and hinder joint and integrated working between health and social care services: a review of research literature. Health Soc. Care Community **22**(3), 225–233 (2014)
14. Wastell, D., White, S.: Beyond bureaucracy: emerging trends in social care informatics. Health Inform. J. **20**(3), 213–219 (2014)
15. Spitzer, W.J., Davidson, K.W.: Future trends in health and health care: implications for social work practice in an aging society. Soc. Work Health Care **52**(10), 959–986 (2013)
16. Coleman, E.A., Berenson, R.A.: Lost in transition: challenges and opportunities for improving the quality of transitional care. Ann. Intern. Med. **141**(7), 533–536 (2004)
17. Hirdes, J.P., et al.: Reliability of the interRAI suite of assessment instruments: a 12-country study of an integrated health information system. BMC Health Serv. Res. **8**(1), 277 (2008)
18. Sampaio, R.F., Luz, M.T.: Funcionalidade e incapacidade humana: explorando o escopo da classificação internacional da Organização Mundial da Saúde. Cadernos de Saúde Pública **25**, 475–483 (2009)

19. Alvarelhão, J., et al.: Comparing the content of instruments assessing environmental factors using the international classification of functioning, disability and health. J. Rehabil. Med. **44**(1), 1–6 (2012)
20. Silva, A.G., Alvarelhão, J., Queirós, A., Rocha, N.P.: Pain intensity is associated with self-reported disability for several domains of life in a sample of patients with musculoskeletal pain aged 50 or more. Disabil. Health J. **6**(4), 369–376 (2013)
21. Cruz, V.T., et al.: Web-based cognitive training: patient adherence and intensity of treatment in an outpatient memory clinic. J. Med. Internet Res. **16**(5), e122 (2014)
22. Nagi, S.: Some conceptual issues in disability and rehabilitation. In: Sociology and Rehabilitation. American Sociological Association, Washington (1965)
23. Silva, C., et al.: Adaptation and validation of WHODAS 2.0 in patients with musculoskeletal pain. Revista de saúde publica, **47**(4), 752–758 (2013)
24. Wade, D.T., Collin, C.: The Barthel ADL Index: a standard measure of physical disability? Int. Disabil. Stud. **10**(2), 64–67 (1988)
25. Lawton, M.P., Brody, E.M.: Assessment of older people: self-maintaining and instrumental activities of daily living. Nurs. Res. **19**(3), 278 (1970)
26. Brorsson, B.E.N.G.T., Asberg, K.H.: Katz index of independence in ADL. Reliability and validity in short-term care. Scand. J. Rehabil. Med. **16**(3), 125–132 (1984)
27. Hawes, C., Vladeck, B.C., Morris, J.N., Charles, D.P.: Implementing the resident assessment instrument: case studies of policymaking of long-term care in eight countries. In: Executive Summary. Milbank Memorial Fund, New York (2003)
28. Marcos-Pablos, S., García-Peñalvo, F.J.: Technological ecosystems in care and assistance: a systematic literature review. Sensors **19**(3), 708 (2019)
29. Mutlag, A.A., Ghani, M.K.A., Arunkumar, N.A., Mohamed, M.A., Mohd, O.: Enabling technologies for fog computing in healthcare IoT systems. Future Gener. Comput. Syst. **90**, 62–78 (2019)
30. Skarlatidou, A., Hamilton, A., Vitos, M., Haklay, M.: What do volunteers want from citizen science technologies? A systematic literature review and best practice guidelines. J. Sci. Commun. **18**(1) (2019)
31. Grossi, G., Lanzarotti, R., Napoletano, P., Noceti, N., Odone, F.: Positive technology for elderly well-being: a review. Pattern Recognition Letters (2019 in press)
32. Queirós, A., Pereira, L., Dias, A., Rocha, N.P.: Technologies for ageing in place to support home monitoring of patients with chronic diseases. In: van den Broek, E. et al. (eds.) Proceedings of the 10th International Conference on Health Informatics, HealthInf 2017, pp. 66–76. INSTICC (2017)
33. Dodd, E., Hawting, P., Horton, E., Karunanithi, M., Livingstone, A.: Australian community care experience on the design, development, deployment and evaluation of implementing the smarter safer homes platform. In: Geissbühler, A., Demongeot, J., Mokhtari, M., Abdulrazak, B., Aloulou, H. (eds.) ICOST 2015. LNCS, vol. 9102, pp. 282–286. Springer, Cham (2015). https://doi.org/10.1007/978-3-319-19312-0_23
34. Saffer, B.Y., Klonsky, E.D.: The relationship of self-reported executive functioning to suicide ideation and attempts: findings from a large US-based online sample. Arch. Suicide Res. **21**(4), 577–594 (2017)
35. Trindade, I.A., Ferreira, C., Pinto-Gouveia, J.: Inflammatory bowel disease symptoms and cognitive fusion's impact on psychological health: an 18-month prospective study. Eur. Psychiatry **41**, S355 (2017)
36. Bible, L.J., Casper, K.A., Seifert, J.L., Porter, K.A.: Assessment of self-care and medication adherence in individuals with mental health conditions. J. Am. Pharm. Assoc. **57**(3), S203–S210 (2017)

37. Boessen, A.B., Verwey, R., Duymelinck, S., van Rossum, E.: An online platform to support the network of caregivers of people with dementia. J. Aging Res. (2017). Article ID 3076859

38. Mbwambo, J., et al.: A customized adherence enhancement program combined with long-acting injectable antipsychotic medication (CAE-L) for poorly adherent patients with chronic psychotic disorder in Tanzania: a pilot study methodological report. Heliyon **5**(6), e01763 (2019)

39. Rosa, A., Martins, A., Costa, V., Queiros, A., Silva, A., Rocha, N.: European Portuguese validation of the Post-Study System Usability Questionnaire (PSSUQ). In: 2015 10th Iberian Conference on Information Systems and Technologies (CISTI), pp. 1–5. IEEE (2015)

40. Lewis, J.: Psychometric evaluation of the PSSUQ using data from five years of usability studies. Int. J. Hum. Comput. Interact. **14**, 463–488 (2002)

41. Wilkinson, S.: Focus group. In: Smith, J. (ed.) Qualitative Psychology: A Practical Guide to Research Methods, pp. 199–221. SAGE Publications (2003)

42. Caplan, S.: Using focus group methodology for ergonomic design. Ergonomics **33**, 527–533 (1990)

43. Cooper, A.: The Inmates Are Running the Asylum. Pearson Education, Upper Saddle River (2004)

44. Queirós, A., et al.: ICF inspired personas to improve development for usability and accessibility in ambient assisted living. Proc. Comput. Sci. **27**, 409–418 (2014)

45. Martins, A.I., Queirós, A., Silva, A.G., Rocha, N.P.: Usability evaluation methods: a systematic review. In: Saeed, S. et al. (eds.) Human Factors in Software Development and Design, pp. 250–273. IGI Global (2015)

46. Martins, A.I., Rosa, A.F., Queirós, A., Silva, A., Rocha, N.P.: Definition and validation of the ICF–usability scale. Proc. Comput. Sci. **67**, 132–139 (2015)

47. Rocha, V.P., Pinto, J.S.: Web platform to support multidimensional assessment of human functioning. In: 2019 14th Iberian Conference on Information Systems and Technologies (CISTI), pp. 1–5. IEEE (2019)

48. Databases, documents and collections-w3resource. https://www.w3resource.com/mongodb/databases-documents-collections.php. Accessed 02 May 2019

49. Bhamra, K.: A Comparative Analysis of MongoDB and Cassandra. University of Bergen, Bergen (2017)

50. Sharding—MongoDB Manual. https://docs.mongodb.com/manual/sharding/. Accessed 02 May 2019

51. MongoDB Advantages. https://www.tutorialspoint.com/mongodb/mongodb_advantages.htm. Accessed 02 May 2019

52. Node.js Introduction. https://www.tutorialspoint.com/nodejs/nodejs_introduction.htm. Accessed 02 May 2019

Digital Media

The Adoption of Digital Marketing by SMEs Entrepreneurs

Claudiu Coman⬤, Maria Mădălina Popica,
and Cătălina-Ionela Rezeanu(✉)⬤

Transilvania University of Braşov, Braşov, Romania500030
catalina.rezeanu@unitbv.ro

Abstract. The new communication forms emerging from the digital field have profoundly changed how companies communicate with their stakeholders. If a few years ago, the attention of companies was centred on websites, at present specialist literature highlights how important it is for SMEs to define and to adopt an online marketing mix and to develop interactive applications. Nevertheless, given that Romania is in the last place in the EU regarding the degree of digitisation of the population and that Google develops a series of programmes for rising digital competencies of entrepreneurs, we intend to scan the attitudes and behaviours of the representatives of companies in the SMEs field regarding digital marketing, to offer clues on how professionals in the digital field and marketing could address this market segment more efficiently. In this respect, we selectively present the results of an exploratory research on the basis of opinion polls applied online. The quantitative data were collected in 2019 through an online survey technique based on a standardised online questionnaire applied to 100 Romanian entrepreneurs from the SMEs sector participating in the project Google Digital Workshop.

Keywords: Digital marketing channels · Social media · Internet marketing

1 Background

The entrepreneurial field manifests more and more openness towards the digital medium and technologies. At present, worldwide, digital marketing channels and tools have started being used by SMEs to promote their business. Alongside a massive investment in digital marketing to reach the market, competition has considerably increased [1]. In spite of opportunities offered by digital environment, most digital interactions between consumers and enterprises are mediated by a small number of platforms.

The research of the relation between digital marketing and entrepreneurship has occurred as a consequence of the difficulty of applying traditional marketing in present contexts, both by large companies and medium-sized or small enterprises (SMEs). The main differences between small and medium-sized enterprises relating to large ones regard resource availability (financial resources, human resources, and facilities) and their strong development potential of competencies [2, 3].

© Springer Nature Switzerland AG 2020
T. Antipova and Á. Rocha (Eds.): DSIC 2019, AISC 1114, pp. 431–441, 2020.
https://doi.org/10.1007/978-3-030-37737-3_37

The combination of proper marketing with entrepreneurial marketing contributes to improving the performances of an entrepreneurial company [3]. Within an entrepreneurial context, digital marketing is an essential factor to take into account when explaining the performance of new companies. Digital marketing is no longer a tendency but a necessity, becoming increasingly important for these companies. In the beginning, many entrepreneurs have underestimated digital marketing and others still underestimate it, even if it has a potential to deeply influence the new globalised world of nowadays.

Masterful use of communication technologies has become essential in the business world and over the last decade, Internet offers more and more opportunities and challenges, especially for SMEs. The Internet allows the improvement of marketing efforts of these companies, facilitating: improvement of efficiency; access to new markets, collaboration between enterprises; customisation of products and services; creation of products together with customers; and development of the relations with them [2, 4].

The new communication forms offered by the digital field have completely changed how companies communicate with stakeholders. If a few years ago, the attention of companies was centred on web sites, at present specialist literature highlights how important is for SMEs to define and to adopt an online marketing mix and to develop interactive applications. Here arises the development of social media platforms and their integration in enhancing business and their performances. For instance, the "e-service quality" concept is based on the idea that an online presence of organisations in a dialogue area with a target market has a major effect on the perception about the organisation and on the market's satisfaction towards the organisation [5–8].

Digital marketing techniques and tools are numerous and complex. Of them, Search Engine Optimisation (SEO) is one of the digital marketing pillars, being part of the organic tools, representing the refining process of a web site, using tools both on pages and outside them so that it is successfully indexed and ranked by search engines [9]. Besides a website, companies can use numerous channels and means of online communication, such as: blogs, e-mails, and social media. Social media has become popular in the present marketing mix in general, and specifically in the promotional mix. Marketing on social networks refers to the process of obtaining website traffic or attention by means of social media platforms, but also for communicating to a large community and listening to that community. Even if social media is a strong and convenient marketing strategy, because of the reduced level of awareness and technique implied by this, most entrepreneurs are not very aware of this concept or they do not take it into account. [10]. Social media marketing programmes usually focus on the efforts to create content that draws attention and encourages readers to share it with their own community of persons in the social network [11]. Paid advertisements conveyed by means of Facebook are standard advertisements that occur in the lateral bar of a user, being conveyed on the basis of interests and demographic information. Advertisements of involvement on Facebook are similar to standard advertisements, but they include an element with which fans can engage in their community, such as a Like or Share button, a video clip, an event, or a survey [12].

Google makes many tools available to entrepreneurs, such as channels and means that can render efficient the procedures in an organisation, which can save time, financial, and space resources. Google Ads is the most well-known and it is considered as a standard in the industry; it allows users to trade in a currency of their choice, it is related to a comprehensive analysis tool and it offers training and certification programmes. Also, Google Ads has at present the best contextual and geographical orientation worldwide, although geography is also offered by Bing ads, Facebook ads, LinkedIn ads, and video YouTube ads [12]. The main types of campaigns that can be conducted by means of a Google Ads account targets the search network with an extension to the display network, only the research network, only the display network, shopping, video and campaigns devoted to the mobile similar applications. Google My Business is a free and easy to use instrument for companies and organisations to manage the online presence on Google, including Google Search and Google Maps. By checking and editing information about a company, the platform helps customers to find a campaign and find out information about it.

A recent research [13] shows that SMEs do not use maximum capacity of new digital tools, such as benefitting from the advantages of opportunities that social media offer. Results firmly encourage researchers to examine SMEs to measure the benefits of using social media as a two-directional communication channel. According to the same source, using social media in SMEs marketing, is a challenge, because there are many cases in which SMEs are not capable of creating interesting content on platforms such as Facebook or Twitter. The conclusion of the study is that using websites and social media tools brings benefits to SMEs through an increase in the awareness and promotion of good relations with customers, an increase in the number of new customers, the capacity of reaching customers at an international level and promoting local business that will improve the image of companies.

In what follows, we intend to highlight the importance that representatives of companies in the SMEs field should give to digital marketing, to offer clues on how professionals in the digital field and marketing could more efficiently address this market segment. In this respect, we will selectively present the results of exploratory research on the basis of an opinion poll applied online in 2019 to a sample of 100 representatives of some Romanian companies that are active in the SMEs field and that took part in the Google Digital Workshop consultancy project.

2 Materials and Methods

2.1 Data Collection Method, Technique and Tools

This study is based on cross-sectional data collected and interpreted by the second author of this paper for the bachelor's thesis called: "The Usage of Digital Marketing Tools among Romanian Entrepreneurs" coordinated by the third author of the paper and secondarily analysed by the first author. The quantitative data was collected in May and June 2019 through the online survey technique based on a standardised questionnaire applied to entrepreneurs participating in the project Google Digital Workshop (convenience sample, N = 100 Romanian entrepreneurs from the SMEs sector) using the Google Forms platform.

The decision to choose a method of inquiry based on a standardised questionnaire applied online primarily took into account the specificity of the studied population, the level of availability, the time they can offer to answer questions, and the need of confidentiality and anonymity. We chose the pragmatic character that the online questionnaire offers as we wanted to surpass the limits imposed by geographical boundaries, but also for an increase in the response rate as a result of anonymity and convenience offered by the online medium.

The questionnaire was drawn up on the basis of operationalisation of the concepts of "adoption of digital marketing", "SMEs' profile", and "entrepreneurs' profile". We operationalised the concept of "adoption of digital marketing" by using five dimensions:

(1) Preference – with sub-dimensions:

- General preference;
- Specific preference;
- Reasons for reluctance;
- General benefits;
- Benefits and specificities.

(2) Competencies – with sub-dimensions:

- General knowledge;
- Specific knowledge;
- Usage abilities;
- Attitudes.

(3) Usage –with sub-dimensions:

- General usage;
- Specific usage;
- Frequency of usage.

(4) Strategy – with sub-dimensions:

- Communication plan;
- Human resources;
- Financial resources;
- Time resources;
- Target market;
- Objectives.

(5) Consultancy – with sub-dimensions:

- Previous experience;
- Perceived necessity;
- Frequency of participation in consultancy sessions;
- Future intentions of participation.

The "SMEs' profile" concept was measured with questions about:

- The company, without it being identified by name;
- The region in which the company is located;
- Field of activity (services, industry, or agriculture);
- Field of activity;
- Turnover;
- Number of employees;
- Experience on the market.

The "entrepreneurs' profile" concept was measured with questions about:

- The position of the respondent in the company;
- The gender of the respondent;
- Age category;
- Marital status;
- Education;
- Origin environment;
- The quality of member in local, national, and international associations of businessmen.

The questionnaire comprised 53 items altogether of which 33 items measure "the adopting of digital marketing techniques", 7 "SMEs' profile" items, and 12 "entrepreneurs' profile" items; the last item aiming at an invitation to make personal observations regarding the study. The tool also comprised an "informed consent", in which we have presented to potential respondents the aim of the research, the development period, the means of choosing a sample, assuring them regarding confidentiality and anonymity. The questions in the questionnaire were mainly of the closed or mixed types, with multiple variants of response, in the end offering to respondents the possibility to add their own commentaries, suggestions, or observations through an open answer.

2.2   Universe of Research

Respondents were selected on the basis of information obtained following consultation within the project Google Digital Workshop, in which the second author of this article took part as a volunteer. The project Google Digital Workshop for SMEs is addressed to entrepreneurs that are starting digital promoting. In 2016, Google started the Digital Workshop in Romania, as a part of the international programme Grow with Google, through which it aims to increase the digital competencies of Romanians, either students, entrepreneurs, or employees in SMEs [14]. In this respect, Google established five hubs within universities in Bucharest (the Academy of Economic Studies), Iasi (Alexandru Ioan Cuza University), Cluj Napoca (Babes-Bolyai University), Constanta (Ovidius University), and Timisoara (Western University), but it also offers consultancy in the digital field in Brasov [14]. Consultancy sessions started on 12 July 2017 along with the opening of the hub in Bucharest [15] and it continues until present.

An important component of the programme is the sessions of preparation in the business field and digital competencies dedicated to local companies, where the Zitec company is a partner. The main objective of meetings between experts in online and SMEs representatives is an increase in the rate of comprehension and usage of the Internet and of solutions available online, given that Romania is in last place among the 28 EU countries regarding the degree of digitisation of the population, according to the DESI top (the Index of Economy and Society) in 2017 [16].

To take part in the project, entrepreneurs subscribe by accessing the website https://atelierul-digital.zitec.com/ and they fill in the data in the form. The website is directed to entrepreneurs by means of paid campaigns targeted precisely for SMEs' market, run through Facebook Ads and Google Ads. Subsequently, they are contacted by one of the consultants of the project and jointly agree on a day and time for consultancy. Until present, over 5,500 entrepreneurs have participated in the project, from various fields of activity and with various experience.

2.3 Sampling

Initially, the sample was projected to be a larger size and to be of a systematic random type, to represent certain key characteristics of the population participating in the project (turnover, field of activity, and geographical area). Initially, the link to the online questionnaire was distributed to 350 entrepreneurs, former participants in the consultancy sessions held by the second author of this paper, within the project Google Digital Workshop. Entrepreneurs were personally contacted after the permission offered verbally at the end of the consultancy session, independently of the project Google Digital Workshop. Nevertheless, as a result of the difficulty of accessing such a population (high refusal rate, non-responses regarding certain data of company identification, certain respondents were eliminated from the database as they subsequently requested withdrawal of participation), we decided to use a convenience sample of only 100 respondents. The reduced volume of the sample is also because it contains only these participants that, within consultancies, have agreed to be subsequently contacted.

On short, we have used non-probability sampling, the selection of the subjects being made on the basis of the availability, this sampling procedure having the advantage of the ease of application and that of economy, but also the disadvantage of not producing generalisable results [17]. The final sample at the level at which the results were extracted has the following profile: six out of ten respondents represent companies that operate in the capital city of the country; eight out of ten in the services sector; the most frequent fields of activity being commercial, education, IT, health, tourism, and industrial sectors; and six out of ten respondents are company administrators, the remaining predominantly general managers, partners/associates, or top management; the experience of the company on the market is, on average, seven years; 52% of respondents are women, and 48% are men; and 42% are between 18 and 34 years old, 40% between 35 and 44 years old, 17% between 44 and 54 years old, and 1% over 55 years old. Over 90% attended higher education, over 90% come from the urban area; more than half are married; and seven out of ten are members in a national or international association of businessmen. It is interesting that, of the total companies included in the sample, 55% operate in the national market, 20% in the global one, 8% in a regional one, 15% in the local market, and 2% do not have a clear estimation of the market in which they operate. We have presented this profile because we started from the premise that the results of the study can be useful to understand and launch hypotheses about the particularities of the populations of entrepreneurs and SMEs with a similar profile.

3 Results

Romanian entrepreneurs participating in the research prefer promotion by means of digital tools to the detriment of traditional classical promotion. Less than half of respondents state that they are to a very large extent aware of the benefits that online promotion generates, many of them considering that this has a great importance in developing the business and contributes considerably to the image of the company, 66% of the questioned entrepreneurs consider that the brands active on the Internet create a more robust relation with the market compared to those that are not active (see Fig. 1) and 61% of them admit they are very open to adopting tools of promoting through digital marketing.

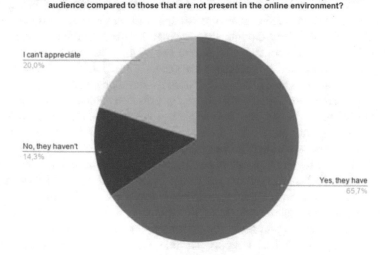

In your opinion,the active brands on the Internet have a stronger relationship with their audience compared to those that are not present in the online environment?

I can't appreciate
20,0%

No, they haven't
14,3%

Yes, they have
65,7%

Fig. 1. The perceived role of the brand online presence in building relations with the target audience

Respondents estimate that the openness of entrepreneurial environment towards digital marketing is an average one. Over half of the questioned entrepreneurs estimate that the main cause of reluctance of entrepreneurs in using digital marketing tools represents the lack of knowing them and almost a quarter report the lack of a dedicated budget. It is relevant in this respect that most respondents estimate they do not know the benefits of digital promoting very well. The reasons less invoked to explain the reluctance of the entrepreneurial environment towards digital marketing are: conservatism and the particularities given by age and education (2%), the fear of public visibility of potential negative feedback (5%), the costs too high (7%), and the difficulty to manage a potentially too rapid development (7%).

Social media prevails in the preferences of entrepreneurs both in terms of promotion and of online communication, followed by an official website and online store. Platforms well known among entrepreneurs are YouTube, WhatsApp, Twitter, Reddit, Instagram, and Facebook, the most used being Facebook and Instagram. Reddit, Instagram, Facebook, and YouTube are considered by most entrepreneurs as being the most useful of the promotion platforms. Conversely, GitHub, Vimeo, Quora, and Snapchat are least used by them and considered the least useful. At the same time, most reluctance is stirred by GitHub and Quora platforms, and the greatest interest for future use is generated by the YouTube platform. Very few of the entrepreneurs questioned can attest to frequent interactions with online promotion tools. To a relatively great extent, they state that they did not interact at all with a certain platform over the last month. Therefore, although entrepreneurs promote using a certain promoting channel (sometimes even appreciating its utility in marketing actions), there is a tendency of not investing time and financial resources personally, adequately, and consistently using it.

Only four of ten entrepreneurs questioned consider online communication as essential in reaching the objectives of a business. The platforms offered by Google, both those paid and the free ones, are not the top preference of entrepreneurs. Regarding the choice between organic growth and paid campaigns, most entrepreneurs consider that both SEO and Google Ads have to be used together to carry out an efficient digital marketing strategy. Regarding usage, the most frequently used digital marketing tools are applied by means of Facebook, e-mail, and official website. The most useful tools are perceived to be those used on Facebook and the official website. Regarding communication with customers, the greatest reluctance is stirred by the tools associated to Microsoft Teams and Twitter platforms, while the intentions of future use are directed to Blog and Google my Business. Among various digital marketing tools and techniques, those associated to Facebook are the most frequently used and considered the most useful. Entrepreneurs manifest the greatest reluctance towards SEO and the greatest number of intentions of usage in the future target Google Ads. The greatest preference in terms of paid campaigns is that for social media.

Most entrepreneurs questioned have a national or international market, digital marketing opening new opportunities by overcoming geographical boundaries with a view to accessing a market as large as possible. Only 53% of the total questioned entrepreneurs have a communication strategy within the company and 35% state that they personally deal with promotion activities of the company. The general manager of the company in most cases manages promotion and, other times it is externalised or it is performed by a specialist within the company. It is interesting that 12% have stated that, in the company they represent, nobody deals with promotion, and 9% that anyone in the company can post online, with product and service promotion content, 65% of the entrepreneurs state that they would be willing to hire a person that strictly deals with promotion and 88% consider that digital marketing implies a high volume of work. Percentages, such as 5% or 10% of turnover, are considered by most people questioned as being the most convenient percentage to invest in digital marketing. On average, entrepreneurs appreciate 9% of the turnover to be the optimum budget to invest in digital marketing.

Most entrepreneurs questioned consider that the main advantages that digital marketing offers are: increase in sales volume (80%); relations development with customers (64%); rapid communication (54%); creation of value for the customer (53%); increase in the engagement of customers (51%); real time feedback providing (49%); overcoming geographical boundaries (46%); and reducing promotion costs (42%). Besides these main advantages, they also gave: solving the problems of customers (34%); monitoring interactions between customers (31%); and the possibility of outlining the market very well (30%). The least frequent advantage given is that related to an increase in brand awareness.

The main reasons for which most entrepreneurs adopt online promotion tools are: an increase in visibility (93%), accessing a market as large as possible (66%), and continuous information of the target market (61%). Beside these main reasons, relatively high frequencies have been also registered for the following: obtaining feedback from the target market (35%) and replacing traditional communication means (30%). The least frequently given reasons are these related to: fund raising, maintaining a relation with the press, adaptation to promotion tendencies adopted by the other entrepreneurs, and monitoring members of the organisation in collective actions.

Most entrepreneurs consider that they have an average level of knowledge regarding digital marketing and a little under half assess an average trust level in their abilities of applying methods and techniques of online promotion. 86% of the respondents have participated in the past in consultancy sessions and 78% are willing to invest in the future in seminars or other formation models by which they can develop their knowledge and abilities regarding digital marketing. Nevertheless, it is worth mentioning that, almost unanimously, those questioned admit that the entrepreneurial environment needs digital marketing consultancy and the best solutions would be monthly participation or once every two months (32%) in consultancy sessions. On average, entrepreneurs estimate that they would need a number of eight sessions per year of consultancy in the field of digital marketing.

4 Conclusion

The aim of this paper was to highlight the opinions, attitudes and behaviours of entrepreneurs in the SMEs sector regarding digital marketing, selectively presenting the results of an exploratory research on the basis of an opinion poll applied online in 2019 on a sample of 100 representatives of Romanian SMEs that took part in the consultancy project Google Digital Workshop.

Results show that, although at a declarative level, the interest of respondents towards digital marketing is raised, nevertheless their perception is that the entrepreneurial environment continues to manifest reluctances, as a result of insufficient knowledge, of lack of trust in their own abilities of applying them, of the lack of allotting a budget devoted to online promotion, of a clear communication strategy, and of a specialist to manage it. Therefore, the demand for consultancy services in the field of digital marketing is very high. The most well-known promotion channels, used and appreciated, are those of the social media type (mainly Facebook, Instagram, and YouTube), in second place being the official website and e-mail. For the future, entrepreneurs intend to orient also towards YouTube and blog. Although the tools offered by Google, both those paid and those free are not the top preferences of the entrepreneurs questioned, there is an openness towards using Google Ads and Google my Business in the future and reluctance towards SEO.

The entrepreneurs questioned are aware of certain general benefits of using techniques and tools specific to digital marketing, but these are rather those connected with sales promotion techniques, that have clear effects in a short term. More complex advantages that require constant engagement and that have strong effects only in the medium and long term, that belong to the field of branding and PR, of constructing fame, of managing communication with the internal market, the press and the community are less well-known and assumed. Therefore, the consultancy services in the field of digital marketing should orient towards forming competencies of identifying and using the most relevant channels of digital communication (not only the most well-known) within a communication strategy for the long run, with communication programmes in the medium term, in which the most appropriate financial, human, and time resources are allotted to create and maintain conversation with various stakeholders and to efficiently manage the feedback offered by them.

The results of this study must be interpreted with caution. Being based on a non-probability sample, they cannot be generalised at the level of Romanian SMEs, Romanian entrepreneurs or participants in the consultancy project Google Digital Workshop. Nevertheless, the study can be representative of those entrepreneurs in the SMEs sector that overlap with the profile of the sample presented in detail in Subsect. 2.1 of this paper.

This paper may inspire future research in the field of new media, as it provides a complex operationalisation of the concept of adopting digital marketing, as well as new hypotheses regarding the opinions, attitudes and behaviours of entrepreneurs in the SMEs sector about digital marketing. Future studies could orient towards investigating how consultancy services in the field of digital marketing can stimulate an attitude change among entrepreneurs in the SMEs sector, so that marketing strategies acquire consistency with the specificities of new media and assume in the long term a coherent, efficient, authentic, and constant conversation with all relevant categories of market. Such a conversation supposes overcoming a one-dimensional vision according to which digital marketing is that cure-all that increases sales in the short run though sell promotion tactics by deepening the knowledge of particularities of channels, techniques, and tools and training skills of their strategic use in compliance with recent theories in the field of branding and PR.

References

1. Chaffey, D., Smith, P.R.: Digital Marketing Excellence: Planning, Optimizing and Integrating Online Marketing. Routledge, London (2017)
2. Rua, O.L.: Entrepreneurial Orientation and Opportunities for Global Economic Growth. Advances in Business Strategy and Competitive Advantage. IGI Global, Harshey (2018)

3. Gilmore, A., Carson, D., Grant, K., Pickett, B., Laney, R.: Managing strategic change in small and medium-sized enterprises: how do owner-managers hand over their networks? Strateg. Change **9**(7), 415–426 (2000)
4. Ansari, A., Mela, C.F.: E-customization. J. Mark. Res. **40**(2), 131–145 (2003)
5. Hristache, D.A., Paicu, C.E., Ismail, N.: The impact of the image of the organization in terms of the online communication paradigm. Theoret. Appl. Econ. **21**(3), 67–74 (2014)
6. Hosu, I., Culic, L., Deac, M.: Comunicarea online - provocări manageriale [Online communication - managerial challenges]. Transilvanian J. Adm. Sci. **35**(2), 19–28 (2014)
7. Udo, G.J., Bagchi, K.K., Kirs, P.J.: An assessment of customers' e-service quality perception, satisfaction and intention. Int. J. Inf. Manag. **30**(6), 481–492 (2010)
8. Batool, A., Bano, S., Khan, M.A., Akhtar, M.N., Naeem, A., Batool, B.: Technology as a black-box in e-business and its impact on customer satisfaction: major corporate sector of Islamabad areas, Pakistan. Far East J. Mark. Manag. **1**(1), 28–53 (2011)
9. Dodson, I.: The Art of Digital Marketing: The Definitive Guide to Creating Strategic, Targeted, and Measurable Online Campaigns. Wiley, Hoboken (2016)
10. Samarasinghe, G.D., Suwandaarachchi, C.M., Ekanayaka, E.M.S.T.: Impact of social media on business performance: empirical study on apparel fashion brand retailers in Sri Lanka. In: Student Research Conference on Marketing (SRCM), Department of Marketing Management, University of Kelaniya, Kelaniya (2017)
11. Mudaliar A., Chava M.: A study of the novel innovation: "social media" - as a form of advertising in the framework of digital marketing. In: Khullar, L., Kavishwar, S.M., Deshpande, S. (eds.) Proceedings of International conference on Business Remodelling: Exploring New Initiatives in Key Business Functions, pp. 14–22. Tirpude Institute of Management Education, Nagpur (Maharashtra), India (2018)
12. Stokes, R.: eMarketing: The Essential Guide to Marketing in a Digital World, 5th edn. Red & Yellow (2013)
13. Taiminen, H., Karjaluoto, H.: The usage of digital marketing channels in SMEs. J. Small Bus. Enterp. Dev. **22**(4), 633–651 (2015)
14. Năstase-Lupu, S.: Google te ajută să deprinzi competențe digitale pentru jobul în HR [Google helps you learn the digital skills for the job in HR] (2019). https://www.wearehr.ro/google-te-ajuta-sa-deprinzi-competente-digitale/. Accessed 08 Aug 2019
15. Abrihan, R.: Înscrieri Gratuite: Google lansează cursuri de Design Thinking pentru antreprenori [Free registration: Google launches Design Thinking courses for entrepreneurs] (2019). https://www.startupcafe.ro/idei-si-antreprenori/google-atelierul-digital-design-thinking.htm. Accessed 08 Aug 2019
16. Andriescu, V.: Zitec și Google lansează primul Atelier Digital din Iași [Zitec and Google launch the first Digital Workshop in Iasi] (2017). https://start-up.ro/zitec-si-google-lanseaza-primul-atelier-digital-din-iasi/. Accessed 08 Aug 2019
17. Babbie, E.R.: The Practice of Social Research, 14th edn. Cengage Learning, Boston (2015)

Digital Platforms as Social Interaction Medias: Virtual Sex Risks

Lidice Haz[1]([⊠]) [iD], Ivette Carrera[2], Freddy Villao[1],
and Ginger Viviana Saltos Bernal[3]

[1] Universidad Estatal Peninsula de Santa Elena, La Libertad, Ecuador
victoria.haz@hotmail.com
[2] Universidad de Guayaquil, Guayaquil, Ecuador
[3] Escuela Superior Politécnica del Litoral, Guayaquil, Ecuador

Abstract. The widespread use of internet has transformed communication and information exchange processes. The technological development has promoted the growth of virtual spaces that encourages virtual social interaction. There are many digital platforms dedicated to meet people and establish relationships since professionally to affective aspects. In this sense, affective relationships include sexual relationships, friendship and romantic love. The main goal of this research is to identify the psychosocial factors that encourages people to establish virtual relationships, particularly sexual erotic relationships; through the analysis of the causes and implications of this behavior. In general, this paper describes the risks of virtual sexual relationships and suggests security mechanisms to minimize them.

Keywords: Cybersex · Virtual space · Social interaction · Virtual sex · Risk · Online dating · Virtual relationships

1 Introduction

The Information and Communication Technologies (ICTs) promotes the development of virtual spaces, which main purpose is the social interaction. The widespread use of internet has changed drastically the social interaction between people that have access to several digital platforms for professionally, personal and entertainment aspects [1].

Internet has had a deep impact on society. The worldwide web has allowed to decentralize the information, thanks to the www, thousands of people can access to a lot of information, resources and online services in an easy and immediate way. Some companies and individuals have adopted the use of weblogs, forums, social networks and other digital platforms to share their ideas and communicate with other people [2, 3].

The use of information technology has allowed to pass time and locations limitations creating complex environments. Individuals can interact in a virtual world in order to meet and communicate with other people no matter where they are. Communication through the web increases the level of risk to which the internet user can be exposed. The ICTs have become in powerful tools that used in malicious ways can affect strongly people's lives, especially for those who frequently use digitals platforms for online dating, virtual sex and pornography [1, 4].

© Springer Nature Switzerland AG 2020
T. Antipova and Á. Rocha (Eds.): DSIC 2019, AISC 1114, pp. 442–453, 2020.
https://doi.org/10.1007/978-3-030-37737-3_38

In this regard, ICTs have transformed the economy, society and culture. They have even influenced in the sexual behavior of people [5]. In fact, Internet allows to create virtual spaces in order to establish sexual relationships through adult websites, meet people by webcam or chat, search erotic and pornography material [6]. However, since more than a decade these topics are the searching with the higher rate of recurrence on the Internet [7]. The growing numbers of users on social networks and specifically on dating websites or sexual websites has been a great of interest for researchers since these websites present a huge number of risks for their users [1].

The digital platforms and online content available are numberless, being pornography the topic with higher prominence. Around 14% of the searching on Internet includes sexual terms, 35% of the downloads is pornography material. This high demand represents between 1 and 97 billions of dollars on annual profits for material production and diffusion in the pornographic industry [8, 9]. According to the statistics published by Pornhub (one of the most pornographic websites visited), their servers share each year around 3.732 petabytes of pornographic material, that is equivalent to 118 gigabytes of pornography per second (around 95 movies in high definition) [10]. The availability of a lot of tools used to access to Internet like computers, tablets, virtual and augmented reality glasses and others, the facility of looking for sexual content online, the anonymity that these websites guaranty and their free cost make Internet a suitable tool to begin experimenting and exploring sexuality [6].

Sexual-erotic interactions through internet exposed to the cyber user to risks and threats such as loss of privacy, identity theft and others. These threats are recurrent when people keep in touch with strangers through the web [1].

Some of these behaviors can become in pathological patterns and may result in a cybersex addiction. This could become in an addiction to sex and internet together [11]. This research states the psychological factors that encourage people to have sexual affective relationships through internet. Furthermore, it determines the causes and implications of this behavior and the technological risks presents in these virtual spaces.

In order to address this subject, this research is classified in four sections: introduction, cybersex, virtual spaces of sexual erotic social interactions and finally conclusions and recommendations.

2 Cybersex

Sexuality is one of the most complex aspect of human being. This implies physical, biological, emotional, cultural and social conditions. Anyone of these elements can influence in others and cause consequences in chain. The sexuality has been analyzed from different disciplines, creating various theories and points of view; aspects such as fertility, homosexuality, gender, pornography and their roles have received totally opposite approaches depending on the context from which they have focused [12].

Virtual spaces reveal the reality of who we are and what we want to be. The XIX century was characterized by the hidden sexuality, XX century by the sexual revolution and XXI century represents sexuality displayed. This century shows sexuality in a spontaneous way, a sexuality without conditions. The human being in virtual spaces shows inclinations to experiment and interact with other people their deeper desires looking for sexual pleasure [1, 11].

Cybersex or virtual sex is the act in which two or more people interact with explicit messages describing a sexual erotic experience using technological resources [11]. This activity generates a change in the human being behavior, in their social and cultural roles. The Internet Era opens a parallel space, a "virtual community or virtual space" that promotes freedom of thoughts and actions without conditions between their users [1]. Virtual spaces combine the real with the imaginary through games, text messages, meetings, videos, sexual proposals, experiences from other genders. In brief, there are a wide range of options where everything is possible and valid in the network taking advantage of the anonymity.

2.1 Characteristics

Cybersex o virtual sex is a manner of having sex without penetration and physical contact where two or more people share sexually explicit messages using a technological platform. The information is transmitted by various types of media such as email, instant messaging, chat rooms, webcam, online gaming and others. Cybersex occurs between strangers or people who have just met by Internet [11]. There are several digital platforms with exclusive chat rooms and live video conferences for finding people who wish to have virtual sex from everywhere of the world. The information (images, videos, audios) transmitted can be real or fake. In these role-playing games, the players fake that they are having sex. They can be passive players where they just watch and listen or can be active players where they describe and responds to the partner(s) with the purpose of stimulating their desires and sexual fantasies. The quality of the virtual sexual encounter depends on the player's capacity of representing images or situations that satisfy to their partners or themselves [13, 14].

This kind of interaction generates in some people chemicals reactions mostly dominated for the dopamine, which produces passionate emotions such as excitement, rewarding sensation, pleasure, desire, freedom [12]. This allows to people feel unusual sensations giving them the capacity of experimenting sexual acts that they would not try in real life [15].

2.2 Advantages and Disadvantages

The availability of sexual content and the ease of access to them on Internet increases their demand between the population. Cybersex can be used as a sane way of expressing sexuality but it exposes to risks that can causes physical and psychological impacts in those who practices it [13].

From an objective point of view, Cybersex has some advantages in sexual life. Its main advantage is that there are not risks of catching sexual transmissions diseases and unintended pregnancies since cybersex does not requires physical contact. People can experiment freedom of thoughts and sexually feelings in a safe way [16].

People with disabilities or chronic diseases such as VIH can find on virtual sex a safe way of feeling sexual pleasure with freedom and without transmission risks. Cybersex also helps couples in traditional relationships when for external reasons are geographically apart, giving to them the option of keeping in touch and preserve their sexually intimacy.

Cybersex also allows to create role-playing games to stimulate fantasies and describes sexual situations. Experiences that can be limited in real world by physical or social factors and fear of rejection. For some people, cybersex provides them a virtual space where they can learn and explore their desires in a safe way and deal with their fear of rejection. Even for those who have phobias related to traumatic sexual experiences, cybersex could help as a therapy in successive interactions [18].

In relation with cybersex disadvantages, the privacy of the participants is the risk most latent in these virtual environments. Furthermore, users who desire to trespass cybersex experiences to reality with other people away from their virtual interaction, can find disappointing the encounter since sexual chemistry could not be real. Also, it is possible that these activities generate cybersex addiction, which would replace the physical contact with other people; disturbing the social interaction capacity and can became in a habit or an exclusive way of sexual interaction [18, 19].

Finally, cybersex represents an ethical and moral dilemma in the society. For traditional couples is necessary to determine if cybersex with other people is an infidelity or not. Although there is no physical contact, these sexual activities could affect the confident and respect in the relationship if there is not a deal about these situations. Moral implications of cybersex between couples and individuals is an ongoing discussion depending of the actual society. But like in all kind of sexual interaction, the discussion should be done with honesty and questioning possible threats [17]. Cybersex establishes sexual erotic interactions with known people and strangers in a virtual space where their actions generate positive or negative consequences between the implicated parts. The decision of having cybersex should be taken by their own choice in response to their desires of pleasure but without forgetting that in the other side of the screen there is a human being who also feels and suffers.

2.3 Psychosocial Aspects

The psychosocial term refers to the relation between the individual and their social environment. In the technological subject, it can be interpreted as the relation between the human being and the use of information and communication technology [20]. The human behavior is affected by the context or the virtual space in which is exposed. Internet and virtual spaces can be raised in two different social contexts [21].

1. The less social context consists in looking for information through network, surfing on the www accessing to different contents and images.
2. The use with more social context involves the interaction or communication exchange with other people connected to the net from all over the world through chats, email, video conferences and others.

In regards to social contexts, the cyber user shows different behaviors and attitudes depending on the influence that the virtual space exercises. Digital platforms as a social interaction media maintain to their users active who are connected during hours without having other kind of social relationship away from virtual space, which can cause isolation from society [1]. This analysis identifies the main motivations and behaviors exposed by the individuals who establish sexual erotic interactions.

According to the scenery previously described, it is necessary questioning Why do people prefer to establish sexual erotic relationship through virtual spaces? Which effects are evident on people behavior in those who use these medias for sexual pleasure? How is the sharing information used in these medias? These are some questions that should be addressed to identify behavioral patterns and security information criteria that should be applied to minimize the risks on these platforms.

Previous research suggests that leading motivations are looking for companionship, avoiding loneliness, escaping from reality or finding pleasure [13, 16, 19]. These motivations can become in factors of risk and generate an addiction to the uncontrollable use of internet and virtual interaction spaces. The impact with more incidence is the influence on the individual mood [22]. The isolation in this virtual spaces eases the adversities rejection and avoiding daily responsibilities including anxiety social disorders or depression. Below, in Fig. 1 is described the main causes and consequences from sexual virtual relationships.

 Causes

 Consequences

•**Looking** for intense experiences

•**Social affective Isolation**

•**Dissociative disorders** between virtual and real world

•**Child sexual abuse** stories

•**Social abilities deficit**

•**Low self - esteem** or body image rejection

•**High introversion** o excessive shyness

•**Negative emotionals states**

•**Disabilities** or congenital physical limitations

•**Sexual dysfunctions**

•**Low frustration tolerance**

Physical changes: general deterioration, sleep rythms alterations, variable weight and tiredness

Alterations in mood: ansiety, sadness, crankiness, impatience, isolation or intolerance

Accademics or working difficulties : decline in performance, delays, absences or disputes.

Troubles in family or social relationships: frequently fighting, mistrust, secrecy or loss of friendship

Fig. 1. Causes and consequences of virtual sexual relationships.

Predisposed individuals to be involved in cybersex are those who shows social abilities deficit, this leads to a lack of social relationships since they prefer virtual relationships over real relationships, which can be described as fake [13]. The results are obvious. People find in this environments satisfaction and responses to their psychosocial necessities, neglecting their job, familial and social responsibilities. The regular and uncontrollable use of these virtual spaces are addictive behaviors where individuals invest time, money and health [23].

For other researchers, controlled and sporadic cybersex is not wrong for the user while it does not interfere in their daily activities and it does not affect their performance, self-esteem or auto control capacity [24]. In other worlds, it is possible that some people find in virtual spaces those things that in "real life" cannot obtain such as friendship, emotional and physiological reactions from a sexual relation.

The parameters described above do not represents a real dimension of the psychosocial behaviors in individuals. The main risk is the time that those people invest in behaviors detached from their responsibilities [23, 24]. Nevertheless, there is no one who can determine behavioral social rules. Currently, there are several ways of interaction that had influenced in society as well as in ethical and moral values that for years had conducted the society behavior. The individual has the right of looking for alternative medias, which promote their wellness and increase their self–esteem [17, 19]. Virtual spaces like interaction medias are socially accepted and used according to the necessities of their users [1]. It is important to learn or re learn new ways of behavior and do not fall in states than can be described as behavioral disorders or addiction caused by the excessive use of virtual communication medias or Internet.

3   Social Interaction Virtual Spaces

Social interaction virtual spaces have changed the concept of classic social relation and free time use. In this virtual medias, users look for keeping touch with known people or strangers. The interaction is done by messages, audios and videos. Social networks generate a debate related to data privacy, which is the most meaningful risk in these medias [25].

Virtual spaces are medias established in different contexts with the purpose of social interaction using a digital platform [1]. These spaces create an exchange between people, groups, religions, professional relations and others. It is an open system in constantly development which involves people who have the same necessities and troubles and their principles are pleasure, satisfaction and sharing in the virtual world [21].

3.1   Digitals Cybersex Platforms

Nowadays, adult websites are one of the technological trends that had changed social behavior. Its main goal is the institution of sexual erotic relations without distinction of gender, race, languages or any other variable that could restrict the interaction and communication.

Pornographic industry is growing exponentially. The use of digital platform produces great economics profits. The popularity of pornographic material in Internet is showed by statistics. According to respectable news sources and researching organizations, pornographic industry produces around 6,754.634.84 dollars per second which means that approximately 65,891.918 internet users are watching pornography and 880,640 internet users are looking for pornographic keywords on searching engines. Between 1998 and 2015 the numbers of pornographic websites growth from 14 to 528 millions [26].

Cybersex digital platforms offer a wide range of options according to the necessities of their users. There are websites with online transmission through webcams. The sexual and erotic activities can be filtered by categories according to user preferences. Generally, these activities ranges from nudes, striptease, dirty talks to masturbation with sex toys. The eroticism can be just between women, men, transgender or couple of artists. The virtual interaction environment can be public or private where private environment requires paid subscription. Bellow in Table 1, websites with the largest audience ot users are shown.

Table 1. Websites of sexual erotic activities.

Website	Year of foundation	Main office	Global Alexa rank 2018
Chaturbate	2011	U.S.	114
CAM4	2007	U.S.	886
BongaCams	2012	Canada	58
LiveJasmin	2001	Rumania	38
ImLive	2002	U.S.	18145
ManyVids	2014	Canada	1362
XNXX	1997	Francia	112
XHamster	2007	Chipre	72
Flirt4free	1996	U.S.	984
YouPorn	2006	U.S.	175

Internet pornography seems infinite and hard to understand. The statistics suggests that pornography industry has a huge impact in internet. Table 2, displays some measurement parameters about internet pornography [26].

Table 2. Internet Pornography Statistics.

Description	Measurements
Pornographic websites	4.2 million (12% of total websites)
Pornographic pages	420 million
Daily pornographic search engine requests	68 million (25% of total search engine requests)
Daily pornographic emails	2.5 billion (8% of total emails)
Internet users who view porn	42.7%
Received unwanted exposure to sexual material	34%
Average daily pornographic emails/user	4.5 per Internet user
Monthly pornographic download (peer-to-peer)	1.5 billion (35% of all downloads)
Daily Gnutella "child pornography" requests	116,000
Websites offering illegal child pornography	100,000
Sexual solicitations of youth made in chat rooms	89%

(*continued*)

Table 2. (*continued*)

Description	Measurements
Youths who received sexual solicitation	1 in 7 (down from 2003 stat of 1 in 3)
Breakdown of male/female visitors to pornography sites	72% male - 28% female
Worldwide visitors to pornographic web sites	72 million visitors to pornography: Monthly
Internet pornography sales	$4.9 billion

According to statics from global newspapers and organizations, it was observed that the users who sign up and visit cybersex websites are more often men between 20 and 65 years old, upper middle class and professionals. The population that mostly works in cybersex are women between 18 and 40 years old. Related to the user nationality, there are users from all over the world. However, there is a greater presence of American and European continent.

3.2   Risks and Technological Threats in the Use of Cybersex Platforms

The use of platforms as a means of communication or for information searching is inevitable. The ease of access to cybersex websites and their anonymity make it attractive among population. This practice has become a current sexual dynamic; therefore, it is important the evaluation of risks and threats to avoid their exploitation [24, 27]. Risks and threats with higher incidences are listed below [1].

- It is not possible to guarantee user privacy on internet. The ease of intercepting webcam transmissions by a hacker is quite high, what could trigger an invasion of people life and became a victim of blackmail or extortion.
- Sexting consists in sending sexually explicit text messages among known people or strangers. The risk on this practice is the non-authorized spread of information shared by SMS. This creates an emotional and cultural harm to the reputation.
- Downloading malware: digitals cybersex and pornographic platforms are suitable spaces to spread malware. Their purpose is to infect and damage as many computers they can. Computers infected can be victim of information theft such as credit cards, personal information and others or involuntary control like zombies.
- Impersonation or identity theft is the illegal appropriation of other person identity. The thief can do financial, migratory and civil transactions through the illegal use of another person identity.
- Grooming, consists in behavioral patterns or actions that an adult adopts in order of gaining the trust of a minor. Its objective is to systematically and deliberately harass to minors to obtain sexual benefit through internet.
- In addition to this regard, there are psychologic and physical risks that can harm people's health. Continuous and uncontrollable cybersex practice and the use of pornography can create addictions and develop predilections of sexual aberrations as zoo philia (sex with animals) or necrophilia (sex with dead) and other similar topics available on internet.

3.3 Security Mechanisms to Reduce Security Risks in Cybersex Platforms

To reduce the materialization risk during the practice of cybersex, some security mechanisms are exposed [1, 27].

- Use tools to block information sent from the computer source. The purpose of these tools is to avoid revealing personal information. This mechanism is very useful when users fill out online forms and wish to buy using credit card.
- Avoid to download information through links that redirect to illegitimate emails, social networks or instant messaging applications in order to prevent the installation and execution of hidden programs that can control the device or steal information
- Evaluate frequently the computer security configuration with the objective of detecting flaws in firewall configuration, verify the antivirus and operating systems updates
- Verify the website security certificate. This can assess the website legitimacy and the use of cipher data
- Do not give personal information such as address, telephone number, financial data, family information and others. To communicate with strangers, it is advisable to use especially created addresses for this purpose.
- Avoid recordings during the cyber sexual transmission. These recordings can be stolen by thirds parties and be broadcasted massively in websites. In this aspect, the use of application that facilitate image and video exchange are an alternative for "sexting" with strangers.
- Do not show the face in the images and videos where naked body is shown. It is possible that these images are trafficked by cyberspace. Also, avoid to show in these images distinctive marks that allows to identify a person for example tattoos, moles, freckles, dressing clothes or daily places
- Do not hurry. Establish data of interest and related context concerning to the person with whom want to practice cybersex. Although, this type of link responds immediate searches, do not let desire generate irresponsible actions. Also, ensure if it is really convenient to continue the experience and share information in the virtual sex chat room
- Do not promote or participate in virtual sex practices with minors. It is important that parents supervise their kid actions during the use of virtual spaces and know the danger of inadequately activities that put in risk security privacy and intimacy of minor.

The mechanisms above described can be applied as security controls that will help to minimize the risk. The main purpose of these controls is to avoid the violation of the privacy and intimacy of the user, who for any reason practice cybersex. In this sense, it is important that users be aware of the type of information they provide and with whom they share it during the interaction in the cyber space [24].

4 Conclusions

The use of virtual spaces for sexual, erotic and social interaction has progressively increased over the last years. This fact could be described as a strength. People are adapting to new behaviors, forms of communication and relationship with others. The responsible use of virtual spaces is an advantage for the individual as long as their incursion and permanence helps to minimize the state of anxiety derived from feeling of loneliness.

However, the excessive and uncontrollable use of virtual spaces represents a high risk for physical and psychological health. The most eminent risk is the alteration of individual social behavior, resulting in a psychological disorder or even addiction.

Cybersex and pornography have become in an attractive and common reality among people. There are those who discover aspects of their sexuality through these types of practices. It can even be a sexual therapy proposal between couples. In addition, it has advantages such as the possibility of anonymity, privacy, accessibility, easy interaction, sex without risk and the lack of commitment.

It is relevant to keep in mind that during these sexual practices there are technological threats that can described as high risk situations. The dangers of exposing yourself virtually with strangers or even with known people, share images that can be exposed uncontrollably in cyberspace, download information and application with hidden malware, access to insecure digital platforms or websites are some of security risks that exposed people privacy and intimacy. In this regard, it is necessary to apply security mechanisms that help to decrease the user exposure to these threats and raise awareness among internet users about the information and level of trust they provide to strangers on the web.

5 Recommendations

In future works, it is important to further investigate factors of the human behavior related to the use of internet and virtual sexual practices. Besides, it should be designed and developed new scales that measure more accurately levels of addiction or hobbies generated by cybersex and pornography, as well as levels of impact according to the mode of cybersex consumption (chat, websites, email, forums, webcam) or demographic aspects as gender, age, socioeconomic status, educational level, among others should be identified and, also it is highly recommendable to evaluate the technological, social, cultural and economic impact created by sexual digital platforms.

References

1. Haz, L., Acaro, X., Guzman, C.J., Espin, L., Molina, M.F.: Digital platforms as means of social interaction: threats and opportunities in online affective relationships. In: Antipova, T., Rocha, A. (eds) Digital Science, DSIC18 2018. Advances in Intelligent Systems and Computing, vol 850, pp. 417–425. Springer, Cham (2019)

2. Ledo, M.V., Vidal, M.N.V., García, L.H.: Redes sociales. Educación Médica Superior **27**(1) (2013)
3. Tobón, S., Guzmán, C.E., Silvano Hernández, J., Cardona, S.: Sociedad del Conocimiento: Estudio documental desde una perspectiva humanista y compleja. Paradigma **36**(2), 7–36 (2015)
4. Lee, Z.W.Y., Cheung, C.M.K.: Problematic use of social networking sites: the role of self-esteem. Int. J. Bus. Inf. Taipei City **9**(2), 143–159 (2014)
5. Toranzo, F M · Aspectos psicosociales de la comunicación y de las relaciones personales en Internet. Anuario de psicología/The UB J. Psychol. **32**(2), 13–30 (2001)
6. Cardenas Amaya, J.P., Martelo, V., Jesus, D., Arevalo Moncayo, M.P.A.: Las TICS como medio para el cibersexo (2016)
7. Freeman-Longo, R.E., Blanchard, G.T.: Sexual abuse in America: Epidemic of the 21th century. Brandon: Safer Society Press (1998)
8. Corazza, O., Blanchard, G.: Should problematic pornography use be considered an addiction?. Research and Advances in Psychiatry (2018)
9. Castro-Calvo, J., Ballester-Arnal, R., Potenza, M.N., King, D.L., Billieux, J.: Does, "forced abstinence" from gaming lead to pornography use? Insight from the April 2018 crash of Fortnite's servers. J. Behav. Addict. **7**(3), 501–502 (2018)
10. Silver, C.: Pornhub 2017 year in review insights report reveals statistical proof we love porn (2018)
11. Wéry, A., Billieux, J.: Problematic cybersex: conceptualization, assessment, and treatment. Addict. Behav. **64**, 238–246 (2017)
12. Lucas, D., Fox, J.: The Psychology of Human Sexuality. The Psychology of Human Sexuality. Noba Textbook Series: Psychology. DEF Publishers, Champaign (2018)
13. Sánchez, N.F.: Redes sociales virtuales, ¿fortalezas o debilidades? un análisis psicosocial relacionado con el cybersexo y la soledad-Virtual social networks, strengths or weaknesses? A psychosocial analysis in relation to cybersex and loneliness. Hamut'ay **3**(2), 42–54 (2017)
14. Tettey, W.J.: Globalization, cybersexuality among ghanaian youth, and moral panic. In: Globalization and Socio-Cultural Processes in Contemporary Africa, pp. 11–37. Palgrave Macmillan, New York (2015)
15. Veale, H.J., Sacks-Davis, R., Weaver, E., Pedrana, A.E., Stoové, M.A., Hellard, M.E.: The use of social networking platforms for sexual health promotion: identifying key strategies for successful user engagement. BCM Public Healt (2015)
16. Sulima, M., Lewicka, M., Skorek, A., Roworth-Stokes, S., Bakalczuk, G.: Cybersexual activity. EJMT **3**, 12 (2016)
17. Knauss, S.: Transcendental relationships? A theological reflection on cybersex and cyber-relationships. Theol. Sex. **15**(3), 329–348 (2009)
18. Guillén, V., Alberola, E.G., Cortell, M.: Debate: Reflexiones acerca de la utilización de las Tecnologías de la Información y la Comunicación (TICs) en Psicología Clínica: eficacia, ventajas, peligros y líneas futuras de investigación. Información psicológica (116), 121–132 (2018)
19. Castro Muñoz, J.A., Vinaccia, S., Ballester-Arnal, R.: Ansiedad social, adicción al internet y al cibersexo: su relación con la percepción de salud (2018)
20. Wallace, P.: The Psychology of the Internet. Cambridge University Press (2015)
21. Serrano-Cobos, J.: Tendencias tecnológicas en internet: hacia un cambio de paradigma. Profesional de la información **25**(6), 843–850 (2016)
22. Weinstein, A.M., Zolek, R., Babkin, A., Cohen, K., Lejoyeux, M.: Factors predicting cybersex use and difficulties in forming intimate relationships among male and female users of cybersex. Front. Psychiatry **6**, 54 (2015)

23. Madrid, L.N.: La adicción a Internet. http://www.psicologia-online.com/colaboradores/nacho/ainternet.htm. Accessed 12 Aug 2019
24. Battaglia, P.: Conoce las ventajas y desventajas del Cibersexo. https://sexontologico.com/cibersexo/. Accessed 13 Aug 2019
25. Dominguez, D.C.: The social webs, typology, use and consumption of the webs 2.0 in today's digital society. Documentación de las Ciencias de la Información **33**, 45 (2010)
26. Ropelato, J.: Internet pornography statistics. 2006. Age **18**(24), 13–61 (2019)
27. Fire, M., Goldschmidt, R., Elovici, Y.: Online social networks: threats and solutions. IEEE Commun. Surv. Tutorials **16**(4), 2019–2036 (2014)

Digital Public Administration

A New Form of Public Services Intended for Citizens of the Slovak Republic

Martina Jakubčinová[✉]

Faculty of Social and Economic Relations, Alexander Dubcek
University in Trencin, Trencin 91150, Slovakia
martina.jakubcinova@tnuni.sk

Abstract. The issue of transforming public services is a pressing issue today.
Moreover, it is precisely the interperson in the citizen-state relationship that is
represented by public services that should have the highest priority. As a pro-
ducer and provider of these services, the state should emphasize their qualitative
shift. The citizen as a user and customer has to be an inspector and driving force
to achieve this goal. For this reason, we have focused on the current status of
public service issues in the Slovak Republic. We focused our attention on the
results of the eGovernment evaluation, which are freely available in the EU
databases. The important key was subjective attitude of the client using the
services. Based on this, it is possible to carry out evaluations and modifications
and thus point out both strengths and weaknesses. The results of the questionnaire investigation and common user opinions were helpfull in our research.
Through analysis, exploration and comparison, we tried to bring the reader
closer to the issue and to present a possible modification of the public service
portfolio.

Keywords: Citizen · eService · Public service

1 Introduction

The sophisticated use of technical-information technologies, the arise of digitalization
or virtual reality is an essential part of present age. The age called Information age. The
boom caused by technology merges into social life and behavior. This demanding
transformation process of a non-linear nature requires a change in the paradigm of
perception of time, space and the world, thus breaking the boundaries and limits of
obligatory thinking. The existence of modern information and communication technologies, their enormous variety and potential, offer to individuals, society and the
State new opportunities for meeting needs (Gupta and Jana, 2003; Moon, 2002;
Snellen, 2002). However, we must also keep in mind that we can use ICT for civilization, not the other way round. It must also be acknowledged that it is not only ICT
but also changes in economic and social values or the transition from regional to global
markets that need to be addressed in this context. For these and other reasons, we have
decided to pay attention to the usability and modernization of public services provided
by the State to its citizens in the form of public goods. This is also related to improving
transparency, which is very sensitive to the public (Welch et al., 2005, Snellen, 2002).

T. Antipova and Á. Rocha (Eds.): DSIC 2019, AISC 1114, pp. 457–469, 2020.
https://doi.org/10.1007/978-3-030-37737-3_39

By building a network of sophisticated public services, the State shows interest in citizens. By applying new technologies and means to activities related to citizen-State or State – citizen agenda, there are significant shifts in this issue. The diametrical systems difference breakes stereotype and sterile environment concept. Advances in public services administration electronization bring added value that society should be able to exploit (Weerakkody, 2012, Mosse and Whitley, 2008). In connection with above mentioned, it is also important to point out that this process could not be carried out without the existence and participation of the society, respectively their citizens. The electronization of this area has the potential to build good relations between government authorities, citizens and businesses by facilitating, refining and stream-lining interactions (Weerakkody, 2012, Lee et al., 2005). As the literature suggests, the quality of public services depends on ICT infrastructure and eGovernment readiness in the country (Mukamurenzi, 2019, Sharma, 2015, Weerakkody, 2012, Martin and Rice, 2011), interaction with citizens (Robbins, 2015, Braithwaite, 2007, Needham, 2006, Moon, 2002) as well as the scope and maximum availability of services (Gouscos et al., 2002). In Slovak republic, the last round of changes were important in transition from a complicated and unclear public administration system to a simple, active and transparent one. Using the new area and reality, the State can communicate with client more efficiently and flexibly. With the launch of the ESO (Effective, Reliable, Open Public Administration) reform in 2013, the State seeks to fulfill the notions of eAdministration and a positive perception of "mandatory" bureaucracy. For these reasons, we had decided to present and compare indicators of the European Commission study of 2014–2018 with the results obtained by us in the relevant years, in particular through a questionnaire survey. Through the questionnaire, we asked respondents about their agenda problem solving experience. The reasons for this activity are the efforts to monitor and compare the situation in time, point out the strengths/weaknesses and evaluate the results. However, in this article we will try to illustrate only some of the results. At the same time, we will try to draw conclusions on eGovernment. These can serve as a backbone for the successful achievement of goals and the development of an innovative vision for the functioning of public eServices, as eTransformation is a key element of success in removing barriers and reducing administrative burdens.

2   State as a Public Service Producer

The State represents a particular form of social establishment. At the same time, it represents a political, economic and social constitution based on the application of specific management tools. The State has to regulate governance practices, integrate citizens' interests in public policy-making, harmonize relations and links between subjects, especially in state-citizen relationship. These activities were pointed out in ancient times by Aristotle, "the State as an apparatus for fulfilling the common good, is to establish a good life for citizens by imposing rules and making laws (constitution)". This idea is complemented by the fact that "a community of any kind can achieve order only if the ruling element is represented by authority". At the same time, another interesting idea of Aristotle, in the work of Zarri (1948), is that the State "is being created and functioning to satisfy the bare needs of life". In a modern sense, the State is a political, economic and social player that emphasizes the elimination of external and internal undesirable phenomena and processes. Through public authority, it organizes the population (Boguszak, 1968), and is also referred to as the "guarantor of the established order in the State" (Kútik and Králik, 2015). The State should protect not only its own economic interests (Habánik et al., 2016, Habánik and Koišová, 2011), but also the interests of citizens (Karremans, 2017, Gilens and Page, 2014), guarantee the protection of people's lives and property (Howard-Hassmann, 2013, Králik and Kútik, 2013), to provide social security and to provide public services (Ochrana, 2001, Stiglitz, 1997).

In the State´s interest should be a comprehensive solution to optimize processes leading to quality improvement of citizens life/society and economic growth of the country. For this reason, attention should be paid to the qualitative side of public services, namely to their electronic form. The European Commission has launched an eGovernment Action Plan 2016–2020 (2016–2020) to address this issue and its challenges. The Commission calls for the principles of digital-by-default, cross-border by default, once-only principle, inclusive by default; privacy & data protection; Openness & transparency by default (European Commission, 2016). It also recommends respecting the principles of good governance (transparency, legality, impartiality, participation, etc.). These goals can be achieved through a sophisticated concept of universal access point and its general acceptance.

2.1 Citizen as a Natural Personnel Substrate of the State

In a well-organized State, the citizen has the exclusive right to participate in public life. It is therefore desirable that both sides respect each other and work together to meet the objectives and expectations. We can say that many of the governments have made progress in this area (European Commission, 2018, Transparency International, 2017). A new study on eGovernment services in the EU reveals that online public services are becoming increasingly accessible across Europe, 81% being now available online (eGovernmnet Benchmark 2018). While trying to move forward in eServices, improve user accessibility and connect to the single the digital market, the problem arises with the qualitative characteristics of these services; e.g. simplicity, transparency, reliability and speed (European Commission, 2018). Within the period 2016 and 2017 and the scoring goals achieved for eight life events measured on the four reference levels (user centricity, transparency, cross-border mobility and key prerequisites), Malta, Estonia, Austria, Latvia and Denmark were more than favorably assessed. In particular, the top-five scoring countries have managed to make public services widely available online, in a mobile friendly manner and with a strong focus on citizen and business users. At the same time, as illustrated in Fig. 1, there is room for growth and development of the issue in all countries under review. The fact that the low-performing countries (compared to 2015 and 2016) in this area are reaching up digital and electronic leaders in this area is also pleasing.

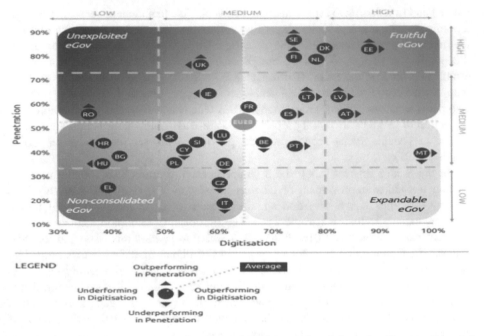

Fig. 1. Overall eGovernment performance (absolute and relative) in Europe (biennial 2016 + 2017 averages). Source: European Commission, 2018

Looking at the results of the Slovak Republic (on figure refered as SK), we can say that the level of digitization (63%) and penetration (53%) is moderate low. Based on eGovernment Benchmark 2018, we can conclude that Slovak portals and services belonging to them are not functional enough (62%), but user characteristics are not user-friendly (52%), digital context characteristics are relatively low compared to similar countries (55%). In view of the results, observations and objectives of the ESO reform (an effective and reliable open administration), the approach of the State has to be more vehement. It is not enough just to modernize and lean the state administration, increase transparency and openness or support the socio-economic growth. It is also necessary to build a well-functioning eGovernment, which is based on sophisticated eServices and office aparatus. The quality of the office aparatus, or building a network of professional personnel equipment, is not a problem of the Slovak Republic alone. Good administration points, as well as the need to share new knowledge, are a problem throughout Europe, which has led the EU to patronize this issue (eGov conference, ePractice.eu and others).

2.2 Citizen in the First Place

The principles of modern democracy (Whelan, 2018, Vančura, 2012, Terchek and Conte, 2000) enhance the role of the citizen. It becomes a center of action and interest, which has not been so far.

Focusing on and anticipating citizens' needs is now crucial. States strive to meet expectations and increase citizens satisfaction with their rights and obligations (Holomek, 2015, Doherty, Horne and Wootton, 2014, Holomek, 2015). At the same time, they are adapting this area so that people with disabilities can also use their services comfortably. They support the functioning of this platform by reducing the administrative burden for players, especially citizens. They are actively approaching the introduction of new technologies and the use of ICT-based public services. This also increases transparency and reduces the possibility of corruption at all levels (Beblavý et al., 2001).

Looking for new interactions between partners for so-called holistic maps. This relatively new category of relation translates into quantitative and qualitative indicators that reform the entire society of the State. Restrictions regarding office hours or paper support are no longer justified. Citizens are allowed to implement selected mandatory and voluntary agendas electronically, without waiting periods or visits the authorities. As an example we can mention Sweden, Malta, Estonia or Denmark, so-called eStates. From Table 1 presenting the achievements in eGovernment, we can see that all EU countries are working to change the status of public services. However, eStates were able to cope best in transformation process. But the remaining countries are no worse at all. Most of them should focus on citizens cross border mobility. The reason is low to zero possibility of using eID and weak eDocuments apparatus. Nevertheless, issues of social, property or health nature can be addressed in the EU through registers and unrestricted access to information. The truth is that only the eStates managed to work independently and completely automatically.

It follows from the foregoing that it is natural to turn the 'terile' environment of the public administration and sector into a 'fertile' environment; flexible, creative, fully functional, reliable and trustworthy.

3 Results

Since 1993, the Slovak Republic has been trying to bring freedom and democracy, development and progress to its citizens. By realizing the values and responsibilities associated with it, it offers to citizens more and more space and opportunities to participate. On the other hand, the State is very well aware of what represents the status of an organizational unit. This is particularly evident in public administration and public services. It seeks to design these issues as optimally as possible, taking into account the opportunities and expectations of stakeholders. The reforms that have affected this area are a clear picture. Mention may be made of the change in the nature of administration and the organization of local state administration (1996–2007), decentralization and modernization of public administration (1999–2019) and computerization of public administration, which are currently in place.

The strategic objectives of the Slovak Republic in the field of eGovernment are conceptually organized into four groups: Satisfaction of citizens, Businesses and other public authorities; Electronic public administration services, Effective and efficient public administration and increased competences of public administration (eGovernment in Slovakia, 2015). Fulfillment of these objectives significantly reinforces the

Table 1. eGovernment performance across policy priorities.

State	User centricity		Transparency		Citizens cross border mobility		Business cross border mobility		Key enablers		Country average		Individual interaction with PA	
	2016	2017	2016	2017	2016	2017	2016	2017	2016	2017	2016	2017	2016	2017
Austria	91	92	79	76	78	59	90	64	71	85	81	77	60	62
Belgium	85	85	73	67	73	53	58	47	58	60	69	62	55	55
Bulgaria	70	71	39	43	30	28	60	59	23	25	44	45	19	21
Croatia	62	64	43	46	58	34	38	45	12	19	42	41	36	32
Cyprus	67	72	50	45	60	42	77	76	58	44	62	55	38	42
Czech rep.	70	76	55	61	55	45	57	61	39	48	55	58	36	46
Denmark	95	93	71	68	68	58	100	84	91	88	85	78	88	89
Estonia	88	91	83	84	74	69	83	73	85	90	82	81	77	78
Finland	90	93	63	66	92	75	63	71	52	66	72	74	82	83
France	86	87	62	64	54	57	65	63	44	45	62	63	66	68
Germany	90	87	75	61	65	37	81	68	64	52	75	63	55	53
Greece	69	75	27	37	60	38	43	47	8	16	41	42	49	47
Hungary	68	68	26	33	40	13	30	38	33	47	39	39	48	47
Ireland	85	85	53	55	70	66	72	74	25	23	61	60	52	55
Italy	82	87	54	54	38	27	78	67	61	56	62	58	24	25
Latvia	88	87	72	76	71	63	99	89	72	82	80	75	69	69
Lithuania	85	89	83	86	52	32	70	64	84	86	74	71	45	48
Luxembourg	80	80	50	50	21	51	63	64	52	51	53	59	76	75
Malta	98	97	94	95	97	87	100	89	98	99	97	95	45	46
Netherlands	88	92	70	69	97	70	71	62	82	78	81	74	76	79
Poland	75	79	53	51	59	34	46	40	35	40	53	48	30	31

(continued)

Table 1. (*continued*)

State	User centricity		Transparency		Citizens cross border mobility		Business cross border mobility		Key enablers		Country average		Individual interaction with PA	
Portugal	93	93	69	70	53	48	100	78	70	71	77	72	45	46
Romania	60	63	48	44	73	32	21	18	5	12	41	33	9	9
Slovakia	**72**	**75**	**33**	**37**	**56**	**26**	**39**	**54**	**53**	**57**	**50**	**48**	**48**	**47**
Slovenia	81	80	51	44	70	60	52	49	43	38	59	54	45	50
Spain	88	90	78	73	41	37	78	79	77	73	72	70	50	52
Sweden	93	89	74	67	79	71	95	76	81	67	84	74	78	84
Un. Kingdom	82	80	59	59	51	53	92	90	22	22	61	60	53	49
EU28	**81**	**82**	**60**	**60**	**61**	**48**	**68**	**50**	**53**	**55**	**64**	**59**	**55**	**59**

User Centricity – indicates to whatextent (informationabout) a serviceisprovided online and howthisisperceived.

Transparent Government – indicates to whatextentgovernments are transparent regarding: theirownresponsibilities and performance, theprocessofservicedelivery and personal data involved.

CrossBorder Mobility – indicates to whatextent EU citizenscan use online services in another country.

KeyEnablers – indicatestheextent to which 5 technicalpre-conditions are available online: ElectronicIdentification (eID), Electronicdocuments (eDocuments), AuthenticSources, ElectronicSafe (eSafe), and Single Sign On (SSO).

Source: processed using by European Commission (2017, 2018), Eurostat (2018) launch of the ESO reform (MV SR, 2014). The aim of this reform is to enable all citizens to benefit from new service platforms, to reduce the administrative burden for citizens and businesses through transparent procedures, or to create and link registers to be used for legal actions. However, this sophistication of services occurs very slowly, inflexibly and unnaturally. This is confirmed by the experience of the lay and professional public, but the State negates this. (NR SR, 2014, MV SR, 2014, Piško, 2013).

As an example we can mention the reduction of administrative burden in relation to the Social Insurance company. Eventhough the Act no. 177/2018 on certain measurements on reducing the administrative burden by using information systems of public administration have been in force for one year, only now is this Act being fulfilled. Documents such as the Certificate of entitlement to parental allowance, a copy of the birth certificate of a child or a Letter of Ownership of real estate do not have to be submitted to the Social Insurance company in paper form. Similarly, parents do not have to send paper receipts to the labor offices to prove their entitlement to child benefit. The reason is the extension of the platform of the public administration information system to include the Central Register of Children, Pupils and Students and the Central Register of University Students.

On contradiction, also point results of the European Commission (see Fig. 2); DESI - index of digital economy and society. This index makes it possible to evaluate and compare EU countries in terms of digital economy and society. Based on given results, we can confirm that there is a slow progress in this issue in the Slovak Republic. However, the positive result of Slovakia is that its value of Digital Public Services almost doubled in comparison with other years. In this context, it is important to note that the Slovak Republic is still below the EU average. A significant contribution to this, is the minimal progress of use of public electronic services aimed for natural persons. While these are in the first place in the viewfinder of the State, in the first phase of the eServices transformation, more attention was paid to the citizen – legal persons, i.e. legal entity. Currently, work is under way to improve the environment of eServices for natural persons.

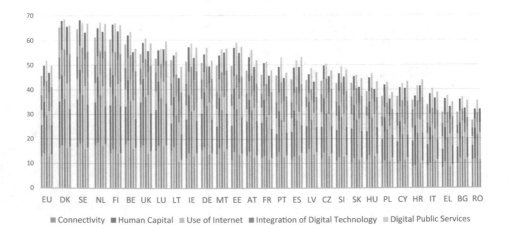

Fig. 2. DESI in time procedure 2014–2018. Source: processed using by European Commission (2014–2018)

We came to similar conclusions based on our own investigation. We collect our data systematically; at regular two-year intervals. The current base of respondents consists of 637 citizens actively implementing agendas with the state. However, as shown in Fig. 3, there is little progress in services. This is due to the above mentioned - focusing the State on citizens – legal persons. At the same time, we can state that citizens' expectations are not met. 74% of respondents do not perceive significant quality improvement of services. Their opinion is based on insufficient quantitative transformations of services (73% of respondents), despite the fact that they can use the assisted service, 660 integrated service points (IOM) or 49 client centers. They also talk about inadequate time allocation for the agenda (68% of respondents). The results of our findings suggest that the average waiting time at the workplace is 1 h 52 min. This was negatively supported mainly by waiting times at traffic inspectorates. In the case of client centers - special workplaces, this time is estimated at 32 min. However, the reports of the Ministry of Interior of the Slovak Republic speak only of waiting times in client centers, namely 19 min intervals and up to 91% satisfaction of clients of the given centers. Differentiation may be due to sample size and interval enumeration methodology. Nevertheless, we believe that these intervals should be lower in both cases. Respondents also point to shortcomings of eServices such as inflexibility (43% of respondents), low added value (69% of respondents), user subordination (84% of respondents) and especially their offer (82% of respondents). It is possible to request e.g. for social benefits, assistance in material need, care allowance or personal assistance. When using eServices, we are confronted with great concerns and too high expectations of the respondents, which, however, have changed over time, but we are not convinced that for the better. In 2014, 61% of respondents were convinced that they would be able to use the eServices system 24 h a day, 7 days a week, without visiting the office, almost immediately by launching the reform. However, in the active use of eServices and eID registration, it was only around 0.5% of respondents surveyed. The rest of the respondents did not have such a license and did not plan to remedy in the near future (3–6 months). This was mainly due to the lack of coverage of eServices, their range, offer and functionality. A certain reversal can be observed in 2016, when the status of eServices improved in terms of functionality and platform expansion, but this status was not rated positively. In the last reporting period, the situation improved, slightly in the assessment and number of eID-registered clients. Compared to the previous year, the number of registered clients is 18.62%. We can see the problem in the presentation and slow expansion in the range of eServices.

Citizens are also discourage from acquiring a new form of ID with an electronic contact chip – entry to the world of eServices. In the context of eID, the issue not only the usability of eGovernment services but also the amount of its procurement (the administrative fee of an eID for a still valid ID without an electronic chip – 4.50€) has been very resonant. Our opinion is that in the case of the first issue of eID to citizen - natural person, it should be for free. This would increase interest in this method and the possibility of communication with the State and its authorities (Table 2). To increase its usability, besides reducing the burden on the citizen in the area of documentary evidence, an increase in the tempo of the agenda's solution by the State would be achieved. At the same time, this may be one way of reversing the ongoing trend of low levels of online public-government interaction.

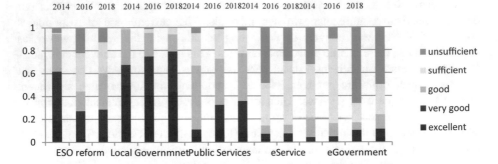

Fig. 3. Questionnaire results

In the field of eServices can be positively evaluated pace and quantity of publishing documents and information through its own websites, but also through the common portal slovakia.sk. Through it, the citizen can actively apply for interaction with the authorities. However, the basic prerequisite for receiving, notifying and submitting the agenda is to activate the eBox. It is conditioned by eID with verified electronic signature. The results of research indicate low growth in number of people using eEnvironment and eID disposition (growth by 6%). However, on the other hand, by launching and updating the public administration portals, citizens spend more time preparing for contact with the authorities. While in 2014, according to our findings, they spent an average of 7 min on this activity, in 2018 the value raised to 16 min. Documents (235 samples) or information that they find on the relevant portals will be prepared and filled in advance at home by 32% of respondents.

Subsequently, they will visit the Office and follow the agenda personally to the desired end. This can be viewed positively in the context of citizens' interest in the issue. On the other hand, there is a lack of understanding the idea of reducing the burdens. Therefore, the State should urgently address and involve in the eServices scheme not only selected, but all services that fall within its field of competence. The rate of change in thinking and application of new technologies and processes in trusted governance should be strengthened. At the same time, this environment should be monitored in much wider dimensions. In addition to fulfilling the declared aims, the individual institutes should be interconnected in such a way that automatic data collection takes place without the citizen's intervention in the system. Moreover, which is a trend in other States, the portfolio of services is expanding into areas related not only to work but also to leisure activities. Therefore, it is possible to use eID for traveling, visiting museums, galleries and libraries as well as many others. We believe that the process of transforming eServices in our environment should also be taken in this direction. We would therefore like to make our contribution to the State and its authorities to act more flexibly on this issue. By the end of the implementation period of the ESO reform (2020), we can hardly speak of its 100% fulfillment. We are less than halfway to our goal. While the results of our inquiries speak of client confidence in the eSystem, its slow pace of expanding the range of services may pose a threat that could undermine that confidence. We positively evaluate the first "swallow birds" that

Table 2. eInteraction with the state

Month	2017				2018			
	eSubmitting		Activation of eSubsets		eSubmitting		Activation of eSubsets	
	Number	$\%^*$	Number	$\%^{**}$	Number	$\%^*$	Number	$\%^{**}$
January	81118	4,23	19156	0,39	252197	4,12	58785	1,19
February	63638	2,98	21333	0,44	243404	3,96	61454	1,24
March	55900	2,36	23655	0,48	265543	4,12	64430	1,30
April	89966	3,51	25627	0,52	249693	3,75	66521	1,34
May	157373	5,67	27773	0,56	373791	5,48	68239	1,38
June	159779	5,02	31794	0,64	286999	4,21	70923	1,43
July	153067	3,67	41688	0,85	297276	4,05	73395	1,48
August	186776	4,12	45319	0,92	323913	4,30	75397	1,52
September	224654	4,55	49365	1,00	unavailable		unavailable	
October	209458	3,97	52730	1,07	unavailable		unavailable	
November	199875	3,69	54149	1,01	unavailable		unavailable	
December	174145	3,11	56011	1,14	unavailable		unavailable	

* Calculation based on amount of Active eBoxes
** Calculation based on amount of Total eBoxes
Source: processed by statistika.slovensko.sk

really disburden clients - the Social Insurance company or the Office of Labor of Social Affairs and Family. The news in their area will certainly be positively evaluated in our further inquiries. Therefore, we recommend involving all institutions, including self-governments, in the systems. Along with this, rebuild the network of passive contact points into actively integrated contact points. This will save time that clients spend in institutions while waiting for their agenda.

4 Conclusion

Evaluation of activities in such a wide area is difficult because of its growth, albeit slow. The results of the evaluation institutes we used in the research are somewhat similar to ours. On the one hand, this pleases us, but on the other it grieves.

The results and findings of our activity point to a number of problem areas concerning eGov and eServices in the Slovak Republic. As the first one we can mention the slow and inflexible transformation of the new form of services. The current state of affairs indicates that a citizen - legal person is moving faster in this context than a citizen - natural person. This is due to the obligation of business entities to communicate with the state through this environment. The number of eServices, which are based on electronic form, is growing and the paper equivalent does no longer exists. This is missing in case of a natural person. They do not have such an obligation, which is logical, especially given the low level of eServices intended specifically for this client. It would therefore be appropriate to continually address both forms for the citizen. The question also arises as to whether it is necessary to levy an administrative fee from citizens who nevertheless express this kind of communication with the State. And last but not least, it is necessary to promptly solve the whole range of the issue, from thinking the basic article of this transformation - bureaucrat, from clerk to the highest, which is the State itself. The results, which are not in accordance with the declared ones, insufficient communication and participation with users or the dislike and disinterest of citizens, undermine the authority and trust that it should be seek first. Partnership and learning from eAuthorities, building a range of benefits and real participation and perception of the citizen at both levels are therefore appropriate. But for the completion of electronic services, political will is needed, only it can start the process.

References

Röstl, A., Hollander, P.: Aristoteles: Aténska ústava. Kalligram, Bratislava (2009)

Beblavý, M., Sičáková, E., Zemanovičová, D.: Transparentnosť v štátnej správe. CPHR TIS, 33 p (2001). http://www.transparency.sk/wp-content/uploads/2010/01/030807_trans.pdf. Accessed 01 Aug 2019

Boguszak, J.: Teorie státu a práva. Díl II. Orbis, Praha (1968)

Braithwaite, V.: Responsive regulation and taxation: introduction. Law Policy **29**(1), 3–10 (2007)

Doherty, T.L., Horne, T., Wootton, S.: Managing Public Services – implementing Changes. Routledge, New York (2014)

European Commission (2015). eGovernment in Slovakia. p. 45. http://portal.egov.sk/en/content/egovernment. Accessed 01 Aug 2019

European Commission (2016). eGovernmnet benchmark 2016. p. 80. https://publications.europa.eu. Accessed 01 Aug 2019

European Commission (2017). eGovernmnet benchmark 2017. p. 190. https://publications.europa.eu. Accessed 01 Aug 2019

European Commission (2018). eGovernmnet benchmark 2018. p. 56. https://ec.europa.eu/digital-single-market. Accessed 01 Aug 2019

Eurostat (2018). Population statistics. https://ec.europa.eu/eurostat. Accessed 01 Aug 2019

Gilens, M., Page, I.B.: Testing theories of american politics: elites, interest groups, and average citizens. Camb. Univ. Press **12**(3), 564–581 (2014)

Gouscos, D., Laskaridis, G., Lioulias, D., Mentzas, G., Georgiadis, P.: An approach to offering one-stop e-government services - available technologies and architectural issues. In: First International Conference on Electronic Government, EGOV 2002, pp. 113–131 (2002)

Gupta, M.P., Jana, D.: E-government evaluation: a framework and case study. Gov. Inf. Q. **20**(4), 365–387 (2003)

Habánik, J.: a kol.: Zmeny v hospodárskej štruktúre regiónov SR v rámci sociálno-ekonomického rozvoja. TnUAD, Trenčín (2016)

Habánik, J., Koišová, E.: Regionálna Ekonomika a Politika. Sprint, Bratislava (2011)

Holomek, J., Klierová, M.: Verejná správa (organizácia, kompetencie, elektronizácia). TnUAD, Trenčn (2015)

Howard-Hassmann, R.E.: Reconsidering the Right to own Property (2013). https://www.tandfonline.com/doi/full/10.1080/14754835.2013.784667. Accessed 13 Apr 2019

Karremans, J.: State interests vs citizens´preferences: on which side do (Labour) parties stand?. EUI Ph. D. theses, Department of Political and Social Sciences (2017). http://cadmus.eui.eu/handle/1814/45985. Accessed 12 Apr 2019

Králik, J., Kútik, J.: Kontrolný Systém a Jeho Subsystémy vo Erejnej Správe. Aleš Čeněk, Praha (2013)

Kútik, J., Králik, J.: Krízový manažment a verejná správa. TnUAD, Trencín (2015)

Lee, S., Tan, X., Trimi, S.: Current practices of leading e-government countries. Commun. ACM 48(10), 99–104 (2005)

Martin, N., Rice, J.: Evaluating and designing electronic government for the future: observations and insights from Australia. Int. J. Electron. Gov. Res. 7(3), 38–56 (2011)

Moon, M.J.: The evolution of e-government among municipalities: rhetoric or reality? Public Adm. Rev. 62(4), 424–433 (2002)

Mosse, B., Whitley, E.A.: Critically classifying: UK e-government website benchmarking and the recasting of the citizen as customer. Inf. Syst. J. 19(2), 149–173 (2008)

Mukamurenzi, S., Grönlund, Å., Sirajul, M.I.: Improving qualities of e-government services in Rwanda: a service provider perspective. EJISDC (2019). https://onlinelibrary.wiley.com/toc/16814835/0/0. Accessed 12 Apr 2019

MV SR: Bilancia reformy ESO po dvoch rokoch (2014). http://www.policiasr.eu/?tlacove-spravy&sprava=bilancia-reformy-eso-po-dvoch-rokoch. Accessed 12 Apr 2019

Needham, C.E.: Customer care and the public service ethos. Public Adm. 84(4), 845–860 (2006)

NR SR: Odpoveď na interpeláciu č.4 (2014). www.nrsr.sk/web/Dynamic/Download.aspx?DocID=436415. Accessed 14 Apr 2019

Ochrana, F.: Verejný Sektor a Efektívní Rozhodování. Management Press, Praha (2001)

Piško, M.: Kaliňák reformu ESO prechválil, živnostníci sa stále nachodia (2013). www.domov.sme.sk/c/6873776. Accessed 14 Mar 2019

Robbins, G., Mulligan, E., Keenana, F.: E-government in the Irish revenue: the revenue on-line service (ROS): a success story? Financ. Account. Manag. 31(4), 363–394 (2015)

Sharma, S.K.: Adoption of e-government services. Transform. Gov. People Process Policy 9(2), 207–222 (2015)

Slovensko Štatistiky. https://statistiky.slovensko.sk. Accessed 14 Mar 2019

Snellen, I.: Electronic governance: implications for citizens, politicians and public servants. Int. Rev. Adm. Sci. 68(2), 183–198 (2002)

Stiglitz, J.E.: Ekonomie Veřejného Sektoru. Grada, Praha (1997)

Terchek, J.R., Conte, T.C.: Theories of Democracy. Rowman & Littlefield Publishers, Lanham (2000)

Vančura, P.: Teorie Demokracie. Trinitas, Svitavy (2012)

Welch, E.W., Hinnant, C.C., Moon, M.J.: Linking ctizen satisfaction with e-government and trust in government. J. Public Adm. Res. Theory 15(3), 371–391 (2005)

Weerakkody, V.: Technology Enabled Transformation of the Public Sector. IGI Global, Hershey (2012)

Whelan, F.G.: Democracy of Theory and Practice. Routledge, New York (2018)

Zarri, J.: Aristotle's theory of the origin of the state (1948). http://www.scholardarity.com/wp-content/uploads/2012/10/Aristotles-Theory-of-the-Origin-of-the-State-DRAFT-2-PDF.pdf. Accessed 13 Apr 2019

Problems of Use of Information Technologies for Management of Social and Economic Development of the Municipal Unit

Akifieva Larisa Vladimirovna$^{(\boxtimes)}$ ⓘ,
Bolshakova Yulia Aleksandrovna ⓘ, Ilicheva Olga Valerievna ⓘ,
Kuchin Sergey Valentinovich ⓘ,
and Belousova Olga Alexandrovna ⓘ

Nizhny Novgorod State Engineering and Economics University,
Nizhny Novgorod Region, Knyaginino 606340, Russia
laraakif@mail.ru

Abstract. There are considered problems of use of information technologies in the public and municipal administration in the article. Development of legal and technological base of digital government is investigated. There are considered directions of informatization of municipal management, introduction of electronic document management. Examples of use of end-to-end digital technologies, cloud computing, social networks, «The Internet of things» are given and the problems connected with data application of technologies are specified, key problems are allocated. Solutions of the revealed problems with the help of adaptation of foreign and Russian experience of the megalopolises which are based on use of the following types of functional services are offered: information services, portal of electronic services, communication services of the platform of electronic polls and referenda of citizens.

Keywords: Informatization of municipal management · A municipal unit · Information technologies · Social and economic development · «Smart City» · Digital government · Functional services

1 Introduction

Large-scale use of modern information systems and technologies in activity of municipal authorities is caused by the positive impact of this process on realization of concrete functions of local self-government: rendering municipal services to the population, expeditious informing inhabitants on results of the activity, the organization of «feedback» with participation of the population, data exchange between structural divisions and departments, the organization of document flow and data handling. Use of opportunities of modern information technologies in municipal management allows to perform informing mass audience, to involve active inhabitants in public discussion of development problems of the municipal unit and development of their decision versions, to lower expenses of local self-government.

T. Antipova and Á. Rocha (Eds.): DSIC 2019, AISC 1114, pp. 470–479, 2020.
https://doi.org/10.1007/978-3-030-37737-3_40

2 Use of Information Technologies in Municipal Management

The purposes of forming and development of an informational society are improvement of quality of life of citizens, ensuring competitiveness of a municipal unit and a territorial subject of the Russian Federation, development of economic, socio-political, cultural and spiritual spheres of life of society, improvement of a system of municipal management on the basis of use informational and communicative technologies (further ICT). Today active development and broad use of ICT in activity of local government authorities is one of important factors of increase in the level of social and economic development of the municipal unit, efficiency of municipal management.

Freely made OECD Better Life Index 2019 on the basis of over two thousand the countries of indexes personally made by inhabitants shows that on the basis of 11 fundamental aspects of quality of life of a modern person of the Russian Federation takes the last places (8) among such countries as Japan, Great Britain, the USA, Sweden, Germany, etc. [1], nevertheless, the Russian Federation for the last 3 years increased this level on 3 points [2, p. 128].

According to Gavrilov A.E., it is necessary to do a great job on integration of society and state for effective and mutually productive interaction. The program of realization ща such updating of the existing mechanism of realization of will of people and the system of management demands creation of the universal productive complex communication medium which will provide: an available information site in any time and place; politicy of openness and publicizing of the events by freedom of information sources; publication of current and interesting events by authors or those who are fond of these events; adaptation of contents of a resource under properties and interests up to the specific user; possibility of expression of the opinion and offers; a possibility of interaction of users according to modern communication methods; means of forming of petitions, statements and programs of a social significance; obligatory attention of the state and municipal bodies to the offered public projects [2, p. 133].

In the last decade the problem of informatization of municipal management processes purchased special importance. It is connected as with strengthening of a role of information and communication channels in the world in general, so with updatings of questions of information technologies implementation in the standard legal framework of the public and municipal administration.

Starting point of administrative reform realization on information and telecommunication infrastructure development and modernization of the municipal unit became the following documents:

- the federal target program «Electronic Russia (2002–2010)» approved by the Resolution of the Russian Federation Government from the 28.01.2002 № 65;
- the strategy of development for the informational society in the Russian Federation approved by the Presidential decree of the Russian Federation on February 7, 2008 № Pr-212 (nowadays it has lost the legal force in connection with the adoption of the new Strategy for 2017–2030);

- the State program of the Russian Federation «An informational society (2011–2020)», approved by the Resolution of the Russian Federation Government from 15.04.2014 № 313.

General purpose of above-mentioned programs was improvement of quality of life and professional activity of citizens of the Russian Federation, improvement of operating conditions of the organizations and enterprises, development of social and economic capacity of the state and increase in its competitiveness by means of use of information and telecommunication technologies, including, in activity of bodies of the public and municipal administration [3, p. 167].

Later basic provisions of federal documents found the reflection at the level of territorial subjects of the Federation in the form of similar regional programs and concepts of information policy with a further specification and transformation within each municipal unit of the region.

For example, in strategy of social and economic development of Nizhny Novgorod Region till 2035 there is planned creation of project office for assistance to development in the region of the leading companies of digital economy (DE) within the Central Committee of Scientific and Technical Documentation (TsKNTD), the realization of pilot projects with the use of end-to-end digital technologies in Nizhny Novgorod region («Smart City», «The Digital Enterprise», «Creation of the Centers of Collective Access to Supercomputer Clusters» and so on) and others [4].

In addition to legislative initiatives on digitalization of economics it is necessary to tell also about other factors which objectively influence this process. Today the quantity of the workplace equipped with computers and having Internet connection grows (Fig. 1) [Regions of Russia. Socio-economic indexes/Federal State Statistics Service [5] (see Fig. 1).

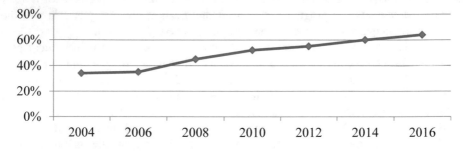

Fig. 1. A share of the workplace equipped with Internet connection, %

Special attention is paid to informatization of local government authorities and increase in efficiency of municipal management by means of the use of information technologies, this level of the power is brought much closer to the population and is a basic point of forming the whole management system in the country, and, at the same time having a set of special problems, beginning with the Internet of the population of the municipal unit from the lack of a possibility of access to the network, finishing with information illiteracy of local government officers, in particular, in rural settlements.

The main directions of informatization of municipal management are introduced in Fig. 1 (see Fig. 2).

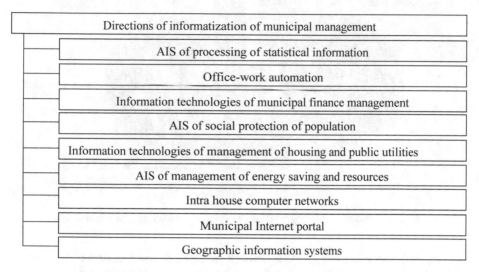

Directions of informatization of municipal management
AIS of processing of statistical information
Office-work automation
Information technologies of municipal finance management
AIS of social protection of population
Information technologies of management of housing and public utilities
AIS of management of energy saving and resources
Intra house computer networks
Municipal Internet portal
Geographic information systems

Fig. 2. Directions of informatization of municipal management

Various tools are developed and used for improvement of the main directions of social and economic development of the municipal unit today. Digitalization is one of strategic directions of development of municipal management.

At the moment the interest in the strategy «Smart City» and its realization grows, as the purpose of «the smart cities» is increase in efficiency of city services with use of the modern information and communication technologies (ICT).

In Fig. 3 leaders in development of the concept «Smart City» on federal districts are introduced (see Fig. 3). Apparently from Fig. 1, the main leaders are concentrated in Volga Federal District, and in two districts leaders in development of the concept of «Smart City» are absent – the Far Eastern Federal District and the North Caucasian Federal District [6, p. 27].

At implementation of the concept of «Smart city» there are used: intellectual management of lighting in the city, a traffic congestion control system, the common system of city security, the automated transport system, an electronic payment system of city municipal services, the electronic system of receiving various public services, the system of full the complete Internet covering, the system of climatic control in municipal and nation-wide constructions and many others. This list can extend in process of introduction of new subsystems and accession them to a single system (see Fig. 3).

For example, the Smart City (Settlement) Project (according to the Strategy of social and economic development of Nizhny Novgorod Region till 2035) assumes increase of quality of citizens' life due to the use of a number of digital services covering various spheres of life of Nizhny Novgorod city became the pilot site for their

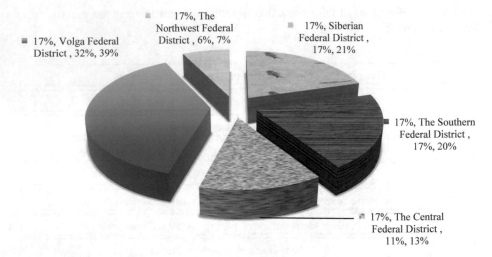

Fig. 3. Leaders in development of the concept of «Smart City» in districts

working off. The first version of the concept assumes the navigation and information system (management of passenger traffic, school buses, cars of a system of housing and public utilities, monitoring of movement of transport) and introduction of an automated system of control of payment of journey, work of service «112», the systems of video fixing of traffic regulations and video surveillances, weight control and forest fire observation. Besides, the list of opportunities of «Smart city» includes development of a system of providing the state services in electronic form and project implementation directed to involvement of inhabitants in process of management of the region. Constant development of the concept «Smart city» with participation of inhabitants for determination of the following services which need to be created [4] is planned.

Due to the state on January 1, 2018 in Russia there are 2777 multipurpose centers and 10558 small offices of the multipurpose centers in the places with small population which provided the population with public services whose quality gradually grows due to the population survey results (Table 1) [7, p. 38].

Table 1. Indicators of quality of rendering state services in the Russian Federation, 2011–2017.

	2011	2012	2013	2014	2015	2016	2017
Indicator of satisfaction of citizens In quality of providing state services, %	74,6	75,5	77,7	81,2	83,3	82,9	86,4
Average time in the queue, min.	There are no data	55	There are no data	There are no data	35,7	22	18,7

However, «Smart cities» have a number of problems which need to be permitted for their further effective functioning. We will notice that problems of these cities are connected with unwillingness of citizens to provide data as they are afraid that the information systems solving city problems except necessary data will obtain also their personal information. This problem is connected with the fact that people interact insufficiently among themselves.

The information and technology basis of the digital government is made by the following trends: cloud computing, Big Data, analytics, mobility, social networks, «Internet of things», cyber security [6, p. 26].

We will note that one of the problems of «Internet of Things» technology is the cyber security. It is very important to ensure information security as because of cyber attacks these technologies will not be able to work, and, therefore, there will be a decline in quality of life of citizens and level of their interaction to bodies of municipal authority. Use of services of information security used on public cloud platforms is also effective from the economic point of view because of rather small cost and can be an exit from this situation.

Office software and e-mail is applied in all state procedures. In this sphere software products including cloud solutions are not protected from information leakages, besides, instruments of control of its distribution are absent. An example of office package for use in the public sector is «My office» which can overcome the specified problems.

Besides, realization of import substitution of information and communication technologies in a public sector is important. According to the Information Security Doctrine ensuring steady and smooth functioning of information infrastructure belongs to national interests in the field of information technologies. For realization of national interests in the field of information technologies it is necessary to develop domestic hardware and the software.

For today technologies of information processing are rather developed and continue to be improved therefore it is possible to note that processes of a mediatization and computerization are well under way. To tell the same concerning intellectualization process, according to us, prematurely as capabilities of people to perception and generation of information in many respects depend on properties of their brain and have biological limits. Therefore today process of intellectualization is expressed more likely in understanding of a role of information in the modern world, creation of the corresponding precepts of law, i.e., is shown indirectly, in acceleration of processes of a mediatization and computerization [8, p. 103].

Today basic standard, material and technological and organizational conditions for active use of information technologies in practice of municipal management are created as a result of carrying out administrative reform that is expressed in availability of official, regularly supported and updated websites, a possibility of receiving many municipal services in an electronic form, work of the multipurpose centers of provision of services, the organization of electronic document management and interdepartmental interaction at municipal units and many municipal enterprises.

Last several years public authorities and local government authorities have tasks of creation of conditions for improvement of quality of rendering public and municipal services, cutting of costs for the maintenance of government and increase in efficiency of its activity with use of modern information technologies.

The main function of any public institution is acceptance of management decisions. At the same time, basic data for decision making, necessary information and decisions are presented in a document type and the volume of technical work with documents is often quite big.

Introduction of electronic document management is now one of priority tasks of authorities which successful realization will allow to provide transition to the better level of work of government and local government officers. Automation of work with the circulating documents in authorities is essential for increase in efficiency of activity of both separate organizations and a management system in general. The systems of electronic office-work and document flow promote creation of a new organization culture in authorities, that leads to facilitating work of local government officers.

The accurate organization of document flow, competent preparation, timely and high-quality execution of the made decisions are guarantee of outstanding performance of management activity, the adjusted system of labour processes, and those lead to high authority of power structures, trust to them from citizens.

They had planned to perform transition of public authorities to electronic document management by 2010, then - by 2015. It is planned in the strategy of development of informational society in the Russian Federation approved by the Russian President on February 7, 2008 No. Order-212 that the share of electronic document management between public authorities has to be 70% in the total amount of document flow by 2015.

But the achievement of the planned results was not reached in the specified terms. It is confirmed by representatives of expert community.

Despite obvious achievements in this sphere last decade, some problems of application of information and communication technologies in municipal management do not lose the relevance, it is necessary to designate among them:

- preserving of inequality in logistics of structural divisions of local government authorities, heterogeneity and incompatibility of the used hardware-software providing;
- insufficient training for work with modern information technologies, absence of specialists in information technology use and development among staff of municipal bodies;
- low level of informing population on opportunities of use of the Internet for receiving municipal services and information about work of local government authorities;
- inequality in ensuring access to municipal services in an electronic form and information on work of local government authorities for inhabitants of rural and remote territories;
- low financial potential of local self-government for information technology use and development of management and independent solution of problems arising in this sphere;
- insufficiently developed information environment of interaction of municipal authority and population as an integral part of development of civil society and democracy.

The last problem is especially relevant, it does not allow fully to realize the potential of the idea of local self-government as initiative mechanism of the decision citizens of problems of territory and population.

When use of Internet services and mobile applications for forming communications with population is delivered on quite a high level in developed countries, the mechanism of participation of population in the course of acceptance of management decisions on development of certain administrative and territorial educations is only formed today in Russia, without relying on a sufficient regulatory framework and practical experience.

3 Results

According to Shamarova G.M., decision the problems interfering the increase in efficiency of integration to information technology development in the sphere of municipal management requires implementation of a number of actions directed on: improvement and development of a regulatory framework in the sphere of information technologies and protection; forming of standards and recommendations in the sphere of use of information and communication technologies (further - ICT) in municipal management for the purpose of introduction and increase in efficiency of use of ICT; development of a training system of specialists in ICT and the qualified users; ensuring access of local government authorities, employees, budget institutions to the Internet that will allow to use information effectively and to perform communication transmission of media of data and to perform exchange of documents in an electronic form; further distribution of electronic document management in local government authorities that will promote increase in efficiency of structural divisions interaction of an municipal unit administration as among themselves, and with the organizations and also the population of the municipal unit and others [9, p. 75].

According to us, for activization of a role of the population in the decision and discussion of problems of municipality it is expedient to adapt foreign experience and experience of large Russian megalopolises which are based on use of the following types of functional services:

– information services. They serve to provide the necessary information on the city to citizens, to establish information exchange between various departments of municipality and other suppliers of data and also between different groups of their suppliers – from city community to business and researchers;
– the portal of electronic services – the crucial element of a system of online service of the population allowing citizens to send requests, to keep track of the status of their processing and to receive directly public services via the web interface or the mobile application without physical presence at public institutions. Experience shows that this service is the most popular with the population of large municipalities as it is directed on much wider target audience in comparison with other projects;

- communication services. One of the most widespread formats is the interactive platforms allowing inhabitants to control work of municipal services: to report about problems and to monitor reaction to the complaints. Crowdsourcing projects of municipal administrations are also effective, i.e. interactive platforms for collecting opinions, ideas and offers from population. Scales of such projects and a format of realization of their results in municipality management depend on specific tasks and a subject. For example, collecting opinions on local projects on improvement belongs to crowdsourcing projects as exchange of ideas on the long-term strategy of municipal development;
- platforms of electronic polls and referenda of citizens. They represent interactive services for the formalized collecting of opinions of the population on various issues of development. The municipal administration forms the agenda/guide for discussion or poll, and the population is invited to vote and answers to the questions posed.

Change of strategic approach to interaction of municipal authority with the population from perception of citizens as passive observers and users of information and services to their perception as active actors in processing of data on municipal management, development of concrete management decisions, introduction of offers on improvement of the territory, etc. is advisable.

Communication services in municipal management are gradually displaced to professional competences today: the resource of crowdsourcing (made by a large number of people) information is used for the purposes of planning and management of municipal development. However, it is important to understand that complete knowledge of the population of managerial activity, understanding of management processes and planning of municipal development from population is necessary for establishment of high-quality dialogue between population and local government authorities and views of territorial community from municipal administration, that is forming of uniform language of communication.

4 Conclusion

Thus, in the Russian Federation own information and communication infrastructure is being created which will allow to provide digital sovereignty and to observe national interests in information security field. As it was stated above, a positive trend is transition on domestic hardware and software. We will notice that in the case of refusal of the foreign software the way out can be the use of free software, but development of domestic software has to become the priority direction. It is also necessary to expand implementation of social networks in public and municipal administration because the use of this tool will allow to improve interaction with citizens. Development of professional services of other types in interactivity degree can become the direction of improvement of professional communication services: transactional and informational ones. The separate group is made by the professional and civil communication services which are of special interest for the developed municipalities by experience of improvement of the existing projects and start of new initiatives. These services which along with requirements satisfaction and the rights of citizens contain the instruments of feedback starting crowdsourcing of the information which is valuable to municipal unit development management.

References

1. http://www.oecdbetterlifeindex.org/ru/#/11111111111, 28 Sept 2019
2. Gavrilov, A.E.: Use of modern information technologies in the concept of strategic management by the municipal unit. In: Modernization of the Russian Economy: Forecasts and Reality. Collection of Scientific Works of the International Academic and Research Conference, pp. 128–135 (2017)
3. Grigoriev, I.S., Gusev, M.N.: Information technologies in municipal management: experience of application in scales city districts. Scientists of a note of Tambov office of ROSMU, no. 10, pp. 167–174 (2014)
4. https://2035.government-nnov.ru. Accessed 23 Jan 2019
5. http://www.gks.ru/wps/wcm/connect/rosstat_main/rosstat/ru/statistics/publications/catalog/doc_1138623506156, 28 Sept 2019
6. Gaysinsky, I.E., Nikonenko, N.D.: Development of information and communication infrastructure of the public and municipal administration within digital sovereignty. Public and municipal administration. Scientific notes of SKAGS, no. 2, pp. 26–29 (2017)
7. Kurilo, A.E., Prokopyev, E.A.: Development of digital technologies in the system of public and municipal administration. Problems of market economy, no. 2, pp. 35–44 (2019)
8. Shamin, E.A., Generalov, I.G., Zavivayev, N.S., Cheremukhin, A.D.: Essence of informatization, its purpose, subjects and NGIEI Objects. Bulletin **11**(54), 99–107 (2015)
9. Shamarova, G.M.: Information technologies in municipal management: problems and prospects of the development. Practice of municipal management, no. 12, pp. 68–75 (2013)

Smart Village. Problems and Prospects in Russia

Anatoly E. Shamin⬚, Olga A. Frolova⁽⊠⁾⬚, Irina V. Shavandina⬚,
Tatyana N. Kutaeva⬚, Dmitry V. Ganin⬚, and Julia Yu. Sysoeva⬚

Nizhny Novgorod State Engineering and Economic University,
Knyaginino, Russia
ekfakngiei@yandex.ru

Abstract. Until recently, the agro-industrial complex of Russia as a branch of economy had low business processes automation. At the present stage, digitalization and automation of the maximum number of agricultural processes of all forms of management is a priority task of the state. The strategic objective is to maximize automation of all production cycle stages to reduce losses, increase agricultural business productivity, optimal resource management, rural infrastructure digitalization, digital and financial literacy of rural population. Based on an analysis of economic literature, it was concluded that intensive introduction of digitalization and the Internet of things into agriculture transforms the less influenced IT industry into a high-tech business due to explosive growth of productivity and reduction of non-productive costs, providing security of regions and the whole country.

Keywords: Agro-industrial complex · Digitalization · Agriculture · Agricultural organizations · Forms of management · Way of life · Efficiency

The time will come when science surpasses the fantasy
Jules Verne

1 Introduction

Western countries are successfully modernizing their economies, developing innovative technologies at an accelerated pace, dominated by artificial intelligence, big data analysis, automation and digital platforms. By 2020, according to the experts forecast, 25% of the world economy will move to digital technologies introduction, allowing the state, enterprises of various industries and society to function effectively.

However, the level of digital technologies introduction in agriculture is still low. World leaders in digital technologies implementation are IT-companies, media, finance and insurance. In real production and logistics, the digitalization level is much lower, and the lowest is in agriculture (in our opinion, the main deterrent factor is peculiarities of agricultural production under modern conditions). Russia ranks 15th in the world in terms of agriculture digitization.

© Springer Nature Switzerland AG 2020
T. Antipova and Á. Rocha (Eds.): DSIC 2019, AISC 1114, pp. 480–486, 2020.
https://doi.org/10.1007/978-3-030-37737-3_41

2 Agriculture Digital Transformation in Russia

Russia has one of the largest agricultural potentials in the world. With only 2.2% of the world population, it has 8.9% of world arable land, 2.6% of pastures, 20% of world freshwater and 8.3% of mineral fertilizer production. Such a natural resource potential allows production of almost all major agricultural products. Nevertheless, our country is one of the main countries among food importers [1].

It should be noted that the key subjects in economic relations to enhance the process of introducing digital economy elements into agriculture are government agencies and agricultural commodity producers. Based on currently formed economic structure of agriculture, one can single out agricultural organizations, households, peasant (farmer) farms (Table 1). These data indicate that over the past 17 years, the agricultural production share in agricultural organizations (+9.9%) and peasant (farmer) farms (+9.2%) has significantly increased. At the same time, households production declined by 19.1%, which from the point of view of the economy digitalization prospects, we consider as a positive phenomenon, since it is agricultural organizations that are more susceptible to new technologies introduction, they have necessary resources (labor, financial, managerial) for digital innovations introduction, and households are among the most conservative and low-sensitivity to new technologies [2].

Table 1. Production of agricultural products by farm categories in Russia for 2000–2017

Years	Households of all categories		Agricultural organizations		Households		Peasant farms	
	Billion rubles	%	Billion rubles	%	Billion rubles	%	Billion rubles	%
2000	724,4	100	335,6	45,2	383,2	51,6	23,6	3,2
2005	1380,9	100	615,6	44,6	681,0	49,3	84,3	6,1
2010	2587,8	100	1150,0	44,4	1250,4	48,3	187,4	7,2
2015	5164,9	100	2657,1	51,4	1932,8	37,4	575,0	11,1
2016	5505,7	100	2890,4	52,5	1951,1	35,4	664,2	12,1
2017	5119,9	100	2818,6	55,1	1665,8	32,5	635,5	12,4
Deviation of 2017 from 2000	+4377,5	–	+2483,0	+9,9	+1282,6	−19,1	+611,9	+9,2

The potential of introducing advanced information technologies in Russian agriculture is enormous. And prospects for the digital economy development, including the agro-industrial complex, is one of the topical issues for discussion in the scientific sphere as well as at the state level. Presidential Decree No. 203 of May 9, 2017 "About the Strategy for the Development of the Information Society in the Russian Federation" [3] identified the main state priorities in this area. In July 2017, the program "Digital economy in the Russian Federation" [4] was approved, the implementation of which should ensure the digital technologies introduction into the society daily life, and this is especially important for rural residents. The Strategy for Scientific and Technological

Development of the Russian Federation, approved by Decree of the President of the Russian Federation of December 1, 2016 No. 642, stated the need to develop technologies that among other things will ensure the transition to highly productive and environmentally friendly agriculture, development and implementation of rational systems the use of chemical and biological protection of plants and animals [5].

Agriculture digital transformation is the transformation of economic activity in agriculture through the digital tools introduction - technologies and platform solutions designed to generate, process, deep analysis and broadcast analysis results in the form of numerical information about objects and subjects of the agricultural economy for subsequent adoption justified management decisions, ensuring technological breakthrough in the agro-industrial complex and achieving significant (at times) productivity growth on farms.

It should be emphasized that the digital products and technologies introduction will be the catalyst for a tangible growth in labor productivity in modern agricultural enterprises.

The basic characteristics of the domestic agricultural sector give an objective idea of an overall level of industry digitalization in comparison with other sectors of the economy. Thus, in the Forecast of agro-industrial complex scientific and technological development of the Russian Federation for the period until 2030, developed by HSE, automation and computerization are low in the vast majority of peasant farms, medium-sized agricultural enterprises, agricultural production cooperatives and personal subsidiary farms. Only the digitalization level of large agricultural holdings is estimated by the developers of the Forecast as high due to export orientation of this category of agricultural producers.

The agriculture digitization, a smart village, processing of large amounts of data, the Internet of things, cross-cutting digital technologies - all these terms are increasingly used to predict the further development of Russian agriculture.

A survey of agricultural commodity producers in 2018–2019 made it possible to formulate typical problems of an agrarian business that can be solved with the pass-through digital technologies introduction in the industry (Table 2) [6].

Table 2. Systematization of problems and end-to-end digital technologies to solve them

No	The problem voiced by agricultural producers	End-to-end digital technology to solve it
1	The lack of an operational systematization process of collected industry indicators and the subsequent generation of reports; lack of array of master data required for agribusiness	– big data
2	Access to aggregated industry information, advanced analytics	– big data
3	Agricultural land monitoring technologies; the use of satellite imagery and unmanned aerial vehicles in agriculture	– industrial internet – wireless technology

(*continued*)

Table 2. (*continued*)

No	The problem voiced by agricultural producers	End-to-end digital technology to solve it
4	Digital technologies for agricultural engineering	– industrial internet – wireless technology
5	Economical products and software solutions for marking farm animals	– components of robotics and sensorics – distributed re-estra systems
6	The use of cross-cutting digital technologies in agriculture: new opportunities overview; availability of a structured digital technology catalog	– all known digital technologies
7	The need to restore work of information and consulting services for the AIC; possibility of obtaining operational agricultural consulting	– technologies of virtual and augmented reality – wireless technology

3 Results

Smart village, what do we mean by this phrase? Back in 2012, the CEO of parallels research, defined the concept of a smart village as creating conditions in which people would have comfortable living and working not in large metropolitan areas, as a kind of habitat for people who could work without leaving home. On the other hand, the concept of a smart village implies the creation of individual houses that fully provide themselves with energy, using alternative sources, a kind of ecovillage.

In our opinion, "smart village" is the end-to-end digital technologies introduction:

- production digitalization, processing and sale of agricultural products in rural municipal areas;
- rural infrastructure digitalization;
- digital and financial literacy of the rural population.

Factors hindering the "smart villages" development and the digital economy formation in agriculture are proposed to be classified into external and internal. External factors include:

- the natural and biological basis of agriculture;
- shortcomings of legislative regulation (for example, secrecy of some aerial photography data, lack of clear rules for the use of unmanned vehicles);
- underdeveloped infrastructure (intermediary, legal, banking and other services) ensuring the digital technologies promotion and implementation;
- the digital technology market underdevelopment;
- insufficient financial support from the state (for example, difficulties in obtaining government subsidies for the digital technologies introduction of precision farming).

Among the internal factors it should be noted production, economic and other, due to the economic activity of an enterprise. Production factors can be formulated as difficulties in predicting final results due to high production risks in agricultural production; uncertainty of the timing of the digital technologies introducing process; dependence of production process on natural and biological factors. The composition of economic factors includes the uneven formation of stocks of their own financial assets; high cost; payback period duration of digital technologies; lack of information about markets; economic risks presence; low demand from consumers for digital technologies for agriculture.

Factors associated with an enterprise business include a shortage of qualified personnel; lack of opportunities to attract investors, scientific organizations and others [2, 7–19]. Considering that one of the Russian scientists' tasks in this regard is to study experience of using domestic digital technologies in agriculture, calculating their economic indicators introduction and use, comparison with foreign analogues, wide popularization of the best Russian practices. As part of implementation of this direction, the laboratory "Business Solutions in Digital Economy" was created in the State Budgetary Educational Institution of Higher Education "Nizhny Novgorod State Engineering and Economic University".

The laboratory performs research and development work on the contractual relations basis with any legal entities and citizens in order to provide self-speaking services in field of adaptation and implementation of professional software products in production activities.

The objectives of the Laboratory:

- organization and conduct of search, innovation and applied research in the field of adaptation and implementation of professional software products in production activities of an enterprise;
- the teaching skills and competencies development that enhance their competitiveness in the modern labor market;
- implementation of targeted integrated development, at the customer request, in the field of adaptation and implementation of professional software products in production activities;
- development of solutions in the field of organizational and managerial, production and technological tasks, based on results of the digital technologies and the automation introduction of individual business processes in enterprises.

The main tasks of the Laboratory are:

(1) combining theoretical and practical training:

- the professional skills formation in the course of solving applied research tasks in the framework of practices conducted on the laboratory basis;
- familiarization with commercial and other business activities;

(2) the personal qualities formation and development: autonomy, creative attitude to work, an ability to make decisions independently, team-working, an ability to resolve conflicts, communication skills;

(3) the students of an integrated system of skills and competencies formation and development that allow them to successfully realize themselves in their chosen profession.

As part of the work of this laboratory, business contracts were signed for the automation of individual business processes and the end-to-end technologies introduction of agricultural organizations in Nizhny Novgorod region.

It remains an open question how to make new digital developments reach agribusinesses and they are interested in introducing them. In this regard, it is proposed to intensify government support for the digital technologies introduction and use in agricultural production. To do this, it makes sense to concentrate more attention on encouraging agricultural producers to apply digital innovations without stopping work on invention and creation of these technologies based on their implementation indicators (Table 3).

Table 3. Indicators of digital efficiency

No	Indicators
1.	Automation of business processes in an organization, %;
2.	Funds aimed at automating business processes in an organization (to the total amount of funds received), %;
3.	Annual update of licensed programs;
4.	The proportion of costs directed at training (retraining) employees to work with software products;
5.	The proportion of employees who use software products for the production processes computerization;
6.	The share of internal workflow conducted through automation

4 Conclusion

Digitalization and "smart villages" are beginning to come to the agricultural sector of Russia and are one of the main vectors of its development. The strategic task facing the industry and what developers are interested in, and most importantly, professionals who are engaged in the developments implementation in practice is how they will be integrated into agricultural production and rural infrastructure. Such integration is impossible without the artificial intelligence use, systematization of large databases, as their volumes grow exponentially.

Thus, "smart agriculture" allows for the maximum automation of agricultural activities, increased yields and product quality, will have an impact on the labor productivity growth at agricultural enterprises, will reduce the enterprises unit costs for business administration, reduce the share of material costs in the agricultural products unit cost in Russia.

References

1. Tsvetkov, V.A.: Agro-industrial complex of Russia. The current state, necessary and sufficient conditions for overcoming the crisis, no. 3, pp. 14–16 (2017)
2. Nemchenko, A.V., Dugina, T.A., Likholetov, E.A.: Digitization as a priority direction of economic development of agrarian production. Bulletin of the Altai Academy of Economics and Law, no. 4, pp. 118–123 (2019)

3. Presidential Decree of 09.05.2017 N 203: About the Strategy for the Development of Information Society in the Russian Federation for 2017–2030. Collection of Legislation of the Russian Federation, 15.05.2017, no. 20, Art. 2901
4. Order of the Government of the Russian Federation of 28.07.2017 No. 1632-p "About approval of the program "Digital Economy of the Russian Federation. Meeting of the Legislation of the Russian Federation"", 07.08.2017, N 32, Art. 5138
5. Decree of the President of the Russian Federation of 01.12.2016 No. 642 No. A of the Strategy for Scientific and Technological Development of the Russian Federation. Meeting of the Legislation of the Russian Federation, 05.12.2016, N 49, Art. 6887
6. Ganieva, I.: The Digital Transformation of Agriculture in Russia: Consolidation of the State and Agribusiness. Achievement of science and technology, no. 4, C. 5–8 (2019)
7. Korska, M.A.: Development of the methodology of economic research problems of agriculture. New science: theoretical and practical view, no. 11-1S.97–99 (2016)
8. Fedorova, O.V., Krutikov, V.K.: New approaches to the effective solution of problems of managing the innovative development of the agro-industrial complex. AIC: economics, management, no. 11, pp. 40–45 (2016)
9. Polulyakh, Yu.G., Adadimova, L.Yu., Oydup, T.M., Chupikova, S.A.: Sustainable development as a single category of problems in the agricultural sector and rural areas. In the collection: food security, import substitution and socio-economic problems of the agro-industrial complex. Materials of the international scientific-practical conference C. 335–339 (2016)
10. Edrenkina, N.M.: Solving the problems of development of small forms of managing in the framework of the priority national project "Development of the agro-industrial complex" in Siberia. Achievement of science and technology of agriculture, no. 4, C. 11–13 (2008)
11. Ryzhenkova, N.Ye., Chepik, D.A.: Development of the innovative potential of the agro-industrial complex: problems, directions. Economics of Agriculture of Russia, no. 6, C. 53–57 (2014)
12. Berezina, N.N., Rakitina, I.S.: Innovative development of the agro-industrial complex and the problems of its financing. In the collection: economics and management: problems, trends and prospects. C. 5–9 (2009)
13. Chernova, A.A.: The development of the agro-industrial complex: problems and solutions. The strategy of sustainable development of the regions of Russia, no. 29, C. 49–50 (2015)
14. Stenkina, M.V.: Problems of Information Support of Agrarian-Economic Research in the Digital Economy. Agrarian economics: state, tasks for the future. VIAPI named after A.A. Nikonov, pp. 281–283 (2018)
15. Larina, T.N., Breeders, N.D.: Potential and prospects for the development of "digital" agriculture in Russia. Agrarian economic science: state, tasks for the future. VIAPI named after A.A. Nikonov, S. 283–285 (2018)
16. Balabanov, V.I., Tsvetkov, I.V., Zhogin, I.M.: Methodology for assessing the effectiveness of introducing information technologies in agriculture. Agrar economics: state, tasks for the future. VIAPI named after A.A. Nikonov, S. 283–285 (2018)
17. Nilsson, D., Meng, Y.-T., Buyvolova, A., Kopyan, A.: Unlocking the potential of digital technologies in agriculture and the search for prospects for small farms. www.worldbank.org
18. Schwab, K., Davis, N.: Technologies of the Fourth Industrial Revolution (translation from English), 320 s. Eksmo, Moscow (2018)
19. Shamina, O.V.: Statistical assessment of informatization of the economy of Nizhny Novgorod region. Bulletin of NIERI. no. 4(71), pp. 93–100 (2017)

Digital Technology and Applied Sciences

Optimization of the Processing of Functional Materials Using Gamma Irradiation

V. Oniskiv[✉], V. Stolbov, and L. Oniskiv

Perm State National Research Polytechnic University,
29, Komsomolsky Avenue, Perm 614990, Russia
Oniskivf@gmail.com

Abstract. A particular problem of optimizing the process of irradiation with gamma quanta of products made of Polyethylene in order to obtain heat-shrinkable products with a shape memory effect is considered. The statement of the problem of optimizing the placement of ionizing radiation sources in an industrial installation is presented. As a criterion of optimization, the parameter of "lost power", which is an excessive dose of radiation, which is not necessary to achieve the required properties of degree of shrinkage of the material, is proposed. The main problem of optimization is the need to analyze an extremely large number of options for the placement of radiation sources. The results are obtained using Monte Carlo methods and combinatorial optimization. Both approaches, performed independently of each other, lead to almost the same solution. Some results of optimization, which have real practical application and allow to significantly reduce the exposure time, are presented. It is possible to expand the scope of the proposed method. It is important to solve the optimization problem for several irradiation chambers simultaneously.

Keywords: Functional materials · Gamma irradiation · Optimization algorithm

1 Introduction

The functional materials with shape memory effect are most widely used in technology and various technologies of modern production [1, 2]. In particular, the electrical, aerospace, automotive, cable [3], construction industry [4] and biomedicine [5] are the largest consumers of these products. The use of radiation modification of materials to achieve the effect of shape memory and obtain other properties of materials is now formed in a successfully developing industry in many countries. The growth rate of production in this industry is from 7 to 10% per year, and the volume is measured in tens of billions of dollars [6, 7, 13]. In radiation technologies, as a rule, radiation with a quantum energy of 10 keV–10 Mev (x-rays, gamma rays, accelerated electrons) is used. More high-energy particles, such as heavy ions with an energy of up to 100–200 Mev, are successfully used in technologies for modifying the surface of metals and alloys in order to achieve high microhardness, obtaining nanoporous track membranes of different structures.

The procedure of gamma irradiation of the material is widely used to create molecular - crosslinked polyethylene [8, 9]. Ionizing gamma irradiation creates excited molecules and chemically active radicals in the material, which are the main element for obtaining new bonds. There is no need to use special additional chemicals.

© Springer Nature Switzerland AG 2020
T. Antipova and Á. Rocha (Eds.): DSIC 2019, AISC 1114, pp. 489–498, 2020.
https://doi.org/10.1007/978-3-030-37737-3_42

It is important to note that gamma radiation does not initiate nuclear reactions and does not induce residual radioactivity. The technology of creating additional inter-molecular bonds using ionizing gamma irradiation from CO60 (cobalt 60) or Cs137 (cesium 137) sources has not only a number of certain advantages associated with the quality of the product (high shrinkage coefficient, high elasticity, low shrinkage tem-perature), but also some design features. In particular, the radiation dose required to produce the cross-linking effects should be of a well-defined value and should be achieved as much as possible simultaneously at all points of the irradiated material.

In real production, for the purpose of guaranteed elimination of defects (insuffi-ciency of insoluble gel fraction in the material), exposure time is calculated by the minimum value of the dose rate. It depends on the location of sources of ionizing radiation (SIR). Optimal placement of point SIR is the subject of this study. It should be noted, that gamma radiation is isotropic.

In existing industrial installations, the location of sources of ionizing radiation with a slight error can be taken as a straight line. Irradiated materials are in rotating sealed containers, with the axis of rotation parallel to the straight line along which the sources of gamma rays. The locations of the radiation sources are fixed in space and cannot be changed for design reasons. Several radioactive elements of different power can be placed at each location of radiation sources. The order of radioactive sources is a control parameter in the optimization problem.

2 Problem Statement and Heuristic Solution Algorithm

2.1 Construction of the Objective Function

There are N Gamma radiation sources and M points of their locations $(N \gg M)$. Let denote by A_i - activity of the i-th radiation source. It is known for every i at the moment $t = 0$ and is denoted as A_{0_i}. Considering isotropic nature of gamma radiation and its dependence on time (half-life), the exposure dose rate $P(x, t)$ at the point with the coordinate x will be determined as:

$$P(x,t) = \sum_{i=1}^{N} \Gamma_c \cdot \widehat{A_{0_i}} \cdot 2^{\left(-t/T_{1/2}\right)}/r_i^2, \tag{1}$$

where: Γ_c – ionization gamma constant, $T_{1/2}$ – is the half-life, which for Co60 is 5.2 years, r_i – is the distance from the i-th source to the point with coordinate x, i - is the summation index for all sources from 1 to N. At each moment of time t some distribution of irradiation power is realized, for which it is possible to find the mini-mum value of the power:

$$P_{min}(t) = \min_{0 \leq x \leq l}(P(x,t)) \quad \forall t \in [0,T]. \tag{2}$$

This value is the basis for calculating the exposure time of the irradiated material. Due to the uneven distribution of radiation sources, some of the material receives an excessive dose, which can be conventionally called "lost" and which is shown in the form of a shaded area W in Fig. 1.

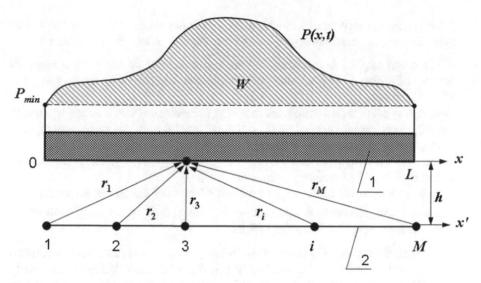

Fig. 1. Scheme of the process of irradiation of polymer material (1) using radiation sources (2)

Then the optimal placement of N sources of ionizing radiation should provide a minimum value of W – "lost" power:

$$W = \int\int_{00}^{Tl} (P(x,t) - P_{min}(t))\, dxdt \to min, \tag{3}$$

where: T - is the period of operation of the radiation unit between two successive recharge procedures, L - is the linear size of the irradiation zone. Note that the value of the functional W can be normalized by L and T.

2.2 Problem Solving Algorithm

When considering the special case when all sources of IR in the formula (1) are distributed in such a way that the radiation activity is uniform along the length L, it is easy to obtain an analytical solution of the power distribution of the form for some fixed moment t: $p(x) = {}^{c}/_{h} \cdot \left(arctg\left((L-x)/_{h}\right) - arctg\left((-x)/_{h}\right)\right)$, where h is the distance from the source line to the material line and c is a constant.

Respectively $p_{min} = {}^{c}/_{h} \cdot arctg\left(L/_{h}\right)$ and of course $p(x)$ is symmetric about $L/_{2}$. The presence of such solutions suggests the possibility of reducing the optimization problem to the search for parameters of a continuous function of the activity distribution $A(x,t)$, given, for example, in the form of a polynomial and ensuring the fulfillment of the condition (3) in the presence of an integral restriction on $A(x,t)$, the essence of which is to provide conditions for the total activity of IR sources. It should be noted that even if there is an effective algorithm for constructing such a function, it is impossible to avoid solving the integer problem of the activity distribution of discrete point sources of IR in accordance with the solution which found.

Taking this into account, the minimization of the functional (3) was carried out by solving the integer optimization problem with a number of technical limitations:

1. It is allowed to place in one location with the number of several gamma radiation sources of different activity. Also permissible and the absence of sources in some places locations.
2. A complete set of radiation sources designed to be charged into a single irradiation chamber may contain several smaller sets of sources with the same activity. For example, the complete set of N sources may contain N_1 sources with activity A_1, N_2 sources with activity A_2, etc. Thus $N_1 + N_2 + \ldots + N_k = N$. Typically, the number of types of sets (k) is small, no more than 3–4.
3. Due to technological features, the problem of optimal placement of sources has symmetry with respect to the point $x = L/2$. This feature significantly reduces the computational cost of solving the problem.

In general, if there are N sources of A_1 and M points of their placement (locations) on the segment of length L, the number of possible placement options is extremely large – M^N and this fact should be taken into account when creating an optimization algorithm which can have a very serious computational complexity. As a rule, no more than 3 A_1 IR sources can be physically located in one location point at the same time. Thus, the number of possible combinations at one location point is significantly reduced. In addition, it is necessary to exclude from the computational procedures variants of calculations W by formula (3) with the number of placed IRS different from N. The optimal solution is sought in the subspace $N_1 + N_2 + \ldots + N_k - N = 0$. Other options, when not all IRS are placed, obviously does not make sense to consider.

Thus, the problem of discrete optimization can be formulated as follows: to find such an optimal distribution of ionizing radiation sources in locations $x_j, j = 1, 3M$, where the objective function at a given time interval from 0 to T reaches a minimum:

$$W(x_j) = \int_0^T \int_0^L \left(\sum_{i=1}^N \Gamma_{CH} * A_{0(i)} * \frac{2^{\left(-\frac{t}{T\Pi}\right)}}{r_i^2(x)} - P_{min}(t) \right) dxdt \rightarrow min, \quad (4)$$

under the following restrictions:

• in one tube can be no more than 3 SIR, the number of tubes in the chamber (locations SIR) is M;
   * the total number of sources of different types is N;
   * the distribution must be symmetrical with respect to the middle of the chamber.

It should be noted that problem (4) belongs to combinatorial optimization problems [10] and is np-complex. To solve such problems usually apply heuristic algorithms [10], which include genetic [11], "ant" [12] algorithms, Monte-Carlo, etc. All heuristic algorithms aimed at reducing the feasible set of solutions to problems in combinatorial optimization and accelerating the sorting of the remaining options.

In this case, we apply the heuristics based on the analytical solution of the problem under the assumption of continuous distribution of radiation sources, which is given above. As it was shown, at uniform distribution of activity of sources on length $A(x)$ the function of power of irradiation $P(x)$ has a maximum on the middle (graphs 1 on Fig. 2) and the power loss is quite large. To achieve a minimum of power losses, it is necessary to place the sources so that their activity at the ends of the segment $[0, L]$ increases, and the power function tends to a constant (graphs 2 in Fig. 2).

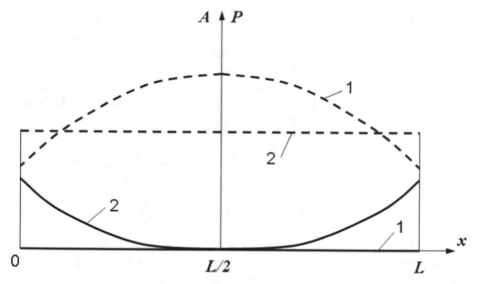

Fig. 2. Qualitative picture of radiation power distribution $P(x)$ (dotted line) at continuous distribution of activity of sources $A(x)$ (solid line).

Now we consider an eristic algorithm for approximate solution of the discrete optimization problem (4), built on the basis of the analytical solution of the continuous problem.

The first step in the work of the algorithm is to search by combinatorial method for a set of admissible variants of source distributions in locations taking into account the limitations of the optimization problem. As noted above, the number of acceptable options is very large. Therefore, it is necessary to limit this set reasonably so that the search for the optimal distribution does not take too much time.

That is why the second step of the algorithm is to take into account the following heuristic conditions:

* the distribution of sources should be such that their activity gradually decreases from the edge to the middle of the segment $[0, L]$, which corresponds to function 2 in Fig. 2.
* the distribution of sources should be such that the approximating function of their activity is sufficiently smooth at a given interval, which means that from the admissible set of solutions it is possible to exclude options in which the total activity of sources at two consecutive locations would be maximal or equal to zero.

The last step of the algorithm is to sort the remaining variants of the source distribution by the value of the target function and search for such variants in which it is minimal.

3 Results

As an example solution, we consider a variant of placing in the chamber 48 of radiation sources A_1, 24 of which are at the time of charging in the camera activity $A = 4{,}54 \cdot 10^{13}$ Bq each (in the diagram marked figure 1), while 24 of the remaining activity of each is $A = 7{,}23 \cdot 10^{13}$ Bq (marked by number 2). In any point locations the number of placed sources should not exceed three. You want to place all sources in 34 places locations. The scheme of typical placement of *SIR* is presented in the form of Table 1 (figure 0 indicates the absence of *SIR*):

Table 1. The location of the IRS of two types in locations of the irradiation chamber

2	2	2	2	2	2	2	2	2	2	2	2	1	1	1	1	1	1	1	1	1	1	2	2	2	2	2	2	2	2	2	2	2	2
0	0	0	0	0	0	0	0	0	0	0	0	0	1	1	1	1	1	1	1	1	1	1	1	0	0	0	0	0	0	0	0	0	0
0	0	0	0	0	0	0	0	0	0	0	0	0	0	0	0	1	1	1	1	1	0	0	0	0	0	0	0	0	0	0	0	0	0

The distribution of power losses in this arrangement is shown in Fig. 3. The corresponding distribution of activity of sources is shown in Fig. 5.

Distribution of lost power that is shown in Fig. 3, contains a characteristic "power sink" at the ends of the irradiation zone. This pattern of distribution is very typical of real production.

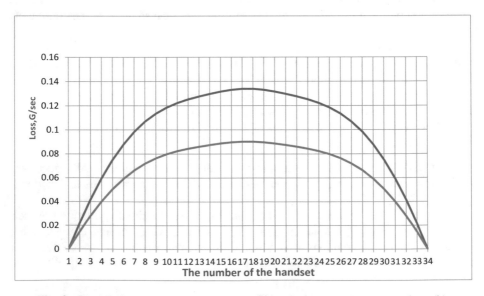

Fig. 3. Instantaneous power loss (at $t = 0$ - blue graph, at $t = 3$ years - red graph).

One of the partial solutions of the optimization problem was obtained by the Monte Carlo method. The input parameters of the problem, of course, were the same. The algorithm is constructed in a classical way. IR sources were randomly placed in 34 locations, the value of the target function (4) was calculated, the best placement option was remembered. The analysis of more than $45 \cdot 10^6$ variants resulted in the following placement:

2	2	2	2	2	0	2	0	0	1	1	2	1	1	1	2	0	0	2	1	1	1	2	1	1	0	0	2	0	2	2	2	2	2
2	2	2	1	1	0	0	0	0	0	0	0	0	0	1	0	0	0	0	1	0	0	0	0	0	0	0	0	0	1	1	2	2	2
1	2	1	1	0	0	0	0	0	0	0	0	0	0	1	0	0	0	0	1	0	0	0	0	0	0	0	0	0	0	1	1	2	1

The value of the solution lies in the fact that it justified the above heuristic hypotheses. However, the use of the Monte Carlo method involves significant computational and time costs and requires additional assumptions.

Optimal, in the sense of minimum power losses, the placement of 48 sources obtained as a result of calculations is as follows (Table 2):

Table 2. The location of the *SIR* of two types in locations of the irradiation chamber

2	2	2	1	2	2	2	1	1	1	1	0	2	2	1	1	2	2	1	1	2	2	0	1	1	1	1	2	2	2	1	2	2	2
2	2	2	1	1	0	0	0	0	0	0	0	0	0	0	0	0	0	0	0	0	0	0	0	0	0	0	0	0	0	1	1	2	2	2
1	1	0	1	0	0	0	0	0	0	0	0	0	0	0	0	0	0	0	0	0	0	0	0	0	0	0	0	0	1	0	1	1

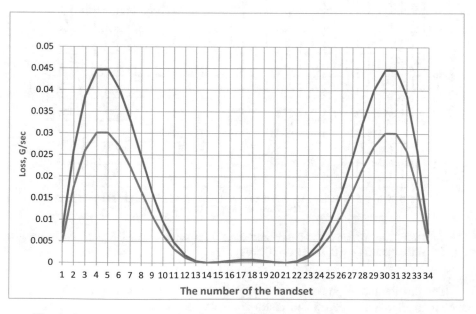

Fig. 4. Instantaneous power loss (at $t = 0$ - blue graph, at $t = 3$ years - red graph)

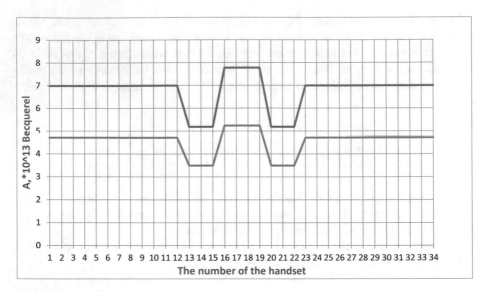

Fig. 5. Distribution of activity in the typical placement of sources. Top graph corresponds to activity at time $t = 0$, lower $t = 3$ years

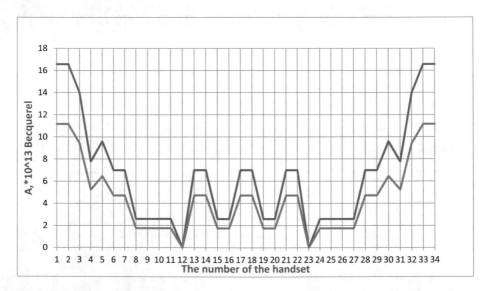

Fig. 6. Distribution of activity at the optimal location of sources. Top the graph corresponds to activity at the moment $t = 0$, lower $t = 3$ years.

The distribution of power losses along the length corresponding to this variant is shown in Fig. 4. On Fig. 6 the distribution of activity of radiation sources corresponding to this variant is given. It should be noted that the values of power loss parameters in the two compared variants actually differ almost at times, in addition, in the optimal variant, as expected, the above-mentioned effect of "power sink" at the ends of the irradiation zone disappeared.

4 Conclusion

In order to assess the effectiveness of the solution obtained, it should be noted that at time $t = 0$ the exposure time is calculated for the standard version, based on the power $P_{min} = 0.22$ G/sec, and for the optimal distribution $P_{min} = 0.27$ G/sec. At time $t = 3$ years, these values are $P_{min} = 0.14$ G/sec and $P_{min} = 0.18$ G/sec, respectively.

Thus, the overall efficiency of the proposed source placement option is in the range of 22–28%, which is undoubtedly very attractive economically, since it does not involve any additional costs in equipment or technology. The effect occurs as consequence of reducing the required exposure time of the material. At the same time, the number of irradiation cycles for a period of 3 years increases by approximately 25%.

It is seen from Fig. 6 that the optimal distribution of source activity along the length of the chamber is close to that given by the heuristics obtained from the analytical solution of the optimization problem.

The direction of further research is related to the optimization of the irradiation process not in one chamber, but in several, between which the exchange of radiation sources is allowed. This problem has significantly greater computational difficulties and the numerical procedure must be extremely efficient.

Acknowledgements. The reported study was funded by the Ministry of Science and Higher Education of the Russian Federation (the unique identifier RFMEFI58617X0055) and by the EC Horizon 2020 is MSCA-RISE-2016 FRAMED Fracture across Scales and Materials, Processes and Disciplines.

References

1. Knyazev, V., Sidorov, N.: Irradiation polyethylene in engineering. Chemistry (1994)
2. Aguilar, M., Roman, S.: Smart Polymers and Their Application, 1st edn. Woodhead Publishing, Cambridge (2014)
3. Zhilkina, N., Larin, Y., Vorob'ev, V.: Investigation of the effect of gamma radiation on physico-mechanical characteristics of polymeric materials to protect materials to protect the casing of optical cabeles. Cabels Wires **3**, 11–15 (2004)
4. Gitman, I., Klyev, A., Gitman, M., Stolbov, V.: Multi-scale approach for strength properties estimation in functional materials. ZAMM Zeitschrift fur Angewandte Mathematik und Mechanik **98**(6), 945–953 (2018)
5. Gitman, M., Skriabin, V., Sotin, V., Batin, S.: Methods for complex assessment of operational life of the functional material in hip replacement. Report 1. Russ. J. Biomech. **21** (4), 310–318 (2017)
6. Anthony, J., Marshall, R., Walo, M.: The evolution of and challenges for industrial radiation processing. Radiat. Phys. Chem. **94**(1), 141–146 (2014)

7. Chmielewski, A., Al-Sheikhly, M., Anthony, J., Marshall, R.: Recent developments in the application of electron accelerators for polymer processing. Radiat. Phys. Chem. **94**(1), 147–150 (2014)
8. Forster, A.L., Tsinas, Z., Al-Sheikly, M.: Effect of irradiation and detection of long-lived polyenyl radicals in highly crystalline ultra-high molar mass polyethylene (UHMMPE) fibers. Polymers **11**(5), 924 (2019)
9. Mahlis, F.: Radiation physics and chemistry of polymers. Atomizdat (1992)
10. Court, B., Figen, Th.: Combinatorial Optimization: Theory and Algorithms. MCNMO (2015)
11. Gladkov, L., Cureychik, V., Cureychik, V.: Genetic algorithms (electronic resource). Fizmatlit (2010)
12. Stovba, C.: Ant algorithmis. Exponenta Pro. Math. Appl. **4**, 70–75 (2003)
13. http://sdamzavaz.net/1-46978.html. Accessed 20 Nov 2018

Indoor Positioning and Guiding System Based on VLC for Visually Impaired People

Simona Riurean$^{(\boxtimes)}$ ⓘ, Monica Leba ⓘ, and Andreea Ionica ⓘ

University of Petroșani, Universitatii str., No 20, 332006 Petroșani, Romania
sriurean@yahoo.com

Abstract. More and more innovative devices with the wireless communication technology based on radio frequency (RF) embedded makes our everyday live easier and offer a healthier life behavior due to improved local or remote transmission capabilities. Internet of things (IoT) has already become reality, therefore the RF space becomes increasingly congested. Alternative technologies as those based on visible light or IR comes to unload the overcrowded RF spectrum and bring new applications (such as optical camera communication – OCC) and local-based services since they allow to be embedded in small devices with Light Emitting Diodes (LED). The Indoor Positioning and Guiding System (IPGS) described in this paper consists of a number of optical transmitters (oTx) with LEDs embedded into each illumination fixture and a low cost, wearable device as optical receiver (oRx) with solar panel, intended to be used by Visually Impaired People (VIP) when accessing public institutions as schools, museums, indoor stations as underground metro or airports with a dedicated infrastructure to support IPGS. The oRx, wear on by VIPs, receives the optical signal piggybacked by light in a form of a sound message when accessing crowded zones where the dedicated infrastructure is installed. The technique used for IPGS setup is the optical Received Signal Strength (oRSS). The specific, particular scenario described here is the main entrance hall of our university.

Keywords: VLC · OCC · Solar panel · LED · IPS · oRSS

1 Introduction

New technologies transform the way we usually think, therefore our everyday behavior as individuals and society is significantly different from one decade to another. Some of the new technologies provide extensive economic, social, well-being, safety and health benefits for humankind, far beyond anything their creators have ever dreamed of. Optical wireless communication (OWC) technology, although not very deep investigated more than a decade ago, promises to be one of this kind, since intensive research efforts by both academics and companies have been noticed during the last 10 years with significant progresses in wireless data communication piggybacked by illumination.

Most of the innovative technologies applied in special situations, as the additional support and guidance for Visually Impaired People (VIP) is, can be of mutual benefit for both visitors and institutions that agree to invest in these indoor positioning and

© Springer Nature Switzerland AG 2020
T. Antipova and Á. Rocha (Eds.): DSIC 2019, AISC 1114, pp. 499–512, 2020.
https://doi.org/10.1007/978-3-030-37737-3_43

guidance systems (IPGS). Positioning systems used outdoor are (most of them) based on Global Positioning System (GPS) and are useless indoor where the signal from GPS is fading or becomes absent. Indoor positioning systems are based on radio frequency (RF) signals, infrared (IR) and visible light (VL) signals.

The IPGS addressed here aims to be a very handy, low cost and useful tool for the visually impaired people as visitors of indoor public places, especially in schools, museums, hospitals, airports, metro underground stations, rail stations, commercial buildings, industrial locations and so on. This paper presents in its first part, the GPS and AGPS technologies as well as wireless communication technologies based on Wi-Fi that are used both outdoor and indoor. The OWC technologies such as VLC and OCC are the most used technologies for current Indoor Positioning Systems (IPSs) developed worldwide.

The IPGS described in this paper is based on VLC wireless technology that sends guiding sound messages by visible light to the VIP who need to find easier and faster their destination into indoor crowded public places consisting of buildings with many, difficult to find or access, end points. The vocal messages are heard by the VIPs solely, who wear the special designed optical receiver (oRx) when their location is under one of the access points (AP). The network of APs indoor has the oTx embedded into the lighting network being a dedicated infrastructure to support IPGS. When the oTx and oRx are in Line of Sight (LoS), the optical signal communication become possible simultaneously with the recorded sound message. In this way, the VIP is aware of both his/her own position in building as well as the nearest surrounding potential destinations.

Beside the novelty of the idea addressed in this paper, namely the IPGS for VIP, the technology used is also new with tremendous perspective of wide applications in different areas. The system is described at theoretical level then the system simulation and prototype development with various tests are also presented.

2   Conventional Positioning System Technologies

The GPS technology, allowed by U.S. Department of Defense to be used for civilian in 1980s, is an already worldwide spread useful navigation tool outdoor where there is sufficient LoS with satellites.

GPS signals, although very useful outdoor, are too week or not traceable indoor and therefore they are not accurate or useless for indoor positioning.

GPS implemented in smart phones has transformed user's behavior from simple mobile device owner into dependent on dynamic map-based way-finding experience [1]. An assisted GPS (A-GPS) technology, also known as A-GPS or AGPS, consists in a smart device that communicates with 4 or more satellites to determine anywhere on earth its exact location coordinates (latitude and longitude). Although it works on the same principles as a GPS, the sources of triangulation information are radio signals from satellites and assistance of servers. As long as the device has a clear LoS to the satellites, it works during any weather. The A-GPS technology, follows the principle of Differential GPS (D-GPS) that provides extra-assistance data from cellular network which is used to reduce acquisition time, improve positioning accuracy and provide communication facilities, thus enhancing the performance of the standard GPS in devices connected to the cellular network. The additional D-GPS reference stations are integral part of the cellular infrastructure by additional signaling procedures between network and terminal. Even if an A-GPS device determines location coordinates faster because they have better connectivity with cell sites than directly with satellites, the location determined via A-GPS are slightly less accurate than GPS [2].

There is a variety of wide-area hybrid location alternative technologies to GPS for positioning and guiding and therefore, the market has shifted significantly over the last years being oversupplied with new smart devices, technologies and dedicated applications. There are also alternative location technologies capable of supporting both indoor and outdoor location, not needing to hand off to another technology when moving anywhere between, being designed to meet ubiquitous consumer. The IPS are developed and optimized for the indoor environment that usually consist of dedicated beacons or location units and/or the processing of signal. The smartphone OS vendors Google and Apple, as well as GPS manufacturers like Broadcom, offer the cellular technology referring to all forms of cellular positioning and location technologies. The wide area Wi-Fi location is currently used to augment GPS in challenging environments, while also providing indoor location without the need for dedicated infrastructure [3].

iBeacon, as well as Eddystone, based on the Bluetooth Low Energy (BLE) proximity sensing are used as applications of IPS. iBeacon transmits an Universally Unique Identifier (UUI) picked up by a compatible application. The identifier together with several other bytes sent are used to determine the device's physical location, track the owner's movement, or trigger a location-based action on the device (such as a check-in on social media or a push notification). With the help of an iBeacon, a smartphone's software can approximately find its relative location to an iBeacon indoor. Under development at the Bluetooth SIG, Bluetooth proximity positioning technologies from companies (like Wireless Werx) as well as the Bluetooth 4.0 location protocol layer, iBeacon is a solution that requires a dedicated infrastructure to support positioning [4]. Although not strictly a positioning technology - there are ways to create a tracking solution using Near Field Communication (NFC) that is seen as core link between location and end-user engagement [5]. Audio location is an old technology that has received significant interest, as well [6].

3 Optical Wireless Communication Technologies

3.1 General Description of OWC

As the radio spectral resources have become increasingly overwhelmed, OWS has raised interest due to enormous prospective for wireless (and potentially mobile) bandwidth resources. OWC technologies have been a subject of a poor research interest for many years (from the 1970s till 2010). During the last decade, on the other hand, an incredible intensive R&D work and innovative applications in the OWC area has been noticed worldwide.

OWS consist of Infrared (IR), Visible Light Communication (VLC), Optical Camera Communication (OCC), Light Fidelity (LiFi) and Ultraviolet (UV) (Fig. 1) [7, 8].

The most important benefits of OWC consist in immunity to EM interference, small components compared with RF, no harmful to human health (with some restraints regulated by EN 62471:2008 standard and European Directive 2002/95/EC) and the potential for high security from using highly directed beams. The most recent approach relay on cost effective solutions and highly energy-efficient.

Near IR (NIR) used in OWC applications, has wavelengths from 750 nm to 950 nm, is invisible to the human-eye. TV remote control (which uses IR light) is one of the most successful innovation from the early days embedded in wide commercial used device that is still worldwide used.

VLC uses wavelengths in visible light, spreads from 380 nm to 780 nm on EM spectrum in a simplex communication topology and is defined by standard IEEE 802.15.7:2011 Short Range Optical Wireless Communications where PHY and MAC layers are described. The optical transmitter (oTx), consists of a Light Emitting Diode (LED) and the optical receiver (oRx) can be a PIN (positive intrinsic negative) or an APD (avalanche photodiode) photodetector (PD). Smart LED technology is known for generating important energy savings and overall improvement in lighting quality.

Due to the fact that LED lighting can be rapidly modulated or fast turned ON and OFF, adding an intelligent driver, the system is able to power LEDs in a manner that produces Morse code-like light patterns at a very high rate that is perceived by human eye as a continuous light.

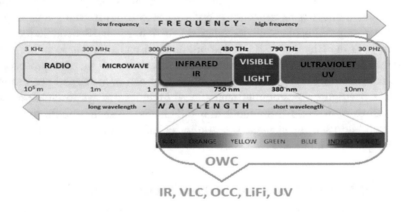

Fig. 1. Part of the electromagnetic (EM) spectrum with OWC emphasized

The VLC concept has been for the first time launched and demonstrated in the Japan's M. Nakagawa Laboratories in 2003, therefore a VLC ID System Development Kit was available on market on 2012.

The OCC is a practical form of VLC that uses a camera or an image sensor as an oRx. The oTx consists of a LED, or an array of LEDs that send On Off Keying (OOK) modulated optical signal. According to IEEE 802.15.7r1, the revision of 802.15.7 targets the commercial usage of VLC systems defining the PHY layer, namely the compatibility between a several types of cameras with different image sensing sampling rates, constant/changeable frame rates and various resolutions.

Strengths of OCC system refer to the use of built-in cameras of smart devices resulting in an overall cost of the system dependent only on software that controls of the built-in camera, therefore it can easily operate as a multiple-input–multiple-output (MIMO) system by extracting the light source from an image sensor functioning in a single-input–single-output (SISO) or multiple-input–single-output (MISO) mode. OCC is stabile over different changes of communication distance, has high quality of Signal to Noise Ratio (SNR) and low complexity of signal processing.

Future challenges of the OCC system refer to a large amount of shot noise due to the random arrival of photons and a high AWGN resulting in background noise coming from natural (sun) or artificial light (other light sources) [9].

LiFi is a MIMO, full duplex communication (usually download VLC and upload IR), high speed data wireless communication between mobile smart devices and other networked devices via LED lights fixtures used to carry data [10]. Ultraviolet wavelengths use very short-range non-line-of-sight links to transmit data.

3.2   Principles of VLC

The general diagram of the IPGS based on Intensity Modulation/Direct Detection (IM/DD) is presented in Fig. 2. Signals from the oTx and oRx are underlined as well as the optical signal indoor with ambient light.

LED is used as light sources in VLC systems aiming to emit the required optical signal. The optical signal transmitted by LED experiences, invariably, free-space diffusion to the receiver consisting in one or more PDs. The PD collects both the desired optical signals and other lighting signals known as Additive White Gaussian Noise (AWGN) with a negative effect on the VLC communication quality. In front of the PD is usually used an optical filter with the aim to decrease the effect of AWGN produced by ambient light. The ambient light consists on both natural light (coming from sun) and different other artificial lights (LEDs, incandescent bulbs, halogen lamps or fluorescent bulbs). An optical non-imaging concentrator is also used in order to improve the Signal-to-Noise Ratio (SNR) at the active area of the PD. Characterization of a VLC channel is done by its optical Channel Impulse Response (CIR), that is used to investigate and attempt to attenuate, as much as possible, the effects of channel distortions.

Both experimental measurements and computer modelling has been already described in different research papers on the channel characterization of indoor and outdoor systems [11].

The VLC system (Fig. 2) consists of the optical transmitter (oTx), optical channel and optical receiver (oRx). In the Electrical Domain (ED), of the oTx, data is processed by the Electrical Modulator (EM) and then by the Optical Intensity Modulator (OIM). Signal is represented here by an electrical voltage or current ($s(t)$). In either case (voltage or current), the electrical power is proportional to $s(t)^2$. The OIM generates an optical signal with intensity of $s_i(t)$. The optical power is proportional with $s(t)$. The signal $s(t)$ can have only real and positive values, therefore the modulation techniques usually applied in radio communication have to be modified. On the optical channel with impulse response *h(t)*, an AWGN will be added and interfere with the optical signal sent by the LED to the PD. Here, the signal is represented by the optical intensity.

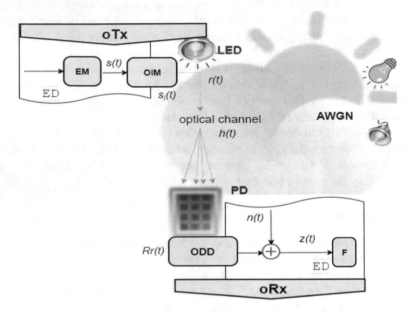

Fig. 2. The general diagram VLC system based on IM/DD

The Optical Direct Detection (ODD) of the oRx, consists of the active area of the PD, converting the signal from the optical to electrical form, therefore the signal:

$$r(t) = h(t) \otimes s(t) \tag{1}$$

will became again an electrical one Rr(t) (voltage or current).

In the electrical domain (ED), the PD's noise signal

$$z(t) = Rr(t) \otimes n(t) \tag{2}$$

where:

n(t) - shot noise.

Shot noise refers to fluctuations of the photons number sensed in PD according to their occurrence independent of each other. Shot noise is one of the main noise source in the optical wireless link and arises fundamentally due to the discrete nature of energy and charge in the PD. A matched filter (F) detection is the last item before the final data out is completed.

4 Indoor Positioning Systems Based on OWC Technologies

4.1 IPS Principles

The adoption of VLC and OCC technologies, mainly in the retail business, enables sellers to augment customers' experience by enabling location-based interaction on mobile devices, which can drive in-store deals. These systems use intelligent driven LED lights, that sends data to the image sensors of a smart device, providing an effective method to enable location and navigation within indoor environments.

Due to the many advantages of OWC, different methods already applied for IPS based on RF are applied in IPS using LEDs. They cover different techniques with specific algorithms:

- Received Signal Strength (RSS) with algorithms (i) trilateration, (ii) fingerprinting and (iii) proximity [12];
- Angle-of-Arrival (AoA) – has the main disadvantage of high cost but, on the other hand, has a very good accuracy with algorithms (i) triangulation, (ii) image transformation. AOA algorithms base on the estimated angles of arrival of signals from multiple LEDs in order to determine the position of the mobile device [13];
- Time of Arrival (ToA) – require an accurate synchronization between the receiver (PD) and emitter (LED) with algorithms (i) trilateration, and (ii) multilateration [14];
- Time-Difference-of-Arrival (TDoA) – requires synchronization between LEDs, and has high cost for the installation of the positioning system with algorithms (i) trilateration, and (ii) multilateration [9];
- Image or vision analyses.

4.2 IPS Early Implementations

The USA company, LVX System [15], during 2011, claimed to be the first company that launched a VLC commercial product.

Oledcomm Company was one of the first company that implemented in France, during 2012, the VLC as an IPS in museums and offices and then, in Lille Carrefour in 2015 [16].

During 2013, a South Korean supermarket uses a guiding light to point out discounts on the smartphone. Emart, the app on the smart phone sorts out automatically the lenses and the lights, so shoppers don't need to worry about compatibility issues. Shopping carts have a special place for the smartphone [17].

Acuity Brands demonstrated in 2014 both at Lightfair and Lux Live in London, the ByteLight IPS, using LED lighting based upon intelligent drivers from eldoLED. Using Lumicast VLC technology from Qualcomm, a smartphone was able to determine its position relative to LED ambient light fixtures to within a 10 cm radius, as well as deliver an accurate orientation of the direction the user was facing. Using light signals received from the lighting fixture, the system proved high accuracy.

Next year at Lightfair the addition of Bluetooth technology was shown to provide a 15 cm radius of detection when the phone was "listening" while in a pocket or purse, and this capability was integrated with Lumicast technology under the ByteLight services platform. The USA's company, Qualcomm presented in 2016 its developed IPS project based on VLC Lumicast. Qualcomm and Acuity Brands collaborate to commercially deploy this project in more than 100 US retail locations [18] (Fig. 3).

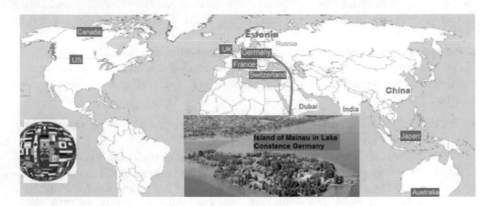

Fig. 3. IPS based on VLC worldwide R&D and implementations

The Japanese company Outstanding Technology, has also launched on 2012 its own product named Commulight for wireless communication, achieving high security and accuracy positioning information platform for smart phones or tablets using LED lighting. Their device consists in a dongle having an USB or a socket that plugs into any device of 3.5 mm jack and a sensor that senses relevant location-based information from the LED lighting fixtures that transmit data. Commulight is designed to give accurate, real time information to visitors, regarding the exhibits in a museum, or to send coupons to the customers in a supermarket [19].

OWC systems can be applied in different areas such as medical facilities [20], industry [21] or aviation.

5   Description of the IPGS for VIP

The general representation of the IPGS (Fig. 4) consists in the oTx embedded into the illumination fixture and the oRx used by the visually impaired actor.

The oTx is a low cost, high quality electronic system integrated into the illumination network as an access point with a pre-recorded unique message describing the present location and all the nearby possible destinations to be accessed into the building. It consists in a Printed Circuit Board (PCB), an electronic board with a microcontroller (MCU Board) and a proximity sensor as well as a Sound Storage Smart Device (SS-SD).

The PCB is designed to modulate the LED light by IM/DD using the OOK technique. The MCU board with the proximity sensor turns the light ON allowing to send the sound message pre-recorded on the SS-SD. The sound will be piggybacked by illumination only after the proximity sensor senses movement and therefore turns the light ON and triggers the recorder message to be send by light.

The oRx consists in a band of low cost flexible solar panel (SP) with the role of photodetector that can be embedded either into high quality headsets (HS) or an HS with an electronic board as sound amplifier (PCB-Amp) as well as into a frequently worn cloth' collar.

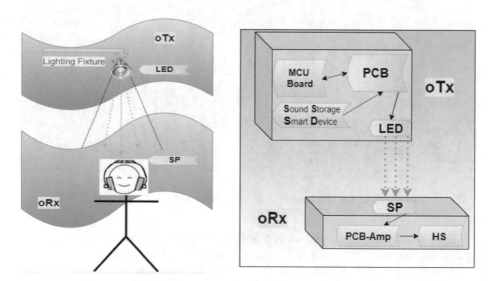

Fig. 4. The general representation of the IPGS with visually impaired actor (left) and a block diagram of IPGS (right)

5.1   Simulation of Both oTx and oRx Modules of the IPGS

The IM/DD technique used by the oTx embedded into the AP integrated into the illumination network is a suitable method to communicate data indoor based on VLC systems, due to its low cost and simplicity [22].

At the oTx module, when the electronic circuit is opened and LED ON, triggered by the proximity sensor, the recorded message is send by optical signal. Here, the signal changes from the electrical into optical one. The optical signal sent by LED is conveyed by light to the oRx and then is changed back from optical to electrical signal by the solar panel. The Arduino Proteus library was used (Fig. 5) in order to simulate the functionality of the oTx with electronic circuit, Arduino board and proximity sensor.

The vocal message regarding the VIP location and nearby destinations are able to be send due to a digital to analogic (D/A) signal converter circuit and the LED with the default optics embedded resulting a signal ready to be send indoor. Channel's default characteristics, as the way of the optical communication signal, can convey sound by light/piggybacked by illumination, to the oRx.

The oRx module, with a flexible solar panel, as the receiver, converts the signal from analog to digital (A/D) form, due to the circuit that drives and converts data received from the oTx.

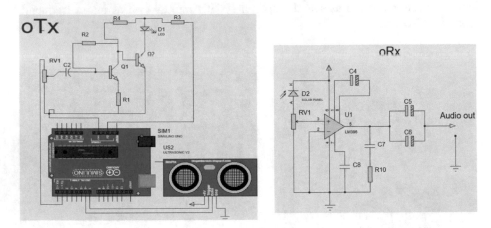

Fig. 5. Simulation of the oTx module (left) and oRx (right) with support of ISIS Proteus

Prior to purchase hardware, when possible, designing, testing and simulating the system is preferable to be done using one of many available useful applications.

In order to do this, Proteus ISIS application has been used due to its advanced, powerful tools, including more than 800 types of microcontroller, support many processor families along with lots of embedded peripherals being a reliable solution for circuit simulation and PCB professional design allowing to create a proper PCB for the VLC system.

Arduino programs are written in Proteus Visual Designer (PVD) using handy flowcharting methods and schematic Arduino shields that are easily placed on schematic capture design space, then the entire Arduino system can be simulated, tested and debugged in this software. The entire Arduino system has been simulated, tested and debugged with software assistance.

The prototype developed to test the VLC functionality and maximum distances for communication is presented in Fig. 6 where the speaker with the solar panel can be seen on the left and the oTx electronic circuit on breadboard on the right. The recorder message is sent by light and played by the speaker. The maximum distance achieved for a clear communication is 3.5 m.

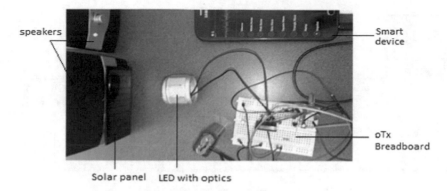

Fig. 6. VLC system tested with oTx on breadboard

5.2 Optical Tx Network Map of IPGS

The IPGS is based on received optical signal strength, being the easiest method with high accuracy to create de optical local area network (oLAN) of APs. There is, however, a main drawback since creating the optical map is time consuming being necessary to be developed for each indoor configuration that is possible to become outdated with any environmental change. Different mobile oRx read the optical received signal strength (oRSS) differently, and VIPs' presence in LoS between APs and oRx affects the RSS as well as any tilt of the solar panel. Three types of misalignment (longitudinal, lateral and angular) of the SP have to be taken into consideration here that have direct effect on the optical power level at the SP area.

Although fingerprinting is one of the most accurate positioning methods, the accuracy can be significantly decreased because of many factors such as: oRSS variation due to higher ambient lighting, environmental changes, multiple VIP presence on oRSS active area attenuating or even blocking the optical path. Such factors will affect the light behavior since the objects indoor as well as the obstacles, their color and type of material will result in reflection, refraction, or diffraction of light. Taking into consideration all intrinsic characteristics of the optical channel, is therefore important. To develop the database by surveying all the site locations to evaluate the oRSS, becomes compulsory, even if it is a time consuming process.

This IPGS model considers the indoor light behavior based on optical path loss (oPL) to adapt oLAN configuration according to the particularities of a specific example and also takes into consideration occurrence of environmental changes.

Different topologies regarding the position of both oTx and oRx, Line of Sight (LoS) or Non Line of Sight (NLoS) have to be also considered in several scenarios of the IPGS. In different scenarios, the optical signal coming from LED, is possible to suffer from severe attenuation, therefore the optical signal is non-linearly distorted at the SP surface. The optical power penalties associated with the optical channel are commonly separated into oPL and multipath dispersion that expresses itself as the Intersymbol Interference (ISI).

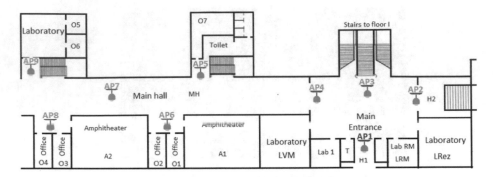

Fig. 7. oTx indoor map for the IPGS. An example.

The positions established for the oTx indoor (Fig. 7) are represented by AP (AP1 … AP9) with corresponding number, starting from the main entrance into the building where amphitheaters and laboratories for students, are. The active area where optical signal strength is powerful enough so the VIP receives the vocal message, into the proximity of each AP, is determined according to ambient light and light behavior due to surrounding objects, their color and type of material.

6   Conclusions

The VLC system considered here has a simplex communication structure with a custom-build infrastructure of APs with the oTx embedded. Each oTx sends guiding sound messages by visible light to the VIP's oRx device, who need to find easier and faster their destination into indoor crowded public facilities (here a university main hall is given as example) with difficult to find and access, end points. The vocal messages that are particular pre-recorded on each AP, send by light, are heard by the VIPs solely, who wear the low-cost, handy, special designed oRx, when their location is in LoS or into a well-established proximity of the AP. The network of APs indoor has the oTx embedded into the lighting setup being a dedicated infrastructure to support IPGS. When the oTx and oRx are in LoS, the proximity sensor turns ON the LED (embedded into the oTx) and the communication of the recorded sound message is simultaneously piggybacked by the optical signal. In this way, the VIP is aware of both his/her own position in building as well as the nearest surrounding possible destinations. Description of the entire IPGS system as well as the hardware prototypes with their simulations are presented here. A short presentation of the optical channel and the influence of the objects indoor onto the optical path with oRSS and the optical power at the receiver are also analyzed.

In addition to guidance, since the light path and irradiation range can be settled by the design and the light's distribution can be controlled, embedded into the lighting equipment, it is also possible to combine the IPGS proposed, with a security system, such as establishing the access authority attached to the place.

References

1. Kashyap, P., Samant, A., et al.: An assisted GPS support for GPS simulators for embedded mobile positioning. In: Proceedings of SPIE-IS&T Electronic Imaging, San Diego, 2–6 August 2009 (2009)
2. van Diggelen, F.: A-GPS: Assisted GPS, GNSS, and SBAS, 1st edn. Artech House, Norwood (2009). ISBN 1596933747, 978-1596933743
3. Nur, K., Feng, S., Ling, C., Ochieng, W.: Geo-spatial Inf. Sci. **16**(3), 155–168 (2013). https://doi.org/10.1080/10095020.2013.817106
4. Faragher, R., Harle, R.: An analysis of the accuracy of bluetooth low energy for indoor positioning applications. In: Proceedings of the 27th International Technical Meeting of the Satellite Division of the Institute of Navigation (ION GNSS+ 2014), Tampa, pp. 201–210 (2014)
5. Motlagh, N.H.: Near Field Communication (NFC) - a technical overview. Thesis Master of Science Degree in Telecommunication Engineering (2012). https://doi.org/10.13140/rg.2.1.1232.0720
6. Hirvonen, T.: Classification of spatial audio location and content using convolutional neural networks. In: Conference: AES 138th Convention, Warsaw, May 2015 (2015)
7. Leba, M., Riurean, S., Ionica, A.: Li-Fi - the path to a new way of communication. In: CISTI 2017 - 12ª Conferência Ibérica de Sistemas e Tecnologias de Informação. IEEE Xplore Digital Library (2017). https://doi.org/10.23919/cisti.2017.7975997
8. Riurean, S., Antipova, T., Rocha, Á., Leba, M., Ionica, A.: VLC, OCC, IR and LiFi reliable optical wireless technologies to be embedded in medical facilities and medical devices. J. Med. Syst. **43**, 308 (2019)
9. Chavez-Burbano, P., Yánez, V.G., Rabadan, J., Pérez-Jiménez, R.: Optical camera communication system for three-dimensional indoor localization. Opt. Int. J. Light Electron Opt. **192**, 162870 (2019)
10. Qiu, Y., Chen, H.H., Meng, W.X.: Channel modeling for visible light communications—a survey. Wirel. Commun. Mob. Comput. **16**, 2016–2034 (2016)
11. Zhang, X., Duan, J., Fu, Y., Shi, A.: Theoretical accuracy analysis of indoor visible light communication positioning system based on received signal strength indicator. J. Lightwave Technol. **32**(21), 3578–3584 (2014)
12. Arafa, A., Dalmiya, S., Klukas, R., Holzman, J.F.: Angle-of-arrival reception for optical wireless location technology. Opt. Express **23**(6), 7755–7766 (2015). https://doi.org/10.1364/oe.23.007755
13. Cardarelli, S., Calabretta, N., Koelling, S., Stabile, R., Williams, K.: Electro-optic device in InP for wide angle of arrival detection in optical wireless communication. In: Optical Fiber Communications Conference and Exhibition (OFC), San Diego, CA, USA, pp. 1–3 (2019)
14. Rabadan, J., Yánez, V.G., Guerra, C., Torres, J.R., Pérez-Jiménez, R.: A novel ranging technique based on optical camera communications and time difference of arrival. Appl. Sci. **9**(11), 2382 (2019)
15. http://www.lvxsystem.com/. Accessed 15 Aug 2019
16. https://www.oledcomm.net/applications/museums-exhibitions/. Accessed 15 Aug 2019
17. Millward, S. (2013). https://www.techinasia.com/korean-supermarket-emart-led-lights-smartphone-app-discounts. Accessed 15 Aug 2019
18. https://www.qualcomm.com/news/releases/2016/03/14/qualcomm-and-acuity-brands-collaborate-commercially-deploy-qualcomm. Accessed 15 Aug 2019
19. https://www.engadget.com/2012/07/16/outstanding-technology-visible-light-communication/. Accessed 15 Aug 2019

20. Riurean, S., Antipova, T., Rocha, A., Leba, M., Ionica, A.: Li-Fi embedded wireless integrated medical assistance system. In: Rocha, Á., Adeli, H., Reis, L., Costanzo, S. (eds.) New Knowledge in Information Systems and Technologies. WorldCIST'19 2019. Advances in Intelligent Systems and Computing, vol. 931, pp. 350–360. Springer, Cham (2019)
21. Riurean, S., Leba, M., Ionica, A., Stoicuta, O., Buioca, C.: Visible light wireless data communication in industrial environments. In: IOP Conference Series: Materials Science and Engineering, vol. 572, p. 012095 (2019)
22. Riurean, S., Stoica, R., Leba, M.: Visible light communication for audio signals. In: 12th International Conference on Applied Electromagnetics, Wireless and Optical Communications (ElectroScience 2017), 31 March–2 April 2017 (2017)

Classification of the Type of Hardened Steel Destruction Using a Deep Learn Neural Network

Mariia Bartolomei[✉], Andrei Kliuev, Aleksei Rogozhnikov,
and Valerii Stolbov

Perm National Research Polytechnic University, Perm 614990, Russia
mbartolomey@mail.ru

Abstract. The problem of images' classification of the fracture surface of steel samples by the nature of the fracture (brittle, viscous or mixed) and the proportion of the viscous component in the fracture is considered. The possibilities of artificial neural networks and deep machine learning are studied in the problem of predicting the physicomechanical properties of hardened steels during fracture based on neural network analysis of fracture surface images. A description of the constructed database of labeled digital microfractograms of the fracture surface of metallic functional materials after special thermomechanical processing used in deep training of neural networks is given. It is shown that quite popular deep neural networks VGG and ResNet with satisfactory accuracy solve the problem of microfractograms classification of various steels according to the type of fracture and the share of the viscous component in the fracture. The results of processing the initial information by a deep neural network are presented and a comparison of the obtained results with experimental data is presented.

Keywords: Deep networks · Machine learning · Functional materials · Microfractogram classification

1 Introduction

Microscopic fractographic analysis allows to get a more complete picture of the state of the material and its properties at the macro and micro levels in the process of destruction, and also allows to control a range of physical and mechanical properties without conducting separate tests for each of them. Depending on the composition and structure of the material, as well as on the conditions of processing and operation, fractures can be brittle, viscous and fatigue character [1]. The fracture usually has a complex relief of the fracture surface, including elements such as cleavage, pits, tear ridges, fatigue grooves, facets, etc. Therefore, the analysis of microfractograms of the fracture surface requires the involvement of highly qualified experts in the field of materials science and solid state physics. It should be noted that the result of the analysis is subjective. At present, it is becoming clear that to increase the efficiency and objectivity of interpreting the results of fractography and identifying the properties of materials, it is necessary to apply modern mathematical methods of data processing and

© Springer Nature Switzerland AG 2020
T. Antipova and Á. Rocha (Eds.): DSIC 2019, AISC 1114, pp. 513–521, 2020.
https://doi.org/10.1007/978-3-030-37737-3_44

artificial intelligence algorithms to obtain more objective information about the structure and properties of the material during destruction [2].

The goal of the work is to create, using machine learning methods in the form of a trained convolutional neural network, a universal tool that would make it possible to objectively relate the microfractogram of the material fracture surface to one of the types of fracture and evaluate the share of the viscous component in the fracture.

To achieve the goal, it is necessary to solve the following tasks:

- Set the task of classifying images of microfractograms of the fracture surface
- Form a database of marked digital photographs of the fracture surface of various materials
- Choose the type of convolutional neural network and train it to classify images of the fracture surface in automatic mode.

In the future, the information system will allow to take on part of the work of experts, averaging their readings, and, possibly, completely replace the opinion of experts for non-specialists in metal science.

Literature review [3–7] showed that on the problem described in the Russian segment of science, these networks have never been used to classify microphotographs, but, due to their indifference to the contents of the images, they can easily be used in including on such data. The lack of research on this topic and the divergence of opinions of experts regarding the determination of the properties of materials during destruction by microphotographs makes this task relevant today.

A neural network must operate in two modes: a training mode and an operating mode. For the training mode, data obtained during experiments on metals are used. In the mode of operation, the neural network should provide prediction of the type of destruction by microphotography. Typically, to classify complex images, a deep neural network is used, a network with several hidden layers [3]. Additional layers allow to build abstractions of ever higher levels, which makes it possible to form a model for recognizing complex objects in the real world. In tasks related to image processing, convolutional neural networks are mainly used because of their greatest efficiency [4–6].

2 Formulation of the Problem

The given set of digital images of the fracture surface (microfractograms) of various metal materials subjected to thermomechanical processing is considered. Each image is marked up, i.e. it is assigned by an expert to one of the types of fracture and the proportion of the viscous component in the fracture is indicated. All images are divided into n classes according to the proportion of the viscous component. It is required to select and train a neural network that allows classification of the image of the studied material to one of the given classes with the required accuracy.

3 Database Formation

The database of digital microfractograms of the fracture surfaces of metallic materials was formed on the basis of processing and analysis of the results obtained by specialists of the Institute of Nanosteels of Magnitogorsk State University. Based on the microstructural analysis of the image of the fracture surface, calculated and expert data on the microstructure of the fracture surface of the material are obtained. Each image was also accompanied by information on the fracture mechanism and the proportion of the viscous component in the fracture. In total, about 200 marked digital microfractograms of the fracture surface of various metal structural and functional (specially processed) steels were obtained. Heat treatment and special methods of intensive plastic deformation (torsion under high pressure, rolling, equal-channel angular pressing, etc.) were used as hardening treatment of steels. After processing, the samples were fixed and subjected to hammer blows to determine the toughness and calculate the total degree of deformation [8]. Pictures were taken using an electron microscope with various magnifications from x14 to x1000. Examples of images can be seen in Fig. 1. Pictures were annotated with physical and mechanical characteristics measured during testing.

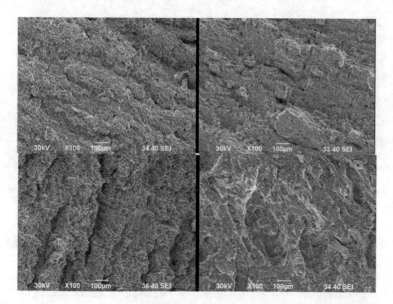

Fig. 1. Microfractograms of the fracture surface with the same magnification.

From the pictures, it can be determined whether the destruction was brittle or viscous. So the Fig. 1 (the upper left photo) shows brittle fracture, and the lower right - viscous. This is seen in the shape of the fracture, brittle fracture is characterized by the appearance of a sharp crack and a flat surface at the place of separation of the atomic layers, and viscous fracture is characterized by the appearance of a blunt opening crack, the crack propagation is slow in itself, the fracture surface is non-smooth, matte, and it scatters light rays [9].

For training and testing neural networks, the number of images obtained experimentally is clearly not enough. Therefore, using special information processing, a lot of images were enlarged.

The first step in creating a multitude of images was the formation of square photos with a side of 256 px. To begin with, the signatures affixed by the electron microscope were removed from all the original images. Then all the original images were cut into such squares. In order, not to lose information from the scraps, the right and lower sides were overlapped as shown in Fig. 2.

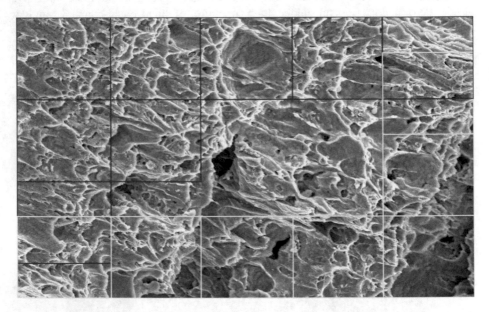

Fig. 2. Images of microphotographs from the training set. The borders of some images after the first pass of the cutting program are marked in red, yellow - after the second.

All images with a microscope magnification from x14 to x1000 were included in the created sample. Training a neural network in a preliminary sample showed that it is difficult for it to distinguish between fragments of images for different classes of different approximations. In this regard, it was proposed to reduce micrographs with a thousand-fold increase in 2 times, thereby making the conditional approximation for them 500 times and not to lose the number of images in the sample. After manipulating the initial data, we obtained the sample shown in Fig. 3.

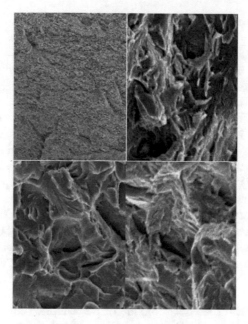

Fig. 3. Cutting images of microphotographs.

In the process of preparing the training set from the available images with the help of various transformations, training and verification sets of 2620 and 1020 photos, respectively, was formed. The training set was divided into classes according to the proportion of viscous component. Table 1 shows the image classes by the share of the viscous component and the distribution of photographs of the training set by them.

After the formation of the database of images of the fracture surface of various hardened steels, the stage of selection and training of the neural network used to solve the classification problem was considered.

Table 1. Distribution of the training set into classes of the viscous component.

Classes share of the viscous component	The number of photos in the training set	The number of photos in the verification set
I class, (0–2%) brittle fracture	440	180
II class, (2–8%) mixed fracture with a light proportion of viscous component	320	180
III class, (8–30%) mixed fracture with a medium proportion of viscous component	740	340
IV class, (30–100%) mixed fracture with a high proportion of viscous component	1120	320

4 Neural Network Selection and Training

To solve the problem, the architecture of the ResNet18 network was chosen as an alternative to the VGG-16 convolutional neural network tested in this area. It was found during training that the ResNet network gives approximately similar results, but the learning process itself takes much less time, so on a sample of 8 thousand images VGG-16 spent 8 h on training, and the ResNet network on weaker equipment and spent the same sample 1.5 h to reduce the time for experiments, as a result, the ResNet network was chosen. A detailed description of the network architecture is available in the source [6]. An untrained version of the network was selected with an input image of 256 × 256 pixels [7]. The following describes the process of preparing the training set and the process of training the network.

As a software platform for implementing the network, the Keras framework was chosen in conjunction with the TensorFlow machine learning library [7].

Deep neural networks were trained using the error back propagation algorithm using the stochastic gradient descent method [5]. The training was carried out on a hardware platform with the following characteristics: the size of simultaneously supplied samples was 16, the number of epochs was 200, each epoch was a complete pass through all images with subsequent mixing of the sample. One computational experiment took 1 h for 150 epochs. Network training lasted no more than 292 iterations. Accuracy calculation was carried out each iteration and is equal to the percentage of correctly recognized images from the test sample formed from the training using random selection. The training was carried out on an NVidia GTX 1080 graphics card with 8 gigabytes of memory. To assess the accuracy of the network, the criteria Top-1 and Top-3 were used (a mistake is considered a situation when among the 3 best options there is no correct answer). Figure 4 shows a graph of changes in accuracy in the learning process. At the time of graduation, the network recognition accuracy reached 98.7%.

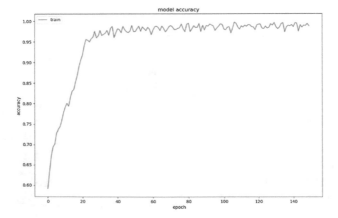

Fig. 4. Change in accuracy during the learning process.

From Fig. 4 it can be seen that the learning process is typical for such tasks and does not have any features. This result shows that the chosen approach to the study of the prediction of the physicomechanical properties of materials, in general, gives positive results, however, the accuracy according to the Top-1 criterion obtained by testing the network on the validation set of images was only 48%, which is insufficient.

Due to the low accuracy of the Top-1 criterion, the magnification interval was reduced in microphotographs. For this, images with a magnification of 1000 times were removed from the training set. After additional preprocessing of the image database, the number of images in the training set was 2117, and 569 in the validation set.

The change in accuracy during the training of the neural network is shown in Fig. 5.

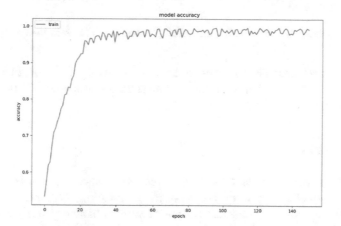

Fig. 5. Change in accuracy during the learning process.

Figure 5 shows that the accuracy of the model during training remained at the same level. However, in this experiment, it was possible to increase the accuracy when testing on a validation set according to the Top-1 criterion only to 60.43%. At the same time, part of the experimental information about images with a high magnification factor was lost. Therefore, images were returned to the training sample with a magnification of 1000 times, but their area was reduced by 4 times, thereby conditionally leading to an increase of 500 times. The reduction was performed using the bilinear interpolation algorithm. Due to the fact that the size of some images has decreased, the number of sliced images in the sample has also changed. So, for example, in the training sample there are already 2347 images, and in the validation - 621. The change in accuracy in the learning process is shown in Fig. 6. The accuracy according to the Top-1 criterion on the training set was 98%, and on the validation set it was 66.81%, which can be considered a satisfactory result for the classification problem under consideration.

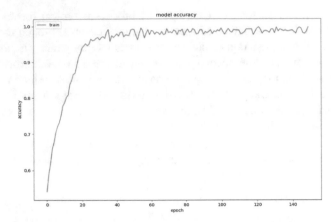

Fig. 6. Change in accuracy during the learning process according to the criterion Top-1.

The shown accuracy allows the use of a trained network as the core of an intelligent system for the integrated assessment of the strength properties of functional and structural materials [10–12].

5   Conclusion

Studies have shown that with the help of convolutional neural networks of deep learning, it is possible to solve the problems of classification of complex images, which are microfractograms of the fracture surface of various hardened steels. However, the accuracy of the classification largely depends on the method of preprocessing the original sample of digital images. The developed recognition tool allows describe with acceptable accuracy the microphotographs of the physical and mechanical properties of steels during fracture and to determine the proportion of the viscous component in the fracture, which, in turn, allows to determine the type of fracture and predict the strength of functional materials.

Acknowledgement. The work was financially supported by the Ministry of Science and Higher Education of the Russian Federation (project identifier RFMEFI58617X0055, as part of the implementation of the international project "Fracture across Scales and Materials, Processes and Disciplines (FRAMED)" under the program "Horizon 2020").

References

1. Kopceva, N., Efimova, Yu., Miholenko, D.: The evolution of the microstructure and properties during heating of ferritic-pearlitic carbon steels with an ultrafine-grained structure formed by intense plastic deformation. VSTU Bull. **7**(9), 85–91 (2011)
2. Gitman, M., Klyuev, A., Stolbov, V., Gitman, I.: Complex estimation of strength properties of functional materials on the basis of the analysis of grain-phase structure parameters. Strength Mater. **49**(5), 710–717 (2017)

3. Schmidhuber, J.: Deep learning in neural networks: an overview. Neural Netw. **61**, 85–117 (2015)
4. Mikolov, T., Karafiat, M., Burget, L., Cernocky, J., Khudanpur, S.: Recurrent neural network based language model. In: 11th Annual Conference of the International Speech Communication Association, Japan, pp. 1045–1048 (2010)
5. Bottou, L.: Stochastic gradient descent tricks. In: Montavon, G., Orr, G.B., Müller, K.R. (eds.) Neural Networks: Tricks of the Trade. LNCS, vol. 7700, pp. 421–436. Springer, Heidelberg (2012)
6. https://neurohive.io/ru/vidy-nejrosetej/resnet-34-50-101/. Accessed 01 June 2019
7. https://github.com/raghakot/keras-resnet. Accessed 29 Feb 2019
8. http://docs.cntd.ru/document/1200005045. Accessed 20 Feb 2019
9. Kolskii, G.: Stress Waves in Solids. Inostrannaya Literatura, Moscow (1955)
10. Sharibin, S., Stolbov, V., Gitman, M., Barishnikov, M.: Development of an intelligent recognition system for complex microstructures on kinds of metals and alloys. Neurocomput. Dev. Appl. **12**, 50–56 (2014)
11. Klyuev, A., Stolbov, V., Sharibin, S.: Visualization of complex grain structures of metals and alloys with identification of their parameters. Sci. Vis. **8**(3), 95–101 (2016)
12. Klestov, R., Klyuev, A., Stolbov, V.: About some approaches to problem of metals and alloys microstructures classification based on neural network technologies. Adv. Eng. Res. **157**, 292–296 (2018)

The Deformation Behavior of Modern Antifriction Polymer Materials in the Elements of Transport and Logistics Systems with Frictional Contact

A. A. Adamov[1] and A. A. Kamenskikh[2(✉)]

[1] Institute of Continuous Media Mechanics, 614013 Perm, Russian Federation
[2] Perm National Research Polytechnic University,
614990 Perm, Russian Federation
anna_kamenskih@mail.ru

Abstract. The deformation behavior of two models of spherical bearing of bridge spans was considered in an axisymmetric formulation as part of the work: with and without grooves with lubricant. Two modern antifriction polymers were considered as sliding layer materials: modified PTFE and antifriction composite material based on PTFE with spherical bronze inclusions and molybdenum disulfide. The physicomechanical and frictional properties of the antifriction spherical layer materials were obtained experimentally by the scientific team of Alfa-Tech LLC and Institute of Continuous Media Mechanics of the Ural Branch of Russian Academy of Science. The approximation of the experiments results the dependence of friction coefficient on the pressure was performed and functions approximating the experiments results with a maximum error of 2.3% were obtained. The approximating functions were used to determine the friction coefficient at pressures >54 MPa acting on the spherical bearing. A series of numerical experiments aimed at identifying the qualitative and quantitative patterns of the deformation behavior of the spherical thin sliding layer materials was carried out as part of the work: contact pressure, contact tangential stress, contact status. Comparison of settlement of numerical models of spherical bearing with a modified PTFE layer and experiments data was performed in the work. It is established that the model taking into account grooves with a lubricant has a minimum settlement error under the frictional properties of the interlayer material declared by the manufacturer, which is approximately 8.44%. The settlement error for all considered bearing models is more than 15% with considering frictional properties of antifriction materials obtained experimentally.

Keywords: Modified PTFE · Composite antifriction material · Contact · Friction · FEM · Polymer properties · Elements of transport and logistics systems · Lubrication · Spherical bearing

T. Antipova and Á. Rocha (Eds.): DSIC 2019, AISC 1114, pp. 522–532, 2020.
https://doi.org/10.1007/978-3-030-37737-3_45

1 Introduction

Bridge structures are responsible units of transport and logistics systems around the world. Increased requirements regarding strength, reliability, stability, durability, safety, etc. presented to them and their elements. The bearing of bridge spans [2–4], which are used to absorb thermal expansion and compression, creep and shrinkage, seismic disturbances, etc., relate to the supporting elements of bridge structures as noted in [1]. The bearings also perceive vertical and horizontal loads from bridge spans. These designs are designed for long-term non-repair periods. Their technical condition has a significant impact on the operation of the bridge structure and can lead to adverse consequences: partial or complete destruction of bridge spans, supports or the structure as a whole. A number of works considers the causes of the bridge structures destruction [5–7] associated with load-supporting elements: destruction caused by loss of elements stability, destruction due to insufficient reliability and breakdown of joints, destruction caused by overloading of supporting structures, etc. Therefore, the study of the bridge structures supporting elements in the framework of the deformed solid mechanics is relevant at the moment. There is an increase in the number of works related to solving problems of analyzing the design features of elements of transport and logistics systems: geometrical configuration and technology of the device of deformation joints [8, 9], bearing [8, 10, 11], bridges spans [12] and other elements. Many works are aimed at optimizing the bearing designs and their elements, as well as the analysis of the supporting capacity, strength, wear resistance of the elements bearing in general and contact parameters in particular [10–15], including the materials from which elements are made. The possibility study of using modern antifriction polymeric materials in the bearing structures [11, 16, etc.], as well as the influence analysis of the structural particulars of the transport and logistics systems elements on the stress-strain contact state of bridge structures are relevant research areas.

2 Problem Statement

A number of topical problems associated with structural particulars of bearings is celebrated bridge-building enterprise: analysis of the effect thickness of a spherical and flat slip layer on the deformation behavior of the unit; influence on the stress-strain state and contact interaction of the geometric configuration of the lubrication recesses in the interlayers (grooves, holes, etc.); influence analysis on the unit deformation contact behavior of the spherical and flat sliding layer location; analysis of the surface treatment of steel plates effect which are contact with antifriction layers on the unit deformation contact behavior ("torn" carving, polished surface, etc.) and others. Two actual problems of critical elements of transport and logistics systems such as spherical bearing are considered in this paper: influence analysis of materials and structural particulars of the spherical sliding layer geometric configuration on the unit stress-strain state as a whole and the parameters of the contact zone in particular.

Friction contact interaction of upper (1) and lower steel plate (2) of the spherical bearing through an elastoplastic polymeric layer (3) is considered in the paper with (Fig. 1b) and without (Fig. 1a) grooves with lubricant (4).

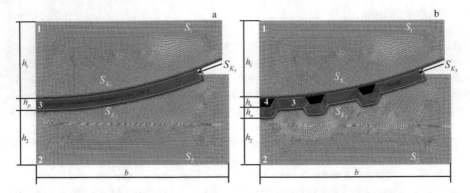

Fig. 1. Finite-element models of the spherical bearing: with (b) and without (a) taking into account the lubricant grooves.

Contact deformation of the spherical bearing with the L-100 (Fig. 1) anti-friction layer manufactured by AlphaTech LLC (Perm) is considered. The spherical bearing is designed for a nominal vertical load of 1000 kN. Two models of the bearing are considered, we introduce the following designations: model A is model of the spherical bearing without taking into account the grooves with lubricant; model B is model of the spherical bearing without taking into account the grooves with lubricant. Typical dimensions of the spherical bearing: maximum height of the top plate $h_1 = 0.03$ m, minimum height of the bottom plate with a spherical neckline $h_2 = 0.02$ m ($h_2 = 0.0179$ m for the model with taking into account the grooves), maximum width of the structure $b_k = 2b = 0.155$ m, thickness of the sliding polymeric layer $h_p = 0.004$ m, thickness of the grooves with lubricant $h_k = 0.003$ m. Constant indentation force and the bending is prohibited on S_1, on S_2 the y-movements are prohibited. The rounding of the sharp edges of antifriction layer 3 is not taken into account in the design.

3 Experimental Part and Mathematical Formulation of the Contact Problem

Previously, the Institute of Continuous Media Mechanics of the Ural Branch of the Russian Academy of Sciences carried out a cycle of experimental studies of the physical and mechanical characteristics of anti-friction materials at complex multistage deformation stories with offloading using a Zwick Z100SN5A testing machine. As a result, the physical and mechanical characteristics of materials were obtained: modulus of elasticity E, Poisson's coefficient v, deformation diagrams $\sigma - \varepsilon$ (Fig. 2). Two materials that can be used as antifriction layers in the design of spherical bearing are considered in the work: modified PTFE (material 1) and antifriction composite material based on PTFE with spherical bronze inclusions and molybdenum disulfide (material 2) with Young's modulus 863.8 and 860.52 MPa and Poisson's coefficients 0.461 and 0.4388 elastic section, respectively [3].

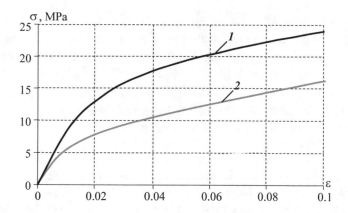

Fig. 2. Deformation diagrams $\sigma - \varepsilon$: 1 – material 1; 2 – material 2.

Further, the scientific team of Alfa-Tech LLC and IMSS of the Ural Branch of the Russian Academy of Sciences prepared special equipment and carried out a series of full-scale experiments aimed at determining the friction properties of polymeric materials [17]. Within the limits of a series of tests, the friction coefficient μ was determined as a function of pressure. Figure 3 shows the results of full-scale experiments with and without taking into account lubricant on the mating surfaces at contact deformation of cylindrical samples with a diameter $0.097 \div 0.103$ m and height 0.01 m by steel plates of the press for the two materials under consideration.

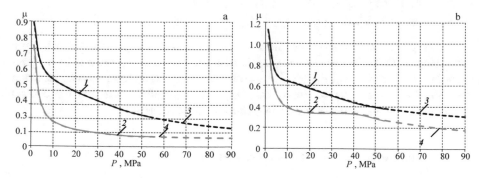

Fig. 3. Dependence of μ on P for material 1 (a) and materials 2 (b): 1 is experimental data without lubricant; 2 is experimental data with lubricant; 3 is approximation of experimental data (1); 4 is approximation of experimental data (2).

The full-scale experiment was carried out in the pressure range of 1.2–54 MPa, and the working pressure range of the spherical bearings can reach 90 MPa. Within the limits of the analysis of experimental data, the approximation of results of full-scale experiments was executed, the approximating functions were selected. Functions approximating experimental results with an error not exceeding 1% are selected for material 1: $\mu(P) = 0,005 + \frac{0,111}{P} + \frac{0,623}{P^2} - \frac{3,57}{P^3} + \frac{3,335}{P^4}$ and $\mu(P) = -0,002 + \frac{1,55}{P} - \frac{17,166}{P^2} + \frac{64,979}{P^3} - \frac{55,745}{P^4}$ for contact with and without lubricant on mating surfaces, respectively. Functions approximating experimental results with an error not exceeding 2,3% are selected for material 2: $\mu(P) = -0,003 + \frac{2,203}{P} - \frac{33,134}{P^2} + \frac{140,289}{P^3} - \frac{124,227}{P^4}$ and $\mu(P) = 0,016 + \frac{1,485}{P} - \frac{15,782}{P^2} + \frac{57,885}{P^3} - \frac{49,101}{P^4}$ for contact with and without lubricant on mating surfaces, respectively. The selected functions are required to calculate the friction coefficient for pressure >54 MPa. In this paper, we consider a spherical bearing of L-100 with a nominal operating vertical load of 1000 kN (~ 55.5 MPa). The general mathematical statement of the problem of contact interaction of two elastic bodies through the anti-friction polymeric layer taking into account all types of friction contact is described in [14]. The problem is considered in the axisymmetric statement; the deformation theory of elasticity was chosen to describe the model of the behavior of the anti-friction layer material. Lubrication is modeled as a low compressible material with $E = 200$ MPa, $\nu = 0.4999$ and $\mu = 0.007$. The contact between the lubricant and anti-friction layer is not taken into account.

The contact boundary conditions are applied to the surface $S_K = S_{K_1} \cup S_{K_2} \cup S_{K_3}$, where the two bodies 1 and 2 are in contact along S_K. The types of contact interaction that are implemented in the problem are given below:

- sliding with friction (friction of rest): $\vec{u}^1 = \vec{u}^2$, $\sigma_n^1 = \sigma_n^2$, $\sigma_{n\tau_1}^1 = \sigma_{n\tau_1}^2$, $\sigma_{n\tau_2}^1 = \sigma_{n\tau_2}^2$, wherein $\sigma_n < 0$, $|\sigma_{n\tau}| < \mu(\sigma_n)|\sigma_n|$;
- sliding with friction (sliding friction): $u_n^1 = u_n^2$, $u_{\tau_1}^1 \neq u_{\tau_1}^2$, $u_{\tau_2}^1 \neq u_{\tau_2}^2$, $\sigma_n^1 = \sigma_n^2$, $\sigma_{n\tau_1}^1 = \sigma_{n\tau_1}^2$, $\sigma_{n\tau_2}^1 = \sigma_{n\tau_2}^2$, wherein $\sigma_n < 0$, $|\sigma_{n\tau}| = \mu(\sigma_n)|\sigma_n|$;
- no contact: $|u_n^1 - u_n^2| \geq 0$, $\sigma_{n\tau_1} = \sigma_{n\tau_2} = \sigma_n = 0$;
- adhesion: $\vec{u}^1 = \vec{u}^2$, $\sigma_n^1 = \sigma_n^2$, $\sigma_{n\tau_1}^1 = \sigma_{n\tau_1}^2$, $\sigma_{n\tau_2}^1 = \sigma_{n\tau_2}^2$,

where $\mu(\sigma_n)$ is the friction coefficient, τ_1, τ_2 are axes designation which lie in a plane tangent to the contact surface, u_n is displacement along a normal to a corresponding contact edge, u_{τ_1}, u_{τ_2} are displacement in a tangential plane, σ_n is stress along the normal to the contact boundary, $\sigma_{n\tau_1}$, $\sigma_{n\tau_2}$ are tangential stresses at the contact boundary, $\sigma_{n\tau}$ is the value of vector tangential contact stresses.

4 Calculations, Results and Discussion

The problem is implemented using axisymmetric finite elements with the Lagrangian approximation and two unknowns in a node. Previously, it was established that numerical solutions for finite element mesh with 8 and 16 elements in the thickness of the layer for the model without taking into account the grooves with lubricant have minor differences [3]. A finite element mesh with 16 elements by layer thickness was chosen for the implementation of the deformation frictional contact problem of the elements of spherical bearing with lubrication grooves, for a better problem solution. Analysis of the deformation behavior of interlayers from two antifriction materials taking into account frictional properties was performed on the selected finite element model. Constant friction coefficient 0.04 declared by manufacturers of modern antifriction materials based on PTFE. A comparative analysis of the frictional properties effect of antifriction materials obtained experimentally and declared by manufacturers on the contact zone parameters for two models of bearing was made therefore: with and without lubricant grooves.

The numerical modeling results are given taking into account the frictional properties of interlayer materials with lubrication. The distribution of the relative contact pressure and the relative contact tangential stress on the main spherical sliding surfaces S_{K_1} and S_{K_2} of the bearing is shown in Fig. 4 for the bearing model A.

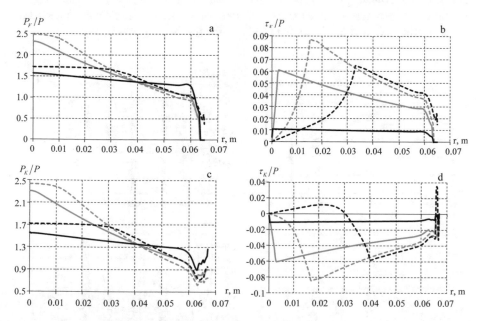

Fig. 4. Relative contact pressure (a, c) and relative contact tangential stress (b, d) at S_{K_1} (a, b) and S_{K_2} (c, d) for the bearing model A: black color is material 1; gray color is material 2; dashed line is frictional properties declared by manufacturers; solid line is frictional properties obtained experimentally.

The distribution of contact parameters is the same for all considered cases: contact pressure has small differences and the contact tangential stress increases in the contact surfaces adhesion zone with a further decrease in the levels of contact parameters in the sliding zone. The "no contact" zone is observed at the edge of the interlayer on S_{K_1} at 2.18 and 2.55% of the contact surface area for material 1 and 2 respectively. The materials frictional properties of the antifriction layer have a significant effect on the adhesion-sliding contact zones distribution: 50.9 and 27.64% of the contact surface area is in a contact adhesion state with frictional properties declared by the manufacturer for material 1 and 2 respectively; the contact adhesion area is much smaller in the case of frictional properties obtained experimentally and is less than 1% for material 1 and about 4% for material 2. Sliding with friction is observed for more than 90% of the contact surface area with frictional properties obtained experimentally. The maximum level of P_K/P is lower by 9 and 5% when taking into account the experimentally obtained frictional properties of materials of the antifriction layer than with the coefficient of friction declared by the materials manufacturers for materials 1 and 2 respectively. Similarly, the maximum level of τ_K/P is lower by 83.64 and 30.2% when taking into account the experimentally obtained frictional properties of materials of the antifriction layer than with the coefficient of friction declared by the materials manufacturers for materials 1 and 2 respectively.

Input lubricant grooves to the model of bearing have a significant effect on the pattern of the contact parameters distribution. The distribution of the relative contact pressure and the relative contact tangential stress on the main spherical sliding surfaces of the bearing S_{K_1} and S_{K_2} is shown in Fig. 5 for the bearing model B.

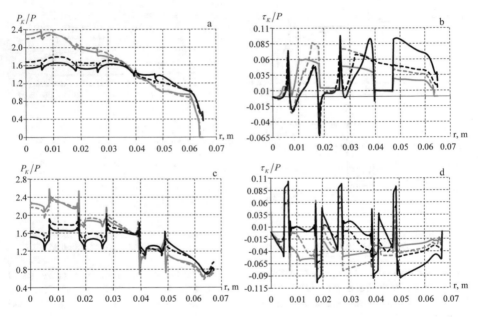

Fig. 5. Relative contact pressure (a, c) and relative contact tangential stress (b, d) at S_{K_1} (a, b) and S_{K_2} (c, d) for the bearing model B: black color is material 1; gray color is material 2; dashed line is frictional properties declared by manufacturers; solid line is frictional properties obtained experimentally.

The input of lubricant grooves in model leads to a change in the pattern of the contact states zones distribution: sliding zones with friction appeared in the antifriction layer central part. The level of contact pressure on average has small differences from the bearing model A. The lubricant grooves have a significant impact on the tangential stress distribution pattern: jumps τ_K/P are observed near the interface areas of antifriction material layer and the lubricant. The frictional properties of interlayer materials obtained experimentally affect of the contact zone parameters: the maximum level P_K/P for the interlayer of material 1 is less by 9.2%, for the interlayer of material 2 is

3.48 and 4.75% more for S_{K_1} and S_{K_2} respectively; the maximum level τ_K/P for the interlayer of material 1 is 23.3 and 49% more on S_{K_1} and S_{K_2} respectively and for the interlayer of material 2 is less by about 30% for all contact surfaces.

The comparison of the numerically obtained values of the bearing draft with the results of a experiments series on the deformation of the real design of the spherical bearing L-100 manufactured by Alfa-Tech LLC with an antifriction layer of material 1 by the vertical load 500–1250 kN was performed as part of the work. Results comparison of the settlement of spherical bearing models with the experiments results at a nominal load of 1000 kN is shown in Fig. 6 for evaluate the numerical simulation results.

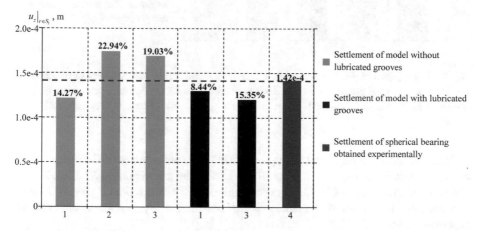

Fig. 6. Settlement of model spherical bearing with layer of material 1: 1 is frictional properties declared by manufacturers; 2 is frictional properties obtained experimentally without lubricant; 3 is frictional properties obtained experimentally with lubricant; 4 is experiment.

It was found that at a nominal vertical load of 1000 kN, the difference in the results of the problem numerical solution for the model taking into account lubricant grooves and the experiment results is minimal for frictional properties declared by the manufacturer and is $\sim 8.44\%$, for the model without grooves is $\sim 14.27\%$. The numerical simulation results of the deformation behavior of the spherical bearing elements, taking into account the frictional properties of material 1, obtained experimentally, differ from the experimental data on the bearing draft by more than 15%. The influence of the antifriction layer geometric configuration on the deformation behavior of the unite can be noted. For example, the input in the model of rounding of the edge antifriction layer of the led to a decrease in the modeling error: at a nominal vertical load of 1000 kN, the difference in the results of the problem numerical solution for the model taking into account lubricant grooves and the experiment results is minimal for frictional properties declared by the manufacturer and is $\sim 7.13\%$, for the model without grooves $\sim 13.05\%$ [17]. A significant reduction in the time of numerical calculation can be noted in this case.

Differences in the settlement of the bearing models A and B with interlayers of materials 1 and 2 are shown in Fig. 7 for work evaluate of the antifriction layer materials.

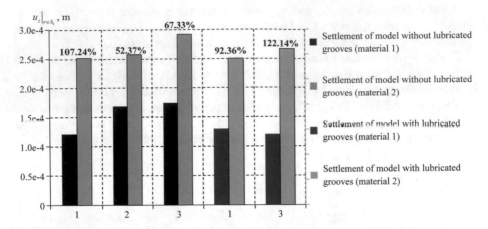

Fig. 7. Comparison of settlement models of spherical support parts with interlayers of material 1 and 2: 1 is frictional properties declared by manufacturers; 2 is frictional properties obtained experimentally without lubricant; 3 is frictional properties obtained experimentally with lubricant.

The spherical bearing draft with an antifriction layer of material 2 is greater than that of spherical bearing with a layer of material 1, by a maximum of 122.14%. This corresponds to the case of frictional contact interaction of the model B elements taking into account the frictional properties of the antifriction layer material are obtained experimentally.

5   Conclusion

An analysis of changes in the contact parameters at a nominal load of 1000 kN was performed for two versions of the bearing model (with and without lubricant grooves), taking into account the frictional properties of the sliding layer materials, obtained experimentally and declared by the manufacturer. The numbers of laws are established in the analysis.

The model without lubricant grooves:

- the level of contact pressure and contact tangential stress is more than 9 and 83% lower, respectively, with a friction coefficient corresponding to contact with lubricant for a layer of material 1 than with a friction coefficient declared by the materials manufacturers;
- the level of contact pressure and contact tangential stress is more than 5 and 30.2% lower, respectively, with a friction coefficient corresponding to contact with lubricant for a layer of material 2 than with a friction coefficient declared by the materials manufacturers.

The model with lubricant grooves:

- the level of contact pressure is less on 9% and the level of contact tangential stress is more on 23.3–49% for a layer of material 1 with a friction coefficient corresponding to contact with the lubricant than with a friction coefficient declared by the materials manufacturers.
- the level of contact pressure is more on 3.48–4.75% and the level of contact tangential stress is less on ∼30% for a layer of material 2 with a friction coefficient corresponding to contact with the lubricant than with a friction coefficient declared by the materials manufacturers.

The series of experiments on deformation of the bearing by a vertical load of 500–1250 kN were carried out. When comparing the problems numerical solution on the deformation of bearing with experiment, it was established:

- the error is about 8% for the model taking into account lubricant grooves and 14% for the model without taking into account grooves (material 1, friction coefficient 0.04);
- the error is about 15% for the model taking into account lubricant grooves and 20% for the model without taking into account grooves (material 2, frictional properties obtained experimentally).
- the draft of bearing with a layer of material 2 is greater than when used as a layer of material 1: for the model with lubrication grooves by 92.36–122.14%; for the model excluding lubrication grooves by 52.37–107.24%.

A significant effect of the antifriction layers materials and the geometric configuration of the spherical bearing on the stress-strain state and contact parameters were established as part of the study. The specification of bearing models is required taking into account in the model a flat sliding layer of antifriction material located in the lower part of the plate with a spherical cut; the three-dimensional formulation of the task, which will allow us to analyze the influence of the lubricant recesses geometry, etc.

Acknowledgements. The study supported by a grant of Russian Science Foundation (project No. 18-79-00147).

References

1. Xue, J.Q., Briseghella, B., Chen, B.C., Zhang, P.Q., Zordan, T.: Optimal design of pile foundation in fully integral. Dev. Int. Bridge Eng. **9**, 3–16 (2016)
2. Akogul, C., Celik, O.C.: Effect of elastomeric bearing modeling parameters on the Seismis design of RC highway bridges with precast concrete girders. In: Proceedings of the 14th World Conference on Earthquake Engineering (2008)
3. Kamenskih, A.A., Trufanov, N.A.: Numerical analysis of the stress state of a spherical contact system with an interlayer of antifriction material. Comput. Continuum Mech. **6**(1), 54–61 (2013)

4. Drozdov, Yu.N., Nadein, V.A., Puchkov, V.N.: Determining relations between earthquake parameters and tribological characteristics of frictional pendular bearings (Seismic Isolators). Russ. Eng. Res. **27**(4), 179–187 (2007)
5. Proske, D.: Bridge Collapse Frequencies Versus Failure Probabilities. Springer, Cham (2018)
6. Kuznetsova, S.V., Kozlov, A.V.: Causes of bridge structures accidents in Russia and the CIS countries. Roads Bridges **1**(39), 204–219 (2018)
7. Ovchinnikov, I.I., Maystrenko, I.Yu., Ovchinnikov, I.G., Uspanov, A.M.: Failures and collapses of bridge constructions, analysis of their causes. Part 4. Russ. J. Transp. Eng. **5**(1), 05SATS118 (2018)
8. Anisimov, A.V., Bakhareva, V.E., Nikolaev, G.I.: Antifriction carbon plastics in machine building. J. Friction Wear **28**(6), 541–545 (2007)
9. Yankovsky, L.V., Kochetkov, A.V., Ovsyannikov, S.V., Trofimenko, Yu.A.: Deformation seams of small structures of small movements: device, repairability, texture. Tech. Regul. Transp. Constr. **3**(7), 6–12 (2014)
10. Becker, T.C., Mahin, S.A.: Correct treatment of rotation of sliding surfaces in a kinematic model of the triple friction pendulum bearing. Earthq. Eng. Struct. Dynam. **42**(2), 311–317 (2013)
11. Choi, E., Lee, J.S., Jeon, H.-K., Park, T., Kim, H.-T.: Static and dynamic behavior of disk bearings for OSPG railway bridges under railway vehicle loading. Nonlinear Dyn. **62**, 73–93 (2010)
12. Ivanov, B.G.: Diagnostics of Damage to the Span of Metal Bridges: A Monograph. Marshrut, Moscow (2006)
13. Saidou Sanda, M., Gauron, O., Turcotte, N., Lamarche, C.-P., Paultre, P., Talbot, M., Laflamme, J.-F.: Efficient finite elements model updating for damage detection in bridges. In: Proceedings of International Conference on Experimental Vibration Analysis for Civil Structures, Lecture Notes in Civil Engineering, pp. 293–305 (2018)
14. Kamenskih, A.A., Trufanov, N.A.: Regularities interaction of elements contact spherical unit with the antifrictional polymeric interlayer. J. Friction Wear **36**(2), 170–176 (2015)
15. Wu, Yi., Wang, H., Li, Ai., Feng, D., Sha, B., Zhang, Yu.: Explicit finite element analysis and experimental verification of a sliding lead rubber bearing. J. Zhejiang Univ. –Sci. A, **18**(5), 363–376 (2017)
16. Peel, H., Luo, S., Cohn, A.G., Fuentes, R.: Localisation of a mobile robot for bridge bearing inspection. Autom. Constr. **94**, 244–256 (2018)
17. Adamov, A.A., Kamenskikh, A.A.: Comparative analysis of the contact deformation of the spherical sliding layer of the bearing with and without taking into account the grooves with lubricant. In: IOP Conference Series: Materials Science and Engineering, vol. 581, p. 012031 (2019)

Methodology for the Formation of Groups of Representative Parts in the Creation of Digital Engineering Industries

Andrey Kutin(✉), Mikhail Sedykh, Svetlana Lutsyuk,
and Vladimir Voronenko

Moscow State University of Technology STANKIN,
127994 Moscow, Russian Federation
aa.kutin@stankin.ru, sedykhmi@mail.ru, sv.lu@bk.ru,
vpvoronenko@yandex.ru

Abstract. A technique is proposed for grouping parts in a multinomenclature production based on a technical and economic analysis of the parameters of products and the manufacturability of parts manufacturing, which allows you to select a representative part of the group and calculate the required number and load factors of technological equipment. Testing the developed methodology at one of the machine-building enterprises showed that the deviation of the machine time for processing the "Plate" workpieces on metal-cutting machines from the estimated average is 12%.

Keywords: Digital manufacturing · Grouping of parts · Technical and economic analysis on the technological processes

1 Introduction

In conditions of rapidly changing requirements for products, given the personification of products dictated by market conditions, the enterprise requires more flexibility of the production system [1]. This complicates the set of requirements for technological and organizational decisions. Changing the technological process under the current conditions, the company sets the task of manufacturing products either with a reduction in costs associated with the stage of technological preparation of production, or with a decrease in the complexity of its manufacture [2].

In this case, the largest number of input parameters should be considered for multinomenclature production. This raises questions about the adaptation of decisions to the technological system and its condition [3]. Any changes in the technological process require a preliminary feasibility study.

© Springer Nature Switzerland AG 2020
T. Antipova and Á. Rocha (Eds.): DSIC 2019, AISC 1114, pp. 533–542, 2020.
https://doi.org/10.1007/978-3-030-37737-3_46

2 The Theoretical Basis for the Development of Grouping Techniques

When developing the methodology for conducting a feasibility study for a new type of product or a new technological process, it was revealed that manufacturability is a criterion that combines technical and economic indicators of products in the current production base of the enterprise [4]. Of particular importance among the parameters of manufacturability are the complexity, material consumption and cost of production.

Under the conditions of existing production, the task is also to ensure the timing of the production order, which also applies to the new process introduced as a result of its partial or complete modernization. In this case, not only testing for manufacturability should be carried out, but also the approval of the adopted production and technological solutions [5].

For conducting a feasibility study of products, processes and forecasting production, an important point is the principle of forming groups of products and the choice of representative parts in groups.

The current grouping techniques and the selection of representative parts [6, 7] are suitable for use in traditional engineering industries in Russia and other countries, but are difficult to apply to digital industries. Therefore, there is a need to develop a new methodology for grouping and selecting representative parts, which allows you to automate certain operations and reduce the work of the designer.

Our proposed method of grouping and selecting representative parts used for digital production consists of the following steps (steps).

The first step is to collect information on the composition of the product range.

First of all, the formation of data of absolute indicators:

1. general product range;
2. annual product name release program;
3. material grade;
4. weight of the part and the workpiece;
5. the number of parts obtained from one workpiece;
6. overall dimensions of the part and workpiece;
7. requirements for precision surfaces of parts.

Then, the formation of data of relative indicators:

1. utilization of the material;
2. weight of material removed per unit of production;
3. planned technological complexity of manufacturing a unit of production;
4. planned technological cost per unit of production;
5. mathematical expectation of the load factor of the workplace.

The mathematical expectation of the load factor of the j-th workplace is determined by the formula [8]:

$$\overline{K^j_{workpiece}} = \frac{1}{n}\sum\nolimits_{i=1}^{n} k^j_{workpiece_i} \tag{1}$$

where $k^j_{workpiece_i}$ – the value of the load factor of the j-th workplace for the i-th part; n – number of work shifts.

This means that all technological processes must be adjusted at the stage of technological preparation of production according to the load factor of technological equipment.

However, it should be borne in mind that consideration of only the technological process with a forecast of the mathematical expectation of the load factor of the workplace may not be enough and will require taking into account the features of the production process as a whole.

The second step is to analyze the design features of the parts and pre-sort the parts. You can use the technique [6], where the principle of dividing many parts into 4 groups is used.

Group 1. Standard parts (parts whose construction is regulated by regulatory documents GOST, OST, etc.);
Group 2. Typical parts - parts close in design to standard parts;
Group 3. Uniform - details of a characteristic uniform design;
Group 4. Special parts - unique parts having a design that cannot be assigned to any of the selected groups.

At this stage of the work, you can enter the classification of assembly units/parts using the classifier of standard products for the first group. Given the requirements for the accuracy of the surfaces of parts, it is determined:
Average value of the accuracy of a part:

$$\overline{T^{part}} = \frac{1}{n}\sum\nolimits_{i=1}^{n} T_i \tag{2}$$

where n – the number of surfaces of the part to be processed,
T_i – accuracy quality value for i-th surface.
The average value of the surface roughness of the part:

$$\overline{Ra^{part}} = \frac{1}{n}\sum\nolimits_{i=1}^{n} Ra_i \tag{3}$$

where Ra_i – value of accuracy quality for the i-th surface.
The only drawback of this technique is the lack of the ability to automate the sorting process today.

The third step is the selection of parameters for the formation of product groups.
The technical and economic analysis of technological processes determines verification by quantitative criteria of accuracy and manufacturability. Accuracy indicators

are the accuracy level and surface roughness of the part, the average values of which were determined at the previous step.

In determining the manufacturability of the design, a special place is taken by labor intensity. The cost of production directly depends on it. Accordingly, knowing the complexity and cost of production in the group of manufactured parts, it is possible to predict the expected indicators of the complexity and cost of products planned for release.

Labor input is a complex indicator, therefore, for its assessment, it is necessary to use simpler parameters. To carry out the grouping process, the number of such parameters should be minimized. It was decided to sort according to the parameters of mass and serial production. To assess the volume of work (laboriousness) of manufacturing products, further introduce correction coefficients that allow you to adjust data within groups.

Sorting by mass and dimensions of parts is carried out into ranges in accordance with the binding to pre-selected technological equipment. Any change in the type of equipment for the formation of individual surfaces of parts will affect the planned load factors of jobs. On the one hand, this complicates the preliminary work of sorting parts and systematizing the source data. On the other hand, this allows obtaining more flexible information for the possibility of taking into account alternative technological processes.

The fourth step is the formation of groups of parts by weight and product release program.

The breakdown of the nomenclature into groups can be carried out according to the principle [7]:

Weight of parts in a group:

$$0{,}5\,M_{max} < M < 2\,M_{min} \tag{4}$$

where M_{max} и M_{min} – respectively, the largest and smallest weight values of products included in the corresponding group.

According to the production program:

$$0{,}1\,N_{max} < N < 10\,N_{min} \tag{5}$$

where N_{max} и N_{min} – respectively, the largest and smallest values of the annual output of products included in the corresponding group.

If these ratios are not fulfilled by this technique, it is recommended to perform additional splitting of products into groups. However, this method does not provide a comparison of groups in the system of indicators of mass and the program for the production of products.

For the relationship of these parameters, the following approach is proposed. It is necessary to build a scatter diagram, and on the abscissa axis indicate the range of masses of production units, and on the ordinate axis - the program for the release of the corresponding name. The nomenclature can be divided into groups according to the principle of mathematical statistics, in which the number of research groups does not exceed the value $k = \sqrt{n}$, where n – total number of samples (total number of items in the product range).

The resulting histogram based on the constructed scatter plot allows you to:

1. zoning the ordinate axis and grouping the abscissa axis;
2. to obtain information on the correspondence of the weighting coefficients k_s according to serial production and k_m according to the mass of products depending on the group.

The fifth step is the formation of groups of parts according to the parameters of accuracy and surface roughness of the parts.

The choice of a representative part of a group can be carried out both from existing parts in this group, and by compiling a virtual representative part. In this case, you can use the concepts of the average value of the quality and the average value of the roughness parameter of the part.

Unlike complex parts, which should contain all the surfaces of the parts of this group and differ in the maximum requirements for the geometric accuracy of the surfaces, such a virtual part gives an average value of the group parameters.

Average accuracy $\overline{T_{repres}^{part}}$ and average surface roughness $\overline{Ra_{repres}^{part}}$ part representatives in group is determined by formulas:

$$\overline{T_{repres}^{part}} = \frac{1}{m} \sum_{j=1}^{m} \left[\frac{1}{n} \sum_{i=1}^{n} \overline{T_{ij}} \right]$$

$$\overline{Ra_{repres}^{part}} = \frac{1}{m} \sum_{j=1}^{m} \left[\frac{1}{n} \sum_{i=1}^{n} \overline{Ra_{ij}} \right]$$

(6)

where j – number of part groups;

m – number of parts in a group

The choice of the value of the accuracy coefficient depends on the average quality of the representative part in the group, and the value of the roughness coefficient depends on the average roughness parameter of the surfaces of the representative part in the group, determined in a similar way.

The average value of the complexity of the virtual part-representative can be determined by the formula:

$$\overline{T_{P_{repres}}^{part}} = \frac{1}{m} \sum_{j=1}^{m} \overline{T_{P_J}}$$

(7)

Similarly determined and the average value of the cost of the part. Then, when assessing the expected laboriousness or cost of a new part, one can take into account the mean square deviation σ of the values of the studied parameter $x_{group.j}$ in the group:

$$\sigma = \sqrt{\sum_{j=1}^{m} \left(x_{group.j} - x_{aver.group} \right)^2 m_{group} \frac{1}{\sum_{j=1}^{m} m_{group}}}$$

(8)

where $x_{group.j}$ – the value of the parameter of new products related to the group,

$x_{aver.group}$ – the average value of the studied parameter of the group

This will significantly reduce the calculation time of labor input and production costs, and knowing the binding of operations to technological equipment, determine the throughput of the technological system and changes in the load factor of machines.

3   Practical Implementation

The practical implementation of the proposed methodology was carried out at the production site of one of the Russian machine-building enterprises for the production of vibration stands of the type BF-45UA-E, BF-70UA-E-T and etc. The 3D model of the vibrating stand BF-70UA-E-T is shown in the Fig. 1.

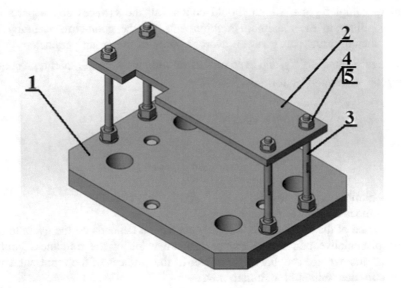

Fig. 1. The 3D model of the vibrating stand BF-70UA-E-T: 1 – Plate, 2 – Pressing, 3 – Rack, 4 – Nut, 5 – Shim.

Step 1. Collection and structuring of information on the composition of the range of products.

The results of the collection of information on the considered product range are presented in the form of a table Fig. 2.

Step 2. Analysis of the design features of the parts and their preliminary sorting into groups.

Parts are analyzed from the point of view of their design features and are recorded in one of 4 groups according to standard execution. Inside each of the 4 groups, parts are also divided depending on the design features, where they are assigned an index, sub-index, etc. depending on the complexity and variety of parts included in the assembly of the product. After preliminary splitting into groups, special parts are selected (Fig. 3).

The name of the Assembly units/Parts	The code of the Assembly units/Parts	The annual program of production, pcs	Parts information								Workpiece information			The amount of parts obtained from the 1st workpiece
			Material grade	Weight, kg	Overall dimensions, mm				Type of variety	Weight, kg	Overall dimensions, mm			
					Length	The width or the diameter of the outer	The height or diameter of the inner				Length	The width or the diameter of the outer	The height or diameter of the inner	
1	2	3	4	5	6	7	8	9	10	11	12	13	14	
Plate	A-13983-1	300	AMg6	3,68	425	425	20	Sheet	10,59	427	427	22	1	
Pressing	A-13983-2	300	AMg6	0,92	420	250	5	Sheet	1,25	422	252	8	1	
Rack	A-13983-3	1200	Steel 45	0,14	250	12		Bar	0,15	250	12		1	
Nut	A-13983-4	2400	Steel 45	0,05										
Shim	A-13983-5	2400	Steel 45	0,015										
Plate	A-13978-1	273	AMg6	3,69	425	425	20	Sheet	10,59	427	427	22	1	
Pressing	A-13978-2	273	AMg6	0,85	420	240	5	Sheet	1,28	422	252	8	1	
Rack	A-13978-3	1092	Steel 45	0,14	250	12		Bar	0,15	250	12		1	
Nut	A-13978-4	2184	Steel 45	0,05										

Fig. 2. The structure of the information on the considered product range.

The name of the Assembly units/Parts	The code of the Assembly units/Parts	The annual program of production, pcs	Selection of products by design features			
			Standard	Typical	Similar	Special
1	2	3	15	16	17	18
Plate	A-13983-1	300				SP1.1.
Plate	A-13978-1	273				SP1.1.
Plate	A-13992-1	160				SP1.1.
Plate	A-13998-1	402				SP1.1.
Plate	A-13994-1	510				SP1.1.
Plate	A-14000-1	350				SP1.1.
Plate	A-14001-1	270				SP1.1.
Plate	A-14010-1	270				SP1.1.
Plate	A-14022-1	108				SP1.1.
Plate	A-14011-1	218				SP1.1.
Plate	A-14012-1	210				SP1.1.
Plate	A-14013-1	247				SP1.1.
Plate	A-14039-1	178				SP1.1.
Plate	A-14023-1	120				SP1.1.
Plate	A-14008-1	300				SP1.1.
Plate	A-13990-1	193				SP1.1.
Plate	A-14041-1	123				SP1.1.

Fig. 3. A selection of parts that belong to a special group.

Step 3. Calculation of the main parameters of grouping.

For parts obtained from sheet material with subsequent machining, it is correct to determine the average value of the roughness and accuracy of the surfaces and conduct further grouping taking into account these indicators. The average values of accuracy and surface roughness of the parts included in the group are determined. The data are entered in the table (Fig. 4).

The name of the Assembly units/Parts	The code of the Assembly units/Parts	The annual program of production, pcs	Choice parts-representative on the basis of ranking of the Assembly units/Parts		
			Weight of material to be removed, kg	Average precision	Average roughness surface Ra
1	2	3	19	20	21
Plate	A-13983-1	300	6,91	14	9,2
Plate	A-13978-1	273	6,9	14	9,2
Plate	A-13992-1	160	2,25	14	13,9
Plate	A-13998-1	402	15,55	14	16,92727
Plate	A-13994-1	510	5,27	14	15,1333
Plate	A-14000-1	350	20,42	14	9,5777778
Plate	A-14001-1	270	2,93	14	17,55556
Plate	A-14010-1	270	14,71	14	9,4125
Plate	A-14022-1	108	15,11	14	10,03
Plate	A-14011-1	218	7,18	14	8,514286
Plate	A-14012-1	210	3,48	14	17,5556
Plate	A-14013-1	247	7,29	14	8,18889
Plate	A-14039-1	178	4,05	14	8,1168
Plate	A-14023-1	120	1,28	14	10,18333
Plate	A-14008-1	300	0.98	14	10.0714

Fig. 4. Determination of average values of accuracy and average surface roughness of parts.

Step 4. Formation of groups of parts by weight and product release program

A breakdown of the nomenclature of parts by weight and production program into 4 groups is carried out. We will sort the nomenclature according to the mass of parts from minimum to maximum, calculate the average mass of parts in the group and the number of parts falling into this group (Fig. 5).

The name of the Assembly units/Parts	The code of the Assembly units/Parts	The annual program of production, pcs	Weight, kg	Group	The average weight of parts in the group, kg	The number of parts in the group, pcs
1	2	3	5	22	23	24
Plate	A-14041-1	123	3,16	I	3,665	874
Plate	A-13983-1	300	3,68			
Plate	A-13978-1	273	3,69			
Plate	A-14039-1	178	4,13			
Plate	A-14013-1	247	4,85	II	5,7275	745
Plate	A-14011-1	218	4,96			
Plate	A-13992-1	160	5,15			
Plate	A-14023-1	120	7,95			
Plate	A-13994-1	510	10,13	III	15,1725	1273
Plate	A-14008-1	300	10,43			
Plate	A-13990-1	193	13,97			
Plate	A-14010-1	270	26,14			
Plate	A-14012-1	210	27,14	IV	31,104	1340
Plate	A-14000-1	350	27,99			
Plate	A-14022-1	108	30,28			
Plate	A-13998-1	402	32,86			
Plate	A-14001-1	270	37,25			

Fig. 5. Sort a group of parts by weight and release program.

Step 5. Determination of the average value of accuracy and roughness of a group of parts.

In each of the groups (column 22), the average value of the accuracy and roughness quality is determined; the data are entered in the table (Fig. 6).

The name of the Assembly units/Parts	The code of the Assembly units/Parts	The annual program of production, pcs	Weight, kg	Choice parts-representative on the basis of ranking of the Assembly units/Parts			Group	The average value of the quality of precise in the group	The average value of surface roughness Ra in the group
				Weight of material to be removed, kg	Average precision	Average roughness surface Ra			
1	2	3	5	19	20	21	22	25	26
Plate	A-14041-1	123	3,16	0,25	14	9,4			
Plate	A-13983-1	300	3,68	6,91	14	9,2			
Plate	A-13978-1	273	3,69	6,9	14	9,2	I	14	8,9792
Plate	A-14039-1	178	4,13	4,05	14	8,1168			
Plate	A-14013-1	247	4,85	7,29	14	8,18889			
Plate	A-14011-1	218	4,96	7,18	14	8,514286			
Plate	A-13992-1	160	5,15	2,25	14	13,9	II	14	10,19663
Plate	A-14023-1	120	7,95	1,28	14	10,18333			
Plate	A-13994-1	510	10,13	5,27	14	15,1333			
Plate	A-14008-1	300	10,43	0,98	14	10,0714			
Plate	A-13990-1	193	13,97	2,01	14	11,175	III	14	11,44805
Plate	A-14010-1	270	26,14	14,71	14	9,4125			
Plate	A-14012-1	210	27,14	3,48	14	17,5556			
Plate	A-14000-1	350	27,99	20,42	14	9,5777778			
Plate	A-14022-1	108	30,28	15,11	14	10,03	IV	14	14,32924
Plate	A-13998-1	402	32,86	15,55	14	16,92727			

Fig. 6. Determining the average value of accuracy and roughness of a group of parts.

Further calculations of the manufacturing time of parts included in the group are based on the "Average value of the accuracy class for the group", "The average surface roughness Ra for the group", as well as the group weight coefficient.

4 Conclusion

The proposed method of grouping parts makes it possible to fairly easily and reasonably sort the parts for the subsequent identification of the part representative of the group, as well as automate this process, which is necessary for digital production. Testing the developed methodology at one of the machine-building enterprises showed that the deviation of the machine time for processing the "Plate" workpieces on metal-cutting machines from the estimated average is 12%.

Acknowledgment. This research was made under financial support of Russian Ministry of education and science, project task № 9.2731.2017/4.6.

References

1. Nyemba, W., Mbohwa, C.: Modelling, simulation and optimization of the materials flow of a multi-product assembling plant. In: 14th Global Conference on Sustainable Manufacturing, GCSM, 3–5 October 2016, Procedia Manufacturing, vol. 8, pp. 59–66 (2017)
2. Fedoseyev, S., Vozhakov, A., Gitman, M.: Manufacture management on the tactical level of planning under the fuzzy initial information. Probl. Manag. **5**, 36–43 (2009)
3. Fedoseev, S., Gitman, M., Stolbov, V., Vozhakov, A.: Control product quality in modern industrial enterprises: a monograph. Publishing House of Perm. nat. issled. Polytechnic, Perm (2011)
4. Losonci, D., Kása, R., Demeter, K., Heidrich, B., Jenei, I.: The impact of shop floor culture and subculture on lean production practices. Int. J. Oper. Prod. Manag. **37**(2), 205–225 (2017). https://doi.org/10.1108/IJOPM-11-2014-0524
5. Kutin, A., Dolgov, V., Sedykh, M., Ivashin, S.: Integration of different computer-aided systems in product designing and process planning on digital manufacturing. In: 11th CIRP Conference on Intelligent Computation in Manufacturing Engineering – CIRP ICME 2017, vol. 67, pp. 476–481 (2018). https://doi.org/10.1016/j.procir.2017.12.247
6. Dolgov, V.A., Kabanov, A.A., Podkidyshev, A.A., Datsyuk, V.I.: Ekspertno-analytical method for determining the composition and quantity of technological equipment for the manufacture of parts of a given production program in the design of multiproduct machine-building enterprises. Vestnik mashinostroeniya **7**, 58–62 (2018)
7. Voronenko, V.P., Solomentsev, Yu.M., Skhirtladze, A.G.: Design Engineering Production, 2nd edn. Drofa, Moscow (2006)
8. Dolgov, V.A., Lutsyuk, S.V., Podkidyshev, A.A.: Organization of the collection and verification of information during the survey of machine-building enterprises. Autom. Mod. Technol. **9**, 43–46 (2014)

Digital Virtual Reality

Virtual Pet: Trends of Development

Daria Bylieva(✉)📧 , Nadezhda Almazova📧, Victoria Lobatyuk📧,
and Anna Rubtsova📧

Peter the Great St. Petersburg Polytechnic University (SPbPU),
195251 Saint-Petersburg, Russia
bylieva_ds@spbstu.ru

Abstract. Information technologies are fundamentally changing modern society. Almost any human activity, including the caring for a pet, is acquiring new formats related to communication in the virtual space. The authors analyzed such a phenomenon as a virtual pet that has been developing since the early 90s of the 20th century on the basis of more than 100 different virtual pet modifications. The most popular among users and purchased more than 1 million times a year around the world are examined in details in this research (video game Petz, Tamagotchi, Furby, Nintendogs, Neopets, My Talking Tom). Thus in this study the evolution of a digital pet is represented. We analyzed how a person models it according to his needs and identified the main trends of the virtual pet development. These basic directions are advanced similarity with a real pet, development of entertaining aspect, virtual pet applying for practical purposes. We proposed four main factors for analyzing changes in the human-virtual pet relationship: touch, game, communication and social interaction concerning the pet.

Keywords: Virtual pet · Digital pet · Virtual reality

1 Introduction

Development of technology changes environment of human existence. Information and communication technologies are being introduced into diverse spheres of human activity [1–5]. The era of virtual pets began in the 90s. In 1995, the video game Petz was released, where dogs and cats are autonomous characters with real-time layered 3D animation and sound, for which it was necessary to care for. In 1996, the first Tamagotchi appeared, a handheld egg with a digital screen, on which a nominal pet hatched. It was necessary to take care of it: to feed and clean, without missing an alarming peep announcing about unmet needs. The character, interests and exterior of the pet, as well as the duration of its life depended on how sensitively the player reacted to all the whims. In 1997, the virtual pet website Neopets appeared allowing the owners of virtual pets living on another planet to actively interact. In 1998, a talking pet Furby was put on sale. In 1999 social dog robots Aibo were produced.

Man's need for a pet is due to many factors. From the psychologists' point of view pets lower stress buffering and provide a kind of social support [6]. Zasloff points out that pets give a person emotional relationship, including love, trust, loyalty, and joyful mutual activity [7]. Special tests like Pet Attachment Survey, Lexington Attachment to

T. Antipova and Á. Rocha (Eds.): DSIC 2019, AISC 1114, pp. 545–554, 2020.
https://doi.org/10.1007/978-3-030-37737-3_47

Pets Scale, Comfort from Companion Animals Scale and other have been developed to assess the attachment and pleasure of owning a pet. Chesney and Lawson in their study compared attitude to live and virtual pets (from Nintendogs) by Comfort from Companion Animals Scale, which indicated that virtual pet does not offer as good companionship as a real cat or dog, but clearly offers some companionship [8].

Some researchers claim that the degree with which children get attached to virtual pets in some cases even matches their affection for real living pets. Similar game presented here intends to teach children between 6 and 12 different social values, e.g. taking care of someone or taking responsibility for your own actions [9]. According to Bloch, from the very beginning Tamagotchi was a toy sanctioned by parent as having educational value and teaching responsibility [10]. On the other hand, research comparing social pet robots with not a living being like a soft toy, showed that the stuffed dog was associated with items rated by adults as relevant to friendship, whereas the virtual dog was associated with items rated as relevant to entertainment [11]. Another study indicates that children engaged more often in apprehensive behavior and attempts at reciprocity with social pet robots, and more often mistreated the stuffed dog and endowed it with animation [12].

The goal of this study is to find out trends of the digital pet evolution, main features of the comparatively easily changeable virtual pet in a historical perspective and how a person models his or her "little brother" to fit his or her needs. In the study, more than 100 different virtual pets were analyzed. Though in the article we will focus on the most popular virtual pets (for example, over a million copies of the video game Petz were sold in a year [13], 30–40 million units of the first models of tamagotchi were sold all over the world in a couple of years [10], in the first three years more than 40 million Furby were sold and it "learned to speak" in 24 languages, more than 6 million units of Nintendogs were sold during a year [8], Neopets currently has over 190 million players worldwide, My Talking Tom has more than a billion downloads in 140 countries).

2 Results

The first version of Tamagotchi was the electronic analogue of a pet. In studies of the time when a toy appeared, researchers focused on its renewability, the cyclical recurrence of life and death, and the possibility of repeating a relationship from the beginning [10]. However, the principal feature that brought together a virtual pet with a real pet and distinguished it from a game was the impossibility to disable the pet or to stop its vital activity as it is customary for computer games. In addition, in the absence of smartphones and gadgets, a unique opportunity to carry Tamagotchi with them, also contributed to the establishment of a close relationship between the owner and the pet.

In the further evolution of the digital pet, two main trends can be traced: **development of similarity with a real pet** and **progress of entertainment aspect**, which is accompanied by a transference of human traits to pets. The third direction is related to **the use of pets for practical purposes** (for instance, a portable virtual pet for asthmatics that measures the lung capacity, and instructs appropriate actions to take [14] or Paro, a robotic baby seal aimed to elicit positive responses from patients and classified as a medical device in the USA [15], robotic dog aid in the social development of children with autism [16]) (Fig. 1).

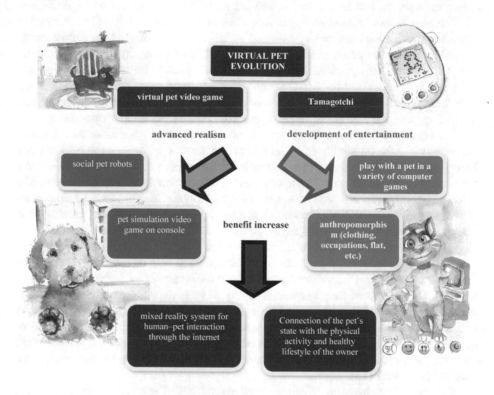

Fig. 1. Three main trends of virtual pets evolution

In the next versions of Tamagotchi, the life of the pet became more and more interesting and exciting. Therefore, the version of Mesutchi and Osutchi (1997) allowed buying male and female characters and to breed youngling. In 1998 Tamagotchi there was the possibility of gathering connected with the risk to suffer from a bear. In 2004 there was a competition in eating rice (in the Japanese version) and bread (in the American version), a competition "who will burst the ball faster". Despite the fact that the number of built-in games is growing, the relative realism of what is happening remains - the pet can really burst the balls or eat rice.

Analyzing relationship between a man and an animal as a meeting of the companion species, Haraway mainly focuses on two ways through which these practical encounters can take place: ***touch*** and ***play*** [17]. It seems to us that for a virtual pet these factors also turn out to be the most important. In addition, communication with the virtual pet plays a special role for interacting with it. All digital pets have negative feedback in the case of proper care absence. Appearing in the first tamagotchi, the peep of the pet who wants some attention, led the digital pet out of the discharge of a technogenic servant who obediently waited for his master to consider it possible to take care of it. Having the ability to state one's own needs, a strange being forced a person to reckon with its desires. There are cases when businessmen postponed business meetings, a person had an accident or refused to fly in an airplane due to the whim of his Tamagotchi, not to mention neglecting work or educational duties [18]. It shocked when the first Tamagotchi boldly avenged to his careless owner by premature death. Virtual and real cemeteries, seeing-off with candles, farewell speeches and gravestones [10, 18] showed the significance of such a realistic final of the virtual pet. Funerals have also been organized for AIBO robotic dogs in Japan, since Sony closed their last tech 'clinic' used to fix them [15]. Virtual pets on the other side of the screen, being neglecting, lose their beautiful appearance, lose various points and opportunities. In Petz, an untidy pet can run away from a careless owner. In addition, an important characteristic is the possibility of social interaction about a pet or with the help of a pet. Pretty quickly, the opportunity to get acquainted with other people's pets entered into a mandatory set of functions. The apotheosis of interaction is achieved on sites like Neopets, where there are private communications, public forums, internal communication, as well as the ability to buy and sell goods, play the stock market, participate in quests, competitions and multiplayer games, etc. Figure 2 presents examples of changes in the main factors of interaction between a person and a virtual pet.

Touch position initially for pets remained "on that side of the screen" could be implemented only with the mouse button. In Petz users can pet, scratch and stroke their animals. The pet immediately reacts in a variety of ways depending on what spot on its body is being petted, how fast, and how the pet feels at the time. Emotion can be expressed through different facial expressions (eyebrows, mouth, ears), styles of movement and body language (sad walks, happy trots, various postures, a variety of tail motions), and sounds (excited playful barks and meows, sad whines and whimpers, yelps of pain, etc.) [13].

The use of touchscreen, which allowed virtual pet to respond to the touch of a finger, brought relationships to a new level. In 2005, a very realistic virtual dog appeared on the handheld video game console with touchscreen and built-in microphone, through which you can give commands (Nintendogs) with the ability to communicate and transfer gifts to other owners via DS's wireless linkup. It can be played with, train, pet, walk, brush, and wash. The dog does not grow old and does not die. However, in the absence of care, it becomes less obedient and more dirty, loses various possibilities, such as participating in competitions, etc.

A separate line of realistic animals make up embodied virtual pet. Robotic dog Aibo first appeared on the market in 1999. Aibo with the exterior of a four-legged robot has a personality, the ability to walk, "see" its environment via camera and recognize spoken commands. Sensors provide to detect distance, acceleration, vibration, sound, and pressure. A special function is the ball game. Aibo can determine the position of a pink ball to go, hit him and head butt it. Aibo can interact with its own kind and even play football. Championships have been held in this sport among four-legged robots. Aibo's exterior refers us to the technique, and behavior to the animal. He has artifactual features (graymetal, flashinglights, and chimingsounds) but it also mimics the shape, motions, and reactions of living dogs [19]. Studies proves children's belief that Aibo liked them. Aibo liked to sit in their laps. Aibo could be their friend, and that they could be a friend to Aibo [20]. Aibo had mental states and sociality, even while stating that the robot did not have biological properties [21]. The new Aibo, launched on the market in 2018, is much less like a robot and more like a dog. He can distinguish faces. A simultaneous localization and mapping camera allow Aibo to map out its surroundings like modern robots - vacuum cleaners do. The extensive variety of movement points makes dog movements very realistic. OLED displays give more expression to eyes. Special application allows you to see the world through the eyes of Aibo and share pictures.

Modern high-tech variants of social pet robots (as well as their simplified counterparts) often include a fur coat, which even more brings perception of them as living beings. Social robot Leonardo developed by the Massachusetts Institute of Technology in the early 2000s, is a fluffy little animal that can make mimic human expression, interact with limited objects, and track objects, learn with a human and independently. Designed for positive emotional response from patients white and fluffy therapeutic robot baby harp seal Paro exhibited to the public in late 2001. The robot responds to petting by moving its tail and opening and closing its eyes, remember names and faces, makes funny sounds and learn actions that generate a favorable reaction.

Today, the market for children has a rich assortment of real-looking fluffy interactive cats and dogs that react to touch, make a variety of sounds and movements. Very interesting is a simplified version of the soothing domestic robot Qoobo pet (2017), where only the following key parts of the animal are left: the furry body, the pillow, and the tail, which has a large library of movement options depending on the touch of the owner. Another development option, on the contrary, emphasizes the technical part of the virtual pet, allowing the child to independently program colors and movements (Unicornbot 2018).

A further increase in realism is possible when using mixed reality technology and ICT to communicate with a real pet. For example, using 3D visualization, haptic sense, remote touch and physical simulation a real poultry and its owner are connected. In the office the owner touches the doll (or virtual 3D live view of the pet) [22]. A game system "Metazoa Ludens" allows you to run away from your own hamster in mixed-reality [23]. You can remotely grow chickens. Netband Internet based system provide feeding and cleaning them via tele robotic and sensor apparatus. You can also just chat with your dog on the Internet [24]. Rover@Home [11] is a system that uses common internet communication tools such as webcams, microphones and speakers to communicate with dogs over the internet. All research in this area emphasize the mutual benefit for a pet, receiving care, and for a man.

Another entertaining direction of research made an opposite step in the 2000s. We observe refusal from realism. A pet acquires more and more anthropomorphic features, such as the ability to wear clothes and even get professions (2006, Tamagotchi Connection Version 4). Now, to entertain a pet you do not need to walk with him or give him toys, it is better to play with him in computer games. In later versions, pets may engage in social interaction. In Tamagotchi P's (2012) this is writing letters and sending mail. Tamagotchi Friends (2014) has a new way of communication, called "short-range communication", which allows two toys to interact with each other when they touch back to back. By doing this, players can visit their friends' tamagotchi, exchange gifts, send messages and earn points.

Some embodied virtual pets also chose the path of humanization. They can also react to touch and voice, but instead of animal sounds, it will tell stories, sing songs, ask questions, etc. It is especially worth noting here Furby, whose popularity was primarily due to the apparent learning ability. Initially speaking his own language "Furbish", it gradually "learned" English.

At this time, games with numerous virtual pets have already taken root in smart-phones. A new generation of virtual pets can be seen on the example of the mobile applications like My Talking Tom series. Interaction through the "separating glass" continues to be sensory. The cat watches the movements of a finger on the screen, purrs, if it is petted, falls from a blow. Tom can be moved and washed with a wash-cloth. You can buy clothes for the cat, various food, interior items for its room by earned coins during the care of the pet (feeding, washing, brushing teeth, sleeping, sending to the toilet, playing computer games). In addition, by paying the game money or by entering the social networks, you can visit Tom's friends or see how the Toms look like in different parts of the globe. The female equivalent of "Tom" is "Talking Angela". There is an opportunity to dance, as well as buy and use cosmetics. In one of the latest versions of Tom, there is possibility to get him a pet (a bird).

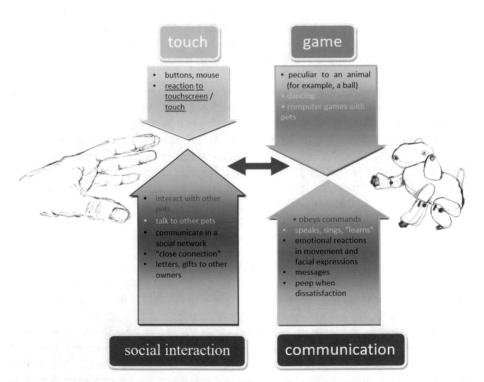

Fig. 2. Four main factors for analyzing changes in the human-virtual pet relationship (examples of the "entertainment" development trend are written in orange, realism is in blue, characteristics for both are in black)

Numerous similar mobile applications report the need to take care of by messages to the owner (differing in this aspect from other mobile games, reminding that they have not been played for a long time, only by more emotional messages about the unfortunate pet). Only the original Tamagotchi could die from inattention. All analogs excluded such a possibility. The game either has levels - age and ends when it reaches, for example, 50 years old (like My Talking Tom) or can last as long as the owner's interest remains.

One of the innovations of My Talking Tom is the stated ability to "speak" (implemented on the principle of repeating spoken words). In a similar game, Mimitos, you can visit the rooms of friends, looking for them by name, and talk by pressing prepared phrases that will be spoken through the dynamic.

Virtual characters are widely used to teach useful skills (for example, in many games it is proposed to brush pet's teeth, thereby teaching children to care for the oral cavity, [25]. A virtual polar bear is suggested to be used as a motivator for energy conservation or to help sick or lonely people (for example, to improve mirror neuron functioning in autistic children [26], rehabilitation for children with Cerebral Palsy [27]).

However, the most popular "useful" direction in the development of tamagotchi is associated with a healthy lifestyle. In American study, virtual pets provided both positive and negative feedback by the pictures of their hosts' breakfast, encouraging proper eating habits [28]. In another research virtual pet contributes to quit eating candy to achieve healthy eating habits [29]. Simplest pedometer based iPhone application linked the achievement of goals by the number of steps with the growth of a small pet "Walkamon". In research of Ramsay, Jin, Maes, and Picard the state of virtual pet was tied to their Fitbit activity data - thus, the more they walked, the fitter, healthier, and happier their digital companion [30]. Singapore My Pet Fitness app (2013) provides you with walking or running coins for which you can buy food and other items for your pet. For teens developed version of the game, with social networking tools. A more complex interactive application for office workers analyzes sensor readings from the seat device and prompts users with their virtual pet to take physical action [31]).

Trend of the virtual pet development as an assistant in the care of physical condition is an interesting phenomenon. When Tamagotchi appeared, he was a stranger from a different technical world who "promotes the introduction of modern technologies" to children, as Bloch notes [10]. At this time, the children were residents of the physical world and guests of the ICT world. Today, the situation has changed. For many people the ICT space is becoming a place of permanent residence for work, study, shopping and recreation, communication and entertainment. Permanent stay in the network sometimes leads to inattention to their physical condition, to the loss of physical activity. Virtual pet becomes the personification of the body, for which it is necessary to take care, to feed healthy food, to provide physical activity, walking, sleeping, etc. In this context, even a peeping fitness bracelet, calling the owner for the activity as he is staying in one place too long, in some sense is an evolved tomagochi. Haraway presented the person as theorized and fabricated hybrids of machine and organism and wrote that people should not be afraid of their joint kinship with animals and machines [32]. Tamagotchi-human cyborg, based on interdependence [33], acquires a new meaning, where an incorporeal thinking substance, which is in a technogenic reality, takes care of tamagotchi declaring the physical needs of the body.

3 Conclusion

The research methodology is based on the analysis of more than 100 different virtual pet modifications. The authors determined main trends of this phenomenon development and identified factors for the analysis of changes in the human-virtual pet relationship. Research methods are represented by the set of different approaches of theoretical and applied scientific knowledge, providing the implementation of analytical tools. In particular, methods of functional, complex and process comparative analysis and synthesis, grouping of actual data were applied. In this study the interdisciplinary approach was used, which allowed us to consider virtual pet as a complex social phenomenon in terms of socio-philosophical problems, the sociology of communication and the Internet. The authors also used such methods as schematization and graphical interpretation of theoretical information and empirical data to reveal the evolution of virtual pets.

The goal of this study was to identify trends of the digital pet evolution, virtual pet's features in the historical perspective and how a person models it to fit his needs. Outside the scope of this study, we leave such questions: why humanity seeks to replace living pets with virtual ones, what people lose and what they get from this replacement.

The main conclusions of the work include the following: virtual pet evolution took place in two main directions. The first trend is an increase in realism, and the second one is development of entertainment, which occurred in parallel with progress in anthropomorphism. However, we can also declare the third direction related to social orientation, along with the entertaining nature. It is virtual pet's assistance with various diseases, such as asthma, autism and others. We also should note the factors indicating changes in the relationship between a person and a virtual pet: touch (from exposure with the keyboard and mouse buttons to touch), game (from games inherent in animals to games inherent in humans as applied to human-virtual pet interactions), social interaction (from interactions with other animals within the game to communication by social networks) and communication (from messages to training and expression of emotions).

Benefit from virtual pets can hardly be overestimated. It is connected not only with the replacement of a real pet if you do not have the possibility to keep it, but also with the development of a person in the process of interaction and learning, with a virtual companion acquisition and formation of the healthy lifestyle idea. Virtual pet can not only be the analogue of the real pet, but also the image of the person's physical part, an anchor in virtual reality for an incorporeal avatar.

Today the speed of technological development is so high that new breakthroughs can be expected, allowing you to create even more technically sophisticated virtual pets. Further studies of new popular virtual pets will show how correctly we have outlined the directions of evolutionary development, and what other human needs they can satisfy.

References

1. Shipunova, O.D., Mureyko, L.V., Berezovskaya, I.P., Kolomeyzev, I.V., Serkova, V.A.: Cultural code in controlling stereotypes of mass consciousness. Eur. Res. Stud. J. **XX**, 694–705 (2017)
2. Shipunova, O.D., Berezovskaya, I.P., Mureyko, L.M., Evseeva, L.I., Evseev, V.V.: Personal intellectual potential in the e-culture conditions. Espacios **39**, 15 (2018)
3. Mureyko, L.V., Shipunova, O.D., Pasholikov, M.A., Romanenko, I.B., Romanenko, Y.M.: The correlation of neurophysiologic and social mechanisms of the subconscious manipulation in media technology. Int. J. Civ. Eng. Technol. **9**, 2020–2028 (2018)
4. Trostinskaia, I.R., Safonova, A.S., Pokrovskaia, N.N.: Professionalization of education within the digital economy and communicative competencies. In: 2017 IEEE VI Forum Strategic Partnership of Universities and Enterprises of Hi-Tech Branches (Science. Education. Innovations) (SPUE), pp. 29–32. IEEE (2017). https://doi.org/10.1109/IVForum.2017.8245961
5. Razinkina, E., Pankova, L., Trostinskaya, I., Pozdeeva, E., Evseeva, L., Tanova, A.: Student satisfaction as an element of education quality monitoring in innovative higher education institution. In: E3S Web Conference, vol. 33, p. 03043 (2018). https://doi.org/10.1051/e3sconf/20183303043
6. Wilson, C.C., Turner, D.C.: Companion Animals in Human Health. Sage Publications (1998)
7. Zasloff, R.L.: Measuring attachment to companion animals: a dog is not a cat is not a bird. Appl. Anim. Behav. Sci. **47**, 43–48 (1996). https://doi.org/10.1016/0168-1591(95)01009-2
8. Chesney, T., Lawson, S.: The illusion of love: does a virtual pet provide the same companionship as a real one? Interact. Stud. **8**, 337–342 (2007). https://doi.org/10.1075/is.8.2.09che
9. Hildmann, H., Uhlemann, A., Livingstone, D.: A mobile phone based virtual pet to teach social norms and behaviour to children. In: 2008 Second IEEE International Conference on Digital Game and Intelligent Toy Enhanced Learning, pp. 15–17. IEEE (2008). https://doi.org/10.1109/DIGITEL.2008.41
10. Bloch, L.-R., Lemish, D.: Disposable love: the rise and fall of a virtual pet. New Media Soc. **1**, 283–303 (1999). https://doi.org/10.1177/14614449922225591
11. Aguiar, N.R., Taylor, M.: Children's concepts of the social affordances of a virtual dog and a stuffed dog. Cogn. Dev. **34**, 16–27 (2015). https://doi.org/10.1016/J.COGDEV.2014.12.004
12. Kahn, P.H., Friedman, B., Pérez-Granados, D.R., Freier, N.G.: Robotic pets in the lives of preschool children. Interact. Stud. **7**, 405–436 (2006). https://doi.org/10.1075/is.7.3.13kah
13. Frank, A., Stern, A., Resner, B.: Socially intelligent agents. In: Dautenhahn, K. (ed.) 1997 Socially Intelligent Agents AAAI Fall Symposium. AAAI Press, Menlo Park (1997)
14. Lee, H.R., Panont, W.R., Plattenburg, B., de la Croix, J.-P., Patharachalam, D., Abowd, G.: Asthmon: empowering asthmatic children's self-management with a virtual pet. In: Proceedings of the 28th of the International Conference Extended Abstracts on Human Factors in Computing Systems - CHI EA 2010, pp. 3583–3588. ACM Press, New York (2010). https://doi.org/10.1145/1753846.1754022
15. Rault, J.-L.: Pets in the digital age: live, robot, or virtual? Front. Vet. Sci. **2**, 11 (2015). https://doi.org/10.3389/fvets.2015.00011
16. Stanton, C.M., Kahn Jr., P.H., Severson, R.L., Ruckert, J.H., Gill, B.T.: Robotic animals might aid in the social development of children with autism. In: Proceedings of the 3rd International Conference on Human Robot Interaction - HRI 2008, p. 271. ACM Press, New York (2008). https://doi.org/10.1145/1349822.1349858

17. Haraway, D.J.: When species meet (2007)
18. Besser, H.: Tamagotchi and Aspectual Shape. http://besser.tsoa.nyu.edu/impact/s97/Focus/Identity/FINAL/ana.htm
19. Melson, G.F., Kahn Jr., P.H., Beck, A., Friedman, B.: Robotic pets in human lives: implications for the human-animal bond and for human relationships with personified technologies. J. Soc. Issues **65**, 545–567 (2009). https://doi.org/10.1111/j.1540-4560.2009.01613.x
20. Kahn, P.H., Gary, H.E., Shen, S.: Children's social relationships with current and near-future robots. Child. Dev. Perspect. **7**, 32–37 (2013). https://doi.org/10.1111/cdep.12011
21. Melson, G.F., Kahn, P.H., Beck, A., Friedman, B., Roberts, T., Garrett, E., Gill, B.T.: Children's behavior toward and understanding of robotic and living dogs. J. Appl. Dev. Psychol. **30**, 92–102 (2009). https://doi.org/10.1016/j.appdev.2008.10.011
22. Lee, S.P., Cheok, A.D., James, T.K.S., Debra, G.P.L., Jie, C.W., Chuang, W., Farbiz, F.: A mobile pet wearable computer and mixed reality system for human–poultry interaction through the internet. Pers. Ubiquit. Comput. **10**, 301–317 (2006). https://doi.org/10.1007/s00779-005-0051-6
23. Tan, R.T.K.C., Cheok, A.D., Peiris, R.L., Wijesena, I.J.P., Tan, D.B.S., Raveendran, K., Nguyen, K.D.T., Sen, Y.P., Yio, E.Z.: Computer game for small pets and humans. In: Presented at the (2007). https://doi.org/10.1007/978-3-540-74873-1_5
24. Rossi, A.P., Rodriguez, S., Cardoso dos Santos, C.R.: A dog using Skype. In: Proceedings of the Third International Conference on Animal-Computer Interaction - ACI 2016, p. 10. ACM Press, New York (2016). https://doi.org/10.1145/2995257.3012019
25. Dillahunt, T., Becker, G., Mankoff, J., Kraut, R.: Motivating Environmentally Sustainable Behavior Changes with a Virtual Polar Bear. Work. Pervasive Persuas. Technol. Environ. Sustain. (2008)
26. Altschuler, E.L.: Play with online virtual pets as a method to improve mirror neuron and real world functioning in autistic children. Med. Hypotheses **70**, 748–749 (2008). https://doi.org/10.1016/J.MEHY.2007.07.030
27. Saimaldahar, D.: Virtual Pet Companion a Digital Console to Enhance the Experience of Children with Cerebral Palsy (2016). http://openresearch.ocadu.ca/id/eprint/694/1/Saimaldahar_Daniah_2016_MDes_DF_Thesis.pdf
28. Byrne, S., Gay, G., Pollack, J.P., Gonzales, A., Retelny, D., Lee, T., Wansink, B.: Caring for mobile phone-based virtual pets can influence youth eating behaviors. J. Child. Media. **6**, 83–99 (2012). https://doi.org/10.1080/17482798.2011.633410
29. Laureano-Cruces, A.L., Rodriguez-Garcia, A.: Design and implementation of an educational virtual pet using the OCC theory. J. Ambient Intell. Humaniz. Comput. **3**, 61–71 (2012). https://doi.org/10.1007/s12652-011-0089-4
30. Ramsay, D.B., Jin, J., Maes, P., Picard, R.: Virtual Pets and Virtual Selves as Exercise Motivation Tools. https://www.davidbramsay.com/public/RamsayAvatar.pdf
31. Min, D.A., Kim, Y., Jang, S.A., Kim, K.Y., Jung, S.-E., Lee, J.-H.: Pretty pelvis: a virtual pet application that breaks sedentary time by promoting gestural interaction. In: Proceedings of the 33rd Annual ACM Conference Extended Abstracts on Human Factors in Computing Systems - CHI EA 2015, pp. 1259–1264. ACM Press, New York (2015). https://doi.org/10.1145/2702613.2732807
32. Haraway, D.: Simians, Cyborgs and Women: The Reinvention of Nature. Free Association Books, London (1991)
33. Gray, C.H.: The Cyborg Handbook. Routledge, London (1995)

Author Index

Printed in the United States
By Bookmasters